The Design Student's Handbook

PEARSON

At Pearson, we take learning personally. Our courses and resources are available as books, online and via multi-lingual packages, helping people learn whatever, wherever and however they choose.

We work with leading authors to develop the strongest learning experiences, bringing cutting-edge thinking and best learning practice to a global market. We craft our print and digital resources to do more to help learners not only understand their content, but to see it in action and apply what they learn, whether studying or at work.

Pearson is the world's leading learning company. Our portfolio includes Penguin, Dorling Kindersley, the Financial Times and our educational business, Pearson International. We are also a leading provider of electronic learning programmes and of test development, processing and scoring services to educational institutions, corporations and professional bodies around the world.

Every day our work helps learning flourish, and wherever learning flourishes, so do people.

To learn more please visit us at: **www.pearson.com/uk**

The Design Student's Handbook

Your Essential Guide to Course, Context and Career

Edited by
Jane Bartholomew
Steve Rutherford

Harlow, England • London • New York • Boston • San Francisco • Toronto • Sydney • Auckland • Singapore • Hong Kong
Tokyo • Seoul • Taipei • New Delhi • Cape Town • São Paulo • Mexico City • Madrid • Amsterdam • Munich • Paris • Milan

PEARSON EDUCATION LIMITED
Edinburgh Gate
Harlow CM20 2JE
United Kingdom
Tel: +44 (0)1279 623623
Web: www.pearson.com/uk

First published 2013 (print and electronic)

ISBN: 978-1-4082-2028-3 (print)
978-1-4082-2030-6 (PDF)
978-1-292-00366-5 (eText)

British Library Cataloguing-in-Publication Data
A catalogue record for the print edition is available from the British Library

Library of Congress Cataloging-in-Publication Data
The design student's handbook / [edited by] Steve Rutherford, Jane Bartholomew.
pages cm
ISBN 978-1-4082-2028-3
1. Design--Handbooks, manuals, etc. I. Rutherford, Steve, 1957- editor of compilation. II. Bartholomew, Jane, editor of compilation. NK1510.D4765 2013
745.4--dc23

2013005152

10 9 8 7 6 5 4 3 2 1
17 16 15 14 13

Cover images: 'Blooming Propeller' Lighting Sculpture by Hsiao-Chi Tsai & Kimiya Yoshikawa, 2012. Image © Tsai & Yoshikawa 2012. All Rights Reserved; Chris Castillo, jewellery and product designer; Alice Palmer; Richard Sneesby; Frederico Zanjacomo; Suet Yi Ceramics.

Print edition typeset in 10/12pt Agenda by 35
Print edition printed and bound by L.E.G.O. S.p.A., Italy

NOTE THAT ANY PAGE CROSS REFERENCES REFER TO THE PRINT EDITION

Contents

Preface

For anyone starting to study design, this text is full of practical and theoretical content and has been put together to provide you with all you need to know. It has been developed to support you from early on in your studies to your first set of opportunities, whether this is going on to postgraduate study, working for a company or setting up your own creative business.

Read this text and gain a greater understanding of the differences between the broad varieties of design disciplines and find out much more about the careers open to you in the industry. There are over 100 stories and career profiles from students, designers and industry specialists, keen to share their thoughts and aspirations with the next generation of designers. It is these deep and meaningful insights that give the text its intriguing and rich dimension.

It takes you through all of the various design disciplines – graphics, fashion and textiles, three-dimensional design, spatial design, interactive media design, theatre, film and television design and craft and will help you clarify your understanding of the scale of the industry and identify where you might see yourself in the future.

The broad range of specialisms have been broken down for you so the types of skills and aptitudes can easily be identified, helping you appreciate the wide range of opportunities available. The principles of designing aesthetically, using colour, improving your drawing and designing skills and communicating design have been written in such a way that you can immediately take from it some practical approaches to help you improve your abilities. The text covers themes that will help you develop a greater awareness of today's contemporary culture, how to design sustainably and inclusively and encourage you to work with the industry as much as possible during your studies. As a successful designer you will have a broader awareness of all that influences potentially great design. This book has also been organised to signpost other important books and websites that will broaden your understanding of the subjects covered.

Devised as a crucial handbook and a core text for all undergraduate students, we recognise that this now reaches out to teachers, career advisors and lecturers too, and we envisage that this could be recommended to all students interested in going on to study design. Students considering coming to the UK to study design may also find this text useful. It provides descriptions of the way that design is taught in the UK and contains many illustrated practical projects and insightful stories written by undergraduate and postgraduate students about their experiences.

Acknowledgements

We would like to thank the many individual students, designers and industry specialists who wanted to tell future students their stories about how they became inspired by this industry. Special thanks go to those contributing authors, from a very wide spectrum of institutions and organisations, who took on the challenge of a whole chapter.

The rich sources of inspiration were captured by visiting trade events, such as 100% Design, D&AD, Milan Furniture Fair, Designers Block, New Designers, Freerange, Origin and Lustre, to obtain a real feel for the whole sector. We therefore would like to thank all of you who remember engaging in conversation about this text and who suggested individuals and organisations to help us in our quest.

This text would not have been possible without the foresight of the publishers, Pearson Education, in commissioning a handbook for budding designers to instil in them the fundamentals and principles of design. Thanks, therefore, go to Andrew Taylor, Paul Stevens and Josie O'Donoghue at Pearson in approaching staff at Nottingham Trent University with this project. Professor Judith Mottram was pivotal in inspiring staff to develop the proposal.

The publisher's anonymous review panel offered diligent and thought-provoking commentary and feedback throughout. Whoever you are, we would like to thank you for engaging with us on this project.

Finally, the last weeks of editing and pulling the initial manuscript together couldn't have been done without the input from Steve Hepworth, the photographer who undertook the picture editing, and the pertinent observations made by Esther Bartholomew in offering the all-important teenager's perspective on the content and writing.

We would like to dedicate this book to all past, present and future students studying design and hope you enjoy being part of this industry as much as we do.

Editors, Jane Bartholomew and Steve Rutherford

Publisher's acknowledgements

We are grateful to the following for permission to reproduce copyright material:

Figures

Figures on pages 4, 121, 123, 124, 126, 127, 126, 127, 130, 129, 130, 131, 135, 134, 136, 138 and 139 from Richard Sneesby; Figure on page 18 from Mike Holden; Figures on page 19 from Katherine Butler; Figures on pages 41, and 42 from Roxy Demetria Eusebio; Figures on pages 53, 314, 398 and 399 from Anna Piper; Figure on page 79 from Eamon Martin; Figures on pages 99, 103, and 105 from Kathryn McKelvey: I would like to thank the following students and graduates of Northumbria University, for their inspiring and creative interactive project work and contribution to the career case studies, Melanie Huang, Katerina Brunclikova, Patrick Niall McGoldrick, Jack Merrell, Alex Steven, Lee Carroll, Andrew Charlton, Dave Barlow, Ross Dixon, James Chorley, Steven Everington, Michael Harmer, Richard Carr, Mike Thomas, Simon Occomore, Laura Straker and Dan Scott; Figures on pages 118, 122, 132, 134 and 137 from Hugo Bugg; Figures on pages 144, 145, and 270 from Harriet Curtis; Figures on pages 148, page 158, page 305, page 305 from Mike Jones; Figure on page 157 from Dan Walker, © BBC; Figures on pages 159 and 162 from Jayne Harvey; Figures on pages 162, 242 and 243 from Hollie Cleaver; Figure on page 165 from Laura Ward; Figure on page 168 from Zosia Stella-Sawicka; Figures on pages 169 and 170 from Dan Walker; Figures on pages 172 and 173 from Dorrie Scott; Figure on page 185 from Sarah Hoyle; Figures on pages 187, 188, and 189 from Frederico Zanjacomo, © Frederico Zanjacomo; Figures on pages 189, 197, 198 and 199 from Land Rover; Figures on pages 195 and 196 from Alexander Taylor; Figure on page 220 from Amy Bicknell; Figure on page 221 from Alexandra Chin; Figure on page 221 from Stuart Brown, stuartbrowndesign.com; Figures on pages 269, 368 and 405 from Chloe Muir; Figure on page 253 from Emma Alderman; Figure on page 300 from NCS UK Limited, NCS – Natural Colour System ®, © property of and used with permission from NCS Colour AB Stockholm 2013. www.ncscolour.co.uk; Figure on page 309 from Laura Thomas; Figure on page 310 from Laura Thomas; Figure on page 312 from Sara Moorhouse; Figures on pages 380 and 403 from Fan Sissoko, Innovation Unit; Figures on pages 401 and 402 from Sally Halls.

Logos

Logo on page 68 from We are SMILE Ltd. www.wearesmile.com; Logo on page 114 from Kathryn McKelvey.

Screenshots

Screenshot on page 78 from Eamon Martin, http://www.artsthread.com/p/emartinmedia/gallery/my-galleries/32133; Screenshot on page 83 from http://www.dancexchange.org.uk/; Screenshot on page 97 from http://www.sarahturner.co.uk/; Screenshots on pages 102, 103, 110, 111 and 113 from Kathryn McKelvey; Screenshot on page 190 from Albert Montserrat; Screenshot on page 234 from www.dweebdesign.co.uk, Craig Foster; Screenshots on page 377 from Kristen Brice; Screenshots on page 422 from Graduate portfolio, http://www.artsthread.com and Imogen Ransley Buxton; Screenshot on page 423 from http://www.artsthread.com/p/elvanotgen/gallery/.

Photographs

The publisher would like to thank the following for their kind permission to reproduce their photographs:

(Key: b-bottom; c-centre; l-left; r-right; t-top)

Alamy Images: Mint Photography 334b, Hugh Threlfall 338r; Crevasse flower vase by Zaha Hadid for Alessi S.p.A., Crusinallo, Italy: 329r; Alice Made This: 239; Arts Thread: Arts Thread 425; Autofil: Jane Bartholomew 214cr; Jane Bartholomew: 5bl, 16tl, 50, 51bl, 117, 214br, 217, 259, 301cr, 301bl, 301br, 310b, 413c, 413b, 414b, Susan Hall 299; Nell Bennett: 181br, 182cl, 267; Elissa Bleakley: 249, 250tl, 250tr; Emily Boniface: 371; Bregenzer Festspiele: adereart 147; Bridgeman Art Library Ltd: Bowl from the El Obeyd period from Syria. Prehistoric civilizations of Mesopotamia, ca 4500 BC. / De Agostini Picture Library / A. Dagli Orti 327, "Casablanca" Sideboard Designed, 1981 (plastic laminate & wood), Sottsass II, Ettore (1917–2007) / Philadelphia Museum of Art, Pennsylvania, PA, USA / Gift of Collab: the Group for Modern and Contemporary Design at Philadelphia Museum of Art & Abet Laminati, 1983 / Studio Ettore Sottsass s.r.l. 334t, Chair (Cesca chair), 1928 (chromium-plated tubular steel, wood & woven cane), Breuer, Marcel (1902–81) / The Israel Museum, Jerusalem, Israel / Gift of the manufacturer 333, Chaise Longue (mixed media), Corbusier, Le (Charles Edouard Jeanneret) (1887–1965) / Museum Thonet, Stuhlmuseum, Frankenberg, Germany / De Agostini Picture Library. / © FLC / ADAGP, Paris and DACS, London 2012, © ADAGP, Paris and DACS, London 2012, 343, Chair (mixed media), Mies van der Rohe, Ludwig (1886–1969) / Vitra Design Museum, Weil-am-Rhein, Germany / De Agostini Picture Library. © DACS 2012 / © DACS 288t, Haystacks at Sunset, Frosty Weather, 1891, Monet, Claude (1840–1926) / Private Collection 303tl, 'Wandle' printed fabric, manufactured by Morris and Co. and Anymer Vallance from 'The Art of William Morris', pub. 1897, Morris, William (1834–96) / Calmann & King, London, UK 330; Adrian Buckmaster: 62, 63; Ken Bushe: 313t; Katherine Butler: 19tr, Dave Williams 19br; Garry Butterfield: 40tl, Squiz Hamilton 40tr; © Estate of Angus Bean / National Portrait Gallery, London: 346; Camera Press Ltd: Caroline Menne 335; Chris Castillo: 9t, 292, 295, 296; Hollie Cleaver: 234t, 244; Richard Collings: 182bl; Hayley Collins: 254; Courtesy the artist and Victoria Miro, London: Grayson Perry A Work in Progress, 2012, Glazed ceramic, H: 60 x dia: 31cms (GP370) © Grayson Perry. Photography © Stephen White 9b; Lana Crabb: 21, 276cl, 276cr, 276br; Nicola Danks: 181cr; Lucy Davies: 261, 400; Leah Dennis: 20tr; Frazer Doyle: 15b; Droog: Bottoms up doorbell for Droog by Peter van der Jagt 344; Roxy Eusebio: 42bl, 43; Eye Revolution Ltd: 338l; Fairline Boats Ltd: 6, 232tl; Craig Fellows: 446tr, 448; Craig Foster: 216; Frederica Cards Ltd: 328; Getty Images: 339, Mondadori Portfolio / UIG 303tc, Popperfoto 332tr; Anna Glasbrook: 251, 464, 465t, 465br, © John Enoch 466; Bathsheba Grossman: Quin Lamp designed by Bathsheba Grossman, photograph by MGX Materialise 336; Sally Halls: © Helen Hamlyn Centre for Design, Royal College of Art 278, 279, 401t; Hannah Lobley Paperwork: 362, 395t, 395b, Dan Lane 397; Jayne Harvey: 160tl; Philippa Hill: 49tr, 56tr, 56cr, 236, 237; Rafael Hoffleit: 283; Mike Holden: 18tl, 18bl, 20cr; Sarah Hoyle: 185br; Oliver Hrubiak: 183, 191tl; Hsiao-Chi Tsai and Kimiya Yoshikawa: 3, 247br; Jodi Hunt: Stonie Reuladair 75tr, 75bl, 75br; Shani Jayawardena: 42cl, 45; Mark Jones: 247bl; Mike Jones: 161; Katrin Klausecker: 271b; Martin Knox: 66; Chandni Kumari: 272cr, 272br, 366tr, 366br; Chris Lamerton: 192; Land Rover: 198cl, 199tr, 200;

Ptolemy Mann: 317; Tony Marsh: 57; Pablo Matteoda: 345; Kathryn McKelvey: 91, 94tr, 94cl, 94cr, 94bl, 100, 100tr, 101, 104, 106, 107, 112, 113t, 114cr; Memphis s.r.l.: Aldo Ballo, Guido Cegani, Peter Ogilivie / Martine Bedin 341; Sophie Minal: 47; Lyndsey Mitchinson: Image: Chris Auld 15tr; Nathan Monk, SMILE www.wearesmile.com: 67, 68tl, 68tr; Alice Moore 39; Jane Moore: 308t; John Moore: 5, 307; Sara Moorhouse: 319tr, 319bl, 319br, 320t, 320cr, 321, 322; Museum of Fine Arts, Boston: Claude Monet, French, 1840–1926. Grainstack (Snow Effect) (detail), 1891. Oil on Canvas. 65.4 x 92.4 cm (25 ¾ x 36 ⅜ in.). Museum of Fine Arts Boston. Gift of Miss Aimée and Miss Rosamond Lamb in memory of Mr and Mrs Horation Appleton Lamb. 1970.253: 303tr; Sian O'Doherty: 46, 52tl, 60tr, 60bl, 61, Dan Staveley / www.danstaveley.co.uk 32tl, 273; Dharma Lounge by Palette Industries: 350; Alice Palmer: Christopher Moore 31; Panasonic: 392; Adele Parsons: 437; Kelsey Pilgrim: 421t; Anna Piper: 360, Steve Rutherford 399t; Anna Pope: 11t, 17, 28, 29tl, 29tr; Megan Randall: 26t, 26b, 26t, 26b; Deryn Relph: © Alick Cotterill 49tl, 256, 469tr, 469b, 470; Steve Rutherford: 6b, 7, 35t, 71tr, 71b, 72, 74, 78b, 85tr, 85bl, 86bl, 86br, 88, 89tl, 89b, 95, 98, 153tl, 177, 180tr, 181tl, 185bl, 193, 205, 208, 209, 210, 213t, 213cr, 213br, 224, 231, 232tr, 233bl, 233br, 255, 258, 260, 262, 265, 285, 303c, 303b, 304t, 308b, 311tl, 311tr, 311br, 312t, 415t, 421b, 445b, 446tl, 453; Science Photo Library Ltd: NYPL / Science Source 332tl; Shutterstock.com: Janaka Dharmasena 289l, Olga Drabovich 288b, nmiskovic 289r, Sergii Rudiuk 287; Alisha Simpson: 313bl, 313br; Fan Sissoko: Innovation Unit: 381, 382, 404t, 404c; Debbie Smyth: 246; Richard Sneesby: 116, 119, 121c, 122cl, 122bl, 123t, 125, 126tr, 128; Beth Snowden: 284; Timothy Soar: 318tl, 318tr; Zosia Stella-Sawicka: 167; Emma Storr: 228, 229tr, 229b; Robyn Swindley: Photographed by Tina Downes 37tr, 47br; Paul T. Dack: 462tr, 462bl; Janet Tan: Stonie Reuladair 76tr, 76b; Alexander Taylor: Peter Guenzel 467tr, Alexander Taylor Studio 196tr, Courtesy of Thorsten van Elten 467bl, 468; The Kobal Collection: 20th Century Fox / Godfrey, Chris 148b; Emily Thomas: 37tl; Phil Thomson: Stonie Reuladair 65; Timorous Beasties: 348; Robyn Townsend: Dave Williams 20tl; Amanda Trimmer: 266, 280, 281; Sarah Turner: 191bl, 337, 471, 472; University of Sunderland: 16tr; Upper Street Events: 48, 178, 411, 48, 178, 411; © V&A Museum, London: 329l; Stephanie Walton: 297cr, 297b; Laura Ward: 160b, 166; Rachel Ward: 175; Hannah Welsh: 184; The Weston Road Academy: 416, 440, 441; David Williams: 18cr; Gill Wilson: 13c, 13b, Adrian Heapy 23, 24; Rebecca Wombell, Harley Gallery: 232cr; www.carphoto.co.uk: David Kimber: 290, 332b; Suet Yi Yip: 13t, 445t; Frederico Zanjacomo: 186; Linyang Zhang: 417, 442, 442tl, 442br.

In some instances we have been unable to trace the owners of copyright material, and we would appreciate any information that would enable us to do so.

What is design?

Part one

1

Introduction

'Knowledge and skill are vital but not as important as passion and creativity. Enjoying what you do and having an enthusiastic approach to design will ensure effective outcomes to any brief. Keeping myself up to date with the design world is integral – I do this by following blogs, subscribing to magazines and looking up exhibitions. Surrounding yourself with great design allows you to explore and understand the industry in greater depth.'

Catherine Perrott, graphic design student

'Blooming Propeller' Lighting Sculpture by Hsiao-Chi Tsai & Kimiya Yoshikawa, 2012

To be creative, to analyse and to make sense of the world through drawing and making is within us all. When we were young we all used 'drawing and making' to express ourselves. As your knowledge and awareness of design develops during your studies, you can begin to appreciate that there are opportunities to make a difference to people's lives. This book will introduce you to the breadth and scope of design as a profession and guide you along the road to a career that contains a wonderful combination of excitement and fulfilment.

One of the surprising aspects of working in design is the huge network of people. According to the most recent research in 2010 by the UK Design Council ('Design Industry Research 2010') there are more than 65,000 self-employed designers working in the UK. There are almost 11,000 design companies employing 82,500 designers and 6,500 design teams in manufacturing companies employing 83,600 designers. There is also the craft and designer-maker aspect of design that includes many small businesses and self-employed designer-makers. According to the UK Crafts Council, in 2011 ('Craft in an age of change' 2012) there were over 23,000 contemporary craft-making businesses. That is a grand total of over 254,000 designers!

In ten years' time you will find that it is impossible to visit a trade fair or exhibition without bumping into people with whom you've worked or studied. This network is crucial to the operation of the design industry. From the same Design Council Survey there is this comment on the nature of design work:

> 'For freelance designers, networking has become increasingly important since 2005. And collaboration is popular – more than half of designers say they work with other businesses.'

So, what can you expect from this book?

The project stories and the career profiles from students, designers and other industry specialists throughout the book will show you the way that individuals think and create. The book provides you with a useful insight into the breadth of subjects available in design education. It also offers you an important set of practical and theoretical chapters that encourages you to debate and appreciate all that it takes to become a successful designer.

Use the book to explore all the design disciplines, not just the ones that you are naturally drawn to. There is also advice on placements, networking and job-seeking. Design is a fascinating, broad subject served by many different companies and organisations, and in order to work in these companies (and with them if you set up your own business) you need to make a success of your studies and make the most of your time as a student.

There is a wide range of practice to choose from. You could become a designer in the industry, work for yourself as a self-employed designer-maker or produce designs to specific briefs for a range of different clients. You would then be operating as a consultant. There is also a lot of multi-disciplinary practice – projects requiring different types of designer or projects where you have to do more than one thing – product and graphics for instance.

Reading about other designers will help you appreciate the commonalities and subtle differences between the disciplines and heighten your awareness of the **cross-disciplinary** opportunities there are to work in other areas.

Some designers are widely experienced and work in more than one discipline. This is known as **multi-disciplinary** working. Self-employed designers, and those working in smaller companies, more often have to work across different disciplines – be responsible for more than just their own discipline. However, everyone at some point will have to work with other designers from other disciplines.

Cross-disciplinary: a designer moving into another discipline. For example, a graphic designer might become interested in designing products.

Multi-disciplinary: this is an opportunity for a designer to have responsibility for working across two different disciplines. For example, a fashion designer might design the clothes and then also design the print for the fabric (therefore working across both fashion and textiles).

There is also the possibility for people with design qualifications to cross into other professions. There is a set of transferable skills that designers need to be equipped with, such as problem solving, lateral thinking and having an eye for detail, which can lead to other careers. Many large companies in the retail sector have teams of buyers who select the product ranges you see in the stores. A knowledge of good design, manufacturing and costing, all of which you will pick up on any design course, is crucial to this role. Others will find that their generic business skills will allow them to move into business and marketing roles in the future.

Once you have graduated you might also consider postgraduate study or teaching. Some students use postgraduate study to delve deeper into their discipline and others use it to cross over and explore other design disciplines. For example, graphic designers might move into interactive media design and furniture designers into interior design. In some cases this can lead to wanting to undertake further research and study for a PhD (Doctor of Philosophy). This will enable you to become a true expert in your field, perhaps with practical and written work published. Other students become very passionate about the written word and develop excellent writing skills, perhaps moving into journalism.

The main contributors to this book have amassed many years of experience through practising, teaching and writing about design. Some of the contributors of the project stories were students when this book was published. Others were just beginning their careers, with some setting up their own companies. What all these people have in common is a desire to tell you about their experiences as a practitioner and offer advice into the profession and practice of design.

We hope that all of the experiences contained in this book will help you to make a success of your studies and help you plan your own career. This is how the text has been structured.

Part 1 – What is design?

Jon Penn's illustration of his Spatial Design proposal

This section contains the various disciplines of design. You will see where you might fit into the big network of design and what possibilities are open to you. Design is a multi-disciplinary profession covering **craft, fashion and textiles, graphics, interactive media, theatre, film and television, spatial** and **3D design**. Each of these disciplines is covered in its own section. In your future you may be working with any of these disciplines and very possibly you will cross over into others at times. Therefore it is important to develop an understanding of the differences in motivation, skills and approaches between these different designers. The process of design is described, illustrated by project stories from industry and from students, so you can gain a greater understanding of what it takes to be a good designer.

Each of the design discipline sections contains an overview and details about:

Today's industry. This provides you with a broad overview of the industry and illustrates the kind of companies that exist and how you might work within the industry.

What courses are there? Describes the range of study opportunities and highlights the differences between them.

Career opportunities. Covers the breadth of possibilities and where they might lead are explored.

What will you be taught? This breaks down the first, second and final year project work.

What is the design process? This is explained through student and industry projects, illustrating the process of designing in the specific disciplines.

What will the final year be like? Read about the students' stories and projects to illuminate the final year experience.

Preparing your portfolio. Summarises the specific requirements for the different disciplines. There is also a comprehensive set of guidelines in Chapter 9 'How to succeed as a design student' and Chapter 15 'So, where is this going to take you?'.

Further resources. This is an important list of books, websites, etc. that signposts you to core texts that you will want to explore.

Project stories and career profiles. These are captivating stories, full of personal insights into either an individual's career or how a student, designer or company worked through a project.

Part 2 – What do you need to know?

Exploring colour and texture

This section contains information on **'How to succeed as a design student'**, **'Being creative and innovative'**, **'Appreciating aesthetics'** and **'Working with colour'** – all essential design skills. Here you will find advice on study skills, how to manage your learning and preparing a CV and portfolio. There will be examples from the real world and education once again, exploring the aspects that designers deal with in aesthetics and colour, mapped out by experts in the field.

Part 3 – What does it all mean?

Tamsin Lakhani's illustration of her 'Fashion Knitwear' collection

This section covers the context in which designers design, with information on **'Design history, culture and context'** and **'Future directions'**. The former defines the theories that influence and steer design through society and time. The latter looks at the part designers are playing in society, now and in the future, and outlines some of the key themes – **sustainability, inclusive design** and **socially responsible design**.

Part 4 – What's next?

John Moore's jewellery

Understanding how to make the most of your time studying is to have an idea of where you might be heading in the future. **'So, where is this going to take you?'** will help you explore potential careers in the industry. If you think you might be interested in self-employment then **'Working for yourself'** will give you a real insight into what it might take for you to consider this as an option. It explores business and finance issues and covers how you can protect your designs by understanding intellectual property rights legislation.

The text illustrates the variety and excitement of the profession of design. We want you to appreciate the opportunities for working in teams with other disciplines, as the following student's experience demonstrates:

Fairline Targa 58

Question: How many designers does it take to design a boat like this?
Answer: Transport designer, interior designer, furniture designer, design engineer and one very lucky design student on placement! Designers from other disciplines were involved, but not within Fairline directly – graphic designer, textile designer, ceramics designer and lighting designer.

The Targa 58 from the company Fairline, one of the world's most famous luxury brands, is a UK-designed and built, 18.07 metre-long motor yacht that will sleep six people plus an optional crew cabin. To bring this and its other products to market is a huge undertaking for this Northamptonshire-based company. Its design facility in Oundle is home to many of the designers listed above. The others are based in the supply companies, who provide items such as furniture, lighting, materials and navigational systems.

Louise Collin was on placement with Fairline during her year out from her Furniture and Product Design degree. As well as seeing first-hand what was required to bring products like this to market, Louise was involved in many aspects of the design work.

Louise Collin on placement with Fairline

Saloon dining table

Her biggest contribution was to the saloon dining table. This was a multi-functional piece of furniture that enabled a clever use of space within the interior. Creating a notion of space, together with using all of the available room within the interior of a boat, are both important factors. Louise had to create a dining table that was large enough for six people to dine at. It also needed to store all the crockery and cutlery on the boat and have room for bottles.

It was also desirable for the table to hinge, to enable easier access to the seating, and of course the styling had to fit in with the overall design of the interior. Louise's solution cleverly packaged all the crockery securely – another vital consideration, as the boat can get up to speeds of 38 knots and the storage must stop any items from becoming displaced.

She was able to develop the idea from initial concept to 3D modelling. This was an important part of the process to prove that it would work and involved collaborating with product and furniture designers and the technical staff in the workshops. Louise could see that this project and the whole placement experience had developed her communication, presentation and negotiation skills – all invaluable for her future as a designer.

There are many more personal stories like this in the book, told first-hand by the students and designers involved. We hope this insight will inspire you.

In summary

Every section contains practical advice, project stories, theories and further resources sections. Following a good read of this text you will:

- understand how simple a good design can be
- see how you can take complete control of your work
- discover the multi-disciplinary nature of design
- appreciate the complex nature of the design industry
- become aware of all the skills you will need to succeed
- appreciate the opportunities open to you by learning from others' experiences.

This is an essential read from the moment you become interested in studying design until you have completed your studies and are into your first years of employment. With a pencil always to hand, feel free to personalise the book: use it; write your thoughts in it; fill it with multi-coloured Post-It™ notes. Studying design offers many wonderful opportunities and making the most of them is the key to succeeding in this industry. Use this to keep you one step ahead, so that you can prepare for what lies in store.

Good luck.
Jane Bartholomew and Steve Rutherford

Craft design

By Professor Kevin Petrie, The University of Sunderland

'Learning how to use materials and tools allows them to become like an extension of my body, which in turn enables me to better communicate with others through the objects I design and make.'

Mike Holden, glass and ceramics graduate

Perspective neckpiece by Chris Castillo, Jewellery and product designer

Craft design is the manipulation of a specific material to create functional objects or artworks. However, craft is so much more than this. The appeal of the work is about the surprise and wonder of how it was made and the intrinsic beauty of the material that the designer brings to life. Traditionally this requires handwork or handicraft, meaning the making of objects using only the hands and simple tools. In the twentieth century the term 'studio crafts' emerged, which relates to crafts practised by independent artists, often working alone or in small groups, as opposed to mass manufacturing in factories. Studio craft includes ceramics, glass-making, metalwork, jewellery, woodworking, textiles and paper-making.

Perhaps what sets the 'craftsperson' apart from other kinds of designer or artist is that they are likely to develop a much more intimate and expert understanding of materials and methods of working them. Therefore, their designs or artworks tend to emerge from this in-depth knowledge of what can be done with a material, for example clay or glass. This gives a great advantage over the non-specialist, who is unlikely to be as able to push the boundaries of a material without this knowledge.

The expression of skill might sometimes take precedence over function, for example through a beautifully turned wooden bowl, the perfect glaze on a pot or a stunning polished and cut-glass form. This combination of the beauty of materials and the skill of the maker can become highly prized and the objects are justifiably sold alongside paintings and sculptures in galleries. The crafts can also cross into other design disciplines, such as interiors, architecture, fashion or art. For example, the Turner Prize-winning artist, Grayson Perry, is essentially using craft skills in pottery to express powerful and, to some people, shocking aspects of life. His work is displayed in major art galleries around the world, so in this case craft meets the art world.

This chapter outlines the opportunities that exist for the crafts graduate and may help you to understand which possible future careers would suit your talents, skills and personality.

Today's crafts world

You might think of 'crafts' as referring to the making of traditional objects, such as functional pottery or

'Total Policing' 2011 (3-metre diameter) by Cate Watkinson, commissioned for Northumbria Police Area Command Headquarters, North Tyneside

baskets. However, the crafts encompass a much broader set of career routes and could be said to overlap with industrial and product design and, as we've already seen, fine art. The beauty of functional craft products though, is that they are accessible and marketable to the general public and have a broad appeal for the domestic buyer.

The term 'designer-maker', or 'maker', has now become commonplace in this field. It usually describes an individual who develops concepts, designs and makes their own work. These kinds of practitioners might make functional items, such as jewellery, pottery, textiles or furniture.

The great advantage of jewellery is that a small studio can be set up at home and the products are easy to transport, making it easier for the seller to display. A maker producing work in clay, glass or wood is likely to have to set up a larger studio, possibly away from their home, due to the scale of equipment required. An alternative to this would be to hire studio space for focused periods of production. For example, a glass-maker might hire a day's access to a glass-blowing studio, with furnaces and other equipment, to produce the initial blown forms. They then might hire a 'cold working' studio, with machinery and resources to finish the work by grinding and

'A work in progress' by Grayson Perry

Courtesy the artist and Victoria Miro, London © Grayson Perry. Photography © Stephen White

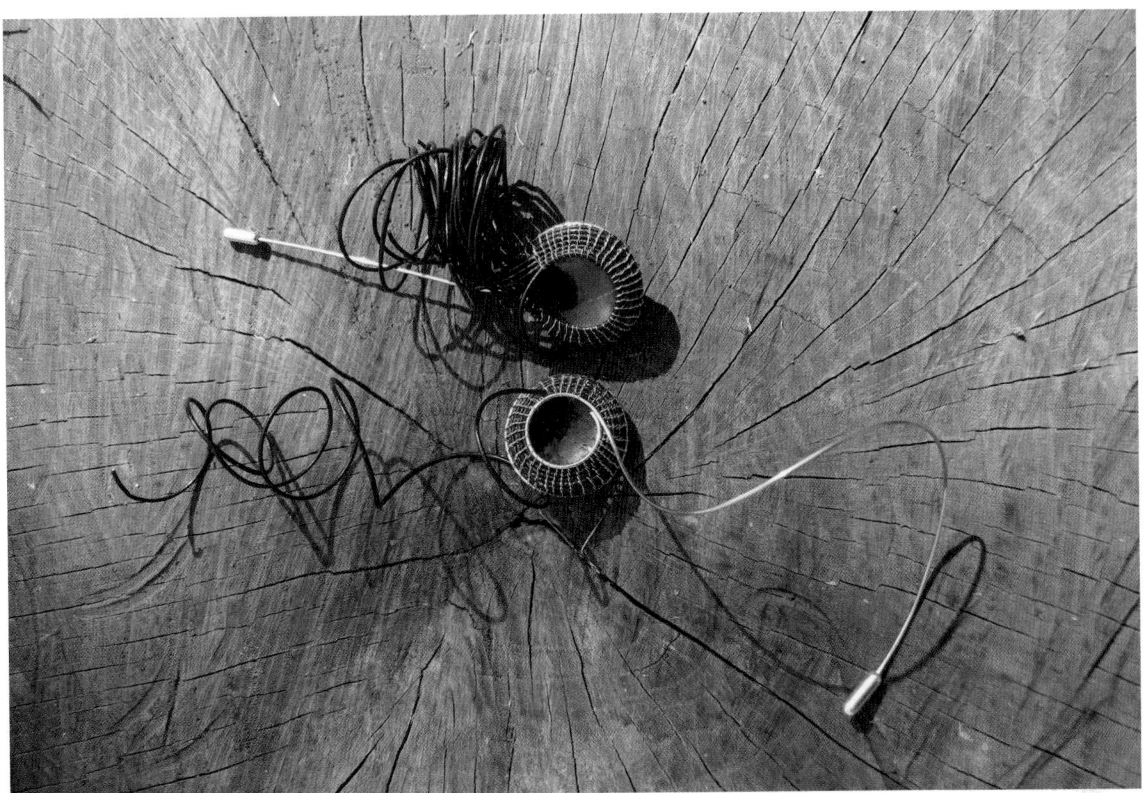

Two pins: basketry-inspired contemporary body adornment by Anna Pope
© Anna Pope/www.annapopedesign.com

polishing. They may even transport bodies of work to another part of the country and pay another skilled craftsperson to finish the pieces. This is common with, for example, the polishing of glass, which is a highly skilled process.

Designer-makers working in this area will often divide their production into two groups: **'bread and butter'** and more **'one-off'** pieces.

Bread and butter: refers to work that forms the staple of the designer-maker's income. This work is often produced in relatively high numbers and sold for a reasonable price. It could include perfume bottles, wine glasses, small ornaments, stained glass for the home, functional ceramics or souvenirs.

One-off: bigger and/or more elaborate pieces, which show a greater complexity of skill or a more experimental approach. The designer-maker may make these pieces to commission or to exhibit alongside the cheaper pieces as a way of attracting interest and developing a reputation for their specialist approach and style.

Wakenshaw Window, Newcastle Cathedral by Cate Watkinson

The realm of the craftsperson also extends beyond the studio to encompass larger-scale work, such as public art and design commissions. For example, a stained glass artist might make windows for churches and cathedrals but they may also develop skills to design large-scale architectural projects and apply them to building facades or street furniture, or produce art pieces for public and private venues.

This approach will involve applying for commissions, which are often advertised in magazines and websites, such as *AN* (*The Artists' Newsletter*). The successful applicant will then respond to a brief and develop an initial idea, which might be presented as a drawing or, as is more common today, as a computer-generated visualisation.

'Home', thrown porcelain and found vintage spoons by Megan Randall
Photograph by Dave Williams

What courses are there?

Craft courses can have different names, ranging from the material-specific such as 'Glass and Ceramics' to broader titles such as 'Applied Arts', 'Contemporary Crafts' or '3-D Design'. There are also courses that deal with more specific aspects of the crafts, such as conservation of wood, metal or architectural interiors, entertainment design crafts, which focus on theatre sets, props and costumes (refer to Chapter 7 for more information on courses in the performance industry), or courses with a focus on promotional and marketing products (refer to Chapter 4 for graphic design).

Specific crafts education has a number of possible strands, including courses based on materials such as glass, ceramics, metals or wood, or broader-based applied arts or contemporary crafts courses. The latter will involve learning about a range of materials and techniques, often specialising at some point during the course. Both types of course largely aim to teach students how to operate as professional designer-makers when they graduate. Some courses might have a 'design' focus, some a 'making' focus, while others take a more philosophical, 'fine art' approach. As the following text goes on to discuss, the generic business skills you will learn can also benefit the transition into other career routes.

COURSE TITLES: **DESIGN CRAFTS; DECORATIVE ARTS; GLASS AND CERAMICS; JEWELLERY; APPLIED ARTS; ACCESSORIES DESIGN**

COURSE DESCRIPTION

As well as the knowledge and experience of working with specific materials, craft-based courses offer a great range of transferable skills that can be taken into other career routes – for example, communication skills, costing, marketing and teamwork. Although there probably won't be many exams or essays, craft courses (like all design courses) should not be seen as an easier option. They require students to work hard, both in terms of 'thinking' and 'making'. Dedicated students who are willing to invest time in nurturing their talent to learn about their material can gain skills that set them apart from other designers. In turn, that can give the crafts graduate some unique selling points and areas of specific expert knowledge in materials, offering a very rewarding, surprising and sustainable career path.

Suet Yi Yip chose to come to the UK to study craft as she wanted to leave behind a stressful lifestyle in Hong Kong. Following graduation in 2011, she now operates as a designer-maker working with clay and wood and says this is the best decision she has made.

Ceramic birds by Suet Yi Yip
Suet Yi ceramics

> *'My inspirations are like ghosts; I do not really know where they come from but they like to "appear" in my head before I go to bed or after some exhilarating exercise. Living in the countryside and being surrounded by nature is subconsciously inspiring me.'*
>
> Suet Yi Yip, Craft Designer

Visit her website to see more of her work: www.suetyiceramics.co.uk.

The ideal crafts student will have an interest in art and design in general and a desire to make things. They will also be keen to learn and, perhaps most importantly, willing to put the time in to understand the materials they choose to work with. It will not be possible to become a skilled maker just by attending the occasional lecture or practical workshop. You will need to use the time allocated outside of taught sessions to practise and develop your own approaches to the materials.

For example, glass-blowing student Mike Holden took real advantage of the learning opportunities offered as part of his course to develop his making skills and contacts. Mike applied for and received university funding to take a class at the Pittsburgh Glass School in the USA. This expanded his skill base and introduced him to his international peers. Mike also undertook a funded internship at the fair trade company Traidcraft, where he designed a logo to be stamped into wine glasses made by the Bolivian company Crisil to separate it from other generic glassware and make it more exclusive. Mike invested the money earned from the internship to travel to Bolivia and test his design with the glass-blowers there. Mike said this 'real world' experience was 'an amazing opportunity to collaborate and share knowledge with Bolivian glass-makers'.

CAREER OPPORTUNITIES

The UK's Crafts Council provides key support for the producers and consumers of the crafts. In the 2010 report 'Craft Matters', they provide a useful overview of the area:

- there are 35,000 contemporary craft makers in the UK, producing a combined turnover of £1bn each year. As well as producing their own work, these makers contribute to many businesses and industries, including film, theatre, dance, fashion and product design.
- over 1 million people visited Crafts Council exhibitions, fairs and events in 2008/09.
- 11 per cent of the UK population visited a craft exhibition in 2008/09 and 17 per cent participated in craft activity within that year.
- thousands of people buy craft; 30,000 visitors to the Crafts Council's 'Collect' 2009 and 'Origin' 2008 events spent over £3m on contemporary craft.

Gill Wilson's paper-making studio. See her career profile at the end of this chapter

The paper-making process

SELF-EMPLOYMENT AND RUNNING YOUR OWN STUDIO

Many graduates working in the crafts set up their own business, often in a studio environment, to produce their work. They may well be supported by grants or loans to invest in key pieces of equipment, such as kilns. They are likely to sell their work through the studio itself or galleries, shops and craft and design fairs. Individuals may purchase items that are for sale or may **commission** a piece of work.

A good website will be key for promoting commissions and is also useful for networking to get your name known. Another option is to run craft classes in your own studio. Refer to Chapter 16, 'Working for yourself', for more information about self-employment.

DESIGNING FOR THE CRAFT INDUSTRY

There are numerous ceramic and glass manufacturers developing product ranges for domestic and contract markets. Interior companies and brands have in-house designers whose job it is to stay ahead of the market place and offer goods that are in keeping with 'new season' trends. This involves plenty of research, and awareness of other products being developed in the company so that there is continuity with the style in a particular season's range. Heals, The Conran Shop and department stores such as House of Fraser and John Lewis will create a specific look and have their own designers develop the ranges, which they will then source or commission from manufacturers. Individual manufacturers, such as Denby or Wedgwood, operate with a team of designers, developing not only three-dimensional forms but also the colour and surface decoration.

'Placement' by Megan Randall
Photograph by Dave Williams

There has been a decline in jobs in the UK's craft industry, for example in the ceramic tableware business, but there is an increased demand worldwide for designers in these fields. The skills learnt through a craft course can take you into this industry and other allied areas, such as product design or interior design. The sensitivity and understanding of materials and your attention to detail, learnt through your craft course, is sure to make you stand out in these sectors.

TEACHING TO SUPPLEMENT YOUR CRAFT PRACTICE

Crafts can be a great focus for teaching – either in schools or at further or higher education institutions. Often those interested in doing art also want to learn craft skills, and this offers the crafts graduate a potential employment route with a reasonable, regular, part-time income. A designer-maker might work in a school to develop a specific project with the children resulting in an artwork for the school. This could be a tiled mural or glass screen, for example. The designer-maker is likely to help the children translate their designs into a craft material, such as ceramic, glass or textile.

As the reputation of a crafts maker develops in terms of their work (as well as their personal qualities), they might well be invited to teach a **master-class** at nationally or internationally recognised centres for the crafts – such as universities, artists' studio cooperatives or at conferences.

This kind of teaching can be very hard work and demanding, as the paying students will want to take as much knowledge as they can from the teacher and get their money's worth. However, this can also offer a wonderful experience for the teacher as it can take you to parts of the world you would not otherwise visit and introduce you to like-minded people. As well as being rewarding, teaching can also be a great

Commission: when a customer likes the work of an individual designer-maker, but desires something slightly different, then they may ask the designer to make a specific piece for them.

Master-class: as a 'master' of a particular material or technique, you will pass on some of your specialist knowledge and experience – often to other professional artists, makers or designers.

learning experience, which in turn helps the development of the teacher's work.

So, the career of a designer-maker is likely to involve what might be called a 'cocktail' of income streams, including selling work, teaching and undertaking commissions. This might be supplemented by other part-time income from allied areas, such as museum work. The skills of the craftsperson/designer-maker might also enable you to move into other areas of work.

1. **Gallery work** – helping the public to learn about the work shown in exhibitions, developing education programmes linked to exhibitions or galleries, selecting and supporting artists and designers showing their work in galleries and helping buyers to select work to purchase in galleries.
2. **Art therapy** – can involve using art to work with children, young people, adults and the elderly to support a wide range of difficulties, disabilities or diagnoses.
3. **Events organisation** – initiating, planning and organising conferences, exhibitions, trade fairs, craft shows and tours.
4. **Environmental planning for local councils** – landscape design, design or commissioning of 'street furniture', such as seating and railings.
5. **Community work** – might include working with disadvantaged groups to develop confidence and creative skills.
6. **Museum curating** – initiation, selection, design, organisation, promotion of exhibitions.
7. **Retail management** – selecting and ordering stock, presenting products for sale, commissioning new products from designers to sell, managing staff and working with customers.
8. **Council arts officer** – working with clients and designers to develop commissions, contributing to the arts policy of a city or region.
9. **Arts administration** – administering funding for artists, designers, organisations and events, promoting and supporting the arts in a city or region.
10. **Publishing** – commissioning publications, working with and advising authors, editing texts and overseeing design.

Lauren England being taught glass polishing by tutor Colin Rennie

Image courtesy of the University of Sunderland Glass and Ceramics Department. Image: Chris Auld

WHAT WILL YOU BE TAUGHT?

Most craft courses will have a similar structure. In the first year you are likely to learn some of the basic techniques relating to materials and processes. For example, if you are studying ceramics you would most likely learn how to build objects by hand, throwing on the wheel, mould making, casting objects in a mould using slip (liquid clay), glazing, firing and some techniques for decoration. You should not expect to master these techniques in your first year, but just get a flavour of how you might use the materials.

Frazer Doyle, ceramics student, opening the kiln to check his finished pieces – always a tense moment!

Image courtesy of the University of Sunderland

First-hand observational drawings capturing form, texture and colour

A glass-blowing demonstration by James Maskrey
Image courtesy of the University of Sunderland

THE FIRST YEAR

In your first year you might learn about the 'science' behind a material. What is glass? What is happening when clay is heated in a kiln? How do different fibres react to different dyes? What are the different kinds of metal? This kind of information can be invaluable, as it will help you to gain a better understanding of why materials behave in certain ways and why things might not turn out the way you expected them to!

You will also learn how to keep visual records of all that inspires you and to understand the importance of research; themes can be useful as a starting point for research. This is something that you will be asked to work on throughout your course. In the wider world, research is about developing new knowledge in relation to a specific question, issue or problem. For the artist or designer, it is about gathering visual information in order to develop an idea of what you want to make. This might involve drawing from life to gather raw material as a starting point, developing this literal information through more imaginative drawings and making models to explore variations on what you might make.

THE SECOND YEAR

In your second year, you will still learn new techniques but will start to develop a more personal set of ideas and approaches to resolving a design brief. This type of work is often called 'self-negotiated'; this means that you will plan your work in response to a brief from a personal perspective, but with your tutor's support.

It is important to remember that each area of the craft field is quite small. This is an important advantage because it means that the ambitious crafts student can be networking internationally in their field by the end of their course. To do this you will have to take advantage of staff contacts and also work hard to maximise your networking chances – for example, through applying for funding to travel to take part in key international events, such as master-classes or exchanges.

Your work will also develop in terms of sophistication and your ability to control the various factors that define your designs. You will become more proficient with scale, weight, surface quality, material choices, colour and aesthetics.

Research can also involve collecting images of other artists' work as inspiration for your ideas, but your own research should be the largest component. A large folder of printouts from the Internet is much less impressive than a folder of 'ideas drawings' done by you. You are likely to have classes in different approaches to drawing and how to develop ideas. Some students find this aspect hard, but it is well worth the effort as good research usually leads to good work. Remember, tutors are always willing to guide and support you – so don't be afraid to ask!

You may also attend lectures covering general art and design history, or craft-based themes. As part of this, you are likely to be asked to give presentations and write essays on a theme. The importance of these aspects should not be underestimated. The confidence you will gain in giving presentations will be invaluable in many professional contexts, such as applying for commissions. For more information about what to expect from studying this subject, have a look at Chapter 13 'Design history, culture and context'.

As well as having lectures on art and design history, you might also be offered lectures by graduates or visiting artists working in your field. Again, don't underestimate the value of this. What you will get

Exploring basketry as inspiration for body adornment by jewellery designer Anna Pope (see her project story at the end of this chapter)

here is a first-hand insight into your chosen profession that might provide you with role models and tips on what to do – or what not to do!

Alongside the making and idea-development skills that you will learn, it is also important to develop other professional skills, such as writing. This is a cornerstone of any profession and could also lead you into a variety of other careers, from journalism and marketing to museums and academia. In order to develop any art and design career, including the crafts, you will need to be able to write succinctly and effectively, and there are many instances when you will need to do this:

- producing a statement about your experience in relation to a job application
- applying for a grant to buy equipment for your studio
- telling others what they will learn in your master-class, or
- giving prospective buyers information about your work in a brochure or on a website.

This last instance is possibly the most important, as the ability to write good promotional material is key to your success. It is not an exaggeration to say that writing can change your life! Good writing can often stem from 'good reading', so keep reading about your subject and think about how others write and communicate. Then try developing your own statements that communicate what your prospective buyers might want to read (not what you necessarily want to write).

WHAT IS THE DESIGN PROCESS LIKE?

You might well be given a brief or a theme to work to. This will help you to develop and focus on one idea that you can explore through learning about the material or process. Examples of a theme might be: 'still life', 'structure', 'contain', or 'interior/exterior'. Alternatively, you might be asked to make a specific object, for example a ceramic cup or a nutcracker out of metal. This framework of a theme or an object is very useful as it gives you a focus, and then your creativity can manifest itself. To paraphrase the twentieth-century Russian composer Stravinsky, 'the more boundaries you give yourself the freer you can be'.

For example, if you think of the 'cup' as a boundary, imagine all the variations that this might take if you consider how you might explore form, scale, texture and colour. In fact, the 'cup' might be used as a basis to express almost any idea or concept. Without some kind of boundary, you might find that you become almost immobilised by the thought of where to start, which then makes the potential to do anything almost impossible. This is often referred to as having **creator's block**.

Glass and ceramics student Mike Holden wanted to create a life-size piece of work as his final project. He undertook research into the exploration of space, form and movement. Alongside the theoretical research, he investigated how to represent movement by using folded paper to create geometric forms. He

Creator's block: a term used to describe the moment in a project when your brain grinds to a halt because it has overheated! In this situation you will need a proper break – a long walk or some form of exercise. This will reset your brain. Leave the work alone for at least a few hours or days if you have to, to unblock your creative flow.

also used polycentric lines to create mathematical curves to track the movement aspect of the project.

The images below show the design process made by Mike leading up to his large sculptural installation.

Geometry dance contextual research into the exploration of movement

This work references Muybridge's famous photography, which recorded movement and combined it with geometry. Rudolf Laban also used this in his theory of **choreutics**.

> **Choreutics**: thought processes, or 'paths', are tracked spatially and are represented as a series of locations, creating flat-sided shapes that track the movement.

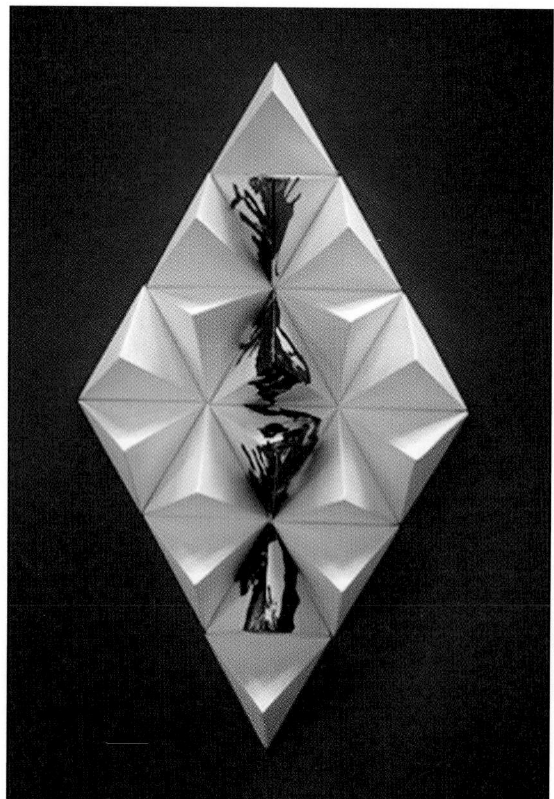

Paper forms – a continued exploration of form using folded paper

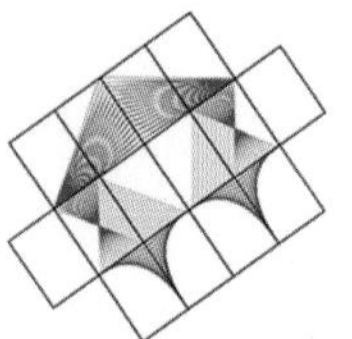
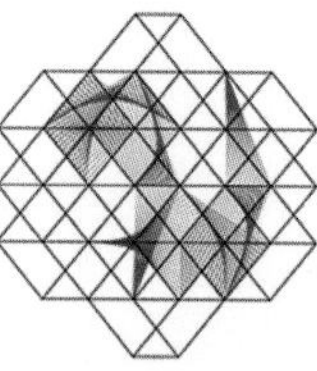

Vector paths, using polycentric lines as a framework

Movement analysis: 'Vector Paths 2012', 1.5m cube by Mike Holden

Photograph by David Williams Photography

As mentioned above, the crafts can be an excellent medium to convey ideas and concepts, as well as to make functional objects. Katherine Butler uses drawing in its broadest sense to develop a body of ceramic work based on the powerful subject of her grandmother's battle with cancer.

In her sketchbook drawing for her 'Malign' installation, next, Katherine starts to explore variations on the theme, and how objects might relate to each other.

Working through her ideas, this is another investigation drawing. At this point in the project, some of the more detailed, developmental drawings could be presented as artworks in themselves and be shown alongside the three-dimensional pieces

Katherine Butler's initial drawing for her 'Malign' installation, charcoal and pencil, A3 size

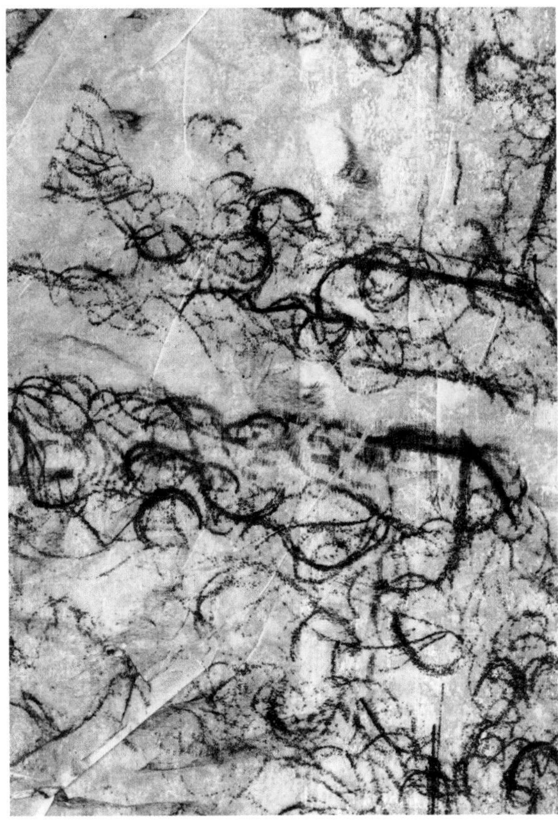

Investigative drawing, charcoal, pencil, acrylic and packing paper

of work to help viewers understand the context for the piece.

The 3D form for the installation piece to be called 'Malign' is investigated by working with cardboard to mock up potential forms. This approach is valuable because it allows you to test and develop ideas quickly before committing them to the more time-consuming craft material – clay in this case.

3D modelling development, cardboard, pen and pencil

Katherine's 'Malign' installation (detail), extruded and slip cast porcelain, 60 × 30cm
Photograph by Dave Williams

This final image was used to promote the British Ceramics Biennial in *Crafts* magazine. This emphasises the importance of taking good quality, professional photographs to promote your craft work. Katherine went on to complete an MA in Ceramics.

NEW TECHNOLOGIES

The use of new technologies is playing an increasing part in the craft field. For example, the 'Flat Pack' design project developed by Colin Rennie is a good example of the combination of a computer-aided

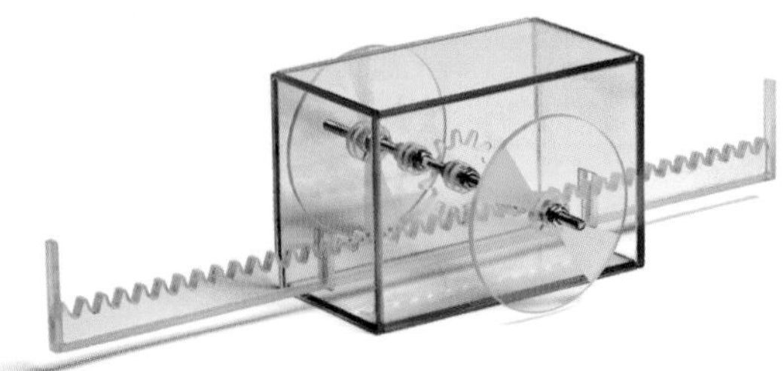

'Flat pack desktop sculpture' by Robyn Townsend, 30 × 12cm
Photograph by David Williams

product design process used with the 'craft' material of glass. The project introduced computer-aided design to students and also combined this with water-jet cutting and glass. Students used the Rhino 3D Design software, which led to the design of the objects via rendering them on screen.

From these designs the students cut 'flat pack' shapes from sheets of glass using the water-jet cutter. Then, they assembled their products from these sheets. The project encouraged students to maximise the use of materials (important in today's economic and sustainability-conscious climate) and resulted in students having demonstrable skills in CAD CAM processes, with a particularly striking and unique output in glass.

One student said, 'The "Flat Pack" project was great as you actually made something real from something virtual created on the computer screen.' This link between designing and making is central to the 'craft' strand of design.

Porcelain slip cast pig heads by final-year student Leah Dennis

Final-year student Mike Holden in the glass-blowing studio

WHAT WILL THE FINAL YEAR BE LIKE?

In the final year you are likely to undertake a major project in which you will make a body of work, or one or more significant pieces. These will usually be presented as part of your final year exhibition. You might also have the chance to show a selection of your work at a national event, such as the New Designers graduate exhibition in London.

This is a great chance to bring together everything that you have learnt so far on your course and to start to operate like a professional. There will be some challenges, such as planning your time to make sure your work is ready on time. You will have to work around your peers with this. For example, you might have to negotiate and plan key things such as booking workshop facilities. It will be vital to plan ahead carefully, and do this by working back from your deadline. Also be sure to add in a contingency plan, as there is often something that doesn't quite go as expected!

As in the 'real world', you will have to work on several things at the same time and not just on making your pieces. Perhaps your cohort will need to fund-raise for a catalogue for your end-of-year show. You might become part of that team and help with the planning and developing the publicity for the show. This means that you will have to make finished pieces much earlier than you might think, so that you can get professional photographs in advance. Other important issues to consider are finding out about how to calculate possible prices for your artefacts and understanding the costs involved. You will also need to be able to write personal statements and press releases about your work. You will need to begin to gather the necessary knowledge early on in your final year in order to be able to prepare appropriately. Take a look at Chapter 16, 'Working for yourself?', for more detailed information and guidance on marketing yourself.

You are also likely to be working on your dissertation or another type of written work that requires considerable research and critical thinking, such as a professional plan or report. However, creative students can sometimes find writing challenging and so can either put off the written work to the last minute or spend too long on it. You should be mindful that both approaches can affect the flow of your studio work. Of course, the best thing would be to start any written projects early and manage your time carefully.

So, the final year can be challenging and it can be stressful but it will also be extremely rewarding (read Lana Crabb's story in Chapter 10, 'Being creative and innovative'). In your final year you should be focused on and confident about what you want to make and be able to bring your ideas and making skills together. It will also be a time when you start to stand on your own feet and make more independent decisions, without as much support from your tutors.

With all this going on, it might be hard to think about life outside college and easier to just focus on progressing through the course. However, you would be wise to think beyond the degree so that you don't lose the momentum gained from experiencing your final year. If you want to be a designer-maker, it is vital to get high-quality, professional-level publicity information, such as business cards, photographs, leaflets and a website, as you will be unlikely to progress without them. Think about forthcoming opportunities, such as competitions, residencies or commissions, that you could apply for. Draw up a list of galleries or outlets that you would like to see your work in and prepare to send them images. Think about how you can keep making your work when you leave college – can you hire facilities or could you work from home? Your tutors can help with all this so make good use of them while you can! Read Chapter 16 to find out more about 'Working for yourself'.

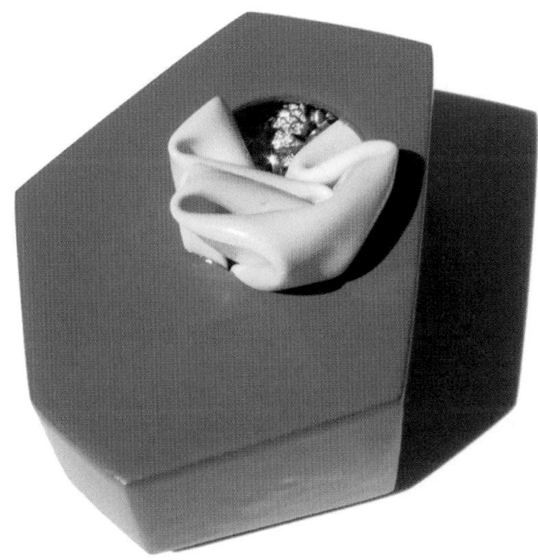

'Peek-A-Boo' gem brooch, painted metal and balloon
Lana Crabb Contemporary Jeweller

Preparing your portfolio

Right throughout your course you should be maintaining a portfolio, which contains a representative cross-section of your work. At some point you might want to apply for a placement, and a portfolio is the main way that the industry would make an assessment as to whether you have the right skills and abilities for the placement on offer.

This portfolio should include drawings as well as pieces that relate to your chosen craft area(s). Drawings are a way of visually demonstrating how you record experiences, and are your chance to show that you are interested in and curious about the world – this is vital for any creative person. Photographs of things you are interested in can also be useful but will not show that you have really taken time to observe and experience things through your own drawn interpretation, so keep these to a minimum.

High-quality images of your work are always necessary, but if you can take actual examples of your craft work to interviews, then this is very useful too. There is more advice on portfolios in Chapter 9, 'How to succeed as a design student'.

Conclusion

The crafts are perhaps one of the more accessible strands of art and design in that most people can understand what they are looking at when they see a ceramic pot, glass vase or piece of jewellery. However, the crafts are also noted as making an important contribution to 'fine art'. The starting point being a material and a function offers the viewer an opportunity to understand the piece more readily.

Craft is an extremely rich area of design to study, with often thousands of years of history to refer back to. In addition to this, there is great potential to combine or blend craft-based subjects with other subjects, such as animation, fashion and computer-aided design. This can yield intriguing results and be the start of something innovative.

It is the passion for making things that unites everyone in this discipline. It is important to consider the diversity of what the crafts can offer, and to think creatively about how you could take those skills into new areas of material exploration or design application. The crafts are about working with a wide range of materials and, through the personal interpretation of you as an individual designer-maker, developing something unique that will appeal to potential customers.

So, perhaps, the crafts are really about people and life as much as they are about materials and objects.

Craft, as you now know, is very diverse. The following stories and career profiles outline the different perspectives of those who have pursued a career in their specific craft disciplines. However, a greater perspective on the variety of craft discipline is on display at the numerous craft fairs held every weekend, up and down the country. Unlike the other design disciplines, designer-makers lay out their work, stand next to it and are only too willing to discuss what they've done with you – the next generation.

Further resources

Books

Adamson, G., *The Craft Reader*, Berg Publishing (2009)
A useful overview of contemporary thinking on the crafts.

Craig, B., *Contemporary Glass*, Black Dog Publishing (2008)
An excellent and richly illustrated book on the diversity of glass as a creative medium.

Hanaor, Z., *Breaking the Mould: New Approaches to Ceramics*, Black Dog Publishing (2007)
An outline of how ceramics can push the boundaries of craft and art.

Seecharran, V., *Contemporary Jewellery Making Techniques*, Search Press (2009)
A modern guide for jewellers and metalsmiths.

Veiteberg, J., *Craft in Transition*, Bergen National Academy of the Arts (2005)
Essays on the transition of crafts into other fields, such as fine art.

Websites

www.acj.org.uk
The Association of Contemporary Jewellers.

www.craftscouncil.org.uk
The starting point for all information on crafts: events, exhibitions, funding and education.

www.designedandmade.co.uk
The information hub for craft designer-makers in the north-east of England. It features a shop, blog and a contacts section for commissioning work.

www.nationalglasscentre.com
One of the UK's best craft centres, in a stunning location. See glass being blown, have a go, then buy something an expert's made in the shop!

Career profile

Name:
Gill Wilson

Current job:
Paper-maker and Gallery Director

Describe what you do

I divide my time between running a gallery and creating textile and paper-fibre pieces of work for the contract art market. Where possible I spend two days a week in the studio working on my own and three days in the gallery. I usually find that commissions are quite time-consuming. In a perfect world I would like to spend more time working on new ideas and developing new concepts and lines of work.

My knowledge and skills as a practitioner have been developed since studying for my degree, largely by making work over several years and stages. I am always researching new developments in my field, going on courses where possible and reading about current developments about craft.

Tell us about your career so far

I studied constructed textiles, and at the end of the course I was awarded a first-class degree and also received a Leverhulme scholarship. I studied paper-making in Japan for three months and this was the beginning of a career studying paper fibres and ideas connected with paper-making. I have worked with paper pulp as a creative medium and have aimed to push the boundaries of what it might possibly represent. I have worked mainly in the contract market, producing art pieces for hotels, and have also worked to commission – developing pieces for private customers. I have always worked with paper as a medium to communicate my ideas.

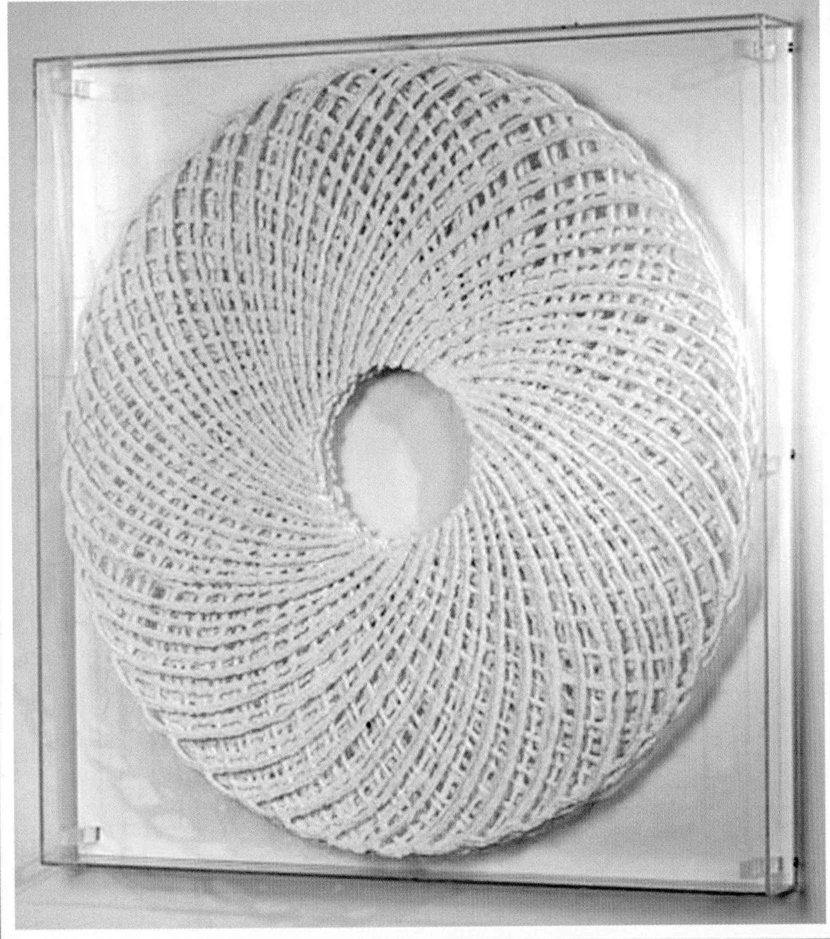

'Fibonacci' by Gill Wilson
Photograph by Adrian Heapy

'Solar' by Gill Wilson
Photograph by Adrian Heapy

The pieces I am currently making are multi-layered and present interesting atmospheric opportunities when lit and can enhance the surrounding environment.

Alongside producing my textile and paper pieces, I have always had other jobs to add stability to my work and life. I have a teaching qualification in the further education sector and have taught on a part-time basis for 15 years. I have also worked for the Arts Council and worked in various galleries.

I now run my own gallery. Although at times this has been quite demanding, I find the contrast between making and selling invaluable. My own paper-work is sometimes quite a solitary activity and the gallery work balances this out as it allows me to work with other artists to curate shows and to interface with the public.

Is there a particular project you would like to tell us about?

I have recently worked on a project for Lebanon – the project consisted of making eight paper pieces in red, cream and black. The work was commissioned for an international exhibition after a curator saw my work at Maison d'Objet – a trade fair in Paris where a wide range of products for interiors are exhibited together so that architects, interior designers and the industry in general can view new work. I am a member of Design Factory, which is an organisation that commercially supports and develops selected designer-makers by promoting their work to a wider audience.

I find it very rewarding to work across several pieces at a time and develop an ambitious set of work that moves my practice on. Often, the demands of very challenging work can bring with it very difficult times, but it also offers a real insight into further developments that might be possible with paper.

What advice do you have for students considering a similar career?

My advice would be to make a well-informed and intelligent appraisal of the proposition and realise that there is no career, only vocation. There can be no career where there is no industry. The only way to make a living in this area is to work hard on the basics, define a range of products and identify a market, wherever that may be in the world. Understand the numbers involved – know your production costs and costs to market. I have always been a great believer in studying business practice to support my creative endeavours, and would recommend that you do the same.

Project story

Company/Designer:
Megan Randall

Project title:
PhD study

Tell us about the project

I studied Glass and Ceramics at the University of Sunderland and my work follows two strands – design-based work and site-specific installations. Although at first glance these two strands of work don't cross-over, the same materials and processes are used in both. One of the installations uses thousands of porcelain vessels. These same vessels can also be sold individually, making them a product. Porcelain is the one 'constant' in my work. I am drawn to combining traditional processes, such as throwing, with newer printing and water-jet cutting techniques. I am very interested in domesticity and how ceramic objects fit in to our day-to-day lives.

I am currently undertaking my PhD, which explores ceramic interventions/installations that are sensitive to specific sites. Most of these installations use documentary methods (photography/video/animation) as a pivotal point of investigation. These works, although very exciting to make, are very unsaleable. My work that sells is used to subsidise the installations I make.

'Teapot'
Photograph by David Williams

I have sold work direct to the public at the Baltic Centre for Contemporary Art in Gateshead as part of the Design Event. I have also undertaken several commissions. These have been organised by myself and through other organisations such as Designed and Made, which is an artists-led network in the north-east of England.

What processes and skills were most relevant to this project?

In my work I use a mixture of techniques and processes, using found objects and creating handmade components that fit together to create a larger piece of work or stand-alone. I tend to always use porcelain for its natural translucent properties.

'Home' installation

What challenges did you face?

Due to having such a diverse practice, and sometimes working up to three different jobs to pay the mortgage, it is sometimes really difficult to get your head in the game: flipping from planning an installation to creating drawings for applications for funding and making time to throw 300 vessels. I tend to have four or five installations on the go at a time, either in the process of making them or documenting outcomes; this sometimes means that projects get dropped midway through making. Making ceramics around work means that there are days (nights) when you are sat up until four in the morning applying ceramic transfers or glueing brooch backs on.

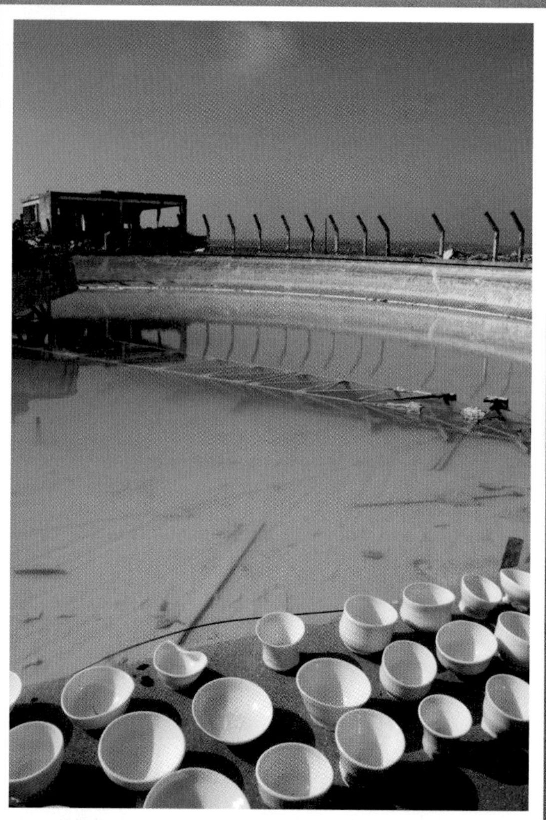

'Steetley Magnesite Works'

What were the highlights of the project?

I think doing my PhD at Sunderland has given me time to think about my work and enabled me to make more (larger) installations. Working with design companies, such as Traidcraft in India and Peru, was an amazing experience and it tested my ability to relinquish the control of making. It was also a fantastic experience to see how ceramic production works. Doing market research for Traidcraft engaged and altered my own practice, making me more aware of what the public are buying.

Project story

Designer/Company:
Cate Watkinson, Watkinson Glass Associates

Project title:
Solar cell seating

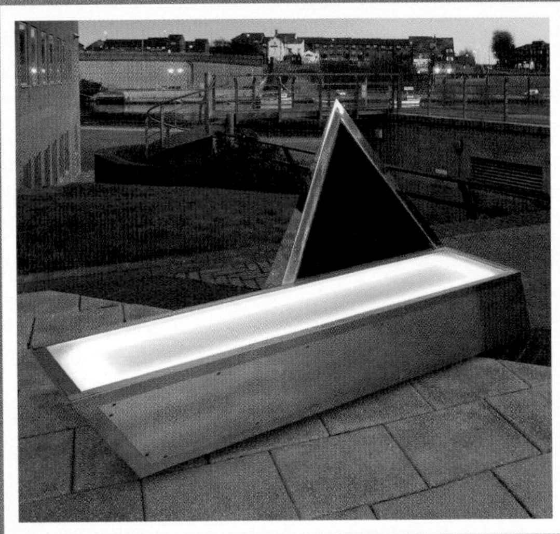

Solar cell seat, University of Sunderland
Photographed by David Williams

Tell us about the project

I have been interested in the idea of using photovoltaic cells (PV – a technology that a device or solar panel uses to produce electricity) in my work for some time. I had designed street furniture for the centre of Newcastle using illuminated glass backs for the public seating.

I had been thinking that it would be an interesting idea to make the seats light up themselves rather than them having to be 'plugged in' to the grid. Designing seats that are portable, easy to locate in town centres and self-lighting seemed to me to be something that town planners and urban designers would be interested in commissioning. The idea stayed in my head until a colleague told me about a Technology Strategy Board grant that would be appropriate for me to apply to for funding a project such as this. I applied and was duly awarded £15,000 to design, build and test prototype seating using PVs. I had no prior knowledge of using PV or solar cells so had to seek help and advice.

In the first instance I went to see a friend who worked with wind turbines and she directed me to the Department of Renewable Energy at the University of Northumbria. With the information gathered, I set myself a task to design and build a seat that had enough PV cells to gather sufficient energy from the sun, which would then be stored in batteries incorporated into the seat. These batteries would power LED lights placed under the glass seat top. A light sensor stored discreetly within the seat unit would then switch the lights on as dusk fell, illuminating the seat.

Many cardboard models later, I came up with a design. Over the years I have built up a team of subcontractors to call upon when I need assistance. I asked one of these, an architect, to draw up the structure I had designed. This was then taken to metal fabricators who made the framework for the seat. The architect directed me towards a renewable energy company that helped me work out how many PV cells I needed to give me enough energy to power the LED lights. The now-specified solar panels were then fabricated by a company in China and the glass elements by local fabricators. The whole piece was assembled for testing in the yard of the metal fabricators and it worked! The seat was transported to St Peter's Campus of the University of Sunderland, where it now lights the way for students returning home after dark.

What processes and skills were most relevant to this project?

Design skills are, of course, important, however there are numerous other skills needed in order to get a piece of work from idea to realisation. I have built up many good working relationships with other companies, of whom I trust the design skills and knowledge and who trust me. Communication is vital for these types of projects, and being a good project manager. It is important to keep each element of the project in hand and know when all the elements need to be brought together.

What challenges did you face?

Initially I did not know anything about PV cells and how they worked and how much energy they could produce. It was a steep learning curve and I was lucky enough to find people who were good at explaining these things to me. Coordinating quite so many different companies was also a challenge. Although the budget appeared generous, with half the money paid up front there was still a good deal of people to pay for their part in the project, so I had be to mindful of the budget and of the deadline. The project had to be completed to a set deadline in order that the second part of the grant would be released.

What were the highlights of the project?

The highlight of the project was seeing the whole piece come together, from its assembly, to its installation, to the first time I saw it switch its own lights on at dusk. Seeing my work out there in the public domain, and doing what I set out to achieve, gave me a deep sense of satisfaction.
www.watkinsonglass.com

Project story

Designer:
Anna Pope

Project title:
Basketry-inspired contemporary body adornment, final-year project

Tell us about the project

In my second year at Loughborough University I had the first opportunity to write my own brief and to begin to specialise in a specific subject area. My brief was to produce a collection of well-crafted, handmade, one-off pieces of body adornment developed from experimentation with already well-established basketry techniques.

The inspiration for this project derived originally from the early stages of the course, when we had the chance to explore a diverse range of products and materials. Projects included constructing stools in MDF and creating batch-produced ceramic vases. Little did I know that the very first studio-based module during the first year would become such a major part of my future work. The task, believe it or not, was to entrap and enclose an egg using only the wood we were provided with. I made the decision to cut a plank down into thin strips and weave them together.

I soon discovered the therapeutic and versatile qualities of weaving, and this spurred me on to research various techniques and materials used by other makers. This was how I came to learn more about basketry – a centuries-old, threatened craft with endless possibilities and strong links to botany.

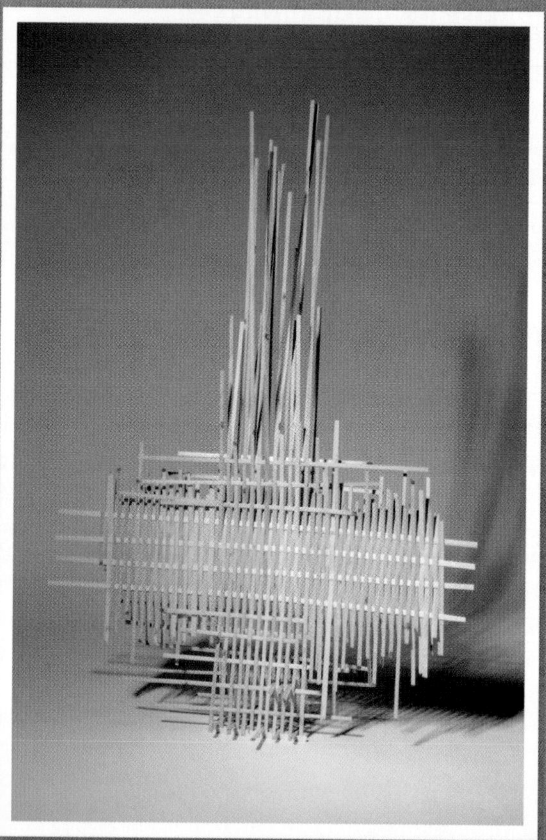

Basketry-inspired jewellery by Anna Pope

A jewellery-based project, early on in the course, helped me realise the pleasure I have when undertaking time-consuming techniques, with precision, on a very small scale. After much thought about what to do for a project in the second year, it dawned on me that I should try combining jewellery with basketry. To my delight this was a successful combination of processes.

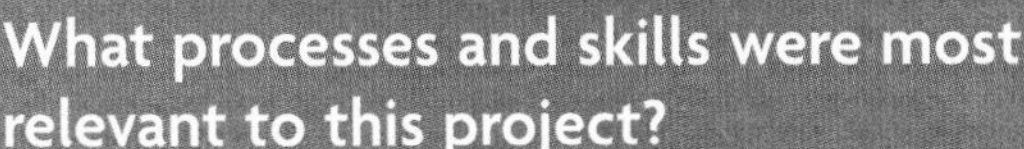

Close-up of enamel and coiling by Anna Pope
© Anna Pope/www.annapopedesign.com

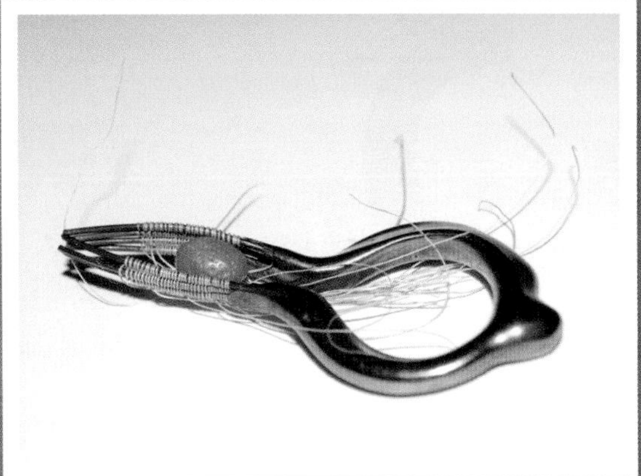

Woven ring on a white background by Anna Pope
© Anna Pope/www.annapopedesign.com

What processes and skills were most relevant to this project?

Through numerous exciting experimentations and by expanding my knowledge I produced two brooch pins created from a traditional basketry technique called coiling. Coiling naturally creates three-dimensional organic forms with a hole at the centre.

This adds an extra element to the pieces, inviting the viewer to look inside. The hole reveals an electric-blue enamel coating that is attached to the domed copper base; the blue hole stands out in contrast with the other materials surrounding it. This was inspired by berries, whose purpose is to attract wildlife. Each piece was individually crafted and made from stainless steel, leather, silver and enamelled copper. These materials were all carefully chosen for their specific qualities.

I intend to develop this concept further, and plan to continue making and set up as a designer-maker – the possibilities to progress with my ideas are endless!

What challenges did you face?

As basketry was not taught on my course, I had to research techniques and practise them myself by following the instructions found. I perceived that this could consume a lot of time and that I needed to devote time at the beginning of the project to understand the process, in order to successfully meet the deadline. At the same time, essays for other modules were required and therefore a timetable was crucial. Part of the project was to write a critical evaluation and I found that this helped me overcome many challenges and kept me focused, helping me to understand my work better. It was frustrating when experiments didn't work, but I kept on trying and eventually found solutions that gave me a great sense of achievement. I learned that things can get tough at times, but I persevered and it all worked out in the end.

What were the highlights of the project?

Learning new skills and discovering exciting and successful outcomes through experimentation. Also, learning traits about myself. I discovered many ways in which materials and techniques could be used and I also developed greater insights into different cultures.

For information on other work that I have produced please visit my website **www.annapopedesign.com**.

Fashion and textile design

By Joyce Thornton, School of the Arts,
University of Northampton

'It wasn't always my plan to have a fashion business. From a young age I thought about going into architecture or fine art. I was always fascinated by making something from scratch – developing ideas, creating patterns and forms. When I studied for my MA at the Royal College of Art, I started making garments and developing innovative construction techniques. This is when I saw the potential for starting a fashion label. It's really satisfying to see the collection coming together and to see the garments being worn.'

Alice Palmer, fashion designer

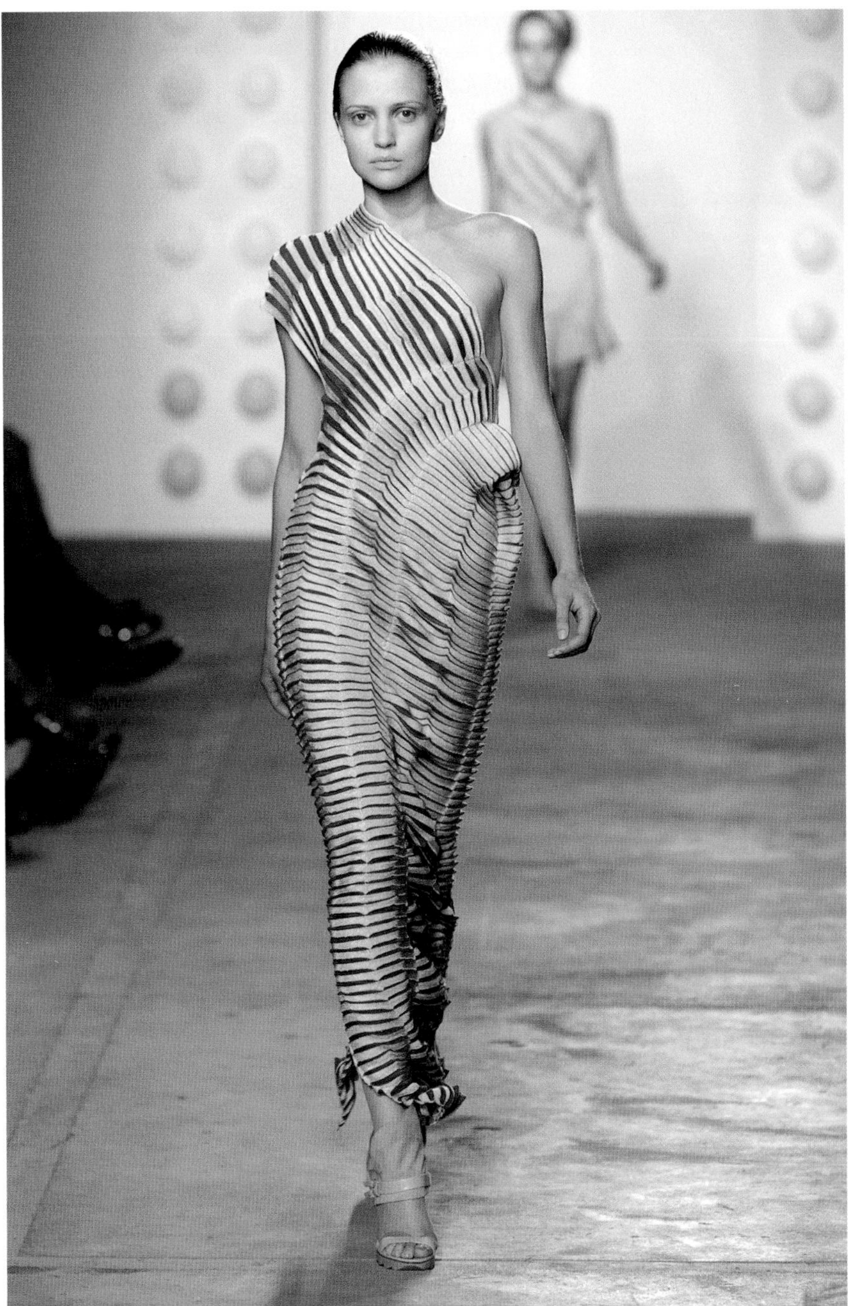

Alice Palmer knitted viscose and silk dress, 'Fossil Warriors' collection, spring/summer 2011
Photograph by Christopher Moore

Fashion and textiles is an exciting area of study, which has enormous appeal to many people with diverse talents and interests. Once seen traditionally as a 'lightweight' subject, there is now increasing respect in both education and in wider society for the fashion industry's complex and diverse business. There is also a growing realisation that the fashion industry makes a highly significant contribution to the economy of the UK, as it does to many other countries worldwide.

Now, more than ever, many young people are engaged in the fast-paced, ever-changing world of fashion and textiles as designers, marketers, salespeople, in the media and in public relations (PR). However, as there are many who choose to study fashion and textiles, the competition for jobs is inevitably fierce. You will have to be both talented and hard-working in order to be successful. The hackneyed phrase of having 'a passion for fashion' is simply not enough to ensure success. However, if, in addition to talent, you are

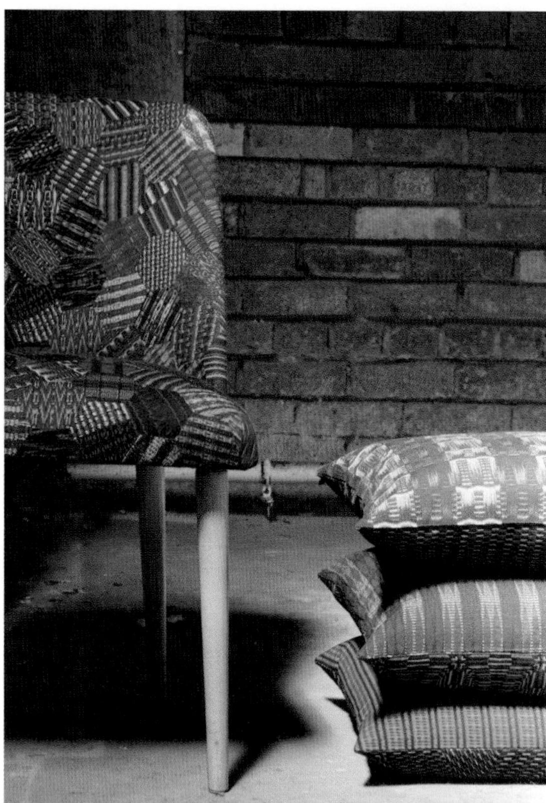

Sian O'Doherty's woven and printed textile collection
Photograph: www.danstaveley.co.uk

prepared to put in the hours and possess determination, dedication and a belief in yourself, the potential for an exciting and rewarding career beckons.

This section outlines the opportunities that exist in fashion and textiles and may help you to understand which possible career path might suit your talents, skills and personality. It details student work and career stories that will help answer questions you might have about studying a degree in fashion, textiles and related subjects.

Today's fashion and textile industry

The fashion and textile industries span the globe and have many facets. Internationally, fashion has become increasingly diverse and is now accessible to a wide range of consumers. A vast array of both established and new brands are available to choose from, and consumers can order with ease from around the UK and throughout the world. This is largely due to the many possibilities opened up by utilising the Internet. For example, many small businesses that traditionally would have had great difficulty financing the overheads and practicalities of running a small store of their own, have found that an online store is a much easier and more flexible alternative.

The fragmentation of the traditional levels of designer, middle market and mass market continues. There have been some big shifts in consumer attitudes; many people now feel completely comfortable creating their wardrobes by mixing high-street buys with designer pieces and adding vintage items, etc. Although seasonal trends still very much matter to the fashion savvy, individual style statements are encouraged, accepted and praised by the fashion media.

Fashion design graduate Anna Lee's final-year work

The opportunities for working in fashion and textiles are many, and the fashion or textile designer is just one member of a large team of creative people who have a role in delivering designs to the public. The industry covers not only the design and manufacture of fabric, clothes, footwear, accessories, interior fabrics, wallpapers and floorcoverings, to name but a few, but also, crucially, the presentation, sales, marketing and promotion of these.

Technology and e-commerce

With the on-going march of technology and the public's embracing of the Internet, the fashion industry has seen some radical and exciting developments in the last 10 to 15 years. For example, it is now accepted that hugely successful fashion retailers may have no physical stores, selling solely online. Companies such as ASOS.com, Mywardrobe.com and luxury e-tailer Net-a-Porter.com are recognised as leaders in this relatively recently created global field.

Their success is part of many purely web-based fashion businesses that are using new and interesting ways of engaging with and selling to their consumers. These relative 'newcomers' create an exciting, interactive online experience for their customers, with video, 3D swivel and zoom features, so that customers can examine their prospective purchases 'up close'. Feedback is encouraged, and initiatives are used to create strong customer loyalty and encourage multiple site visits, including apps, 'personalised' emails and increasing use of mobile messaging, featuring suggested products to complement the customer's latest purchases.

Careers in fashion e-tailing involve styling products for photography and video, writing inviting and accurate product descriptions (which requires fabric and garment construction knowledge) and sourcing and buying products from established brands and exciting new designers throughout the world.

The success of fashion e-tailers has motivated traditional fashion stores to improve and expand their own websites. Retailers have often found that online sales can make up for lower in-store sales, to a large extent.

The development of business-to-business online **fashion forecasting** has become a successful mini-industry, creating roles that require a wide range of skills and approaches.

International catwalk trends also play an important part in this, because it is the fashion designers that other design sectors look to for certain trend influences as it is perceived that they are at the forefront in offering consumers thoughts about what is coming next.

Teams of analysts translate the breadth of inspirations into predicted scenarios or themes (usually 18 months to 4 years ahead, depending on the industry). Companies such as Promostyl, **Première Vision**, Here & There, Moda, Li Edelkort's Trend Union and Peclers & Carlin traditionally have developed trend books for particular forthcoming seasons. These contain a full set of themes, palettes determining colour and texture, fabric/material swatches and product styles. Companies then buy these trend books and, more commonly these days, pay a subscription fee to some of the online fashion forecasting services, such as WGSN, Stylesight, Trendstop, MpD Click or The Future Laboratory. Armed with this information, in addition to their own research, companies are well informed about forthcoming trends and this can then filter into their future product ranges. The industry's embracing of these ever-developing resources has facilitated the increasing speed of **fast fashion**. Many debates around this way of working continue to rage. The opportunities in fashion are now exciting at all levels of the industry.

Luxury and high standards of workmanship are still prized, and many at the top end of the industry continue to thrive. However, at the opposite end of the fashion scale, the product quality and choice at the value end of the mass market has radically improved and continues to attract a broad customer base.

Fashion forecasting: the trend forecasting process involves gathering research and information from a wide range of diverse aspects of modern life and world influences. These might include factors such as the global economic climate, modern technologies and the natural world, combined with textures, colours, art, design, literature, music and film references, which are used to depict a set of conceptual themes for the fashion and interior industries.

Première Vision: as part of this trade fair, one of the smaller exhibitions within (entitled Indigo) comprises many smaller companies, often referred to as swatch studios. They specialise in developing textile designs that are then sold on to textile manufacturing and fashion houses. These are then used as designs to be put into production, or as inspiration for forthcoming collections.

Fast fashion: this is the term used to describe the speed at which new fashions can be conceived/sketched out and turned into collections and be ready for sale in the shops or on a website. Many successful retailers have refined and revised their systems, by using new technologies and advanced computer programs, but also by building flexibility into their production processes. This enables them to switch manufacturers if they have to, depending on what is being requested, and, in some cases, can reduce lead times to as little as three weeks.

Ethical and sustainable fashion

Alongside all of this, an increasing awareness of our wasteful, consumerist society is also picking up speed. This has led to the growth of **ethical fashion** brands.

> **Ethical fashion**: ethical fashion brands form a small but influential sector of the industry. In recent years many small brands have emerged with ethical credentials. These brands appeal to customers who are concerned about the impact on the environment of mass-production processes and the constant fuelling of our consumerist society.

Many companies are making real commitments to reducing the impact of their production processes on the environment. Big high-street retailers have also embraced ethical issues, with companies such as Marks and Spencer in the UK making real commitments to reducing their impact on the environment. However, the ever-present need for making a profit still limits progress. Some brands are forming a small but influential sector of the industry. In recent years many small brands have emerged with ethical credentials. 'Esthetica' is the collective name for the area at London Fashion Week where ethical fashion brands exhibit together to raise awareness of these all-important issues.

Fashion media

The fashion media is undergoing a time of great change; the traditional 'glossy' print media has generally suffered a shrinking readership. At the same time, the range of free online resources has grown extensively, and retailers have embraced new opportunities – developing editorial both for their websites and for new, in-store magazines. Fashion blogging has had an impact, with some knowledgeable bloggers, such as Susanna Lau of Style Bubble, respected for their refreshing, honest and relatively unbiased assessments. Lau is now also involved in many other projects, including styling consultancy (for example, she has worked with Topshop to create an edited range from their main collections). Image-based blogs, such as Scott Schumann's The Sartorialist and photographer Tommy Ton's photo blog for Style.com, have been particularly influential. Success in this field requires a keen eye for style detail, great photography and a clear, engaging, easily accessible format.

A radical aspect of all of this change is that blogs allow and encourage direct feedback from readers and consumers. According to renowned fashion writer Suzy Menkes, this marks a huge change. She famously commented, 'Fashion has ceased to be a monologue and has become a conversation.' More recently, new image-based sites, such as Tumblr, Pinterest and Instagram have opened up yet more online platforms, allowing further possibilities for stylists, designers, journalists and retailers.

Whatever course you do, studying both the practical and theoretical aspects of fashion and/or textiles will prepare you well for the industry in general. Depending on who you become, and how your specific skills develop, you may become a designer working for a multi-national company or set up your own business. Here is a brief description of the job roles that you might consider:

Fashion designer:

creates collections to suit a company's customer profile, working for a fashion brand, a garment manufacturer, an established designer or yourself.

Textile designer:

develops textile designs and fabrics for use in the fashion, interior, graphic, medical and automotive industries, working for textile manufacturers, design companies or as a designer-maker.

Visual merchandiser:

this role demands a clear understanding of the way customers respond to the marketing of products. Visual merchandisers are usually responsible for creating exciting and enticing store window and interior displays for retailers, and are therefore involved in the advertising and marketing of the products.

Buyer:

fashion and textile buyers are responsible for overseeing the development and selection of a range of products aimed at a specific market level.

Sales manager:

this role is about organising and leading a sales team to ensure sales targets are met.

Marketing and PR:

this job involves promoting the brand or the product to the world and making sure it gets good publicity.

Stylist:

develops trend research, visual mood boards and concepts to steer the design team.

Retail manager and assistant:

responsible for managing and motivating the retail sales team and ensuring that customers receive high levels of service.

Exhibition organiser:

works collaboratively with everyone involved in the exhibition and plans and manages the schedule of events.

Curator:

is responsible for acquiring, caring for and developing displays of works of art or artefacts to educate and inspire visitors.

Journalist:

this role is for those that enjoy writing in a creative, inspiring and entertaining way that engages the reader about the content.

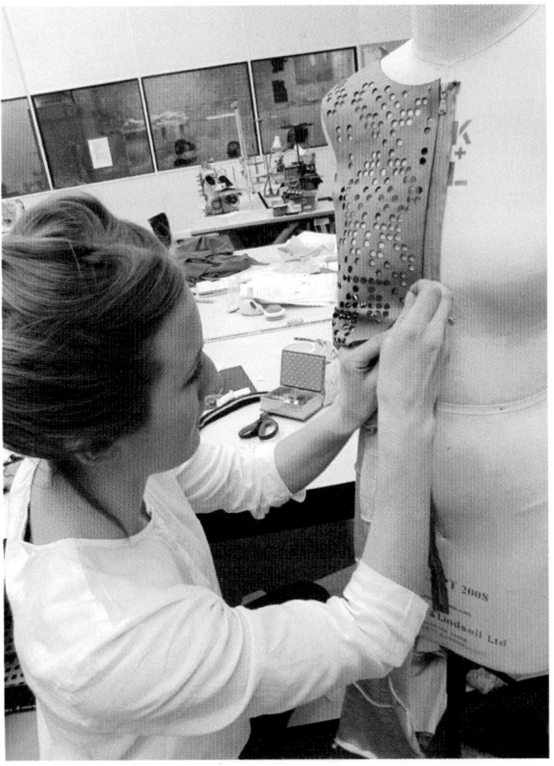

Olivia Williams, fashion knitwear student, working with laser-cut leather panels

What courses are there?

There is a vast choice of degree courses in fashion and textile subjects offered in the UK and abroad. You can choose to study a pure fashion or pure textiles course, or do one that combines both disciplines:

Fashion Design:

teaches you how to develop design concepts for collections, which are then explored through the manipulation of three-dimensional form for womenswear, menswear or childrenswear. Specific projects might be themed for a particular use – for example, sportswear.

Textile Design:

is about developing patterns and textures on and within fabrics for fashion and interior applications. Textile processes include print, weave, knit and embroidery in general terms, but other processes such as laser cutting and felt making are also often available. Not all courses cover all textile processes, so it is worth looking into each course in some detail.

Fashion and Textile Design:

a combined course that covers the fundamental principles of both the Fashion Design and Textile Design courses ensuring that connections are made between designs for textiles and their relationship with three-dimensional forms – clearly two very different sets of design skills. Knitwear Design bridges the gap between these two disciplines and utilises all the skills described above.

Anna Lee combines textiles with fashion for her final collection inspired by folklore and mythology

Anna Lee's fashion designs on the catwalk

Anna's designs include wonderful, intricate, hand-drawn prints that feature dragons, tigers and other magical beasts. Fire and water are also depicted in her illustrations. The intense, bright colours of orange and turquoise dominate, with ink-black linear patterns spread across the rich surfaces of velvet, silk and viscose jersey.

Her fashion collection was shown on the catwalk at Graduate Fashion Week in London in 2010 and made a big impact. Anna was awarded The Zandra Rhodes Textile Award at the Gala show the same year.

Once she graduated, Anna took up the opportunity of an internship at Alberta Ferretti in Italy, where she worked as a print designer before returning to London to study for a postgraduate qualification in Creative Pattern Cutting. She also undertook internships at fashion brands Peter Pilotto and Felicity Brown. She is now working for the menswear brand 'A Child of the Jago', owned by Joseph Corre. Anna Lee commented:

> 'The environment is amazing and I am learning so much. The brand has a distinctive aesthetic – both Punk and Japanese influences are important, and my position as pattern cutter allows me to investigate menswear construction. I'm excited to see where this job will take me.'

There are also many other related courses that specialise in different aspects of fashion, for example, Fashion Marketing/Branding/Communication, Fashion Technology, Knitwear Design, Contour Fashion, Fashion Accessories and Footwear Design, to name but a few.

Fashion Marketing/Branding/Communication:

The course content can vary widely, depending on the individual institution. Some courses have a clear business focus, and will suit students who want to work in promotion, marketing, media and perhaps fashion forecasting, whereas others have a much more creative slant and include aspects of design.

Knitwear Design:

This course deals specifically with fabric construction through knitted textiles, working with highly specialised machinery that will equip you with the skills needed for a career in this field.

Accessory Design:

This course explores a wide range of materials for application to, for example, bags, shoes, hats, belts and jewellery for this growing marketplace.

Contour Fashion:

This is a highly specialised area of fashion design, specifically focusing on developing underwear and lingerie and learning about structured pieces that require under-wiring and **boning**, such as bras and corsets.

Boning: this is a technique used in 'contour' fashion to stiffen the seams of a bodice or corset. In Victorian times 'boning' was made from whalebone, but modern 'boning' is made from lightweight, flexible vinyl.

Fashion Design with Technology:

This course consists of many technical aspects of fabric development, pattern cutting and garment production, alongside contextual studies.

Footwear Design:

This will cover specialist techniques and construction methods used in this section of the industry and will be expected to explore working with a variety of materials, including leather.

Emily Thomas' footwear collection, in collaboration with Dr Martens

Example of a final-year Textile Design degree show
Photograph by Tina Downes

Recent footwear graduate Emily Thomas created a 'girly grunge look' for her final collection. Emily aimed to be as eco-friendly as possible in her work. Supported by the iconic Dr Martens brand and inspired by a laid-back hippy vibe, her successful collection contrasts feminine colour and pattern with Dr Martens' tough footwear shapes. When she visited Lineapelle (a large leather trade fair in Italy) she saw lots of exciting dyeing techniques and interesting effects on leather. This, coupled with her interest in eco-fashion, inspired her to 're-invent' tie-dye techniques for footwear.

'I was fortunate to spend a couple of days a week working in a tannery at the Institute for Creative Leather Technologies, based at the University of Northampton, where I could dye and finish all the leather myself. It took some months to develop different techniques on the leather and then I found one that I was happy with. I used the most environmentally friendly methods available to me in these processes.

My research led me to the company Dr Martens. I've always been inspired by them due to their British heritage and their anti-conformist attitude. These were values that I also shared. The experience of working with them has meant so much to me and I have learnt so much from it about pattern cutting and design. I wanted my collection to be practical, creative and wearable, and working with Dr Martens allowed me to do that. I wanted each item in my collection to be quite different, to go with the nature of the tie-dye leather I was using and to suit the type of customer I was designing my collection for. I feel I have achieved what I set out to do, and I'm very pleased with the outcome.'

Emily Thomas, footwear designer

The more research you do into the actual content of the courses that appeal to you the better. Often courses with the same title can have very different content and expected outcomes. To get a clear idea of the actual work that graduates of a particular course produce, it is a very good plan to visit the summer shows and exhibitions, which are open to the public. In addition, graduate showcases, in particular Graduate Fashion Week, New Designers and Free Range in London, are all excellent opportunities for this research.

In order to get into the industry as quickly as possible following graduation, you will need to demonstrate that you have embraced all the opportunities available to you by taking on work experience or placement during your course. If you choose a three-year course then, as you study for only a maximum of 30 weeks of the year, you have plenty of time in your holidays to build in work experience that can help you find out more about the industry. If you choose a four-year course, then this is often referred to as a sandwich course, with the third year of study based out in the industry on placement.

'My work experience during the holidays has been thoroughly rewarding, giving me an insight into a part of the industry that has enhanced what I'm learning at university.'

Stephanie Aloncel, fashion design student, prior to commencing her second year of study

Historical and cultural context

The study of design context and culture form an integral part of all courses, although the amount of time dedicated to this will vary from course to course. This is usually taught through lectures and seminars, and subjects might include influential movements, such as the Art Deco period or Pop Art, as well as covering aspects of contemporary practice and the history of fashion and textiles.

You will be expected to research and develop essays on specific topics and perhaps give presentations. The main purpose of studying a subject like this is to build your critical awareness skills and gain a greater understanding of the way that learning about context can enhance your practical design outcomes. This part of the course usually culminates in the final year with the production of a dissertation about some aspect of fashion, textiles or a cultural topic that has become of personal interest to you. Refer to Chapter 13, 'Design history, culture and context', for a useful introduction to this.

You may also be offered the opportunity to gain an overview of other aspects of the fashion and textile industry, with some courses perhaps building in to the curriculum an introduction to, for example, business, fashion and trend forecasting, buying, management and sourcing.

COURSE TITLE: **FASHION DESIGN**

Tamsin Lakhani's fashion knitwear collection, Spring/Summer 2013

Fashion Design concentrates essentially on creating collections of garments. During the course you will develop a range of skills, from research and two-dimensional (2D) visual communication skills to honing your practical skills of garment construction in order to achieve a three-dimensional (3D) product. Fashion Design graduates work at many levels of the industry, from designer level through to the high street. The 'flavour' of each Fashion Design course will differ according to the specific elements of the course content. The teaching staff and the links that each institution has with industry will also determine the differences between courses. Some courses encourage students to explore their own creativity without any boundaries, while others maintain a more commercial approach to developing industry-ready designers and focus more on preparing their students to enter the commercial fashion industry.

Fashion Design degree programmes provide the skills that enable graduates to be proficient in designing and creating garments and collections of clothes. Students can study for a BA (Hons) in Fashion Design (a three- or four-year course) or, alternatively, a Foundation Degree in Fashion that is usually a two-year course.

Fashion students need to understand the context in which their work is going to be presented. Gaining an understanding of who their intended customer is, where they are positioned in the diverse marketplace, the importance of branding, marketing and networking and gaining valuable work experience in the industry are all aspects of most Fashion Design courses. Transferable skills are also recognised as being very important, so research skills, working to a defined brief (and to specific deadlines) and presentation skills form a crucial part of most fashion degree courses.

As part of studying Fashion Design you are bound to want to work with interesting fabrics that have wonderful prints on them or have been embroidered, woven or knitted. There are a number of courses that combine studying fashion and textiles together. However, the majority of institutions see these two subjects as separate courses, and will instead offer students the opportunity to work collaboratively, or to have access to the textile manufacturing equipment so that some of the visual effects through the manipulation of textures and images can be realised as part of their fashion collection.

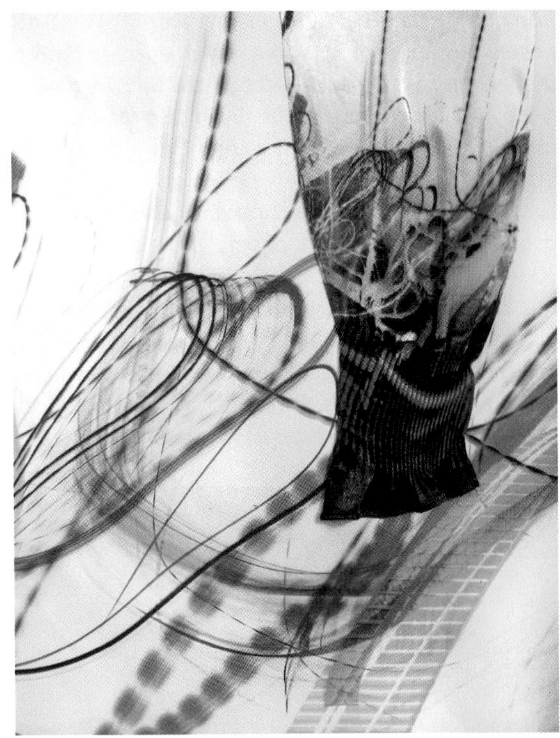

Alice Moore, textile design student

> *'This fabric design was inspired by capturing the movement of the city at night through the medium of photography and CAD. The design was then printed digitally on silk and manipulated by using a smocker pleater. By mocking up a sleeve, this helped visualise the intended fashion context for the design.'*
>
> Alice Moore, textile design student

To apply designs to possible products is an important part of the textile designer's process. Garry Butterfield's story describes how he also wanted to build in new textile developments to his fashion collection.

Garry James Butterfield specialised in menswear design for his fashion collection in the final year. Garry's first year at university gave him a good grounding in aspects of fashion and textile design and also enabled him to sample designing for footwear and accessories. Garry then chose to study Fashion Design for the remainder of his three-year course. In his second year he worked on a variety of briefs and entered a competition run by The Leathersellers

Company, which he subsequently won. Part of his prize was gaining sponsorship, which enabled him to use leather for his final collection the following year.

Garry's inspiration for the distinctive materials he used in his final collection came from a university-organised visit to Première Vision, a large fabric trade fair in Paris. Here he spotted a leather sample that had been laser-etched with a design, giving a distinctive 3D effect. Realising that he was unable to afford the expensive leather he had sourced, Garry applied his creativity and inventiveness, working with the university's laser cutter and the technical staff, to create his own leather designs. Garry's final menswear pieces utilised the rugged textures of goatskin and leather.

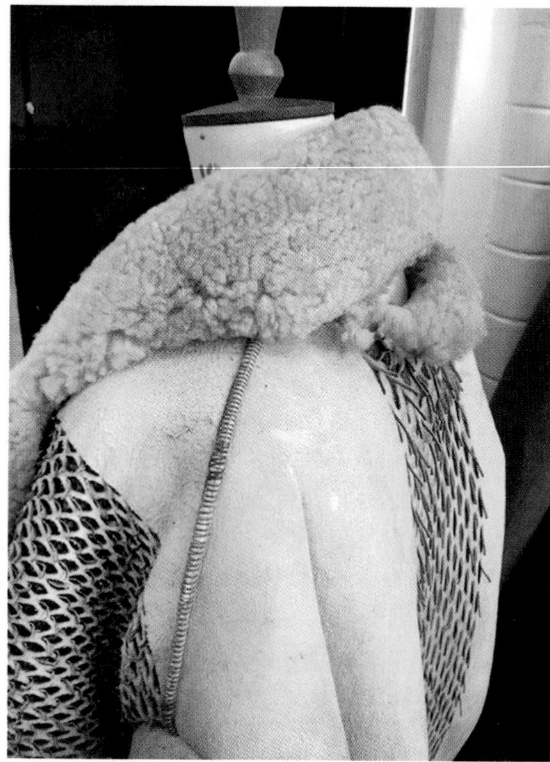

Laser-cut sheepskin by Garry James Butterfield
Photograph by Joyce Thomton

His experimentation involved precise laser cutting into the skins, resulting in rich, decorative three-dimensional effects 'scorched' out of the hair on the surface. Intriguingly, Garry's inspiration for his patterning was drawn from the distinctly urban setting of underground tunnels for tube trains. So successful was the resulting collection, cleverly mixing powerful primitive images with luxurious overtones, that he was chosen from hundreds of hopefuls to be part of a promotional fashion shoot by the famous photographer, Rankin, at Graduate Fashion Week in 2010. He was also shortlisted for the Graduate Fashion Week Gala Show.

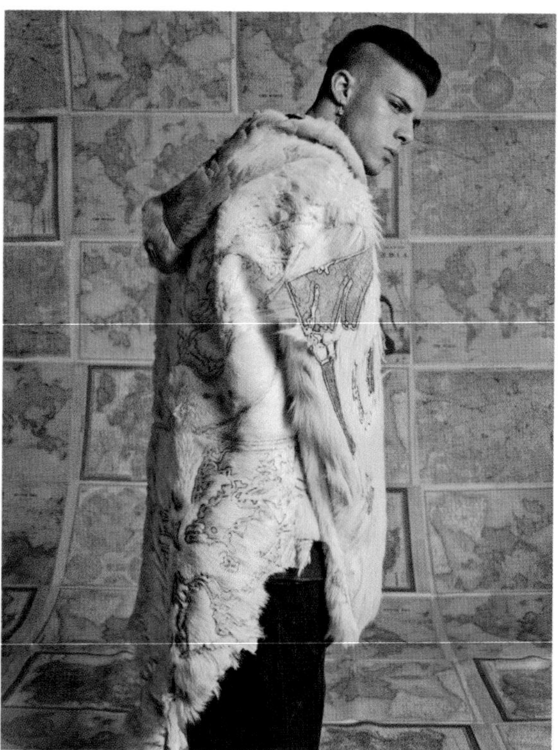

Fashion forms inspired by urban settings, by Garry Butterfield
Photograph by Squiz Hamilton

CAREER OPPORTUNITIES

The big challenge for fashion students and graduates is finding out where real opportunities for eventual employment lie, and this requires both an open mind and some dedicated research. Not all fashion design graduates will find employment as designers, as in the UK there are many more design graduates than fashion design opportunities, but the industry is broad and varied in its job roles where a fashion design degree is essential. However, with the right amount of talent and determination, anything is possible. Womenswear is the biggest market area, but other areas of opportunity include menswear, childrenswear, sportswear, contour, footwear and accessories. Some of the larger garment manufacturers to note are Quantum Clothing Group, Sara Lee Courtaulds, Pentland Group, Dewhirst, Sherwood and Martin International.

Some fashion design graduates also choose to go into fashion buying if they have a keen interest in the business, as well as design. Other creative options for fashion design graduates are in styling, working for retailers or for the media, in visual merchandising, or in fashion forecasting, as described earlier.

Career success in the commercial fashion industry is ultimately linked to the sales of your product ranges. Therefore, understanding the target customer, whichever level you are in, is crucial to anticipating their needs and desires. Graduates entering careers in high-street fashion design can usually expect initially to work alongside an experienced designer as a design assistant for a year or two. This is an important time in which to learn about the company and its ethos and develop a commercial awareness of the fashion calendar and much more. At the designer level, some graduates will seek to work with an established designer in their atelier (designer's studio) for a year or two, before attempting to set up in business themselves.

WHAT WILL YOU BE TAUGHT?

Communicating visually your fashion design concepts is the key component of all fashion courses and therefore developing your drawing skills will be an important feature of becoming a good fashion designer. All courses will also explore drawing using different media and computer-aided design (CAD), so that a creative individual whose drawing may not be their strongest point can develop their own style by incorporating other approaches.

Visual communication encompasses sketchbook work, design development sheets and more 'polished' final illustrations. Technical or 'spec' (short for specification) drawings for garments are also important, so that students learn to be very clear, specific and accurate in their design work. The majority of this work in industry is now done via various computer programs.

Understanding different fabrics and exploring how different materials can be used is a very important part of the course, and you will be encouraged to visit industry-specific trade fairs to develop your knowledge. The biggest commercial fabric fair in Europe is Première Vision, which is held in Paris twice a year in February and September.

You will also learn about the technical aspects of garment construction, and this will give you a solid grounding in pattern cutting and making. Most fashion courses teach you to appreciate the draping qualities of cloth and how to use a mannequin or tailor's dummy to 'model' your ideas by pinning and stitching cloth to simulate a garment, as well as transferring these to the 'flat' pattern-cutting process. Construction workshops usually begin with basic garment construction in the first year and then projects gradually become more complex, with students becoming more competent in

Roxy Eusebio's fashion design project

aspects such as collar construction and lined outerwear, etc., as the course progresses.

THE FIRST YEAR

In the first year of study, you will be introduced to the 2D and 3D skills that are required. You will develop your individual ideas and concepts and use a variety of research methods. Creating sketchbooks and design development sheets and investigating fabric and material properties is part of this process.

Inspiration will be gained from visiting art and design exhibitions, using library resources and specialist trend research materials. Students will also cover illustration and the use of CAD, using software programs such as Adobe Illustrator, Photoshop and other packages.

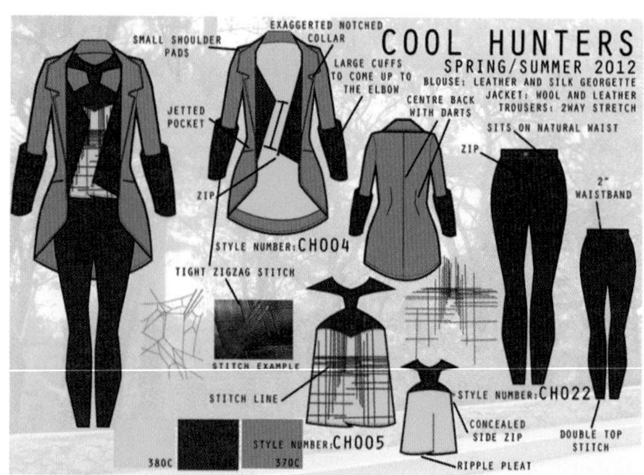

Specification board identifying the colour and detail of each garment by Roxy Eusebio

Inspiration board determining the trend concept for a project by Roxy Eusebio

You will then explore the opportunities, forms and styling of the garments and source fabrics and accessories to bring the collection to life.

Line sheet illustrating the range of silhouettes in the collection by Roxy Eusebio

THE SECOND YEAR

> *'Learning a vast amount of new processes that I've not encountered before has been quite a challenge in my second year. At times it has felt a little overwhelming – but I have learned such a lot in a short period of time and I am thankful that I now have this knowledge to take with me into my final year. One of the main skills I have gained this year is perseverance. I've found that, even if I don't feel completely enthusiastic about a certain brief, I mustn't give up on it – taking baby steps forward will help it all come together in the end. I've also learned not to compare my work so much with other people's. I've realised that everyone works at a different pace and has different styles of working, and as long as I'm doing my best and staying dedicated to my work, then that's all that matters. My advice is don't give up – just keep working hard.'*
>
> Hannah Podbury, second-year student

In the second year, you will build on the skills you have gained in your first year and will also be encouraged to develop your own distinctive design philosophy, or 'handwriting'. Once again, design development is realised through research and the continuation of practical projects involving pattern cutting and making.

Design projects in the second year generally become more demanding, and require students to be much more independent. Practical projects become more technically challenging – for example, students may learn to cut patterns and construct a lined outerwear jacket with lapels.

WHAT IS THE DESIGN PROCESS LIKE?

Design development happens by means of various projects throughout the duration of the course, and will usually take the form of a number of 'briefs'. A fashion design brief is intended to inspire and challenge students, and will normally have the following key elements:

1. A target market – for example, a specific brand or a customer profile.
2. A list of key requirements – for example: a sketchbook with design research and ideas; design development sheets with fabric swatches and annotated notes and details; a specification drawing, with back and front views and other fine details; paper patterns and a 'toile'; a final garment; and reflective evaluation of the process.
3. A deadline for when the work must be completed – you will often be required to present your final work to tutors and fellow students.

Sometimes the brief can be linked to a 'live project', set in conjunction with one of the university's partners in the industry. For example, the brief could be to design an item or a small range of items that would fit well into a particular brand's collection. Students are guided through the various stages of the project, from research to the end product, and the completed body of work (initial sketches, fabric sourcing, samples, pattern, final specification, illustration and finished product) is then assessed. A representative from the industry partner's design team might join the final assessment panel, often giving valuable verbal feedback directly to the student.

Students at the University of Northampton were set the challenge of designing a distinctive garment to illustrate Schwarzcopf's 'Key Looks'. Choosing from their trend-led themes, the students were required to create some initial ideas in their sketchbooks and then develop these ideas, sourcing fabrics and trims before creating their final design.

Roxy Eusebio, winner of the Schwarzkopf live project
Photograph by Caroline Southernwood

The final design had to be presented professionally using CAD software, and then the concepts explained verbally and visually to the Schwarzcopf team. The team viewed the work of the students, listening to all of the student presentations, before selecting a short-list of finalists. The four finalists selected were then required to create paper patterns and toiles (mock-up garments made in plain cotton) before finally constructing their final, finished outfit in their chosen fabric. The Schwarzcopf team then returned to give their feedback to the group and present awards to the winning students.

Second year fashion design student Roxy Eusebio describes her experience:

> 'I think a "live" project is an experience that drives you to do better. Of the initial concepts we were presented with, I was drawn to two of the Schwarzcopf concepts, "Dark Angel" and "Cool Hunters". When I began my research, I found that I was able to find a great many ideas on the "Dark Angel" theme, but I found it was just too easy . . . whereas the "Cool Hunters" theme was much more challenging – and that's what I wanted. Using the "buzz words" associated with the theme, I researched in depth my ideas of what this meant to me visually, and soon found myself really engaged with the concept. I did further research into future trends, using the resources available

through the university and elsewhere, and I began to create ideas for possible garment shapes.

The most enjoyable part was when I realised that I could do so much with this concept – I thoroughly enjoyed the research stage. I am by nature a perfectionist, which hinders me to some extent when it comes to working in my sketchbooks. I have a habit of working slowly and neatly, which can defeat the object of a sketchbook, but I persevered and, once I'd completed it, I felt really excited by what I'd achieved. I love the design development stage, but this time I found I got "stuck" for a while. This often happens, and I've learned to just keep going. Eventually, you get past the "stuck" phase, and then the ideas just keep coming – it really is the most satisfying feeling. Fabric sourcing is one of the most challenging aspects of the fashion design process – it really takes a lot of research to find the right materials, and this is so important.

One of the other really important things I learned on this project is that my time management skills need more work! I found, as it came up to the presentation, I was running out of time, which caused me to panic and created lots of stress. Managing your time has such a big impact on your outcome, and so I've realised that planning each week of a project is really important.

I am scared but excited at the same time. My final major project collection is something that I have looked forward to creating since I started my course. I know there may be stress and tears this coming year, but I also know that the sense of achievement that I'll have at the end will be amazing and priceless to me.'
Roxy Eusebio

WHAT WILL THE FINAL YEAR BE LIKE?

In the final year of study, fashion design students work towards creating an individual collection of garments. The final collection is usually made up of between four and six complete outfits that are usually showcased in a graduate catwalk show. The final fashion collection is an expression of a student's individual creative abilities and should reach a high professional standard.

Tamsin Lakhani's final collection on the catwalk

Tamsin Lakhani describes the inspirations for her final fashion knitwear project:

> 'I derive a lot of my design inspiration from different cultures; living in Africa and India for the majority of my life has influenced my fascination with colour and pattern, as both are rife in their traditional fabrics. To commence my research I started looking at fabrics such as Kente cloth, which I learnt was rich with symbolism – from the geometric patterns used to the choice of colours.

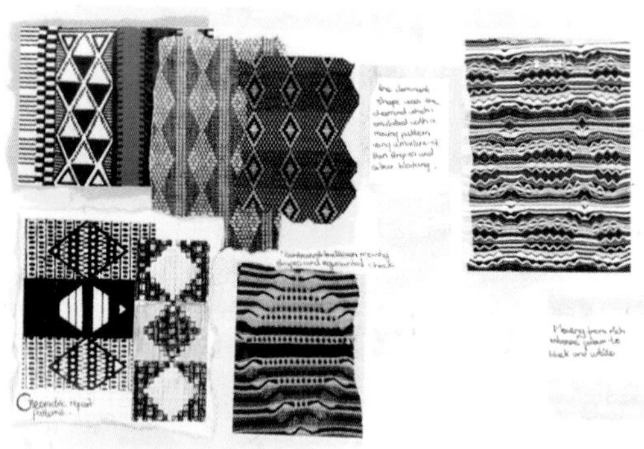

Development board, inspired by Africa and India, by Tamsin Lakhani

I was interested in how the intensity of pattern in a cloth indicated hierarchy, with concentrated pattern for the wealthy and sparse pattern for the poor. This led me to look into the economic situation in Africa during the time of colonisation, focusing mainly on poverty and the hierarchy of power.

Knitted dress design by Tamsin Lakhani

These ideas were translated into my garments by using a saturation of pattern and colour in parts of the fabric, or a change in scale of pattern in order to harmonise the juxtaposition of rich and poor.'

During this long and complex process, you will be expected to use all the skills you have learned over the duration of the course, from sketching to sourcing fabrics and from pattern cutting to creating finished garments, accompanied by technical specs and illustrations.

Fabric and concept ideas board for the project 'The girl with the white balloon' by Shani Jayawardena

The final fashion collection is usually shown in a catwalk presentation using professional models. Some courses offer alternatives to a final catwalk show and have static shows, which suit some collections better, and use film and the Internet to explore exciting new platforms for promoting their work. You will also probably get involved in styling for a photo shoot to showcase your final garment.

Shani Jayawardena's styled photo shoot

Designer: Shani Marion Jayawardena www.girlwiththewhiteballoon.blogspot.com
Photography: Alice Craig Membrey; Model: Hope Osborn; Hair & make-up: Katie Knott

As a culmination of your studies, you will create a professionally presented portfolio of your work as well as a 'digital portfolio' of selected pieces.

COURSE TITLE: **TEXTILE DESIGN**

By Jane Bartholomew, Nottingham Trent University and Joyce Thornton, School of the Arts, University of Northampton

Sian O'Doherty's idea development in her sketchbook (see later in the chapter for her full story)

COURSE DESCRIPTION

This is about the designing and making of fabrics that are then used within a predominantly fashion, interior or art context. All courses will have a slightly different focus, and the types of processes available in each institution will vary. Some courses focus specifically on surface pattern and this traditionally covers printing and embroidery processes in general terms. However, laser cutting, embossing and the development of surface patterns on other materials such as ceramics or metals might also be explored.

'I've always been inspired by the natural and organic movements, colours and reactions within science. I found great inspiration from the mineral collection at the Natural History Museum, as well as studying magnetic fields, Ferro fluid movements and microscopic life-forms. I wanted to experiment with thermo chromic pigments and magnetic materials within the printing process to create reactive fabrics. I also used this

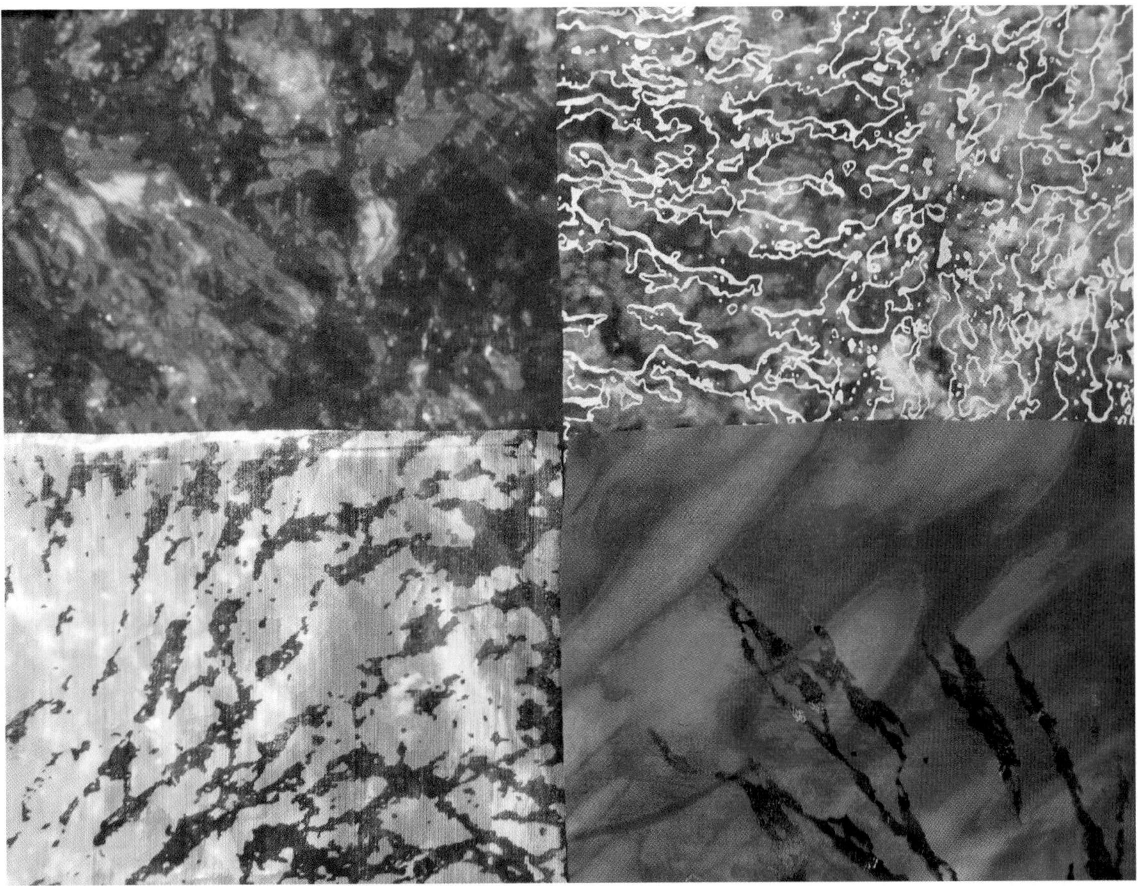

'Star and Galaxy' printed textile designs by Sophie Minal

concept to develop designs by making marks with iron filings and magnets. These designs were then turned into screens for printing. I experimented with using adhesives to attach foils, iron filings and iron pyrite to the fabrics. I also experimented with the way the screens were overlaid with one another to create spontaneous designs.'

Sophie Minal, surface textile design student

Constructed textiles deals with making cloth from scratch using yarns or fibres. These processes include weaving, knitting, lacemaking, tapestry and the construction of non-woven fabrics such as felt and paper. Some courses are focused on specific processes and some cover all of them, allowing you to specialise if that suits the way you work. Others may push for greater levels of investigation by working across the traditional boundaries of these processes using a variety of different media and materials, occasionally referred to as multi-media textiles.

Courses may also have a product focus and determine, as part of each brief, that the textile designs are for a particular product and market level. Others may be broad-based, allowing for each textile to be articulated through illustration, as opposed to prototyping

Drawing, concept and fabric developments by Robyn Swindley
Photograph by Tina Downes

actual products. This may depend on how relevant the actual product application is to the development of the textile design. For example, if you are designing fabrics for light shades, with varying degrees of opacity and translucency built in to the design, then you will want them to be seen in their 'lit' state. Another example might be the development of a printed image that is to be placed on a specific part of a garment. Then, it might be important to make up the garment, to demonstrate how decisions about scale were resolved.

CAREER OPPORTUNITIES

Textile graduates face many of the same challenges as fashion graduates, and many find that work experience opportunities can unlock the door to eventual employment. Emerging textile designers need to be focused and do some serious research and networking to enter the industry. Your graduate show and other major graduate showcases, such as the New Designers event held in London each July, are opportunities for you to meet people from the industry and show them your final year's work.

Careers available following the completion of a degree in textile design are diverse, as the knowledge obtained is not only about the specifics of designing fabrics but also about gaining a deeper understanding of the breadth of the industry. In addition to the fashion and interior companies, there are also the technical, medical, automotive and even building industries that require textile designers.

Becoming a designer is one option, but there are also opportunities to enter buying, styling, merchandising, sales and management within a textile industry context.

New Designers exhibition, Business Design Centre, London 2012

Textile technology and yarn development is another avenue to explore. In addition to this, you may also consider further study that will involve deeper research, or a career in teaching.

As a textile designer based in the fashion industry, you might be creating prints and artwork for in-house collections to be applied to womenswear, menswear, childrenswear or sportswear. The larger employee retailers, such as Marks and Spencer, Topshop, Zara and Debenhams, will employ textile designers as part of their in-house design teams. Some high street retailers also buy designs from swatch studios, where studio designers or freelance design teams will have created ranges for brands to choose from. Working for a print studio you will be expected to create designs to fit in with the specific themes and trends for the season. Textile studios include Palm Studios, Helena Gavshon and Jack Jones Design, among many others.

Other graduates may choose self-employment and operate as a **freelance designer**, or become a **designer-maker** where they create ranges of products and sell them direct to the customer, through retail outlets and galleries or online.

Freelance designer: works either from their own studio or an agent's studio, creating designs for many different clients. The key to successful freelancing is a full book of contacts – something you are unlikely to have when you have just graduated. However, some graduates do find regular work from specific clients through their graduate show and are able to build up their contacts over time. The designs are sold to the client, who can use them in any way they wish. See Chapter 16 for information on intellectual property rights (IPR).

Designer-maker: having become a specialist in a particular process, or set of products, you may want to become self-employed and sell your work through specialist trade fairs, independent retailers and galleries. In textiles, designer-makers usually produce handmade products of high quality (sometimes the products are unique 'one-offs' but more usually they are limited edition pieces from small production runs). Products might include scarves, wraps, throws, cushions and products for interiors. In collaboration with other makers or manufacturers, textile designer-makers can also create clothing, wallpapers, upholstered furniture, light fittings and many other products.

'Retro Rainbow' cushions by Deryn Relph
© Alick Cotterill

Textile designers may also find job opportunities within the graphic design industry in relation to packaging design or brand identity. The application possibilities here are also broad, if you think about stationery, book and CD covers and the print market in general.

Philippa Hill's ideas board (see her story at the end of this chapter)

WHAT WILL YOU BE TAUGHT?

All courses will introduce you to the fact that there are many potential product applications for textile designs. You will learn about the specific weights of cloth that are appropriate for certain items of clothing, for example. Fabrics to be worn on the body need to drape well or have a softness to them if they are to be worn next to the skin. The properties of yarn, and subsequently the fabric, play a very important part in high-performance sportswear. Yarn companies who specialise in technical research develop fabrics with unique qualities. A good example of this is non-iron cotton – a clever invention.

Fashion accessories, from hats to scarves and jewellery to footwear, offer you options to undertake product research. Coats, dresses, ties, suits, bridalwear, lingerie, shirts and nightwear all have potential focus for a project. Designing for interior products is another long list that includes wallpaper, rugs, carpets, upholstery fabrics, furniture, lighting, bedlinen, ceramics and soft furnishings for window dressings and cushions.

Various briefs will be set for you throughout the duration of the course, some determined by staff, others by the industry and others by yourself. You will probably be encouraged to write your own briefs during your final year and may also be encouraged to undertake competition briefs as well. The history of textile design and its relevance to the world we live in today will naturally become part of your studies. Essays and presentations on these broader contextual subjects will be built into your course.

Design development is documented through visual exploration and written reflection in the sketchbook. It is also spurred on by learning techniques and processes, and can be triggered by research. You will also be encouraged to experiment and take risks with your use of colour, texture, scale and repeat, and will probably investigate dyeing methods and textile technology. Laser cutting and specialist focus on footwear or leather work may also be offered at certain institutions, so if you have specific interests then search carefully for the right course.

As with many design courses, an awareness of the market and the customer you are designing for is important, as is the relevance of using certain materials and processes for particular market levels. All briefs encourage high levels of self-motivation and encompass drawing, visual and theoretical research, process development and clear communication of ideas through visualisations. Portraying your designs working within the market sector you have chosen to focus on is an important skill.

You will learn about traditional and automated methods of production in all the disciplines. The industry today requires textile designers to be fully aware of the way fabrics are manufactured. Digital printing is now a very familiar process, with much development clearly visible in our homes, on the high street and in the common use of photographic imagery printed on fabric. However, traditional print techniques still have a very important role to play and there is an important distinction to be made here: as a student, you will need to decide whether you enjoy working with your hands or are happy developing the majority of your designs on a computer.

Here is a brief description of the specialisms available within textile design courses:

Printing:

In print you will learn about traditional block and screen printing methods: as well as digital developments. Your studies will also include a deep understanding of image development, dyestuffs, printing inks and specialist techniques such as discharge and devoré printing, and will encompass an understanding of fabric qualities.

Weaving:

You will learn the fundamentals of designing and making fabrics and gain an appreciation of yarn properties and fabric weights. You will learn how to control colour in the weaving process by applying appropriate structures to enhance effects. Texture and image development play a big part in woven textile design. You will learn about both the hand-weaving and the automated dobby and jacquard weaving processes.

Knitting:

Various structures and yarn types will be explored and you will learn about the importance of the handle and drape of a cloth, as the majority of knitted fabric is to be worn on the body. Imagery can also be captured in knitted textiles and there is a range of automated knitting machines, from jacquard to circular, for you to become familiar with.

Embroidery:

You will learn about a range of techniques that will develop either constructed or surface effects; these include beadwork and techniques generally referred to as 'embellishment'. Embroidery can be created by machine and also by hand. Hand embroidery techniques can be very time-consuming but have always been popular, and are widely used in high-end fashion and on bridal-wear and eveningwear.

Multi-media:

This explores what you might do with materials that are not traditionally associated with textiles, and allows for greater flexibility in crossing the boundaries into other design disciplines, such as ceramics, for example. With this as your focus, anything is possible and it provides you with an opportunity for high levels of creativity and innovation.

THE FIRST YEAR

You will be encouraged to explore and experiment with many mark-making approaches and to work with a broad range of media. Drawing is a key focus of all textile design courses, so there will be a lot of emphasis placed on your ability to record what you see in a variety of different media, from traditional drawing techniques with paint and other mark-making tools to collage or photography.

A drawing workshop

It is very easy to stay in your 'comfort zone' and continue to draw and mark-make in the same style that you did when you first learned to be creative with your drawing. It is also easy to work with the same media. Studying drawing at this level, however, needs to motivate you to be adventurous and attempt new approaches to capturing visual inspiration and experiment with mark-making processes.

TIP

With each new project, set yourself the task to work with a palette of colour that is unfamiliar to you. The industry needs designers who can work within tight parameters across a broad range of palettes, from pastel shades to brights to neutrals.

During the first year you will develop your individual ideas and concepts and use a variety of research methods – creating sketchbooks and design developments and investigating many different techniques and processes to identify which ones inspire you.

Mixed-media experimentation, translating natural forms

The technical aspects of textile design are integral to the specific design processes, and most courses are designed to give students a solid grounding in their chosen discipline – print, weave, knit, embroidery or multi-media. Some courses also offer additional 'open access' sessions to their technical facilities so that you can work across the disciplines if it suits your project. A key focus will be computer-aided design. Training on the use of Adobe Photoshop and Illustrator packages is now essential for working in the textile industry, together with subject-specific software. The packages available will vary between courses and will depend on the machinery being used.

THE SECOND YEAR

In the second year you will build on the skills you have gained in the first year and will probably start to specialise in a process, and on either surface or constructed textiles. You will also begin to develop your own distinctive design philosophy or creative style. You will be building your confidence in developing good quality designs that are suitable for particular applications, with a more intuitive approach to responding to customer-based briefs. The second year is about gaining a greater understanding of the industry, and there might be live projects and visits to companies and trade fairs to help you appreciate how the industry operates.

WHAT IS THE DESIGN PROCESS LIKE?

Drawing with a wide range of media is by far the most important thing to be able to do as a textile designer. Taking the time to observe carefully what's in front of you and capture the imagery, colours and textures with high levels of creativity forms a large part of your studio work. This is the traditional starting point for the development of textile designs and is arguably still the most important skill to have.

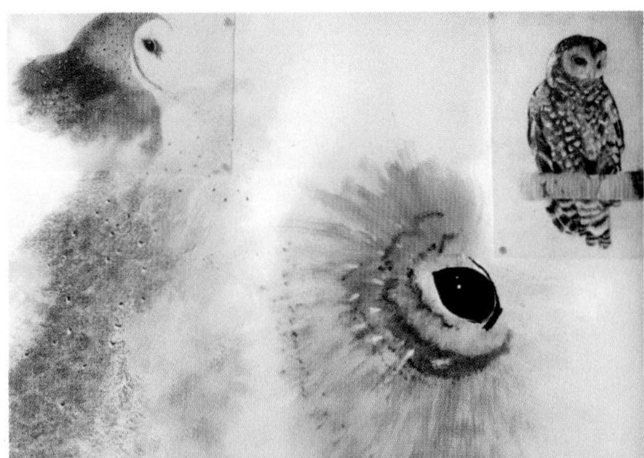

Rachel Kempe's owl drawings

Manipulating the information in your drawings by taking particular aspects of the colour and texture and combining this with the imagery and then

Sian O'Doherty's idea development in her sketchbook

considering the potential ways in which the repeat might work or which scale would look best, is all part of the process development that all students studying textiles will need to work through.

A project's inspiration and starting-point can come from many varied sources. Developing your own theme and undertaking drawings that explore that theme is just one way of starting a project. Visiting an exhibition, reading an article or book, wanting to find out about a specific process or yarn, or identifying a possible gap in the market place for a new product or fabric might all be starting-points for a project.

Whether you are learning about print, weave, embroidery, knit or multi-media textile processes, you are sure to interface with both hand processes and those done by computerised modern machinery. You will also need to undertake material and process research to achieve the qualities in your fabric design that you are striving for.

Plenty of idea generation and in-depth technical experimentation are fundamental to the development of the project. CAD software will also play an important role in designing textiles, and this is often used as a design development tool to help you appreciate how automated equipment might more closely interpret your visual ideas, as well as the design tools in themselves.

Resolving a brief involves the ability to work your technical, visual and creative ideas together and to consider for whom you are designing.

> 'As a conceptual artist, I start with preliminary sketchbook work: drawing, photography, texture and media exploration. I then explore working with different scales and play around with the composition of the pieces. Excessive amounts of technical exploration, experimenting with materials and paying close attention to my drawings helps me achieve my final designs.'
>
> Rachel Kempe, textile design student, specialising in embroidery

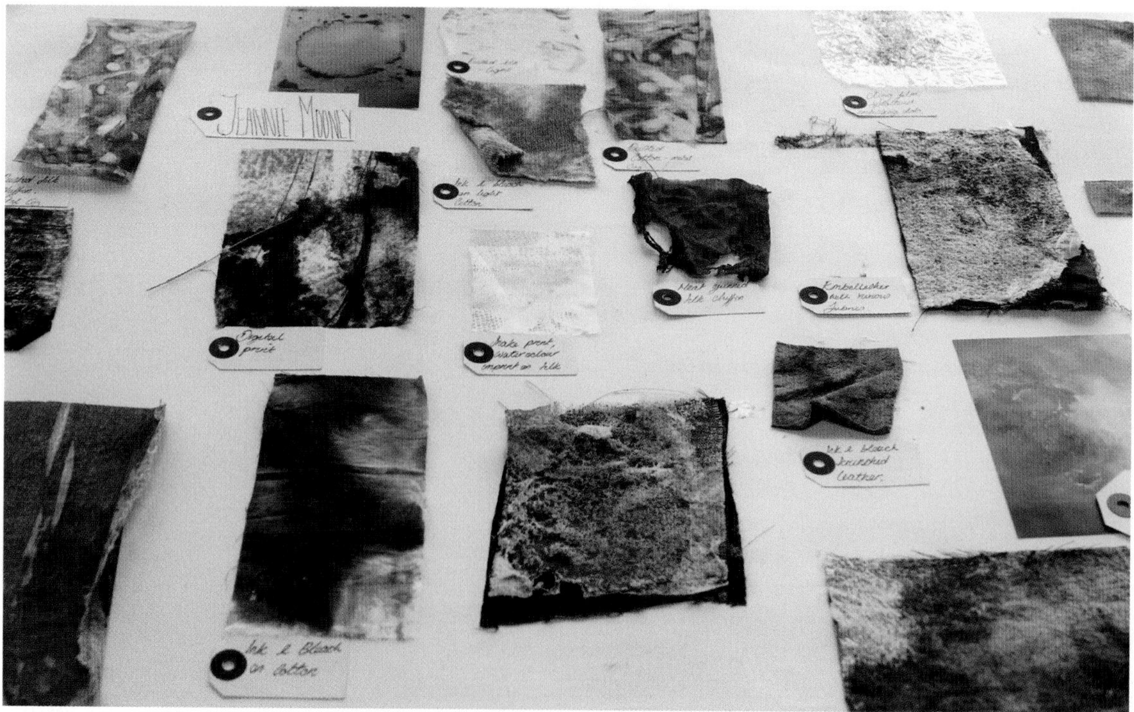

Rachel Kempe's technical investigations into printing processes

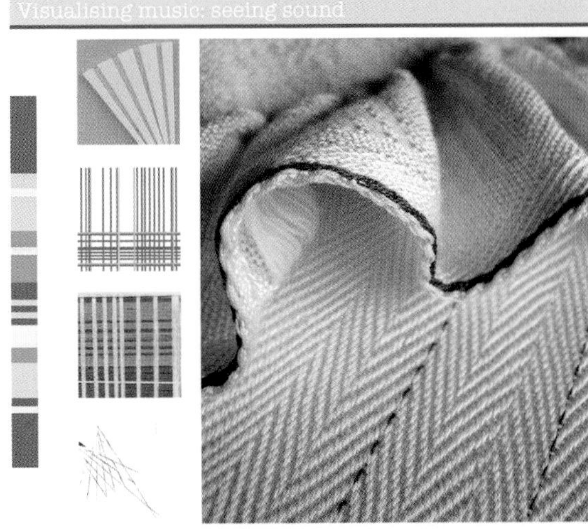

Anna Piper's experimentation with colour proportion in weaving

WHAT WILL THE FINAL YEAR BE LIKE?

During the final year, you will use all of the skills, and more, that you have learned over the duration of the course to draw, research, design-develop, make textiles and visualise your design outcomes with effective illustrations.

In the final year you will probably undertake a couple of projects, enter competitions and try out some new processes. You will experiment with new approaches in the hope that you will arrive at some innovative qualities within your portfolio that will help the industry notice you at the degree show and when you go for job interviews. Aim to create a body of work that expresses your individuality and aim for a high standard of work that is professionally presented.

It is likely that you will define your own projects at final year, so choose the themes and contexts for them wisely. Think about the sort of company you want to work for and the sort of career you are hoping to pursue and imagine what would be best to have in your portfolio that will give you the best possible chance of having a body of work that suits this career focus.

Butterfly canvas design by Rachel Kempe

Preparing your portfolio

During the first and second years of your course you will continue to build the content of your portfolio. Your personal flair and design strengths should be reflected within, and there are a few things that you should always include:

1. **At least one strong example of a successful project** that demonstrates your capabilities of working to a set brief. Prospective employers (even for work experience) will always be interested in how you creatively resolved and worked with the constraints of a project brief. You should also include an example that shows your personal interest in colour and fabric – portfolios should reflect the tactile aspects of fashion and textile subjects, and not just contain flat, two-dimensional artwork.

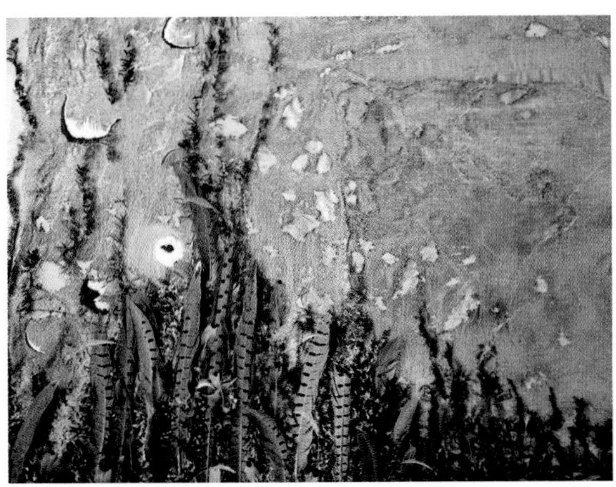

Owl design by Rachel Kempe

2. **Some strong examples of drawing should be included** – perhaps a copy of some excellent sketchbook work that shows your ability to sketch quickly. Some 'polished' fashion or interior product illustrations should also be available so that the viewer can see the context for your design outcomes.
3. **Accurate technical specifications** or examples of working with processes should be included.
4. **Photographs of final garments or products** should always be included as this can demonstrate any styling or layout abilities that you might have. This photography should be presented as professionally as possible.

When taking your portfolio to interviews, you will need to tailor it to suit the company that you wish to work with. Think carefully about what they are looking for and rearrange your portfolio of work to suit the specific client.

In short, your portfolio should be exciting and imaginative, demonstrating your creative flair, but it should also reflect the breadth of your skills and your versatility as a designer.

Conclusion

This section has tackled the breadth of the fashion and textiles industry and has provided a clear starting point from which you can find out more about this subject. It will also have mapped out possible futures for you. Remember that other sections in the book can guide you through your studies and offer tips on how to prepare for placements and the degree show and how to look for a job. The projects and career profiles are sure to have inspired and provided you with a variety of thoughts about what you might do after your studies.

In the broad discipline of fashion and textiles, there is a wide range of practice. Four very different designers have been selected to tell you their stories after 'Further resources'.

Further resources

Books

Aldrich, W., *Metric Pattern Cutting for Women's Wear*, Blackwell Publishing (2008)

Although not for beginners, this is a reference book recommended by many pattern cutting tutors for those who have started to acquire more skills.

Atkinson, M., *How to Create Your Final Collection*, Laurence King Publishing (2012)

Many great examples of fashion-students' work from around the world and useful, practical information are included in this inspiring book.

Blackman, C., *100 Years of Fashion*, Laurence King Publishing (2012)

Tracking the progress of fashion from the turn of the last century, this is an informative and fascinating history.

Bowles, M. and Issacs, C., *Digital Textile Design*, Laurence King Publishing (2012)

An excellent book for budding printed textile designers, this has lots of information, tutorials in PhotoShop and Illustrator programs, and is packed with inspirational images.

Braddock Clarke, S. and O'Mahoney, M., *Techno-Textiles: Revolutionary Fabrics for Fashion and Design*, Thames and Hudson (1999)

Braddock Clarke, S. and O'Mahoney, M., *Techno-Textiles 2: Revolutionary Fabrics for Fashion and Design*, Thames and Hudson (2007)

These are informative books with detailed information on innovative textiles, looking at electronic, scientific and other ground-breaking new developments in the textile sector.

Colchester, C., *Textiles Today: A Global Survey of Trends and Traditions*, Thames and Hudson (2009)

This book gives an informative overview of the many factors affecting textiles today, including the impact of new technologies and issues of sustainability.

Hywel, D., *Fashion Designers' Sketchbooks*, Laurence King Publishing (2010)

Packed with inspiring visual drawing examples, revealing the many differing approaches and styles of successful fashion designers.

Jenkyn Jones, S., *Fashion Design (Portfolio Skills)*, Laurence King Publishing (2011)

An asset to many students since it was first published in 2002, this remains an invaluable overview of the process of fashion design, containing plenty of practical advice.

Leach, R. and Fox, S., *The Fashion Resource Book: Research for Design*, Thames & Hudson (2012)

This is an excellent, contemporary and comprehensive book on this complex topic, which is both informative and inspiring.

McKelvey, K. and Munslow, J., *Fashion Design: Process, Innovation and Practice*, John Wiley and Sons (2011)

This informative book uses many actual examples of student work to explain the design process.

Nicol, K., *Embellished: New Vintage*, A & C Black (2012)

Inspiring contemporary embroidery – a visual delight.

Parish, P., *Pattern Cutting: The Architecture of Fashion*, AVA Publishing (2013)

Clear, informative and inspiring book for fashion designers.

Roig, G.M. and Fernandez, A., *Drawing for Fashion Designers*, Anova (2008)

This includes advice on how to use colour and how to portray fabrics in sketches.

Selby, M., *Contemporary Weaving Patterns*, A & C Black Publishers Ltd (2012)

Well-known woven fabric designer Margo Selby has developed a very useful book to introduce and explain the nature of constructing woven cloth, suitable for those studying to degree level.

Websites

www.texprint.org.uk

The annual Texprint showcase highlights some of the best graduate textile talent in the UK. Twenty-four graduate designers are selected annually for this prestigious showcase, receiving practical advice and mentoring to assist their transition from graduate to professional textile designer. The website details many career profiles of Texprint alumni specialising in knit, weave, print and mixed media.

Project story

Designer:
Philippa Hill, final-year textile design student

Project title:
'I am the Jumper'

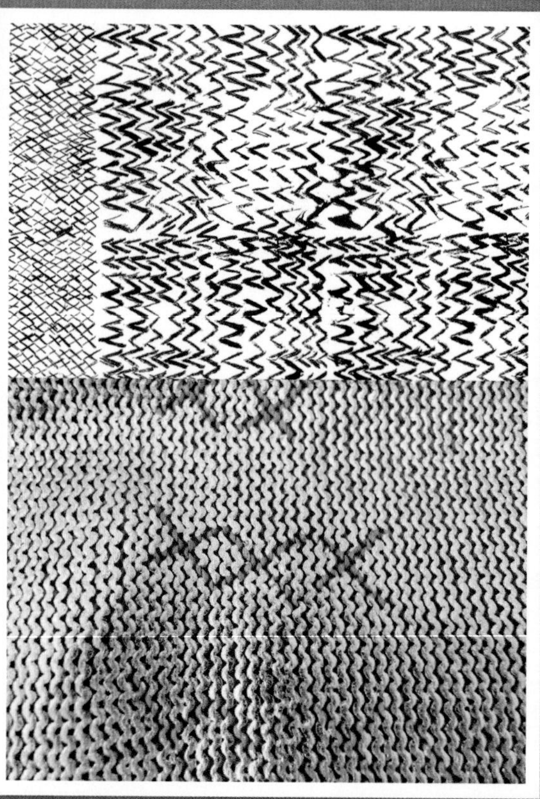

Mark-making and texture investigation

Tell us about the project

'I am the Jumper', my final project, is inspired by traditional fishermen's jumpers and heritage textiles. These jumpers were traditional garments worn by fishermen in coastal communities in the UK. I concentrated on the coastal villages of Fife, Scotland and in Guernsey, since my family originates from the island. My project is dedicated to my Granny and is a very personal project, making me more determined for it to be successful. The fishermen's jumpers are called ganseys and guernseys and traditionally were made by family members for the male fishermen. They were typically navy blue and patterned differently depending on the knitter's skill and imagination in patterning. Each handmade gansey embodies a pattern imprinted with knit into the surface and supposedly tells a story of the fisherman it belongs to. The romantic legend says that a fisherman could be identified by his gansey and the patterning on the front! In my project I aimed to add and subtract from the surface using print techniques and embroidery echoing the knit textures on these jumpers. I explored mark-making inspired by the fishing villages and the vanishing folklore and culture in my drawing.

My collection is aimed at high-end womenswear fashion, for a consumer who dislikes mass-made pieces, keeps garments from her ancestors and likes to associate stories or memories with clothes. She likes time-laboured pieces that will last. She remembers her grandmother knitting or doing embroidery and is interested in her own personal heritage.

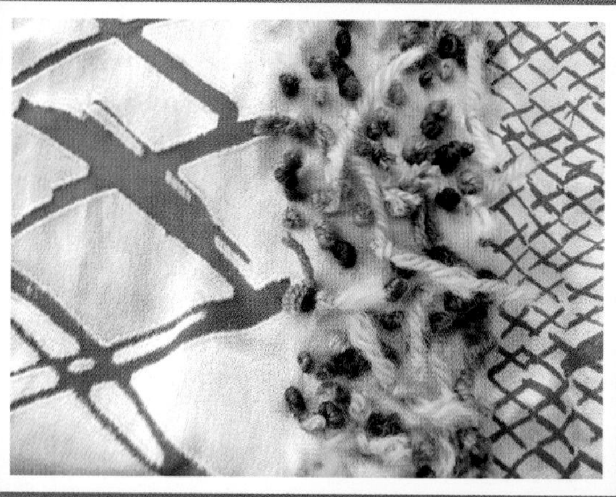

Devoré printing process with embroidery techniques

What processes and skills were most relevant to the project?

I had a variety of knitted fabrics made up by external companies and machine knitters. Specialising in print, I screen printed onto these fabrics using a variety of different binders and print techniques. I was really interested in building upon print room traditions and spent hours experimenting in the print room. Alongside the knits, I used a range of silk and cotton fabrics. I loved finding an 'unknown' devoré fabric and pushing my use of colour and texture with the knit.

I've been passionate about being a textile designer since I was a young child and designed my own fashion catalogue on scrap paper (my mum actually sent it off to a high-street brand's HQ and I spent an afternoon in their design department – but that's another story), and so with determination and a positive attitude I'm sure something brilliant will come along in the future.

A printed length of Philippa Hill's fabric
Tony Marsh

What challenges did you face?

When I started my final year, I found it incredibly hard to come up with a concept for my project. We were given examples of briefs to help us, but I found that I was over-complicating my ideas. I chose to research the history of the fishermen's jumpers for my dissertation and, during my research, I uncovered a lot of depth to the topic and I realised that I could share my research for my dissertation with my studio work. This meant that I could concentrate on one research subject and go into a lot more depth.

What were the highlights of the project?

At the end of my four-year course at Edinburgh College of Art I exhibited my work at the degree show, which was the first major display of my work to the public. It was lovely to read people's comments about my work and listen to their stories about fishermen's jumpers – I have become obsessed!

Project story

Designer:
Jasmin Giles, fashion accessories designer

Project title:
Final-year Degree Project

Tell us about the project

I have a vested interest in surface textures and have always worked three-dimensionally, creating objects that are sculptural and organic. Exploring the different qualities of materials and numerous techniques was what inspired my designs for my degree show collection. I often chose to use materials that I could manipulate and mould, and soon became fascinated with the properties of wax. My intention was to create multiple droplet-shaped beads from wax that I could incorporate into my knitted textile designs. I realised that the degree show was my first opportunity to gain exposure, meet potential clients, develop a network and begin to establish a personal aesthetic. It was my ambition to produce a series of highly finished products.

Jasmin Giles, contemporary jewellery

What processes and skills are most relevant to this project?

I chose to study textile design at Chelsea College of Art and Design because of its versatility and endless possibilities of material quality and production methods. In the first year we had the opportunity to sample four textile methods – knit, stitch, print and weave – and were expected to select one of them at the start of the second year. I felt it important to choose a specialism that was entirely new to me; I wanted to learn a commercial technique and master a skill, so I chose knitwear.

The beauty of all textile techniques is that they are adaptable and can be crossed with one another. By specialising in one method, this does not mean you can't adopt other processes too. This was integral to my development throughout my degree. Having this freedom to work with different processes meant I had the confidence to explore less traditional textile processes.

What challenges did you face?

Going into the final year felt like a huge leap and I was very conscious of the pressure to develop a collection that would represent something very personal and unique.

What were the highlights of the project?

I received a great response at the show and was offered sponsorship from a private foundation to continue producing work and developing ideas. Your degree show is a platform for opportunity and can be hugely beneficial to your success if approached professionally and with enthusiasm.

After graduating, the first goal was to find a creative studio space. I felt it important to have a space where I was around like-minded artists, so as to have a support network and an influential working environment. I have continued to develop new work and have exhibited my collections through trade shows to familiarise myself with the market place and begin to generate a professional network. Through my sponsorship, I have had the experience of pitching my designs to a panel of industry professionals and working to deadlines. I have come to learn that there is no rush to brand yourself too quickly; having the time to focus on making new work and developing ideas is more fulfilling and will satisfy your need to be creative.

Jasmin Giles' fashion accessories

Project story

Designer:
Sian O'Doherty, final-year textile design student

Project title:
Perceived Perceptions

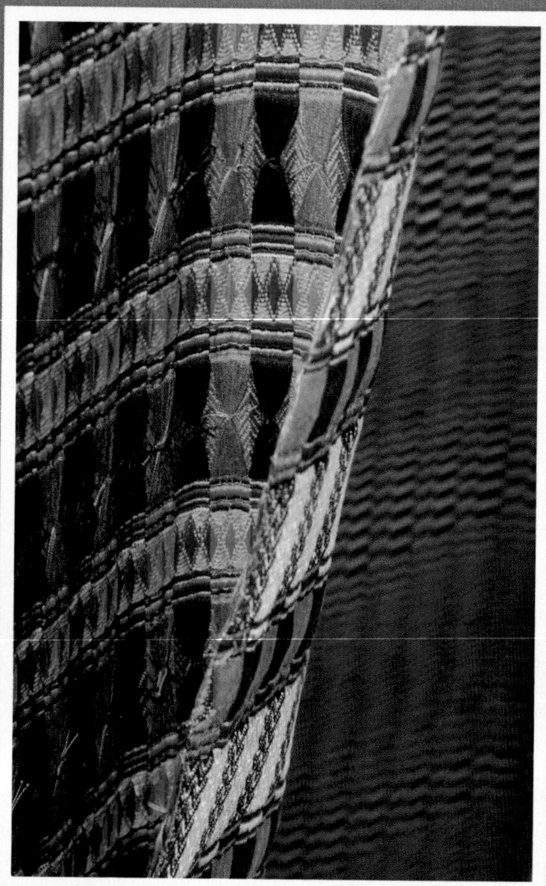

Multi-layered weaves by Sian O'Doherty

Tell us about the project

From the very beginning I was drawn to creating a collection that incorporated optical illusions. My creations, on second inspection, reveal themselves differently, and are not what they first seem. Fundamentally, I didn't want my work to reveal itself in one glance and to be too 'obvious'. My initial inspiration came from Google Earth images – taking references from the patterns and vibrant colours created by estuaries. Initially, I explored with as many techniques and processes as possible, allowing them to respond to and influence one another and fuel further experimentation.

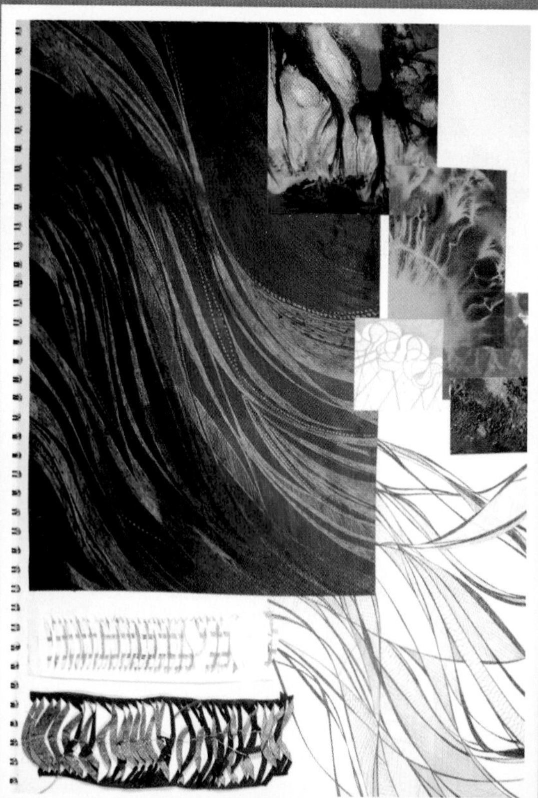

Sketchbook development – oil bar drawings and pen drawings, Sian O'Doherty

As my work progressed I naturally focused on fewer disciplines, developing in-depth technical exploration of multi-layered weave structures, combined with colour investigations and how threads in the fabric could be manipulated.

Keeping in mind my intentions of creating optical illusions, I developed digital representations of my woven creations and developed new print patterns that would be impossible to weave, yet still retaining all of the visual character of a woven cloth. By using modern technology, I managed to further develop my initial weaves by taking them in a new direction.

What processes and skills were most relevant to this project?

Initially all the skills/techniques I had learnt over my degree were relevant to this project. By drawing on numerous techniques and processes I found that they started to respond to one another and allow for

Distorted effects in weaving, Sian O'Doherty

further exciting developments and progressions to occur that you would not have previously expected. I began this project exploring knit, weave, multi-media, hand-cut paper, laser cutting and digital techniques. I kept all my options open, allowing myself to consider every possible outcome – which allowed for a greater range and breadth of development considering both interiors and fashion applications.

What challenges did you face?

Oh my goodness. Where to begin? Time, money, equipment failure, other commitments, self-doubt . . . and these are just a few! There are always challenges; they are unavoidable but they make reaching the end even more worthwhile and satisfying.

On many occasions I literally had to drag myself home, go to sleep, wake up and tell myself 'Get a grip, learn from it and do it again', and I know that I will be telling myself this on many more occasions to come.

As I had already completed a diploma course and redirected my studies, I think I was very focused and determined to succeed. I lived at home where I had space to work without distractions, and living at home also eased financial pressures because any design course, as students will realise, has the added expenses of materials and use of machinery, etc.

Any advice for others?

If I was able to go back to my school days, when the pressures of suddenly deciding what it is that you want to do for the rest of your life are on your shoulders, I would say don't panic and do what it is that makes you happy. I realise that this is easier said than done, as I did not do this myself for numerous reasons and found myself for two years on a sports science degree, believing that it would provide me with good job opportunities. I came to my senses and applied to study textile design at Carmarthenshire College at the age of 23 and it was the best decision I ever made.

What were the highlights of the project?

With any project my fear is coming to the end and standing back and not being happy with its conclusion and knowing that I could have done more and could have pushed myself further. Therefore my highlight was coming to the end and not feeling like this!! I knew I couldn't have done any more in the time that I had, but I was excited about where I could take it next . . . **www.sianodohertyblogspot.com**.

Editors' note: Sian won the 'Fabrics' Award at Interiors UK 2013 just before this book went to print!

Career profile

Name:
Anita Quansah

Current job:
Fashion Designer

Describe what you do

I have created a highly successful creative jewellery brand, harnessing my background in textile design. I am now based in Buckinghamshire but studied at Chelsea College of Art & Design, specialising in embroidery and fabric manipulation. I began to translate my techniques and unique mixes of textures into jewellery, creating one-off, distinctive pieces.

I really look forward to working, and it gives me great joy to take a design that was a concept and then translate that into a distinctive work of wearable art. Most of my day is spent beading and creating complex textures.

Anita Quansah jewellery
Stylist: Sylvia Holden; Photographer: Adrian Buckmaster

Tell us about your career so far

I grew up in a family that is very business-minded and I'm creating an online shop on my website. My late grandmother was a huge influence on me – she was a seamstress and a teacher who trained many women to use their skills to get them back to work. She inspired many women to make something of their lives – and she inspired me.

My cultural heritage has a huge impact on my work. I come from a mixed African background – half-Ghanaian and half-Nigerian. This has exposed me to lots of rich African traditions, particularly in the use of materials and textiles. The bold and vibrant colours from Africa are phenomenal. I incorporate rich African prints mixed with vintage elements and new materials, such as chains, pearls and shells. I believe this mix creates a rare and vibrant new look. I am also influenced by things I see every day: music, art, people from diverse cultures and distinctive styles.

After completing my studies, I was successful in applying to Texprint in 2006, which was amazing (refer to 'Further resources' for more information), and this gave me a great platform to showcase my work and skills to the fashion and textile industry. Through exhibiting in London and at Indigo (part of Première Vision), Paris, I had the amazing opportunity to collaborate with the fashion designer Christian Lacroix. Other prestigious design houses now commission my work.

What inspires you?

My materials are from Africa, Europe and Asia, as well as from flea markets, vintage fairs and stores in the UK. I love to use unexpected elements – reclaimed pieces from vintage jewellery, semiprecious stones, rare African beads and colourful textiles. I weave them together to create strong, expressive, unusual, one-off statements.

I want to continue to maintain my craftsmanship and keep my creative spirit alive, making unique conversational pieces and continuing to wow people. It gives me great joy to know that my pieces are appreciated by so many people, including celebrities. I want to continue to raise awareness of recycling and upcycling. I'm planning a bigger studio – I want to do workshops to start teaching others how to use their creative skills. I also want to take this idea to Africa. My ultimate ambition is to make my brand more accessible and eventually to be recognised across the

Anita Quansah jewellery
Stylist: Sylvia Holden;
Photographer: Adrian Buckmaster

world. I aspire to be stocked in stores such as Harrods, Selfridges, Liberty and Neiman Marcus. I would also love to work with more fashion design houses and couturiers, such as Dior and Jean Paul Gaultier.

What advice do you have for students considering a similar career?

Three words: dream, believe, achieve. This gets me through everything. Hard work and perseverance pays off in the end. In this industry there are a lot of hurdles, but if you are focused and believe in yourself and your product you will stand out from the rest. Love every bit of what you do and enjoy the joy it gives to others too – that is priceless.

To find out more about Anita's work visit the following websites:

www.anitaquansahlondon.com
www.notjustalabel.com
www.shrimptoncouture.com

Graphic design

By Phil Thomson, Birmingham Institute of Art and Design, Birmingham City University

> *'In short: get as much industry experience as you can – intern, freelance, enter competitions, persevere and ask questions. When you land that dream job don't be complacent, soak up everything, surround yourself with the brightest, work incredibly hard and keep asking questions. What's meant for you won't pass you by.'*
>
> James Chorley, art director at AKQA

Greg Howell's experimental posters at 'GASH' – part of a student-led exhibition in a city-centre retailers
Photograph by Stonie Reuladair

Graphic design has its roots in what used to be called 'commercial art', and while there have been a number of attempts to redefine the discipline with the rapid advances in technology over the past hundred years – through 'commercial design', 'graphic art', 'graphic media' and 'media design' – it retains its unique place at the heart of all design subjects. We rely on good graphic design for almost all means of expression where a message has to be conveyed. Indeed, the understanding and practice of the subject is fundamental to anyone involved in the applied arts. If it has a problem, it is a good one – for it exists as a construct, embracing many different skills that, as a generic, are interdependent, and in specific terms have the potential to propel an industrious designer to the heights of success.

In an age where computer software has made designers of everyone and spawned a DIY clip-art culture, it is even more important to make a distinction between basic forms of graphic design and effective visual communication. The difference is in how a designer 'thinks'. A good graphic designer will have discovered semiotics, colour theory or the psychology of perception, or may well have an opinion on John Berger's *Ways of Seeing* (2008) – a book that deals with the cultural issues of art and design and helps us to understand the subject as a visual language, whether it is in advertising (created to 'improve' our lives) or the way in which documentary images of third-world countries are portrayed. That 'thinking' is portrayed very well in this real-life project by Martin Knox:

> 'Cancer Research UK identified 100 of their charity shops that were underperforming and considered what they should do with these. From the outset, our client was adamant that its most important feature should be the connection to the charity. We illustrated that it should be three things in balance, with none being more prominent than another. Those three things were the charity + the product + the experience. And it was on this basis that we conceived 'Wishes' as a business, as a brand and as a retail identity and environment.
>
> We discussed the type of customer we were going to be "targeting". I was not interested in demographics (ABC1 rubbish). I had no thoughts to making this concept appeal to a particular audience. It was important to me that this chain appealed to whoever crossed its threshold; that it was whatever that individual needed it to be for them at that time.
>
> We presented our creative concepts as moods and visuals that illustrated "what the shops would feel like when you entered them". It was only much later that we developed designs illustrating the physicality of the business and brand.

'Wishes', a retail business conceived, designed and implemented by KNOX for Cancer Research UK

My greatest joy came when we launched the first shop in Hammersmith, London. The then marketing director for CRUK came up to me and said "Martin, it is amazing. This is exactly what you showed us at that very first presentation when you showed us illustrations of the sense, feel and spirit of the shops." Twelve shops were launched and then the chain grew to 35.

And, each shop did feel different. We wanted them to have a feeling of the positive personality of the folk running the shops. Visiting the Macclesfield store some months after it opened, I walked in and . . . it was buzzing with energy. There was palpable electricity in the air. This was fully reflective of Eddy – the incredibly enthusiastic and fun manager. Northwich was run by Lesley; she was very calm, caring and gentle. That is what her shop felt like when I walked in.

We "creatives" have immense power over how folk interact with the world around us. With that power comes responsibility, and if you are able to acknowledge that power and take responsibility for using it, only positively, then you are more likely to do amazing work.

Consider an idea. Where is it? What is its structure? Where does it come from? It has no shape or form. It does not exist other than somewhere that is intangible. Consider then, your purpose. To take that idea and make it manifest, to make it real. This looks like magic to me, the work of the alchemist. The alchemist is far more powerful and effective than the designer.'

We are confronted by the work of the graphic designer everywhere – from the moment the face of our alarm clock taunts us in the morning to when we stare despairingly at the 'to-do' list on our calendar as we turn in for the night. What we are dealing with is the organisation of information – visual information – whether it is very specific details in the small print of a legally binding contract or highly persuasive images on television and in the glossy magazines. It's there on a home page, the opening credits of the next must-see film, the seduction of CGI, wherever you look next on Facebook and, most importantly, the instructions you really ought to follow on the medicine packet. See the story on medicine packaging at the end of Chapter 14, 'Future Directions'.

In fact, graphic design keeps us straight on what we want to know and need to know (the Highway Code, for instance) and what we don't want to know (the credit card statement from our bank). That makes it essential to all forms of communication where we find text and image.

Today's graphic design industry

A glance at the leading employers in the fields of graphic design, advertising and related industries reveals a wide range of companies involved in shaping contemporary visual communication, both nationally and internationally.

TIP

It would make a useful study to compare the output from a number of national advertising agencies and studios and also to research your own local and regional companies. As an exercise in networking, it is where it all starts – familiarisation with who the players are and how they might fit in with your own aspirations.

Whether high profile and mainstream or working somewhere in a design service industry, all companies have their unique character and expertise and attract graduate skills and creativity to fit in with their clients' needs. It is worth noting that the labels that employers give these positions may vary – and can be affected by technological advances – yet in almost every case, a basic graphic design education is likely to feature somewhere in the background, even for the most modest of local employment positions.

There will always be opportunities for a confident graduate, ready to take on the world of design, and you could be the one with the new idea. It is definitely a profession where innovation pays off. Take this recent example from SMILE, a forward-thinking young company of recent graduates. Nathan Monk of SMILE describes the project:

> 'In 2011, SMILE began conversations with Birmingham Hippodrome about how members of the public could interact, using their smart phones, with Six Summer Saturdays – an annual season of outdoor performances funded by Arts Council England.
>
> SMILE set about generating ideas for this brief. Our key catalyst was that, in 2011, Birmingham was voted Europe's "most boring city" for the second year in a row. Michael Smith, from BBC's *The Culture Show*, set to contest this title by investigating the city's cultural highlights. He concluded the programme by saying: "If you don't want your culture spoonfed, if you're a bit more adventurous and you're willing to go off the beaten track, it's all here waiting for you discover it."
>
> One of SMILE's proposed ideas was an "Invisible Art Gallery" – the creation of an app that could deliver images, audio and video to the user via **augmented reality**. The notion of self-discovery and exploration that Michael Smith had hit upon was at the core of the Invisible Art Gallery app.

Augmented reality (AR): a live, direct or indirect, view of a physical, real-world environment whose elements are augmented by computer-generated sensory input, such as sound, video, graphics or GPS data. It is related to a more general concept called mediated reality, in which a view of reality is modified (possibly even diminished rather than augmented) by a computer.

Running with the idea of an augmented reality app, the content for the app was developed collaboratively with a group of regional artists with experience of creating digital works. The result is "Zoe's Magic Camera" – an augmented reality app for iPhones (funded by Birmingham Hippodrome) that allows the public to interact in six key spaces in Birmingham city centre. The user gets to see an animated narrative of a young girl named Zoe discovering a variety of creatures in Birmingham.

One of the creatures Zoe discovers

All images © We are SMILE Ltd www.wearesmile.com

The introduction to the SMILE app

Finding your way around using the map

Alongside this are three secret dance performances choreographed by Rosie Kay. These are pilot films, and details of how to view them were released via social networking sites over the duration of "Six Summer Saturdays".

"Zoe's Magic Camera" was launched at The Mailbox in Birmingham on the 6th July 2012. SMILE created the app so that no markers were used in the physical world, meaning that the audience simply needed to turn up to a location and hold their phone up to see the magic take place. It is a fine example of how graphic design *thinkers* can find new creative forms of expression and bring them to life.

www.wearesmile.com

The 'Zoe's Magic Camera' logo designed by SMILE

What courses are there?

If your passion is with technology, you may be more suited to studying the mechanics of the graphic design industry – the software, computer language, programming and attendant make-up skills for web and digital applications, or the inner workings of the camera or projector and what equipment plugs in where. With a curiosity about how it all comes together, you would be exploring the interface between the technical aspects of how the industry functions and the end results. This might be referred to as the HOW axis.

On the other hand, while a working knowledge of some of these elements is essential to new graduates, your bias may be towards concepts, philosophy, ideas, layout and language, in which case there might be more attention paid to the WHY axis. They are not mutually exclusive routes into successful practice of graphic design, but with design courses the important distinction is between training and education. The weighting of content varies, with some offering a largely technical experience, while others provide more theory and a wider context in which the subject can be studied. All courses strive to get that balance just right, but you need to make yourself aware of the nature of the course you are interested in.

The headings in the next section feature as subject pathways, or modules, in one form or another *within* graphic design courses, but there are also a number of discrete degree courses whose title accurately describes what they offer. Some have named awards – for example, BA (Hons) Visual Communication (Illustration) or BA (Hons) Visual Communication (Photography), specialising exclusively in one particular area or another in the final year. It is a simple fact that your talent may not cover all of these subjects, nor will your interest, but most courses provide an early opportunity for experiment and exploration in much the same way that an art and design foundation course is set up – with a short diagnostic period – giving you an opportunity to find out your creative strengths and weaknesses. As you progress, there will also be room for more personal craft skills, such as illustration and photography, to surface, as well as plenty of conceptual thinking – all of which can be put to use in a variety of ways.

COURSE TITLE: **GRAPHIC DESIGN**

This section will deal with the labels and constituent elements that make up most graphic design courses, many of them coming loosely under the banner of 'Visual Communication'. For this subject, 'Visual Communication' and 'Graphic Design' are the most common course labels, with variations that include:

- Graphic Media
- Graphic Communication
- Graphic Design, Illustration & Digital Media
- Design Digital
- Entrepreneurship in Graphics and Digital Media
- Advertising and Brand Management
- Design and Visual Arts (Graphic Design)

It would be safe to assume that those courses with the words 'Graphic Design' squarely in the title lay claim to meeting your specific graphic design needs. However, it makes sense to interrogate the courses themselves in order to find the best fit for you.

COURSE DESCRIPTION

It is certain that all the courses, interspersed with appropriate workshops, will test your knowledge and ability, to one extent or another, in the key components of the subject. Among the key subjects sure to be covered are:

Editorial design:

This is variously known as Design for Print or Publishing Design and covers the appropriate use of layout, grids, pictorial content, etc. in books, magazines, newspapers and websites. An editorial designer will know how to style a publication using the right typographic solutions, photography and illustration.

Typography:

This area of interest deals with visible language, a core skill with relevance to all formats. Sometimes known as Applied Typography, it deals with the design of letterforms, calligraphy, fonts and the nuance of how type has a specific tone of voice and makes connections with a defined audience.

Branding:

You will find this subject given a number of add-on descriptions, such as Lifestyle Branding and Corporate Branding. Essentially it is all Brand Communication and is divided into two areas: commercial – dealing with the relationship between the product or service and its public through persuasion; or non-commercial – corporate identity, the logo, promoting the organisation's values and positioning the product or service in the market place.

Advertising:

Probably the most nomadic of all the sub-routes, you may be working with text and illustration, photography, three-dimensional models and animation all at once and will need to know enough to make the appropriate decisions that will get your ideas across to your clients. Advertising relies on two main skills – the ability to visualise and the ability to write. You need to be a student of human nature, have a vivid imagination and understand both the product and the market it is intended for.

Packaging design:

The best packaging designers are likely to know the length of time it may take for the ink to fade on a wrapper when exposed to extreme conditions in storage. From the visual impact of a label on a supermarket shelf to the nature of vacuum-forming, Packaging Design spans a wide range of skill sets and knowledge. Much more than surface application of a good design, you might just as readily be asked to create a customised container for a new product, setting you problems more akin to the spatial thinking of a product designer.

Information design:

This is the area of graphic design we rely on most – without ever realising it. The subject deals with the activity of making sure all the right kinds of information are in the right place; anything from medical know-how, charts, maps and diagrams to company annual reports, tax return forms and bank statements. It also includes signage, more descriptively known as way-finding, and the world of business-to-business communication.

CAREER OPPORTUNITIES

One of the most interesting aspects of graphic design is that it is called an 'industry', when in fact, very little ever gets made, it isn't tactile and the 'products' are more like services. There is seldom one single discipline attached to one job and this leads to an evolving list of positions and job titles. While many require interrelated skills and knowledge, others may be driven by expertise of a particular software package or have a specific technical application. In general, you are likely to encounter some of the following roles:

Freelance designer:

This term is associated primarily with a one-person operation, handling everything from client contact and briefings to design, preparation of artwork, quotations and costing, buying the print or the advertising space in the press or on television – as well as all that is expected of the business, tax and financing side of being a designer. This is really the most exacting position, and the lifestyle suits some more than others.

In-house designer:

Many large companies retain their own design studios to work exclusively on their own print and promotion, rather than using the services of an advertising agency.

Junior designer:

Whether self-employed or working as part of a team, this describes someone with technical proficiency and the ability to multi-task, usually in relation to completing other designers' work. It may be your first role as a graduate, but there is always the opportunity to have your ideas listened to and your opinions taken seriously.

Digital designer:

With particular computer skills and a recognised track record in a small studio, you may have the sole responsibility for all digital work, in what used to be known as desktop publishing. This means demonstrating prowess in QuarkXPress, In-Design, etc.; and preparing the designer's work as print-ready files.

Art worker:

Skills for this position surround the ability to work with computer programs to a very high level of quality and finish in the creation and execution of design work for production. You will need initiative, an eye for detail and be able to find your way from platform to platform at breakneck speed, with accuracy and diligence in the interpretation of a designer's instructions, using the latest software.

Art director:

This role is concerned with the management of ideas – structuring and steering a design or advertising concept through to completion. The position is what used to be known as a *visualiser*, requiring the use of drawing and software skills that bring an idea to life – and it is one half of a creative duo with a copywriter. Art directors are usually objective, imaginative and lateral risk-takers, well able to interpret a design or advertising brief.

Copywriter:

The word 'copy' in this case sums up absolutely anything that appears in print and in other forms of media using words. A copywriter will be working on anything from the names of products and companies, instructions, directions on packages, slogans, strap-lines and catch-phrases to complicated technical data, captions and annotation. Specialisms might also include the supervising of translations for overseas markets.

Creative director:

The role of a creative director is to take responsibility for all aspects of a design or advertising studio. Concepts, design and production all come under this brief, with the creative director being accountable to the head of the advertising agency or design studio. He or she will have an in-depth knowledge of the clients and be able to manage the flow of work.

WHAT WILL YOU BE TAUGHT?

Student Ruby Hirst devising graphics layouts in the studio

In spite of the fact that core skills feature heavily in the demands of a course, the most fundamental quality you can possess is your own individual curiosity. It is at the heart of all learning, and from the outset of your graphic design education the onus will be on you to make use of every opportunity the course can provide. There will be optional workshops for you to take advantage of, visiting speakers you cannot afford to miss and, to a certain extent, it will be this extra activity that adds value to your experience. The more you enter into academic life, the more confident you will become as you develop your own opinions and ideas on your subject.

The ability to visually record and document the world around you is paramount. Look at signs and advertisements, the logos you wear, the typefaces on wrappers, the flashing images of the games console. Do you know why you like them or don't like them? Everything you see is designed; it is deliberately that way and it is strategic. You need to find out why these design solutions do the job they do (or don't work at all in your opinion) and turn that enquiry into a continual resource for your own ideas.

Gathering evidence of the way you see the world is important. You will hear the practice referred to variously as a sketchbook or a visual diary. The idea is that you demonstrate that you are thinking about things, editing and exploring your subject. This is where you record your thoughts and ideas, first try out what might happen visually and analyse possible next steps. You might hear the process called R&D – research and development. Whatever it is called, the notion of reflection is important, since you show

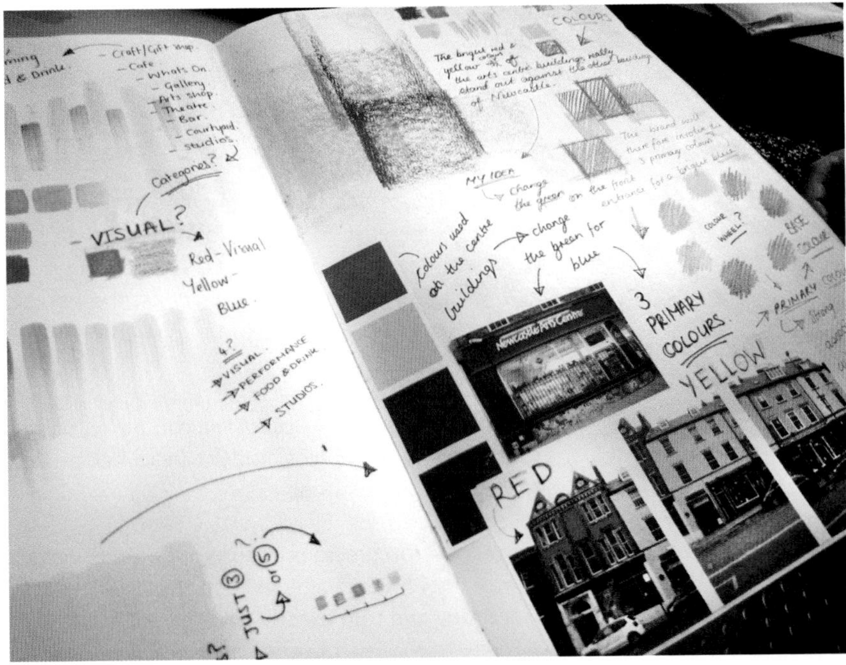

Graphic design final-year student Ruth Wood researches colour for a corporate identity project

your own unique character and creative voice through your thought processes, experimentation and by learning from your mistakes.

Make sure your research does not stop at a highlighter pen and a pile of magazines. From the outset you will be expected to evaluate what you read, put it in a context and, through trial and error, development and final editing, arrive at an informed defence of your own work. Showing the processes through which you have arrived at your design solutions is very important.

THE FIRST YEAR

From early on in the first year of a degree course, you will be introduced to the history and context of your subject and new ways of expressing your ideas in lectures and small group sessions. This will help you to place what you are doing in a historical context as you become familiar with the artists and art movements that have shaped the world of design. The theory may be a well-known advertising phenomenon, such as Trevor Beattie's FCUK campaign, which you might deconstruct to discuss the appropriate use of type, how photographs were edited, what media were employed and what the message was about. Words such as 'brand values', 'viral' and 'ambient' will creep into your vocabulary as you become aware of the inner workings of advertising. You should make sure you know their meaning.

A typical early question might be – how do you brand yourself? That could lead to a project where you are asked to create your own pictorial or typographic identity, which would involve the basic ingredients of text and image and what you might do with them to express who you are. Even before you join a course, the likelihood is that you will have been involved in some level of graphic thinking. There are plenty of unsigned bands whose music is already packaged by covers and posters and by an online presence – and sometimes it is hard to tell the difference between an amateur and a professional result.

Graphic design student Liam Doerr considers various alternative page layouts

Initially on your course, you are likely to find yourself in a period of experimentation, where general problem solving will involve you in drawing, choosing type, illustrating an idea, taking photographs around a theme or proposal and, with whatever technical skills you possess, learning how to communicate effectively with an audience. This will be done in a structured way through lectures, workshops and tutorials, with continuous assessment of your work at the heart of your learning experience. It might be basic logo design and corporate identity, or leaflet, brochure and magazine layouts where grid, focal point and hierarchy of information are introduced. Or it could be how to construct a narrative sequence by illustrating a storyboard for a short, stop-frame animation. Any number of routes can be taken into a world that you actually already inhabit. It will all be designed to help you gradually understand that world as a visual *communicator*.

There will also be technical workshops where you will be introduced to new programs and encouraged to try out ideas in the computer lab. The aim will be to help you understand as many forms of visual communication as possible and give you the intellectual and mechanical tools with which to prepare and show your ideas to a finished, presentable standard. And, if you are shy, you need to get used to criticism and debate, for you will be asked to talk about your work.

By the end of your first year, it is likely that you will have a stronger idea of your areas of interest and your confidence will be growing; however, this can only really happen if you apply yourself to your studies. As with many design subjects, you need to maximise your learning across all years of the course. These are not subjects that can be crammed in the last few weeks of term.

Ruth Wood, a recent graduate of graphic design, has put together this help list for you based on her experiences as a graphics student (read her project story later in the section):

1. Don't be too precious about your sketchbooks. At school I was taught to keep them very neat and tidy and told exactly what to put inside them, that is, 'six observational drawings' and

'five pages of development work'. You can forget all of that at university. Be as messy as you want to be. They're there to help with the creative journey.

2. Doodling, sketching and note-taking can be really worthwhile. I like to carry around a little sketchbook with me most of the time so I can jot down ideas while I'm on the train/bus, in cafés or at the library.
3. Taking breaks helps, especially when you go for walks – they help you think differently about a project.
4. Talking to other people and hearing their opinions and ideas about your project can be extremely useful.
5. Try not to go for the most obvious idea straight away.
6. Try to take advantage of all the university facilities made available to you (I'm a little disappointed with myself for not having screen printed more work).
7. Getting work experience while at university can be very helpful. I was fortunate to get a month's placement at a good branding agency during one summer holiday. It was because of that placement that I have received another two-month internship with them once I've graduated.
8. Work hard and have fun. I've had an amazing three years at university.

THE SECOND YEAR

Your second year will provide you with a more in-depth understanding of the kind of role you might take up. There will be more advanced projects, possibly with commercial settings, where you yourself choose the appropriate means of getting a message across and take more control over the outcome. It will be a time to add to your knowledge of the **semantics** and **semiotics** of advertising, or to test out just how much you prefer working with new technologies.

Through it all runs the principle of problem solving, identifying a route through a visual communication issue and understanding the market or end user you wish to reach. There will be new words and phrases to look up, such as *taxonomy* and *type hierarchy*, and an abundance of technical jargon to master. You will be expected to have a notebook glued to your side, with an appetite for 'finding out stuff' and never letting something go by when the meaning of it was unclear.

INTERFACING WITH THE INDUSTRY

There will be collaborative briefs, so that you get a better feel for team-working, negotiation and a grasp of complementary practice. There will also be a range of national student competitions to enter, such as D&AD (Design and Art Direction) or the RSA (Royal Society of Arts), and 'live' briefs from guest lecturers and leading figures in design. Many courses will also make sure you get first-hand experience in the real world through work placement schemes and internships in advertising agencies and design studios. This part of a course can be fairly competitive and the opportunity might be life-changing. It is all intended to confirm your potential to have a vocation in your chosen area of practice.

You will have a more elaborate CV by now, with proof of communication, networking and organisational skills and a working knowledge of the professional world you are moving towards. If you are a quick learner, you might want to consider branching out on your own while still studying. This is exactly what Jack Nicholl aimed for while in his second year of a Visual Communication degree course. The result is a successful on-going clothes and merchandising label, known as 'hype', that he now manages more or less as a self-employed individual, all the while combining a growing understanding of graphic design with online marketing. He would be the first to admit he has a long way to go, yet his business acumen is attracting attention.

All this leads on to a final year where you will negotiate your projects with your tutors and become more independent. It is a good idea to choose briefs that demonstrate your area of interest and showcase your expertise.

Semantics: the study of the meaning of words. It deals with the language used (as in advertising or political propaganda) to achieve a desired effect on an audience, especially through the use of words with novel or dual meanings.

Semiotics: this is the theory of visual signs and symbols and how they are used as representations with a number of alternative meanings and constructions.

Refining a brand by Ruth Wood. See her story at the end of this chapter

Jack Nicholl's branding company 'hype'

WHAT IS THE DESIGN PROCESS LIKE?

Defining the problem:

As with any form of study, the more organised you can be, the better. With practical subjects it is no different. Usually the design process begins by identifying a problem that needs to be solved – say, the creation of a new name, logo and packaging for a product, the need for a series of advertisements promoting a charity or a title sequence for a film short. The details of this would form the brief.

Looking for answers:

Even before deciding what a particular piece of design might look like, you would need to answer questions such as who is it for (client or end-user) and why (what is it meant to do?), what is it saying (message) and who are you talking to (target audience), etc., and these kinds of issues would continually have to be kept in mind while you are coming up with designs. The starting point is usually the need to solve a visual problem. The natural next stage is to look for visual answers – and there are four main sources for this:

1. **inspirations** – what you experience in life every day from which you draw inspiration – in other words, the things about which you have opinions (it might be politics or music or fashion); these make up your personal taste and style
2. **influences** – other people's artwork and their views on design, current trends and persuasive creative voices
3. **genres** – specific examples of existing work in the same genre you are dealing with, that is, how other designers have solved a similar problem (contextualisation)
4. **your own ideas** – *'off the top of your head'*, which of course, in reality, don't just happen – there will always be a dominant influence or two affecting the source of your so-called 'inspiration'.

All this subconsciously feeds into your research, though it does not in itself provide you with an answer. In a way, it is about you learning the rules that govern that particular form of enquiry, and it is surprising how closely it can be tied in with your personality. Crudely, we all know what we like, but that isn't enough. We need to understand objectively what is going on. The 'inspiration' comes when you mix all of these elements together and create a challenging piece of original design thinking, where suitability for purpose, legibility, aesthetics, information priority and appeal are all in their rightful place. To achieve this you will have to experiment, explore ideas, materials and words, try things out and be prepared to make mistakes.

As an example of how to go about finding a solution, you can see here how open the early stages can be in the explorations made by Jodi Hunt, a second-year student studying Visual Communication. She took on the problem of designing a new label for Heinz Salad Cream and immersed herself in all aspects of what the subject was about – even to the point of cooking with and using the product – all the while recording the progress she was making in her visual journal. To her, it wasn't just about drawing but also experiencing the product, in order to gain an understanding of what it was all about. Her journal, or sketchbook, was a repository, full of possibilities with, notably, nothing resolved too early.

Jodi Hunt's second-year research visual journal
Photograph by Stonie Reuladair

Jodi Hunt's second-year research visual journal showing the results of experiments
Photograph by Stonie Reuladair

Since trial and error is at the heart of creativity, the continual testing of solutions through drawing, painting, cut and paste, masking off, constructing mini-versions of things (prototype/maquette) and generally messing around, is the only way to the satisfactory development of an idea. You need to exhaust all the possibilities, sometimes even the most unlikely and ridiculous ideas. Only then does the editing process begin.

Jodi Hunt's second-year research visual journal
Photograph by Stonie Reuladair

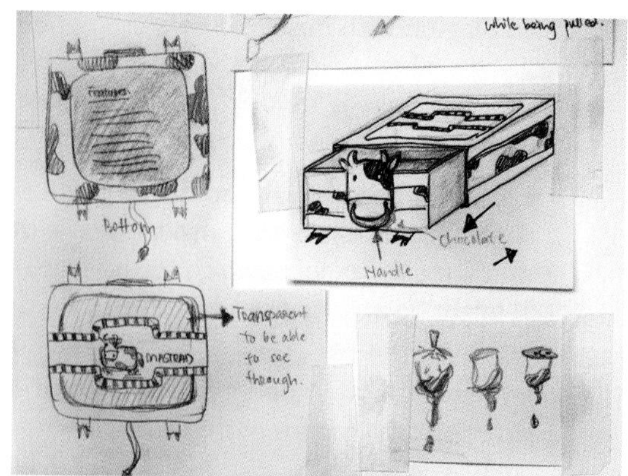

A spread from Janet Tan's research visual journal
Photography by Stonie Reuladair

When you reach a certain point, having created a refined version of what might be a suitable design solution, it will be time to evaluate it. That means asking the questions,

- does it work?
- do people get it?
- do people like it?

Janet Tan, a fellow student of Jodi, looked at how she might redesign a plain, simple, boxed fondue set to inject it with warmth and humour. The process took her through a number of stages of analysing how the items would be packaged, considering functionality versus fun.

Even having achieved a good design solution, so many other elements have to be factored in, such as production costs, shelf-life, contents protection, etc.

You can see in these images how the idea develops. These effective pieces of communication, which you will produce in all your projects, are crucial to taking ideas further and important in demonstrating to others how you think.

Towards the end of your course, you may well be expected to work to the same exacting standards you will meet in life beyond study, though it will be almost impossible to anticipate all the problems and issues you will face. It means getting used to criticism, knowing everything there is to know about what will happen to your designs and maintaining your dignity and sense of humour in the face of overwhelming odds. While coming up with an idea can be a difficult and often solitary experience, the design process is definitely not a lone pursuit. It is worth noting that you will seldom work alone and collaboration is actually to be welcomed.

A mock-up of the design by Janet Tan
Photography by Stonie Reuladair

WHAT WILL THE FINAL YEAR BE LIKE?

In this final stage, usually the third year of a three-year degree course, the idea is that your planned career strategy comes together in work that you have chosen to undertake, all supported by technical back-up and tutorial guidance. The best results are likely to be from an area of personal interest married to what you are good at – that is, an area of practice, whether it is fully digital web applications for a catalogue or the materials, design, three-dimensions and tactile making of the retail packaging of cosmetics. Apart from competition briefs or undertaking industry projects, you will take sole responsibility for your projects and for your choice and mode of study, negotiating your way through the year and bringing together all you have learned. You will be continually comparing, evaluating and contextualising what you produce, hopefully having developed a unique creative signature.

Graphic design is a complex and at times confusing subject, with many levels of ability and understanding associated with it. Thus, the demands of a degree course mean you also have to prove yourself as a design *thinker*, not just a competent designer. The difference can be in the quality of your writing and your ability to express yourself.

TIP

A high level of research, debate and well-considered argument will mark you out as distinctive. It is an area of your study that should start early. The talented and articulate candidate for a design position is more likely to be noticed than the merely talented. Know your subject and how to talk about it.

Your studying and writing skills are likely to be tested. The most common approach is in the form of a well-referenced and annotated dissertation (essay), usually between 6,000 and 10,000 words. However, there are a number of variations to this, such as a 'patchwork essay', in which you weave together your own writing with drawings and sketches and sourced information in a more personal style, probably with a higher proportion of visual content. Another choice might be a 'reflective critical commentary', where the focus of the writing is on your own work, your influences and final outcomes, rather like a well-documented diary. Usually, courses prefer that the research for the theory component is tied in with your first practical 'negotiated' projects in the final year and it is worth working ahead of time – starting your preparation early in the summer – as such modules attract a significant credit rating towards your final classification.

Many students have a 'themed' year, with linked projects, and they develop a considerable body of work as a result, but it is the final major project that usually stands out. There might still be time to get into the computer lab and pick up some new skills and techniques as the pressure mounts, but the focus will be on consolidating your ability as a communicator and demonstrating your employability in whatever sector you have chosen. The issue is not 'I want to be a graphic designer', but what kind of graphic designer, doing what – and where?

In all of this, by default, you are already aligning yourself with the world of work and thinking beyond your three years of study, even if you do not yet possess a wider appreciation of the entire industry. Whether through web design, packaging or graphic novel, you can make it easier on yourself if you know who you will be talking to and demonstrate that you can produce work that is relevant to your 'market' – the kinds of people you'd like to reach. The more you weave the elements of your course together at this stage, with quality and substance, the more likely you are to be recognised by the people whose attention you wish to attract.

If your specialism has been typography, make sure it naturally shines through in your preoccupations with designed fonts or dynamic text-only communication. If it's advertising, make sure your headlines make people laugh or feel guilty.

Your final year is a time for a final student expression of your own vision and creative ability before you face the economic realities of the profession. The best work will come when you eat, sleep, walk, talk and breathe your subject, boring everyone in sight with your newly acquired knowledge. It is also a good way of testing out how seriously you are being taken. You are moving towards the excitement of a final major project – 'going public', in the end, with a graduation show.

There are a number of end-of-year exhibitions and design competitions that showcase graduate work, organised by national bodies such as New Designers, Design & Art Direction, New Blood, Young Creative

Network and The Royal Society of Arts. You should make sure that you give yourself every opportunity of involvement in these initiatives and welcome the exposure. Having your work featured in industry publications, the right editorials and magazine spreads is an excellent way to enhance your reputation.

Preparing your portfolio

A prospective employer will want to take a look at the work you do, yet have very little time in which to view it. That means making your portfolio as concise as possible. This would apply to online portfolios as well as physical ones in portable cases. More importantly, the work should speak for itself – so avoid lengthy explanations and complicated annotation. You should stick to short titles – though, paradoxically, too many captions, especially when the form and meaning are self-evident, are a distraction.

For the world of advertising, you will have your 'book' – a compound term for the joint work of yourself and your co-designer. It will feature your 'scamps' (quick early sketches) for instantly communicating your ideas and a few well-documented, finished or published campaigns. In thinking through the sequence of showing your work, it can be a good idea to keep the best examples till last if it is in print – and being economic and memorable is much preferred to volume as you can easily overstate your case. If you have an online portfolio, make sure it is your work they notice and not the cleverness of the site you are on. Or have your own site. Your digital/web skills might be really important.

A creative director will be looking for three elements, the three 'p's:

1. **personality** in your work (an individual approach)
2. **process** (your methods of thinking and doing)
3. **potential** as a member of his or her team (how you fit in to their company and ethos).

Eamon Martin launched his portfolio online following graduation, care of Arts Thread. An eye-catching set of images and a good, clear, honest description of his skills and ambitions made him stand out. Here is

Graphic designer Eamon Martin's online graduate portfolio

A page from Ruth Wood's physical A2 portfolio

the description about himself that clearly states his skills and abilities:

> Visual Communication: GRAPHIC DESIGN
>
> Visual Communication: ILLUSTRATION
>
> Visual Communication: INTERACTIVE/MOTION
>
> Skilled in all aspects of graphic design. Accomplished illustrator in a variety of mediums.
>
> Experience of web design. Ability to produce creative motion graphics and animations. An eye for detail and quality control. A trouble-shooter and creative problem solver.
>
> An enthusiastic amateur photographer. Great team player, thrives in a group environment. Good knowledge of industry standard office and graphic software. Advanced knowledge of Photoshop. Proficient in Illustrator, InDesign, Flash, Dreamweaver, After Effects, MS Word, Excel and Powerpoint. Working knowledge of 3D Studio Max.
>
> Excellent eye for layout and colour.
>
> Ability to turn around polished design work on tight deadlines while juggling multiple projects.

Eamon Martin's graphic design, inspired by the music of Argentine musician Matias Aguayo

His projects online demonstrate all his skills, from great, highly imaginative illustrations to moving images.

> 'This series of illustrations is inspired by music. While working on this project I was listening to the music of Argentine musician Matias Aguayo. The work uses ink, stencils and old letraset transfers.'

An online portfolio presentation will have a very different feel to a tactile version, but that may well be the long shot that grants you the personal interview. If directing someone to your website, for instance, make sure it is a stand-alone site, as group sites can be misleading and confusing. If you have done your homework, you will already know what might attract attention from the creative director, what kind of work the company is involved in – clients, billing (size of company, what it is worth) – their likely areas of interest, etc., and you will have a good idea of what skills and creativity you might bring to their business.

Conclusion

At every stage, it is the concept and the idea that marks out the level and the quality of the work you will undertake. You simply have to have good ideas and know how to express them to the right people.

It is no secret that contemporary graphic design struggles to define itself. There is a complex and fascinating range of subjects to master, many of which are inter-dependent, and the biggest debate surrounds the name itself. The fact is that whether you consciously and creatively lay out a personal letter, an email or are responsible for designing a global advertising phenomena, as Rick Poyner points out in *Eye* magazine, Issue 82, 'Graphic design is more of a "field" than a true discipline.' This suggests a way of working, a mind-set, even an 'ology' in its own right. It means that specific knowledge and a range of key skills are central to being a graphic designer and there is a need for you to get to grips with these at an increasingly sophisticated level.

Why? Because however much 'graphic' and 'design' have strayed away from their old roots, the generic title for this subject has permeated almost every commercial art discipline in one way or another, making an absolutely fundamental contribution to the way we see the world. Graphic design is integral to all forms of visual media, and more and more driven by the new technologies, which ensures that it is a dynamic and ever-evolving area for study. As such, it is an indispensable part of our daily lives – full of entertainment, surprise and rich rewards for those who master its intricacies.

Further resources

Books

Burtenshaw, K., Mahon, N. & Barfoot, C., *The Fundamentals of Creative Advertising*, AVA Publishing (2006)

This book will introduce you to all you need to know about creative advertising. It deals with how agencies plan their work and produce successful advertising campaigns – with plenty of examples and detailed analysis.

Calori, C., *Signage and Wayfinding Design: A Complete Guide to Creating Environmental Graphic Design Systems*, John Wiley & Sons (2007)

If you are interested in typography on the street and in the workplace, inside buildings and between them, including pictograph language, signage, the accessing of vital public information and an appreciation of the simple business of just getting around (wayfinding), then this book is a perfect introduction.

Crow, D., *Visible Signs: An Introduction to Semiotics in the Visual Arts*, AVA (2003)

Getting to grips with the content of this book is an important part of the designer's brief. In a well-illustrated and easily absorbed style, the author explains the relevance of semiotics to graphic design, advertising and, indeed, all aspects of visual communication, and shows how concepts, meaning and an understanding of the message relate to the theories surrounding signs and signifiers.

Tschichold, J., *The New Typography*, Berkeley: University Of California Press (1928/2006)

Do not be put off by the first date of publication. As one of the giants of early type design, Tschichold's work remains as important today as in those pre-computer years. This book will give you an insight into the thinking behind the design of type fonts, showing the complexity and discipline required to produce such influential work and the political and social context in which it was created. Modern editions are available but the original 1928 one is said to be the best.

Zappaterra, Y., *Art Direction and Editorial Design*, Abrams Studio: Image (2008)

Offering you plenty of real case studies and giving you the opportunity to put your own versions into practice, this book takes you step-by-step through all essential aspects of good editorial design and layout.

Websites

www.adbusters.org

The up-to-the-minute online magazine format of *Adbusters* is a feast for the imagination and one of the most educational and entertaining sites dealing with the current state of the communication industries. It offers debate and argument that will challenge your views and definitely push you towards developing your own, well-informed opinions.

www.brandnoise.typepad.com

The consumer insights thinktank scenarioDNA are behind this very current blog. It can make the difference between you being a designer and a design *thinker* and really ought to be constantly on your list of 'must visit, must contribute' activities when you are in front of your screen. An essential tool in the pursuit of theoretical and contextual studies.

www.dandad.org

The pinnacle of any student achievement is winning an award while still studying. This site gives you all you need to know about how the professional organisation D&AD (Design & Art Direction) works, what it stands for and how you can be a part of the exciting and ever-changing world of current design thinking by entering their student awards for the opportunity to compete for their much-coveted 'Yellow Pencil'.

www.typotheque.com/site/index.php

This is a type foundry with an outstanding pedigree, and companies such as this should become an essential part of the tools and resources you will need to function as a designer. You would do well to familiarise yourself with their work and that of other similar houses.

www.ycnonline.com

In much the same way as D&AD, the Young Creatives Network finds and nurtures new talent for the design industries – in their own words, 'marrying a multi-disciplinary creative studio, an ever-growing talent network and long-standing education programmes'.

Magazines

Campaign

This is the trade journal of advertising from a UK perspective, with the latest inside track on which company holds a particular account and the effect of politics and economics on the 'market place'. The journal runs stories on who shapes which agency, the rise and fall of products and brands and the art of making the whole industry tick.

Creative Review

Describing itself as featuring 'advertising, design and visual culture', this monthly magazine comes in print and digital formats and healthily engages in all aspects of the wider design community – for example, interior design, product design, film, etc. – and its news format makes it readily digestible.

Design Week

For up-to-the-minute facts and opinions, *Design Week* is by far the best channel through which you can understand and keep ahead of the UK's advertising sector. It has the UK's leading design website and specialises in design news and jobs. The range is wide, from top agencies with nationally recognised brands to local 'one-stop' shops and freelance activity, commercial and retail opportunities. It offers a strong vocational view of how you might fit into employment within graphic design.

Eye

Beautifully designed and printed, and with exceptional quality in the editing, this is the magazine to subscribe to for an international review of graphic design. A UK publication with a global reach, it appears quarterly, which makes it accessible and affordable. You will find yourself returning to it as a valuable resource and constant source of inspiration.

I.D.

This magazine is unashamedly about cutting-edge design from every discipline, with thoughtful articles on what drives the latest trends. The coverage features the world's best designers and throws light on their approach to concepts, the products they endorse within their practice and the environments in which they work. Furniture, fashion and innovative product design are all considered.

Career profile

Name:
James Coleman

Current job:
Managing Director at Supercool

Describe what you do

Running a small business, I'm a bit of a jack of all trades. I deal with a lot of the new business, talking to prospective clients, putting proposals and costs together for people. I project manage some of our larger jobs and work with my partner and our developer across various projects. I also look after the finances and policy of the business. Sometimes, if I'm lucky, I also get to design as well; I guess this is around 20 per cent of my time, at most.

I sometimes envy junior designer roles, where the job is wall-to-wall graphic design. While I'm sure it's a very challenging job, and very daunting for a lot of starters, there's a certain clarity in the briefs you receive. When you start working for yourself, you're always considering things outside the client brief – there's a lot of other baggage. Will the client fire us? Will this job lead to more work? Have we charged enough? How much time should we be spending on this work?

Working for myself has helped me to be pragmatic about the creative process. You need to be organised. You also need to make sure you don't become too much of an artist. While it's sometimes tempting to over-engineer and always make things perfect, it's worth remembering that a project is a service – there are always client objectives and always a budget. It's only recently that we've decided to employ other people at Supercool, and being a very small organisation that has always tried to compete with much larger agencies has meant we've had to be flexible, we've often had to learn things on the fly, and sometimes only for a specific job. Understanding what's worth spending time learning or putting in place, what's worth investing in and who's worth wooing are all things we've not necessarily always got right.

Tell us about your career so far

I didn't want to do A levels. I was offered a place on a BTEC course in Design for Media. It included graphic design, photography, illustration, animation and film – it sounded brilliant. Two years of that led me straight onto a Visual Communications degree course at Birmingham Institute of Art and Design. I still had no idea what I'd do for a living, but I loved it. I think I particularly loved the variety of disciplines. When I got halfway through my second year on the degree course, I made a couple of friends during an exchange trip. Skills-wise we were all slightly different, but we loved each other's work. The three of us boldly decided that we weren't going to work for anybody else and instead set up our own business.

A year later we graduated and it became obvious that we'd failed miserably at building up any kind of network, let alone potential clients. One of us dropped out and we were down to two. We had virtually no work for the first six months, and the only thing to show for it was a fine from Her Majesty's Revenue & Customs (the tax man) for failing to submit a tax return.

However, we got lucky. Through a friend of a friend of a friend we met a guy who was starting up his own publishing company. He needed a bunch of magazines and directories designing and asked if we did magazine design – we said 'of course' and got to work. The first magazine was all about interior design. It was an advertising-funded free magazine that got delivered to 20,000 homes in and around Birmingham. We had two months to design the brand and put together the first issue. We worked day and night to get the artwork turned around. It was at this point we realised how unprepared for working life we were. The first issue was eventually sent, and despite it taking two months of solid work between two of us, within a year one of us could do it within just four days.

In my spare time I started to play with the web. We found that most people we worked for wanted websites and started working with developers. We'd put the design together and hand it over. However, this became a very painful process – essentially managing jobs we had very little control over. From that, I decided it would be far easier to do it all myself, so I started to design and build sites for a few of our clients. Looking back, a couple of them were truly awful. However, I persisted and eventually really started to enjoy the process. The sites I was developing started to get bigger and I began to find that I was a much better designer for screen than print.

I've certainly developed far more as a screen designer than anything else over the last five years. This has meant really getting to know how the web works under the hood, what makes a well-coded, accessible, usable and sometimes search-engine-friendly web page. I find the interaction between page and user more interesting – how design changes between devices and how users have the control to affect space and layout. The combination of client interfacing and design processes for screens makes me more of a 'User Experience Designer' than anything else – at the moment at least.

Graphic design can be really difficult. To produce anything that's worth anybody paying for you need to put your heart and soul into it, and when it doesn't work out, when people reject it and you get knocked back, it can feel like the end of the world – working for yourself seems to magnify this as there are often no buffers. So far I've been really lucky that most of the time my clients have been really good, and I get to work with some great people.

Is there a particular project you would like to tell us about?

DanceXchange is a national dance agency based at Birmingham's Hippodrome. They teach and perform contemporary dance, as well as produce their own work and run a number of community-based projects. We've built up a strong reputation within the arts and cultural sector in the region. This really helps you to be in the loop with the right people when opportunities arise. We'd already got a few strong examples of dance-based work within our portfolio so knew we had a good chance. We probably put a

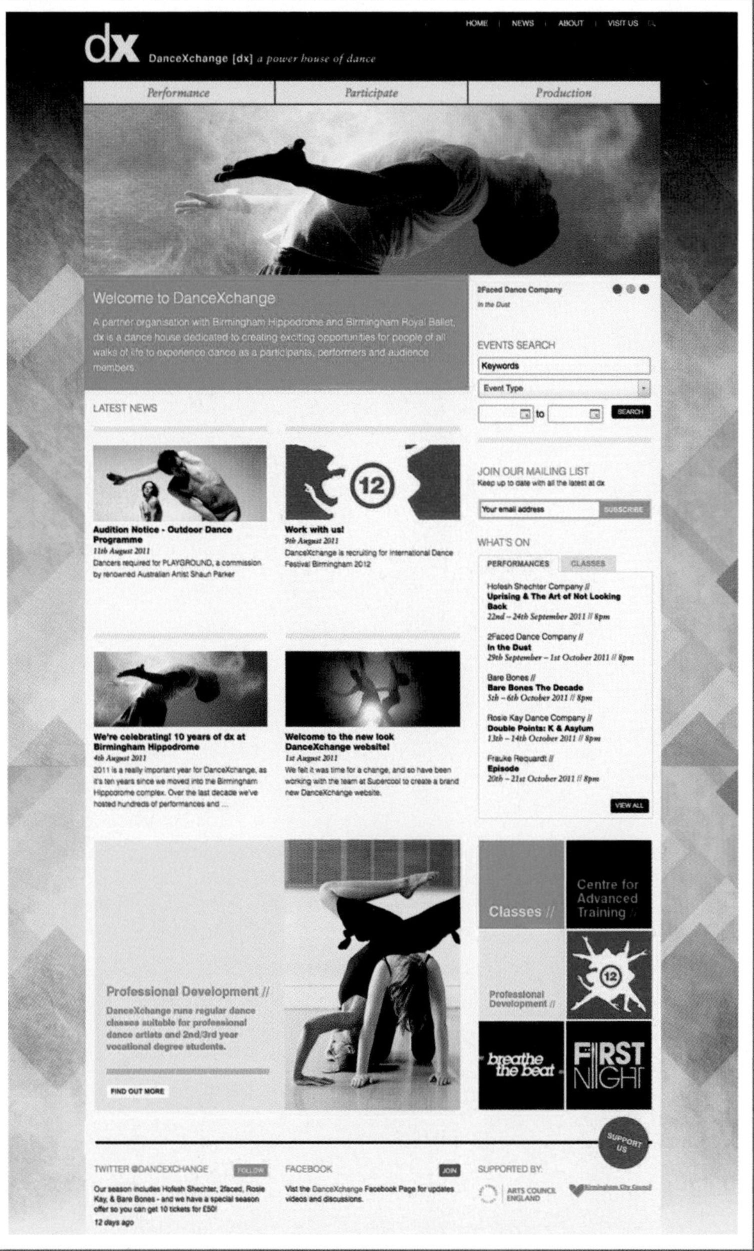

A sophisticated new look for DanceXchange, designed by James Coleman of Supercool

Image: Supercool

lot more into the pitch than the initial job was worth, and I think it was because of that I was so happy when they decided to work with us.

We were commissioned initially to redesign the organisation's season brochure, along with completely redesigning and redeveloping their website. We first carried out a refresh of their brand identity – building on, and making the most of, what was already there.

Having been applied across the entirety of DanceXchange's communications collateral, the new look has been extremely well received internally at DanceXchange, by their board of directors and partners, as well as – perhaps most importantly – by DanceXchange's audience.

What advice do you have for students considering a similar career?

I think my tutors at university were probably right. Had I experienced the real world of design to start off with, I think I would have probably found it easier than going to work for myself. We had a few very difficult first years, not only trying to set up a business but also just doing the work – it was really tough.

And then the second piece of advice from my tutors – build up a network. This is so much easier nowadays given we've got Facebook, LinkedIn and particularly Twitter. You can find the right people and then get to know them and hopefully get across a bit of your personality too in ways that weren't possible ten years ago. When we started, we actually used the *Yellow Pages* to find competitors within the area.

Project story

Designer:
Ruth Wood, final year graphic design student

Project Title:
Newcastle Arts Centre Rebrand

Early colour work in sketchbook

Tell us about the project

This was a self-negotiated project. I was really keen to do a branding project, particularly for something arts related. I finally focused on the Newcastle Arts Centre. I had often visited the centre, and after looking at their website decided their brand could be very interesting to redesign.

I sent an email to the centre asking for permission to take photographs around their buildings. I also asked a few questions about what they do. Their feedback was useful; I discovered that they were very keen to promote local arts and that their target market was 'everyone'. These findings really helped inform my decisions.

After a lot of research I began generating ideas. I had noticed that the brightly painted centre really stood out against the other buildings of Newcastle. This observation led me to focus on the idea that the Newcastle Arts Centre brings colour and creativity to the region. By also using blue for the front of the centre, I decided to use the three primary colours to define the different parts of the centre: red for visual arts, yellow for performing arts and blue for food and drink.

Final year branding project, Ruth Wood

I wanted to emphasise the expressive nature of the arts centre. As it's most well-known for its quality art shop, it seemed a good idea to use the materials available in the shop in the branding. Quite a lot of time was spent having fun mark making with inks, charcoal, pastels, paints, etc. It was finally decided that these marks, combined with the three colours, would be applied across all the branding.

The brand style and logo can be applied to a wide variety of formats – packaging, signage, advertising. For this project I decided to start with what I felt was most important first – signage and the 'what's on' guide. If I still had time after these were created, I would then go on to design the other elements.

As a final touch I worked on some advertising for the centre. This involved applying the bold colours and marks across the more dull parts of Newcastle city, on walls, pathways and staircases. This would hopefully emphasise that the centre really does bring colour and creativity to the local area.

What processes and skills were most relevant to this project?

Research is always vital with every project. I really love researching a topic, particularly when it involves visiting places. I've looked a bit weird at times. One project of mine involved exploring my local library. I got a few odd looks from people as I took photos of bookshelves and signs and took out all the post-it notes and bookmarks left in the various books. Gathering photos, flyers and printed ephemera can be really helpful. You can discover things you didn't notice before.

Mark making was important with this project. I often like to experiment with media and materials when I'm a bit lost for ideas and sick of just being on the computer.

What challenges did you face?

I was initially unsure about how to approach this project. At present the arts centre has a variety of different logos and names for its various parts – its theatre is called 'The Black Swan' and the art shop, 'Details'. I wasn't sure whether I should create different individual brands for each part. In the end, the task was made a lot simpler by just creating one, simple, unifying, umbrella brand for the whole centre.

I find coming up with a strong idea for a project can be really difficult. I can spend a lot of time struggling and stressing to get the right idea. With this project, however, the idea came pretty easily, which was quite a surprise. Such a surprise that I worried that it wasn't right and wasted time trying to take it further. I didn't need to.

I also wish I was more decisive. It's a really useful skill to have. One of my downfalls has been having lots of ideas but not being able to pick which one to carry on with. I would have saved myself a lot of time if I'd had more self-belief and worked straightaway with initial ideas.

What were the highlights of the project?

I find projects a lot more exciting when they're not spent entirely on the computer. This project was fun because it meant I could experiment with different materials and media. I also really enjoyed making the various brand touch-points.

In general I like graphic design because the work you produce can be so varied. It can be 2D or 3D, printed or digital. I love generating ideas and problem solving. I like drawing and making. I also like graphics for some of the barmy things I've ended up doing during my university course – making model trees that will hold lollipops, taking photographs of chewing gum spat on the pavement and cutting more than 500 triangles out of paper.

ARTS
NE1

The final logo for the arts centre

Sketchbook fun!

Career profile

Name:
Judith Doherty

Current job:
Marketing Assistant

Describe what you do

I work in a very busy marketing department for a stone, porcelain and ceramic tile supplier. My role consists mainly of graphic design, and I have contributed largely to many design projects during my short time at this rapidly expanding company. I am predominantly based in the head office in Surrey; however, I occasionally travel into London due to the recent opening of the company's flagship showroom in Clerkenwell in the heart of the design district in central London. Our main customer base is architects and interior designers, and it is essential that our company's brand grows and develops with the ever-changing trends of the very competitive design industry. My role is very varied and no day is the same. I have been involved in the design of advertisements for national magazines, updating the design of brochures, mood boards and other marketing materials used by our sales team and designing promotional material for campaigns and showroom openings. I have also been heavily involved in website design and elements of e-marketing, which we use to promote and launch our showroom openings and sale events. As marketing is such a busy department, it is essential that I am organised, proactive and that my communication skills are second to none. Deadlines have to be met in order for a promotional campaign to be successful – 'time is money' as I remind myself!

Tell us about your career so far

I studied BA (Hons) Furniture Design and then MA Design Management, graduating in 2009 with a new-found love of marketing and high ambitions for a career in that area. To fund my Master's study I worked part-time as a retail supervisor. At the time of my graduation the recession hit and this resulted in graduate jobs being few and far between, so I continued in my current role, full-time, as supervisor and then visual merchandiser. For personal reasons I moved from Birmingham to Woking, Surrey and got a full-time senior visual associate role in Gap Kids in Guildford, while continuing to hunt for my dream marketing job. The role at Gap really helped to maintain my creative and marketing skills as I was solely responsible for the appearance of the store, and I played a key role in promoting the company through my window displays and in-store visual merchandising. My work was highly praised by senior management and I was invited to help merchandise the flagship store in London in preparation for a visit from the CEO, so after 11 months of service I feel I had achieved a lot in such a short time. Now, in my current marketing role, I am using the skills I learnt at university and gaining very valuable experience that will help me progress my marketing career.

Is there a project you would like to tell us about?

Our company's new showroom in Clerkenwell was planned for over five years, long before I joined the company, and this was a big investment for the company. I joined at the right time to start preparing the marketing for the showroom. It was such a pleasure to be involved in this project, and I feel my contribution was very valuable. I was part of the team involved in all the graphics for the showroom, including the invitations, leaflets and online campaigns and the vinyl lettering that is used on the displays and windows of the showroom. I was responsible for designing a guest name-badge system for the events we held and, most importantly, designed and organised the marketing material for the tile sample displays, which proved to be the most popular part of the showroom. The showroom was open in time for Clerkenwell Design Week, which was held in May 2012. This event itself is one of the most extravagant and attractive events held in Clerkenwell each year; it involves design showrooms across this part of London displaying their most inspirational work to some 20,000 visitors who come to Clerkenwell over the three-day event. As we were taking part for the first time, our branding and marketing was hugely important to ensure we made that first impression last. We wanted to appeal to potential new clients and to ensure that our showroom space stood out from the others, and it was great to experience this. In preparation for Clerkenwell Design Week, I created and organised the endless promotional leaflets, e-marketing and facebook campaign visuals, which all helped to bring in over 2,000 visitors to our showroom.

What advice do you have for students considering a similar career?

My previous jobs have not been directly associated with my degrees, yet I gained good experience that I have utilised in my current marketing role. I never lost sight of where I wanted to be and how I wanted to start my career. Even though I had to wait a bit longer to get into a role that I had studied so hard for, I still made the most of the jobs I did along the way. Every experience was a valuable addition to my CV. Marketing is a broad area but can be really competitive. The advice I would give to you now is to never give up trying to get a role you really want. Fight for it, and when you are in that role make sure you are indispensable! That was the advice I was given when I was looking for a role to start my career, and it will be with me always.

Project story

Designer:
Suzana Bašić, graphic design student

Project title:
New York Water Taxi Rebrand

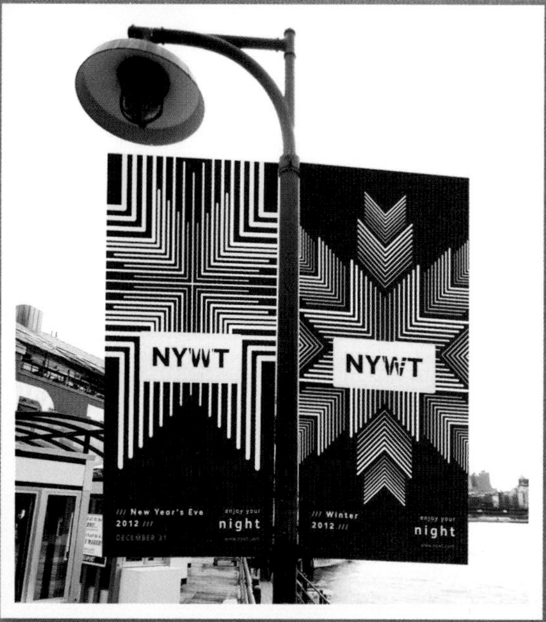

New York Water Taxi identity

Tell us about the project

This four-week, final-year project was derived from a much wider initial brief, entitled 'A grand day out', which I chose to respond to as a result of my visit to New York. Due to it being very much an open brief, I felt that it was necessary to define a subject matter in order to create a clear aim, so to begin the process I recorded my journey to help me consider what direction to take. From this broad overview I gradually pinpointed a topic that I gained interest in by researching it in more detail. The scope in many of the briefs allows you to explore the areas that you most enjoy.

The decision to rebrand the New York Water Taxi (NYWT) was influenced by my experience there; having taken a boat trip I was able to familiarise myself with the brand by obtaining a first-hand account of the service. My choice to focus on branding reflects my interest in this specialism.

What processes and skills were most relevant to this project?

Specific research naturally led onto related areas, and ranged from the branding of cities and tourist attractions to water transport and nautical themed imagery. In considering the purpose and target audience, I evaluated the balance of these core elements in order to underpin the brand identity and so portray it through my design. In any project, I think that this first research stage is crucial in many ways. The subject matter is explored, not only for inspiration or to see what's already out there, but by selectively categorising areas of research; you will understand and visually define the problem.

A number of visual brainstorms drew out the key elements to take forward and adapt within the new design. Iconic imagery, such as the distinctive yellow exterior and black-and-white chequered pattern associated with New York taxicabs, is followed through in the water taxis. I wanted to maintain this link in some way, however with greater emphasis on 'water' to give it a more distinctive look and set it apart from the New York cabs. I produced numerous initial logo designs that continued to change and improve, while implementing them onto the brand touch-points, such as tickets and posters.

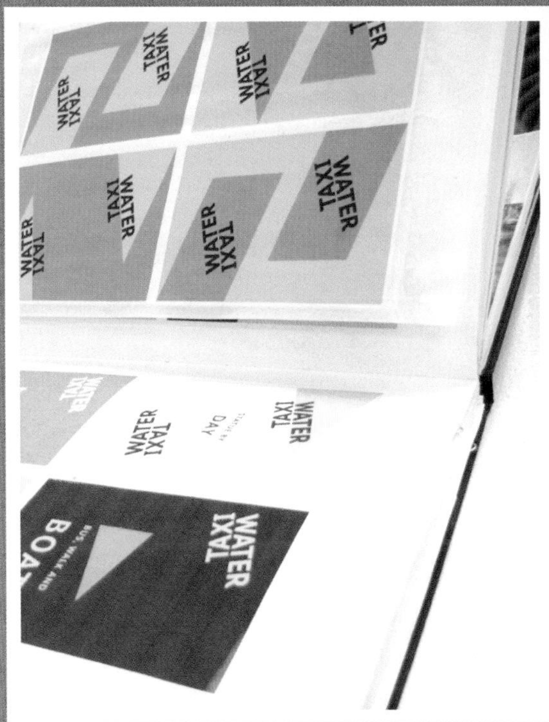

Early concept work

What challenges did you face?

The main challenge that I faced was creating a brand that was as versatile as the service. The boats are primarily a tourist attraction but can also provide everyday or special event transport. They host annual celebrations, EcoCruises, work in conjunction with bike, bus and walking tours and there is even a service that takes you straight to the IKEA store! However, as I had a lot of ideas to begin with, I was able to select the parts that worked well and integrate them. It wasn't until a couple of weeks into the project that I felt the concept was strong enough to apply it to the different platforms.

What were the highlights of the project?

Making judgements was a lot easier due to the process that I had undergone. After trying out a variety of layouts and further development of each outcome, I was able to efficiently refine the final designs while maintaining a consistent visual style in order to depict a coherent brand. In keeping with taxi visual language, the logo is bold and simply highlights the 'W' within the NYWT acronym. It plays a big part in holding the concept together. The window within the letter reveals the 'water' ripples that continually move, allowing the design to constantly change and stay engaging. This arrow-like symbol was influenced by the black-and-white chequers from the existing NYWT branding, with the surrounding yellow rectangle representing the dominant colour of the boat itself. The design suggests efficient transportation and its fluidity caters for the vast array of tours and services they operate. The repetition of the symbol to create patterns and images for promotional material in effect replaces the chequers as the iconic aspect of the brand.

It's always useful to talk to others outside of your project and take on board their perspective, so don't work in isolation. A conversation may just trigger an idea. Finally, trust your instincts.

Rolling out the brand and promoting it

5

Interactive media design

By Kathryn McKelvey, Northumbria University

'Interactive media design gives students the opportunity and freedom to explore a wide range of new media subjects, ranging from UX design (user experience design) to interactive exhibition design and everything in between.'

Oliver Johnston, interactive media design student

Jack Merrell, 'Twitograph', BA Interactive Media Design

We are surrounded by complex yet functional ways of interactivity that have been designed to, hopefully, make our lives simpler. For example, many people take for granted their use of social networking and websites that provide them with quicker, simpler ways to keep in touch with what is going on in the world and ways of buying things. The development of the 'smartphone' has become a welcome extension, with the added facility to download applications for entertainment or information. All of these 'interactions' have to be designed. It is appreciating the future demands of the customer that is pivotal to becoming a successful interactive or interaction designer (the terms are largely interchangeable, although interaction would imply the detail of the communication and interactive could imply the bigger picture of systems).

The media, communication and computing industries have been merging and converging for a number of years and are able to adopt other design disciplines when necessary. This new combination is becoming known as 'interactive media design'. Interactive design is a broad description. It requires a user to interact with products and services that are digital and computer-based. Other associated terms for interactive design are:

- human-computer interaction
- multimedia
- interactive multimedia
- new media
- interface design
- interaction design
- digital media

Some of these terms (multimedia and new media) have become outdated and have been replaced by new terms, such as 'experience design'. The premise in all of these is that the user interacts with text, sound, graphics, animation and/or video. With 'experience design' an emotional, entertaining or satisfying experience is desired. The ubiquitous interactive products we are all aware of are websites and games, and there are particularly rich interactive experiences in these media.

As well as interactive games and web design, this discipline of design encompasses technological developments in social networking and mobile technologies. It also deals with interactive installations. For example, Cadbury World in Birmingham have an interactive floor that works with pressure sensitive sensors. Visitors interact with games about chocolate by stepping, jumping or running across the floor. See www.adammontandon.com/chocolate-infinity for this and more interactivity examples.

During the next few pages you will find out more about the purpose of studying the subject of interactive media design and will gain a greater awareness of the industry and the opportunities within it that will lead to a rewarding career. Further resources are also available in this section of the text – career profiles, project stories, books and websites. Read on to find out about the sort of career opportunities that are available to you as the communication, media and computing worlds collide to provide us, as customers, with well-designed interventions that support our busy, ever-advancing lifestyles.

Today's interactive media design industry

There are some key developments that have defined the interactive media design discipline as it is today:

1. **The development of the silicon chip and Silicon Valley, in California**. Silicon Valley is still the 'hub' for high-tech businesses and innovative technological developments in America.
2. **Steve Jobs – an American pioneer of the personal computer**, designer/inventor and co-founder of Apple Inc.
3. **IDEO, an American-based design consultancy** that is committed to taking a human-centred, design-based approach to help other companies innovate and grow their business. It was founded by British industrial and interaction designer, Bill Moggridge, in 1991. He advocates a user-centred design process and thinks interaction design should be a mainstream discipline. www.ideo.com.
4. **Tim Berners-Lee**, who made a proposal to create an information management system around 1989, and in 1990 successfully communicated between a server and a Hypertext Transfer Protocol (HTTP) client. This was the birth of the World Wide Web.
5. **Massachusetts Institute of Technology (MIT)**, a pioneering American university committed to technological education and research.
6. **Nicolas Negroponte, founder of MIT Media Lab**, which supports the investigation of the human-computer interface.

Interactions occur in many parts of our lives nowadays, we may well take them for granted. They had to be 'designed' and very much had the **user** as the focus. The user sometimes is a customer or consumer using a service or device, or trying to find out some information. The interaction may also be for entertainment purposes, allowing the user to gain new knowledge or be momentarily inspired.

User: anyone who is using a digital service, product, computer or any other digital device. To begin to understand the user, observe a friend or colleague playing an online game. How much experience do they have with using the controls; ask them if they can remember what it was like to learn to use the interface? Look at the menu system on your smartphone; try changing the time and date, how easy is it to find the information you require? Try using someone else's phone, unfamiliar to you, to do the same task. Are there familiar elements in the interface that guide you or do you have to explore, discover and learn new icons and interactions? Think about the journey you have taken – ask someone else to do the same tasks and watch how they go about it.

Here are three examples, some of which you might not have noticed even if you use them, involving 'simple' interactions designed to streamline services, aid information accessibility and provide entertainment.

1. **Services – Oyster cards** – Purchasing Tube tickets in London is made easier by using touchscreen technologies. The 'Oyster' card (known as a 'contactless smartcard') can be purchased to make it possible to take multiple journeys. It is reusable, as you can electronically transfer money on to it and keep it topped up so that you can use it to travel around London. The start and end of each journey requires the owner to place the card in range of an electromagnetic field, on a yellow circular reader at each automated barrier in London Underground stations. These actions are recorded and fares are deducted from the card.
2. **E-commerce – Amazon** – Regular users of Amazon, the multinational e-commerce site, will find the interactive shopping experience very straightforward. Go to the site, select a product area, put in a search keyword, then scroll through the results, clicking and selecting as you see what you are looking for. When you add an item to the 'Basket' and 'Proceed to Checkout' you simply type in your password, then purchase the item. Credit card details are retained to make the whole process simple and effective.
3. **Entertainment – BES Interactive Wall** – This is described as an 'office video wall' and has been developed for the financial services by the company

YDreams. It combines video and art installation approaches and the wall alternates thematic images and video, triggered by the movement of passers-by. It was originally developed for Espaco BES Arte e Fanaca in central Lisbon. For more examples, visit www.ydreams.com.

Often the most successful experiences are those interactions that you hardly notice – except that the experience was pleasurable and worked well, as these examples show. Amazon has taken this further and uses technology on its e-commerce website mentioned above to provide a more personal buying experience; the site remembers what you bought and emails you updates about new products that you might be interested in buying, which is a very clever interactive system that provides a personal service, adding the 'personal touch'. Sometimes these types of feedback system can become irritating, however, there is usually a facility to cancel these notifications in the email.

Through social networking sites such as Facebook, users and companies can create profiles that can be seen online. Users can leave 'status' messages and communicate with each other in statements and comments. Groups are formed where users have common 'likes', and it is these shared interests that bring people together in these virtual worlds of communication. Both the Amazon e-commerce site and Facebook had to be designed carefully, based on an original idea. This required a great deal of development to get it right, development that continues with each launch of the latest version.

Interactive media design requires a combination of concept, design and technical skill in varying measure (dependent upon what is required to create a working prototype that sells the idea). Both are web-based services and require good usability and accessibility for the user. The Internet is a good place to begin learning about interactivity.

Companies can advertise new fashion lines, events and sales on Facebook by targeting customers in banner advertisements and other smaller adverts. The user can play games and view new content from companies. In terms of customers having a great online shopping experience, marketing strategies and brand identities are much more closely linked these days and involve a variety of media platforms, for example smartphones, the iPad and the Internet. This ensures that customers are satisfied with their interactive experience and the company is able to gather intelligence on its customers and ensure that it provides them with a cohesive shopping experience. This is referred to as digital marketing. The interactive designer's job is to design the **web information architecture** that allows the customer to do what they need to do, referred to as **user experience interactions**.

Web information architecture: the information architect designs the organisation and labelling of websites and online communities using primarily 'wireframe' diagrams.

User experience interactions: the user experience architect looks at all aspects of the experience, including the aesthetics, interface and physical interaction – in fact, the whole 'experience'.

The interactive industry cannot afford to stand still, even for a moment, as both designers and consumers are forever chasing a faster-paced, slicker operation to improve the way things work:

- **Cloud computing** – our computer resources will be available to us anywhere and will be accessible by any type of device, whether mobile or fixed.
- **Touch, gesture, voice** – user interfaces will be operated by touch, gesture, voice and our eyes and no longer by wired-in mice and keyboards.
- **Connected domestic digital devices** – all can become connected through the use of the Internet and scalable interfaces (an interface design from a 52″ TV screen can be re-worked and applied to a mobile phone).

'Recognition' and 'gestural' interaction using sensors and triggers is becoming a much broader area of interactivity development and is explored by artists and designers. Movements using the arms or sweeping hand gestures or a pointed finger are sensed by interactive elements, which then react in some way, usually visually, sometimes by sound, to the participant's movement. At the moment this technology is deployed in games technology, such as Xbox Kinect.

There are many directions in which the technical and interactive possibilities are heading, and it is important but very time-consuming to keep abreast of these. Sharing discoveries is very important to the process – following blogs, setting up blogs, inventing the next blog type interaction – these are all in the domain of interactive media design. Jack Merrell's project details how he works and why he has chosen to focus his energies on interactive design:

I am a graduate of Interactive Media Design, I consider myself to be experimental in my approach to design. Following my degree, I decided to continue my studies by doing a Master's course in Multi-Disciplinary Design Innovation, so that I could continue exploring the subject in more depth. Looking back at my final major project in Interactive Media Design, I wanted to create an installation to produce a physical read-out of information in "real time" that would represent a set of data in a new and interesting way to engage and inform a user.

Early on in the research phase, I wanted to create a physical installation. I researched art installation and I looked at machines that could draw, such as Rafael Lozano-Hemmer's "Recorders" installation. I started to construct a device from random mechanical parts from printers and scanners, and used an **Arduino board** to control a servomotor.

Arduino board: an open-source hardware and software microcontroller board that makes the use of electronics easier in art and design interactive installation-type projects.

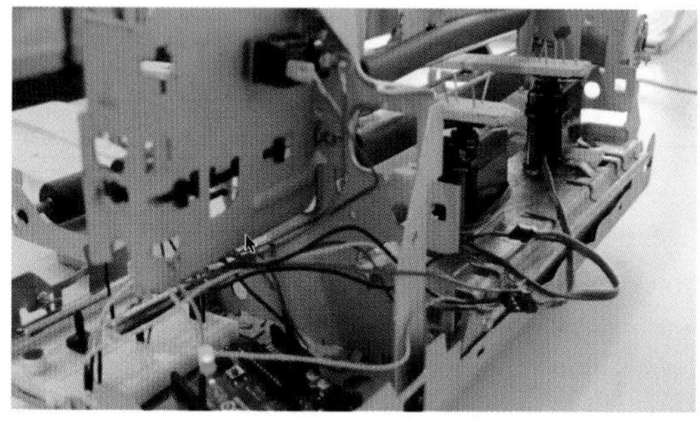

A servomotor

The cardboard arm to support the pens

The mechanisms were then attached to a cardboard arm holding a pen and all hooked up to the computer. Eventually, after testing and reviewing it, I could compare "trending" behaviours in Twitter – the way that numbers of people responded to certain stories. This was a successful attempt to create a new and insightful way of physically recording Twitter feeds.

Simple but effective power and transmission systems

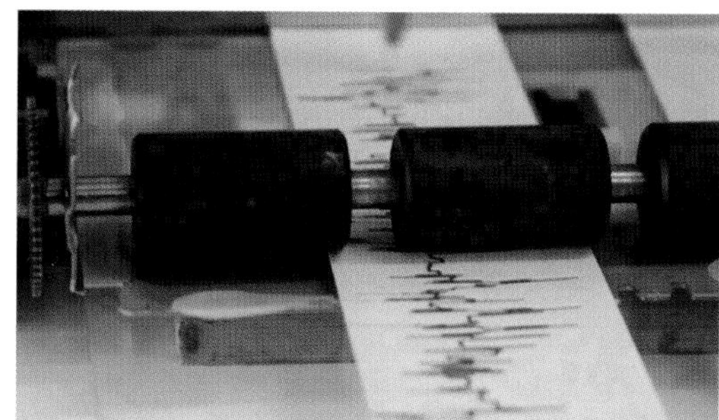

The seismograph

Initially I didn't actually want to map Twitter feeds. Instead, I wanted to see how other data, on a more local scale, might look when graphed over time, such as Wi-Fi signal strength or a local weather report.

Craftsmanship was a hugely important part of my "Twitograph" project, and I enjoyed seeing all of the different materials being brought together to make the machine work! Reusing old pieces of discarded technology was a satisfying part of the making process.

I exhibited this at my graduate show and at the D&AD graduate show in London and it provoked a lot of interest. I received a positive reaction from people intrigued by this mysterious-looking machine. They really wanted to know what it did and how it worked.'

Take a look at Camille Utterback's 'Text Rain'. This is an interactive installation, the text raining down is in the form of a poem separated into single letters. Using a large projection screen and a mirrored projection system, participants can see a black-and-white version of themselves interacting with the falling coloured letters. The letters can be 'held' in suspension where they interact with dark objects and bodies, but when these are removed the letters continue to fall. This work illustrates clearly how gestures and interaction can be recognised within the projections of this art installation – camilleutterback.com/projects/text-rain.

Another more recent example is Camille's 'Active Ecosystem' project from 2011 (camilleutterback.com/category/projects) located in Sacramento International Airport. It involves 14 LCD screens mounted on an elevator shaft. The elevator and human movement within interact with animations inspired by fish, leaves and seeds – these grow and move when the elevator is called and when it stops or moves to a new floor. The animations also change colour dependent upon the time of day and the seasons, where different animated fish will spawn at different points in the year and leaves will fall from the trees in Autumn.

Take a look at the company Imagination; their motto is 'experience is everything'. From a design perspective they work in branding, communications, marketing and in events and exhibition design, where they develop interactive experiences. Visit www.imagination.com/our-work. One particular example is the Ted Baker 'London Calling' 1950s tour of the flagship store in Tokyo. Shoppers were greeted with the TedTaxi when entering the store – half a London cab. In the rear windows of the cab a video shows a journey around London in the 1950s and shoppers can take photographs and share them online to encourage an 'ongoing social media interaction'. An old-fashioned telephone exchange is lit up too and attracts shoppers to the stairwell, where they can play with red, white and blue glowing LED connections, triggering sound effects and conversations from London in the 1950s.

What do you need to be successful?

To be successful in this discipline you will need to develop excellent visual communication/graphic design skills and be able to draw your thoughts out from original concept to finished idea, so that a client can completely understand your thinking.

Laura Verbaten mixing old and new design techniques in the studio

You need to have a keen interest in computers and a desire to find out about new technologies and how they can be applied to new developments in interactive design.

Ideally you would also have an interest in understanding people's desires for interacting more efficiently, so this would demand that you work more closely with end-users to understand the subtleties of what is required. Developing your personal skills and being able to work well in a team will also be crucially important for you, and having a sense of fun and a desire to entertain through developing interesting gaming concepts will help you get on in this industry.

There is a wide range of employment possibilities in the games, web, interactive and interaction fields as designers, usability experts, information architects and user experience architects. This is not an industry that relies on one particular physical location in the world. As communication devices continue to improve, such tools as Skype are used regularly for their videoconferencing facility between industrial collaborators situated anywhere in the world.

What courses are there?

Interactive media design is a young, exciting, fast-expanding sector of the design industry and there are a wide variety of courses to prepare you for it. While the computer is the main tool for experimentation, you may also find courses based in the traditional Graphic Design subject area.

Courses might be either BA (Bachelor of Arts) or BSc (Bachelor of Science), dependent upon the amount of computing and programming involved in the course content. Traditionally, the more programming involved the more likely it is to be a BSc, though simple programming is included in BA courses nowadays, especially where electronic, 'outside of the computer box' interactions are designed. Course titles might include Digital Media, Multimedia or Interactive Media Design. Computer Games Design and Web Design are more specific in their content, with particular markets driving their development. Graphic Design courses will also have a strong digital element, with varying levels of emphasis placed on interactive media. Courses where information systems and computing are mentioned as key components of study are probably to be located within computing departments rather than design schools. Read the descriptions of the course content carefully, as they all have their own focus. The common theme that runs through them all is the development of core skills that the industry needs, such as communication, team working and a keen interest in technologies.

If you are interested in courses that have strong interactive and animation content then remember that these are both rewarding and deep subjects in their own right and may not be equally balanced in any one course. There are usually pathways or options built in to a course, along with your own opportunity to tailor your final projects to suit your interests and help you enter the sector of the industry that interests you most.

It is always possible to continue your studies at postgraduate level (Master of Arts or Science) to gain an even deeper understanding of the subject and to be able to offer a competitive edge to employers. Two-year foundation degrees are also available (equivalent to years one and two of a degree) at further and higher education colleges and can often have more technical content than design in their curriculum.

COURSE TITLE: **INTERACTIVE MEDIA DESIGN**

COURSE DESCRIPTION

> *'Interactive Media Design students will have access to cutting edge technologies, a creative and enticing working environment and the opportunity to explore a wide range of design-based media.'*
>
> Sean Radcliffe, Interactive Media Design student

Many Interactive Media Design courses are graphic design-based, with the principles of typography also being taught. The differences between information, promotion and advertising design will also be highlighted. A key focus will be to build on these skills, through project work, by adapting and designing for the web and games industry (online and gaming consoles), interactive television, mobile technologies and phone applications, or Apps as they are known. Apps are designed for a specific task. The best Apps have a great idea at their core and, if the device has a touch screen, use gestural interactions to operate. These may have more of an 'art'-based and 'experiential' approach to them, as well as the functional Apps that we are all beginning to know.

WEB DESIGN

As a framework for mapping out interaction design, web design is ideal. It involves the planning and design of websites according to the aims of your brief. This includes designing:

Information architecture and site structure:

This is about developing a hierarchy of pages within the website. It is useful to picture this as a set of folders, subfolders and documents within these subfolders. This will help you understand how users will navigate back and forth and across the 'menu' that is available.

User interface:

It is important to develop a 'user-friendly' experience, allowing the user to interact with the software in an intuitive and natural way.

Navigation:

Developing a logical system for the user to get around the site and be able to find what they are looking for with a certain amount of ease is a very important component.

Layout:

The design of this will depend upon whether it is an information or promotional site, but the drivers are all about how a user will want to access the information.

Colours:

Company brands and logos will have specific colours within, and these will need to be considered when exploring colour themes for the site. Designing accessibly is also an important feature of all good websites, to ensure that those with impaired sight, for example, are catered for.

Fonts:

It is best to use the licensed web-safe fonts that are likely to be present on all computers so that the website remains the same across all browsers.

Imagery:

All images will need to be modified so that they are the right size for screen viewing but are also of good quality where necessary. It is important to be aware of the speed at which websites load and to not make the imagery too memory-heavy.

Sarah Turner's self-designed commercial website

These elements, combined with the principles of interaction, design balance, proportion, rhythm, emphasis and unity, will ensure that you create a website that meets the goals of the client and satisfies you as the designer.

CAREER OPPORTUNITIES

An interactive designer may take up a career that crosses several different disciplines. After finishing your degree, you will have a diverse set of skills that could take you in the direction of a web designer or developer, depending on your level of graphic design and programming skills. The designer could design the visual 'front end' of the website and the web developer could work on the programming and server end of the site, making sure that all interactions and transactions worked.

These boundaries are blurred and, depending on the skills and interests you have and the course that you undertake, you will probably migrate towards one end or the other. Job descriptions might have the same title but might vary in content, dependent on the focus of that specific company, so you will need to do plenty of research to find out what the content of the job actually entails. For example, a small to medium-sized enterprise (SME) may expect you to design and develop a prototype for a project's outcome. This is in contrast with a larger company, where you may only be required to come up with the concept as there will be a technical team who will take the initial design, undertake the programming and develop the necessary addition of interactive elements to turn it into a prototype. In this instance you would be expected to liaise closely with the technical team to ensure the idea is realised accurately. Technical skills are important, even if you don't create the prototype, as an understanding of what is possible is essential to getting the most from a design.

Some traditional graphic design agencies now expect graduates to have a degree of web design and interaction design experience. Studying interactive design will mean that you can produce and test prototypes to prove that your concept works.

Job titles found in the digital industries could include designer, project manager, account manager, art director and business development manager. Graduates may also become information architects or user experience architects, where the user is key in the web design or mobile devices design process.

Most interactive projects have elements of video and sound design, and students may want to diversify and cross over into this industry. Other career opportunities are now also growing within digital **special effects** and **virtual reality** sectors.

Special effects: live action footage is combined with computer-generated imagery (CGI), which is added or manipulated to create environments and effects that would be too expensive, impossible to shoot or dangerous to capture on film.

Virtual reality: this is an environment created by software where the user is expected to suspend their belief and accept it as a real experience that involves responding to sight and sound. The environment can be a simulation of a real environment for education and training purposes or an imagined environment for an interactive story or game. The user can interact with the environment.

There is another element of crossover, as companies in the product design industry also need multidisciplinary teams of designers to develop prototypes. For example, interaction designers may also find themselves designing interfaces for domestic products such as microwave ovens: think of the screen options and the order of use of the object. In the future, many products will have touch screens like those on smartphones.

Service design is also becoming a part of the interactive/interaction designer's remit. It requires the organisation of people, infrastructure, components and communication in order to improve the interaction between service providers and service users. The activity of the service suggests patterns of behaviour by the user. Service design can be intangible and tangible – look again at the examples given earlier on in the text to do with the Oyster card, travel and the use of the passport. The service can also have emotional value, in that it should improve our experience. (See the sections on inclusive design and socially responsible design in Chapter 14, 'Future directions'.)

WHAT WILL YOU BE TAUGHT?

THE FIRST YEAR

In the first year you will learn about the principles of interaction and communication. Some Interactive Media courses have a graphic design base to support the 'look and feel', or branding, of a project. Fundamental skills in **typography, grids, information design** and drawing would form the first part of the course.

Typography: the craft of arranging type using typefaces, point sizes, line length and spacing to communicate language.

Grids: a system for arranging blocks of type and imagery to create an underlying consistency in page layout, pioneered by Josef Muller Brockmann.

Information design: the skill of preparing and presenting information for efficiency and effective understanding of the user; this may involve using visual representations of data, such as graphs, pie-charts, photographs and illustrations.

The majority of Interactive Media Design courses focus on the 'front end'. Within the industry, the term 'front end' refers to the interactive and visual spectrum of a product or service, with the 'back end' referring to the computer programming side of interactivity. You will also have an introduction to areas such as sound and user experience (the feeling that a human has when he/she interacts with technology), or human-centred design (where people are at the centre of the design process with the intention of improving designs for them), as these are key components of the subject.

From a technical perspective, you will learn to use and apply Adobe Photoshop and Illustrator software

Jade Crighton demonstrating her breadth of skills

for image creation. Web-design software will also be taught and may consist of Adobe Dreamweaver and Flash. Some courses may take a purist approach and teach mark-up languages such as HTML (Hypertext Mark-up Language) from the beginning. This forms the building blocks of all websites and allows for the webpage structure to include headings, paragraphs, links, images and other objects. There are now 'open source' (free-access software) resources, such as Word Press templates, that are proving to be popular alternatives to traditional tools.

In the first year, most courses have a series of lectures and seminars that look at contemporary design influences or art history. Industry specialists may also be brought in to deliver lectures about their own work. All of this is considered important as it broadens your knowledge and helps you make connections between things you have studied and things you are designing. It is also an important basis for the dissertation and design reports undertaken in the final year. Often your choice of subject for your dissertation will be close to one of your personal interests and be broadly to do with other interests you may have in art, design or technology-based fields. Read more about your final year later in this section.

THE SECOND YEAR

This will build on your skills and experience, particularly in web design, but will also expand into mobile devices and application design (Apps) – a very popular market. Interactive television and uses of social media may also be explored. Games design tends to belong to the second year as well, together with interactive and experimental experiences. You may also encounter more advanced sound design using **MIDI** controllers, **Ableton Live** software and **Pro Tools**.

MIDI: this is an acronym for Musical Instrument Digital Interface – an electronic industry specification that allows digital instruments, computers and other devices to connect with each other.

Ableton Live: a music sequencer, loop-based software tool for composing and arranging, as well as enabling live performances. It also allows for the mixing of tracks.

Pro Tools: this is a widely used professional tool for recording and editing in film, television and music production.

Initial sketches for Melanie Huang's 'Rockwithaheart' project

A key focus will be to undertake deeper study about the 'user', and this will be aided by the integration of live projects and competitions to give the projects a clear direction. You will be encouraged to move away from the computer and explore 'experiential' design and tackle more advanced work in the use of sensors, Arduino boards, **TouchDesigner** software and **processing** – all with a view to producing interactive installations for the arts, entertainment and information industries.

The Arduino board

> **TouchDesigner**: a software tool by Derivative, for creating visual real-time projections, live music visuals, interactive systems or rapid prototyping – really any type of rich user experience.
>
> **Processing**: this is an open source programming language that makes it easier to create interactions, images and animations. It was designed to teach computer programming in a visual context.

Take a look at **processing.org** for a range of examples of what is possible when using Processing. Tutorials are available, as well as the ability to download the open source software. This is free distribution and access to software, made available to anyone who wants to use it for design and development purposes. However, you should acknowledge the company in any outputs you produce that use it.

This second-year experiential design project by Melanie Huang shows the wired Arduino board and her experimentation with coverings for her 'Rockwithaheart'.

The project is called 'Rockwithaheart'. It is an installation that examines the concept of inanimate objects adopting human-like relationships. It works to display an object's ability to extend both our physical and social interactions. Through a variety of sensors communicating with Processing via an Arduino board, Rockwithaheart is able to respond and react to its surroundings, forming a relationship with the participant. Touching the rock provokes a reaction on the relationship meter (in the background of the image below), so the user would have a more visually appealing experience. The coloured, semi-circular bar shows 'neglect' for the rock on the orange side and 'adoration' on the green side.

As you gain skills you are responsible for continuing to develop them. Interpersonal aspects, such as working in teams, sharing workloads and managing deadlines, are

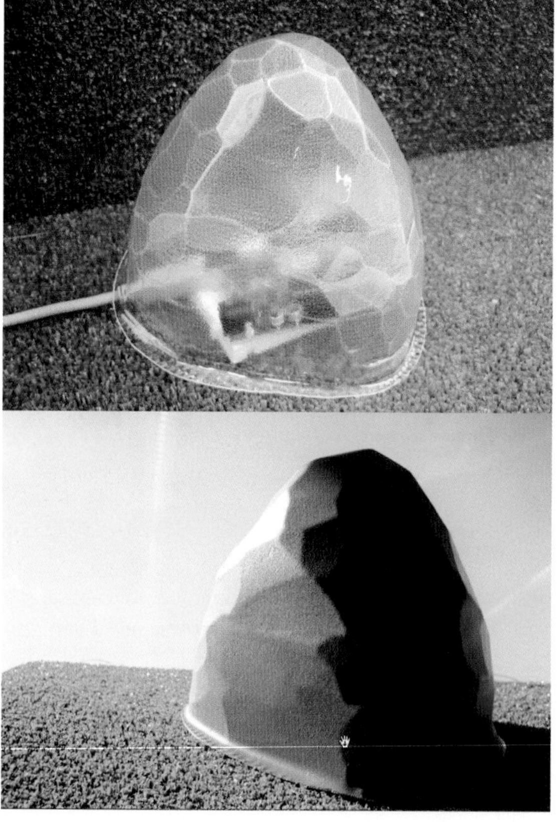

The prototype for 'Rockwithaheart'

Melanie Huang's 'Rockwithaheart' and the relationship meter

important skills that will need your full commitment. Team working may well have begun in your first year as 'ice-breaker' sessions for students to get to know each other, but in later years team working is a much more serious pursuit and required by any employer.

'Live' or collaborative learning projects will occur once you have a good level of skill, knowledge and understanding. These projects require a high level of professionalism as the companies you work with could be very high profile. Or they could be companies not associated with the technology industry but needing its skills – fashion houses, banks, regional councils, publishers – all requiring a web presence, a redevelopment of their current website or a new marketing strategy that uses new technologies rather than traditional marketing methods. The interactive design industry can provide valuable services to any company and you will find this experience of working on live projects invaluable.

You may take a year out from your studies to gain industry experience or have a placement year built in to your course. This might be organised by a member of staff, or you may take the lead yourself in finding a placement. Making this choice will depend on your interests and where you might want to be based. Finding a placement always involves an interview, so you need to prepare yourself by creating a well-designed CV and developing a portfolio that clearly shows the skills that you have attained to date. (Refer to Chapter 9, 'How to succeed as a design student', for information on CV, portfolio and interview preparation.)

WHAT IS THE DESIGN PROCESS?

THE BRIEF

A typical interactive design process will have a brief or assignment. You will firstly need to analyse the brief's requirements and do some initial inspirational research for ideas and concepts, which can be generated by doing a mind-map exercise (refer to Chapter 10 'Being creative and innovative'). Gaining an awareness of the context for the brief by reading around the subject can also provide useful research – use journals and magazines to support this initial investigation. When ideas form, they need to be developed into well-formed concepts. These are usually presented to your peers and tutors to gain feedback. Further research will need to be undertaken as the process develops in terms of design/style, the user and technical skills required to create a prototype.

DEFINING THE GOALS

You will need to begin by defining the goals of the project. Who is the project aimed at? Is there an audience or particular user? It is a useful approach to make up personas (the development of a hypothetical personality to test design functions) or scenarios (profile of how users might interact with a computer system) to help build a picture of the types of detail and functions that need to be included in the research.

DEFINING THE STRUCTURE OF THE PROJECT

The structure of the project would be considered next, where 'wireframes' (a basic drawing that explains the structure of a web design) are designed and the content is organised under headings. Then 'metaphors' may be explored to aid in understanding how the project will function. Ask yourself – will information be in folders or on pages? Will the user scroll through the options on a list or walk down a virtual corridor and go through a door?

BRANDING AND STYLE

The branding or style/aesthetic of the project will then be developed to achieve the desired 'look and feel'. Ideas are explored in layouts, grids and mock-ups. The project is then built and may be user-tested at a number of points during creation. These tests are reflected back and the project rebuilt or altered. Finally, the project is presented to your peers and lecturers for feedback.

A further part of the process might be to create from scratch a brand or style for your projects. This would be through the **style guide**, which is really a set of rules that control how a brand and its logos, colour schemes, typefaces, etc., are rolled out.

> **Style guide**: a graphically oriented publication that shows the development of a logo, for example, and how that logo should be used across divergent media – a series of rules of use. The publication is created for the client to use when establishing or updating a brand.

The style guide is something that in reality is handed to a client, so that everyone in the company can follow the designer's rules. For example, a company website might have a new identity, with a new logo and rules for scaling the logo, working in colour and black and white, colour schemes for 'rollovers', navigation, backgrounds and common web fonts or style sheet specifications.

This is demonstrated here by two solutions to a collaborative learning project by second-year Multimedia Design students. They have designed a new website for WGSN's EDU site, including a new 'community' element. WGSN (Worth Global Style Network) is an online trend forecasting organisation, highly respected in the design world. The company predicts what the trend forecasts would be for the forthcoming fashion seasons in relation to mood, colour, texture and style.

Look at the continuity of the identity and logo positioning in each of the pages. Colour and pattern is used to differentiate between pages in these examples.

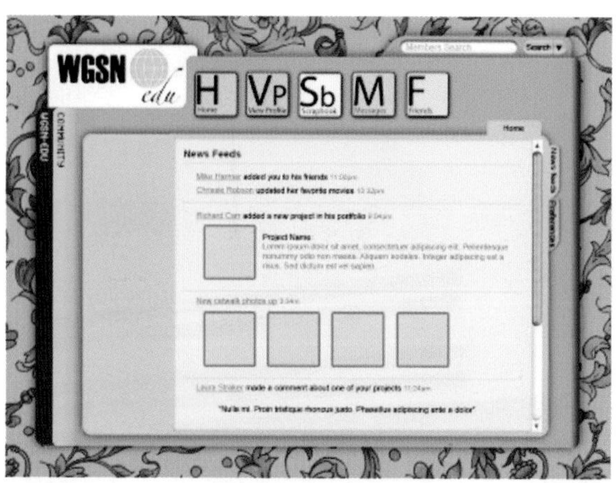

Student project – design 1, image 1

Student project – design 1, image 2

Student project – design 2, image 1

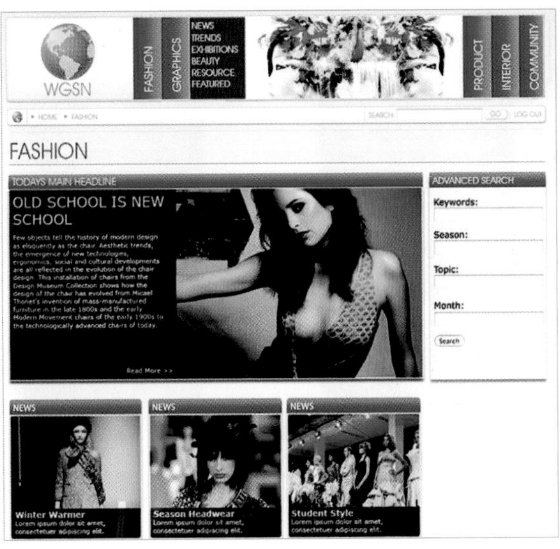

Student project – design 2, image 2

WHAT WILL THE FINAL YEAR BE LIKE?

In the final year you will find that you are more in control of your project choices and the way you manage your time. You should also have a better idea of in which direction you wish to take your career, as this will make the projects you choose more purposeful for you. You will need to take a more objective approach to designing and be capable also of evaluating and critically appraising your own work in relation to the needs of the industry.

You should be able to confidently apply intellectual and design process skills to your solutions. If this 'stepping up a level' happens quite early on in the final year, then this gives you more time to develop more focused, specific ideas and demonstrate a deeper understanding of your subject. A skill you will also need to develop is time management and the preparation of 'critical paths' and 'project time-lines', as these will improve your organisation skills. They describe a process where you have considered how much time each section of the project and the process will take, and have made a plan to incorporate it all so that you know that you can meet a deadline.

The final year is very much about you exploring your individual skills, talents and interests within your projects and you will be able to create more ambitious works, preceded by research and idea development that will allow initial concepts and further experimentation to occur. It is likely that you will then do some specific product and market research related to your chosen concept.

You will be encouraged to write your own brief for the projects, which should focus on your design strengths and skills and hopefully reflect your personal career choices. The prototype will usually be a completed, interactive project. A report of some sort usually accompanies it and this will give you the opportunity to explain and critically justify your design approach and the research methodology undertaken to support the design.

PRESENTING YOUR WORK

You will also need to think about the best ways to communicate the design outcome via a visual and verbal presentation. A good presentation will include the design proposal, research and investigation, the intended audience/end user, content and design requirements, prototype development and your evaluation of goals reached and its suitability for the intended market. (Refer to Chapter 9, 'How to succeed as a design student', for more tips on this.)

MAJOR PROJECTS

Major projects are often in an area where you may feel you need to learn more technical skills or learn about a whole new interactive subject. Your tutors and technicians will support you in realising your ideas, so don't be scared to experiment. The final year is about being professional and ready to enter the industry, so collaborative projects and competition briefs will provide you with great experiences.

Illustrated boards for an interactive submission to the D&AD Awards, by Katerina Brunclikova, interactive media design graduate

You should also try and enter national and international competitions, such as the Royal Society of Arts (RSA) bursary and the Design & Art Directors (D&AD) competition. If you are shortlisted for any of the competitions then you might be invited for interview to talk about your work in more detail so that the judges can understand more about the thinking behind the project and to view other work in your

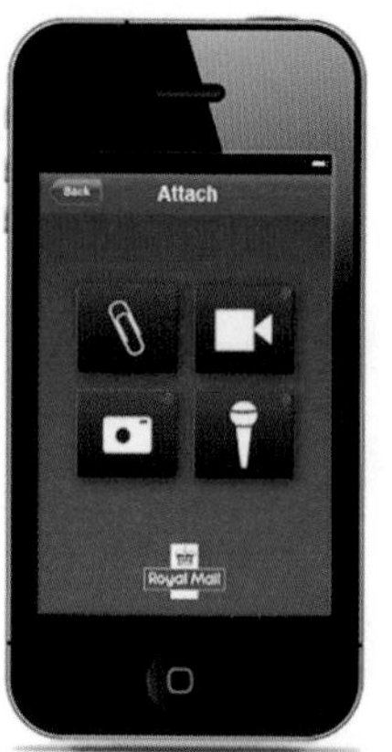

RSA entry QR (quick response) code stamp and smartphone application for Royal Mail by Alex Steven

portfolio. These competitions have briefs set by high-profile industry professionals and provide opportunities for students to produce a well-rounded portfolio by dealing with contemporary design issues. They are a fantastic opportunity to get your design skills in front of the industry, as in this example:

> 'This brief was set by the Royal Society of Arts. The brief was to update the postage stamp idea for Royal Mail. Often the RSA brief is to simply design a new set of traditional postage stamps. I wanted to develop a more emotional approach to sending letters and this became my focus. I took a more contemporary approach and utilised my interactive design skills and technological understanding.
>
> I wanted the stamp to be a "smart" stamp, a stamp that could hold media such as video messages and pictures within it using QR codes or similar technology, and also provide instant information to the postal service in order to speed up the processing of post. My idea was to design an application for a smartphone that would allow the receiver to scan the QR code and unlock any messages contained within. The application would also allow the sender to purchase stamps, upload messages and imagery, check the weight of the letter and attach an image to the QR code, so that it is aesthetically pleasing on the letter front and is still readable by a code reader.'

Most design courses have a 'degree show' or 'showcase' at the end of the final year. This is often a celebration of the students' work and is open to the industry, the public and friends and family. Interactive design offers great experiences for visitors to play with new ideas and technology. Screen-based work, such as website design, has to be compelling to use. Experiential, 'out of the computer box' work is more immediate as it may work with sensors and movement and be visible.

Many courses also exhibit undergraduate work in London so that students get the opportunity to find out what their peers are doing (and how they compare with each other) and also to show their work to potential employers. The best showcases for Interactive Media Design students are at the internationally renowned exhibitions New Designers and D&AD, both in London. Graphic design, illustration, animation, motion graphics, interactive design and advertising are all showcased there. Both exhibitions offer awards that are held in high regard within the industry, and both shows offer you the opportunity to network with people from the industry. You might be invited to consider a placement or even a job offer at one of these events.

Not all institutions attend these shows and not all students get to attend. Some institutions pay for the stand and therefore select work. Some students get together and raise funds to have their own independent presence. The benefits are obvious though, and many final-year students aim for their work to be selected for these shows and being noticed once they get there, as in the following final-year project.

The work shown here, by Patrick Niall McGoldrick, Interactive Media Design graduate, is a small part of a very large personal project done in final year – a major research project for a product design company. It looks at how the television exists with other

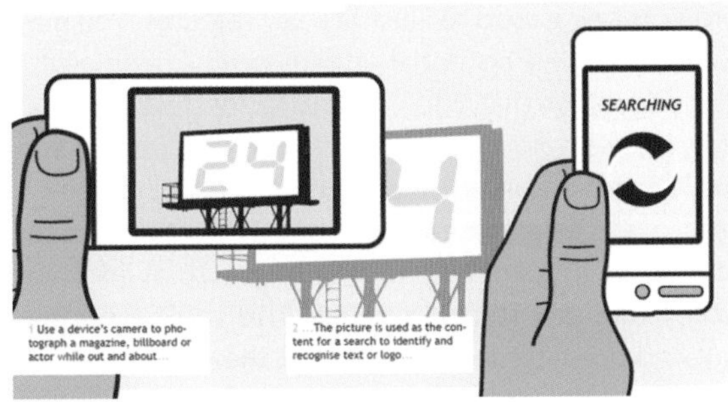

Smartphone search facility by Patrick Niall McGoldrick

> *'I began researching competitors and new technological developments. I then looked into user personas and trends in television, mobile computing, desktop PC and smartphone. Predicting the users' needs and interests really helped me tailor the product and the service.'*
>
> Patrick Niall McGoldrick, interactive media design graduate

devices in the domestic environment: how to share and switch between media; how to deliver different usage modes and work with a flexible and scalable user interface.

The image above indicates how a mobile phone's camera could photograph a billboard or some such object. The picture is used to search for information. The image below shows an HDTV with 'chat' commencing and making use of the high-quality screen. The call can be assigned back to the smartphone if the user needs to move away from the TV.

Preparing your portfolio

Your portfolio will probably contain some static printouts, or screenshots, of your interactive work (have a look at the screenshots of Dave Barlow's experiential design project later in the text), perhaps incorporating DVDs within the visual presentation. Graphic design agencies may well prefer this method of presentation (see also Steven Everington's D&AD boards). In applying for the role of junior interactive

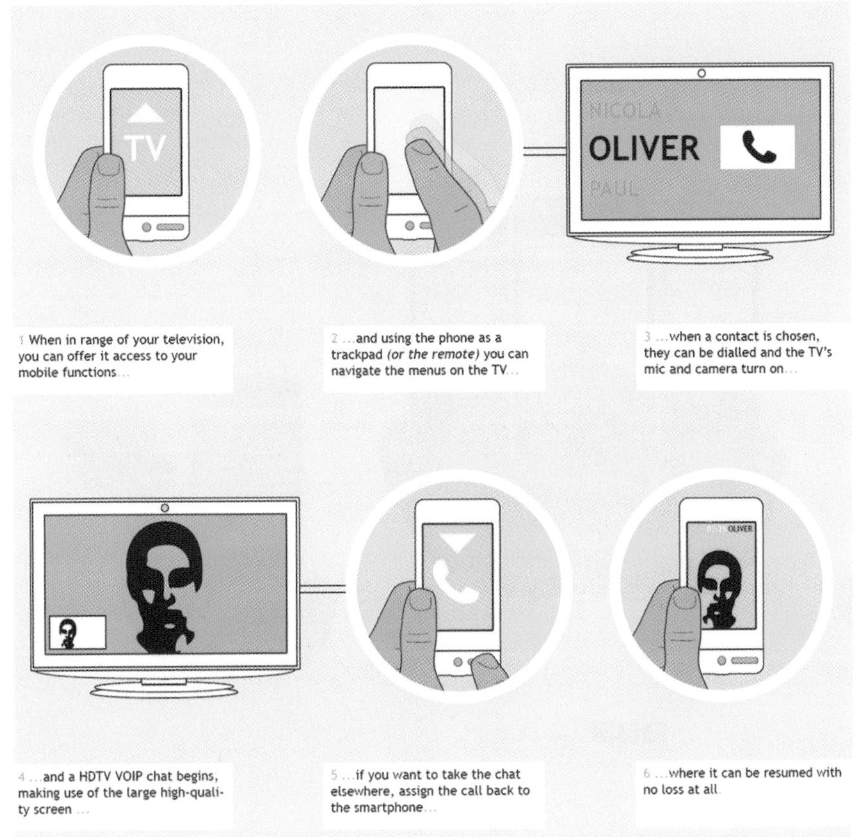

Phone/TV chat system by Patrick Niall McGoldrick

designer, you would need to show a range of interactive work, with perhaps a graphic design project included in the portfolio. Just as important, you would also be expected to have a personal website that contains your projects and demonstrates your design and communication skills.

It is worth finding out what presentation modes a particular company might prefer before you arrange to show them your work.

Here is a presentation board by Steven Everington, a graduate of Multimedia Design, for submission to the D&AD Awards. The BBC set the project brief, which was to design an interactive 'widget' – part of an interface or application that allows a user to access a service or function. These boards now form part of his visual portfolio and show strong graphic design skills.

It is also worth considering a show-reel that captures and demonstrates alternative interactive projects being used to illustrate key features. You may want to show a potential employer an example of a games project that you could offer on DVD, so that they can sample it and interact with it in person and explore its usability. Employers often like to see a graduate's design process and your project documentation can contain the broad architecture of information that supports the project. This can be printed out and presented visually so that the viewer can get a quick and easy feel for your design process and levels of ability. It is rather like a very smart sketchbook but with a lot of thinking, decision making and reflection built into it. (There is some further portfolio advice in Chapter 9, 'How to succeed as a design student'.)

This next project by Dave Barlow, a second-year Interactive Media Design student, is one where experience is key. The focus of the project was on exploring physical movement, in real space, to trigger and manipulate live music and visuals with gestures. More explicitly, the idea was to project visual elements following X, Y and Z coordinates, like a graph, along a tracking system that gives the effect of following the performer around the space. The software used was TouchDesigner. Dave has recorded this interactive

Steven Everington's D&AD competition entry board

Still frames from Dave Barlow's experiential design project

demonstration, or experience, to capture the gestural interactions. These images are screenshots of the video footage and are used in his physical portfolio to explain the stages of interaction.

Conclusion

Interactive media design is a very exciting and ever-responding industry, requiring a fresh, novel approach to design. You will need a wide variety of skills, particularly drawing, and be able to demonstrate a real interest in understanding customer needs. Good research and analytical skills will also help you progress quickly.

The actual depth of the subject is really up to you. There is so much rapid change and development in technology that it can feel overwhelming to identify where you fit in; however, once immersed in it you will learn fast.

Further resources

Books

Cancellaro, J., *Exploring Sound Design for Interactive Media*, Delmar (2005)

This book covers sound design for interactive art and installations from linear to non-linear sound. It ranges from acoustics to creating and mixing, to music for all types of digital media.

Frick, T., *Managing Interactive Media Projects*, Delmar (2007)

This book moves from the idea to design, development and production, to the maintenance of the project, through research, planning, making, user testing, reviewing and specifying the technical needs. It looks at any digital media, including websites, DVDs and CDs.

Maeda, J., *The Laws of Simplicity: Design, Technology, Business, Life*, MIT Press (2012)

An important book that offers ten laws for balancing simplicity and complexity, without sacrificing comfort and meaning in products.

Moggridge, B., *Designing Interactions*, MIT Press (2006)

A well-known book that looks at interfaces and good interaction design through interviews with 40 of Moggridge's heroes, who make the computer accessible to all. It includes a DVD and a website – www.designinginteractions.com.

Özcan, O. and Yantac, A.E., *Creative Thinking for Interactive Media Design*, Lulu.com (2009)

This book suggests that interactive media design balances art, design and technology, and discusses design education as needing to plan for the beginner through to the more advanced designer level, especially as it encompasses a broad range of media, such as television and mobile devices.

Pruitt, J. and Adlin, T., *The Persona Lifecycle: Keeping People in Mind Throughout Product Design*, Morgan Kaufmann (2006)

This book illustrates the best ways to create and use personas in design work. There are a range of industries showing how they use personas in the design process and so enhance the user experience.

Ware, C., *Information Visualization: Perception for Design*, 3rd Edition, Morgan Kaufmann (2012)

A practical guide for interaction, graphic and web designers, as well as data analysts. This book looks at visual perception and related principles through art and science, resulting in greater clarity and persuasiveness.

Wiedemann, J. (Ed.), *Web Design: Interactive and Games*, Taschen (2008)

This book is about interactivity and the Internet. It is not just concerned with clicking on the hyperlinks but includes the full experience of Flash, video and sound, and has case studies of the best interactive sites available at the time of publishing. It also illustrates online gaming for a full interactive experience.

Websites

www.aaronkoblin.com/work/rh/index.html

An artist and designer who specialises in digital technologies and data.

www.cadburyworld.co.uk/CADBURYWORLD/Pages/Welcome.aspx

The information website where you can buy tickets for this interactive experience.

camilleutterback.com/category/projects

Interactive artist, Camille Utterback, shares some of her art installations.

www.fastcompany.com/design/2011/50-most-influential-designers-in-america

The Fast Company list of the top 50 designers in America; these include a number of interactive designers – it's worth checking out their work.

www.nexusinteractivearts.com/work

Nexus describes itself thus: 'Nexus Interactive Arts works in interactive moving image. It shares the same high values for storytelling and design through collaboration with the Nexus film and animation division, but brings technology as an exciting partner.'

processing.org

This is a very useful open source programming environment for anyone who wants to create images, animations and interactions. It can be used like a software sketchbook to teach the fundamentals of computer programming visually. Students, artists, designers, researchers and hobbyists can use Processing for learning, prototyping and production.

www.sciencemuseum.org.uk/visitmuseum/jamesmay.aspx

Visit the Science Museum in London and use the virtual reality application to explore.

www.ted.com

Talks that offer free knowledge and inspiration about **T**echnology, **E**ntertainment and **D**esign.

theneverendingwhy.placeboworld.co.uk/

Interact with this animation for the band Placebo.

www.webmonkey.com/2010/02/information_architecture_tutorial

John Shiple provides a comprehensive view of the information architecture design process.

www.thewildernessdowntown.com

An interactive film by Chris Milk for the band Arcade Fire.

www.wired.co.uk

Wired magazine discusses how developments in new technology affect culture.

Project story

Designer:
Andrew Charlton, interactive media design student

Project title:
Game – War

Tell us about the project

I wanted to show that computer games are a burgeoning art form and that they are capable of dealing with much more complex subjects than the majority do at present. It was my plan to create a game that elicits a much wider range of emotions from the player.

My initial research was to look at current art-type games, such as The Path (a short horror game inspired by tales of Little Red Riding Hood), Passage by Jason Rohrer (which encompasses an entire life), Flower (a Playstation version of a poem that intends to create a relaxing and soothing experience) and The Graveyard (an interactive painting game).

I decided to create a game that showed the real horror of war. It was set in the trenches of World War I and ultimately the main character would always be killed in no man's land.

What processes and skills were most relevant to this project?

I had to research and work out who my key audience was – I realised it was for a small audience familiar with first-person shooter games (FPS), for players who were over 18 and who would have some life-experiences and non-game players who would appreciate an emotive experience.

The style and mood of the game was inspired by WWI poets such as Wilfred Owen and Siegfried Sassoon and harrowing photographs of the war. It was clear that the environment was going to be critical so I researched trench structures, no man's land, the conditions of war, weaponry and equipment to give the game authenticity.

I had to design the graphical user interface (GUI) or heads-up displays (HUDs) – the part of the display that gives the player information about game play. Common aspects of war game GUIs are indications of ammunition, health and armour, compasses or maps and gun crosshairs. I wanted to include these only to persuade players, early on, that this might be a normal game.

What challenges did you face?

I needed to learn Cinema 4D for modelling objects in the game and Unity 3D to build the game. I already had extensive knowledge of the Adobe suite of software to create textures and other imagery, but the 3D world was very new to me.

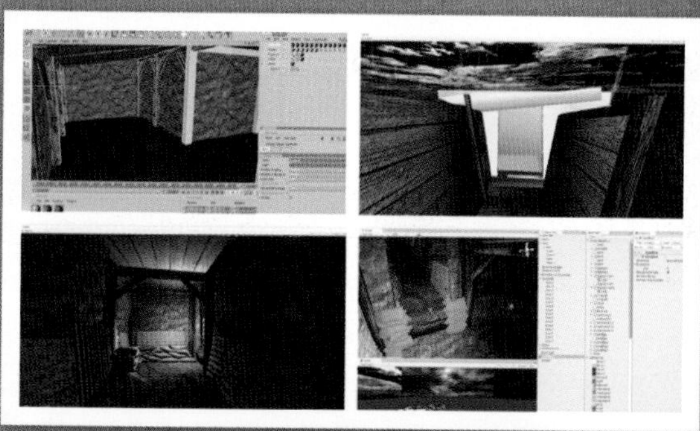

Building the 3D environment for the game

The game's realistic environment (skybox)

What were the highlights of the project?

I needed to create an initial prototype, to test the game. I did this by modelling the trenches, dugouts, terrain, items and equipment and creating a skybox that looked very dramatic to fit the mood. Feedback was mixed as the subject matter didn't appeal to everyone, but this was what I expected.

This was no ordinary game though. I created a 'death' animation that brought to an end the game play and allowed for reflection on what had just happened. Sound was very important to the experience, so I edited sound tracks and sound effects to create a sombre mood. I set out to elicit different emotional responses from my game, to have a dramatic but thought-provoking and reflective end to the game. I believe I achieved that.

Final death scene

Project story

Designer:
Lee Carroll, interactive media designer

Project title:
QTrip bus information system

A bus stand with the multimedia screen displaying updates to bus arrivals and news headlines

Describe your project

I would like to tell you about my final project at university. It was a 'service design' project that utilised a number of technology platforms. It started from an idea to promote bus usage. I wanted to convince the public to use public transport instead of the car and to simplify the experience of using a bus or train. I wanted to develop an interactive timetable and website and reward users for using public transport because of global warming and its connection to car usage. The main aims of the project were to:

- convince users to use public transport
- communicate environmental issues in a fun way
- develop an online social network to create a competition for best environmentalist
- reduce confusion and frustration for current public transport users
- incorporate new revenue streams to allow public transport companies to reduce fares
- make the public transport experience more enjoyable.

What processes were most relevant to the project?

The process I undertook was to define who my audience was: commuters; the elderly; business people; school children and parents; students; shoppers; people without cars; people with cars who wish to use them less; environmentally conscious people; single parents and young mothers. My next task was to develop scenarios of users to work out which functions were required from the service. I also wrote a 'diary of a bus user' for a full week to gather research on changes of use of bus travel at particular times and days. I looked at competitors in terms of existing solutions, including existing transport services as well as companies using Apps and websites, plus environmentally friendly companies of any type.

So, I intended to replace static timetables with an interactive timetable accessed via multimedia screens at bus stops. These would notify passengers of the current bus position and any delays due to traffic build up. I realised I could also maximise advertising space by selling banner space on the screen to national newspapers. The idea was to show main headlines and then say where the story would be continued (newspapers could sell more and this advertising revenue could help to reduce the cost of bus travel). The website could house information about carbon emissions and make users aware of tips to reduce their individual carbon footprint.

The phone App, including QR code to scan as a travel ticket

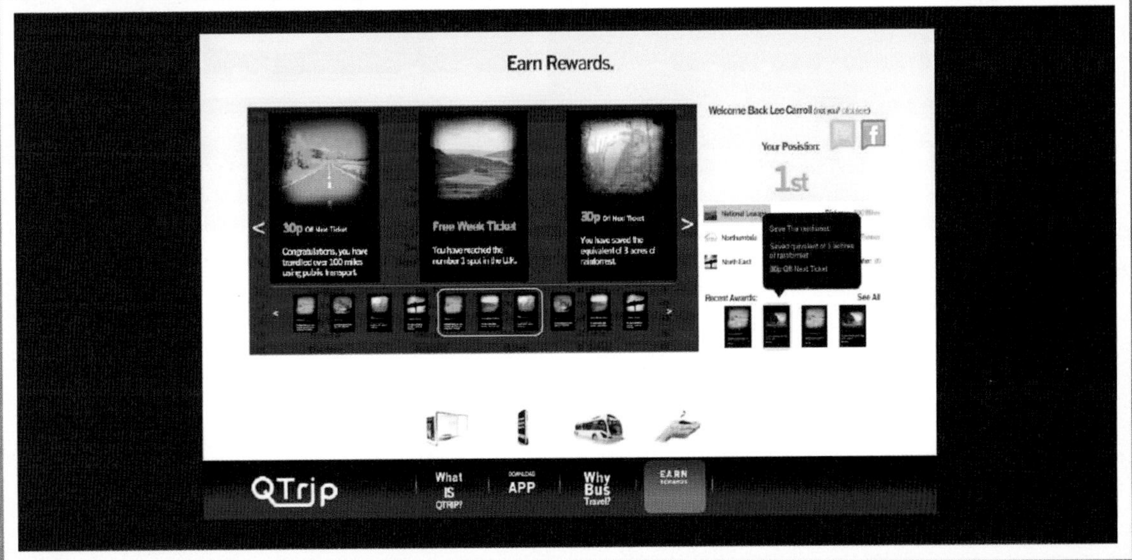

The rewards scheme on the website

The primary function of the service would be to notify passengers of their desired bus arrival instead of standing and waiting at a bus stop for long periods of time. This would be done by downloading my App from the website.

The user would start the App and select the bus and bus stop. For example, if they lived 10 minutes away from a bus stop and were at home, they could set their App alarm for 11 minutes and arrive at the stop with minimal waiting time. A bonus to bus travel would be the 'measured' amount of time spent on the bus and then the user would be notified about what they have saved by using the bus, in terms of rain forest protection and less carbon emissions. This part of the service could allow the user to upload their results to the website, where a national league would show who is doing the most to help prevent carbon emissions, and rewards could be earned.

What challenges did you face?

Of course, in rolling a big project like this out there would be a long list of questions to answer, demonstrating how broad the issues can be within graphic design:

- Should these screens be solar powered, scratch resistant and graffiti proof?

Company identity for this service – QTrip

- Could I use GPS for bus tracking?
- Could the user pay for tickets with their mobile phone?
- How would a route be chosen – on the App on a smartphone?
- Could I use the Google Maps infrastructure for the positioning of buses?
- How would users use their rewards for being environmentally friendly?

What were the highlights?

Designing the functional requirements of the project, such as the multimedia screen interface, information streaming and adjustable light levels. I designed the service and tested the initial prototypes on each platform, then made changes from feedback given from a number of sources – lecturers, friends, peers and bus passengers.

Branding/livery on a bus
Photograph by Lee Carroll

Spatial design

By Richard Sneesby, Chartered Landscape Architect
Formerly Course Leader, University College Falmouth

'Understanding my building took ages. As a listed building built on a hill, with one part of the building connecting to the other through what would have been windows, it felt very confusing. However there is something very satisfying in knowing your building back to front and upside down. It can give you reasons for doing things, as well as strong evidence for choosing not to do others. Knowledge of your site and building will give your project gravitas and give you confidence in the structure of your design.'

Harriet Fleur Curtis, interior architecture & design student

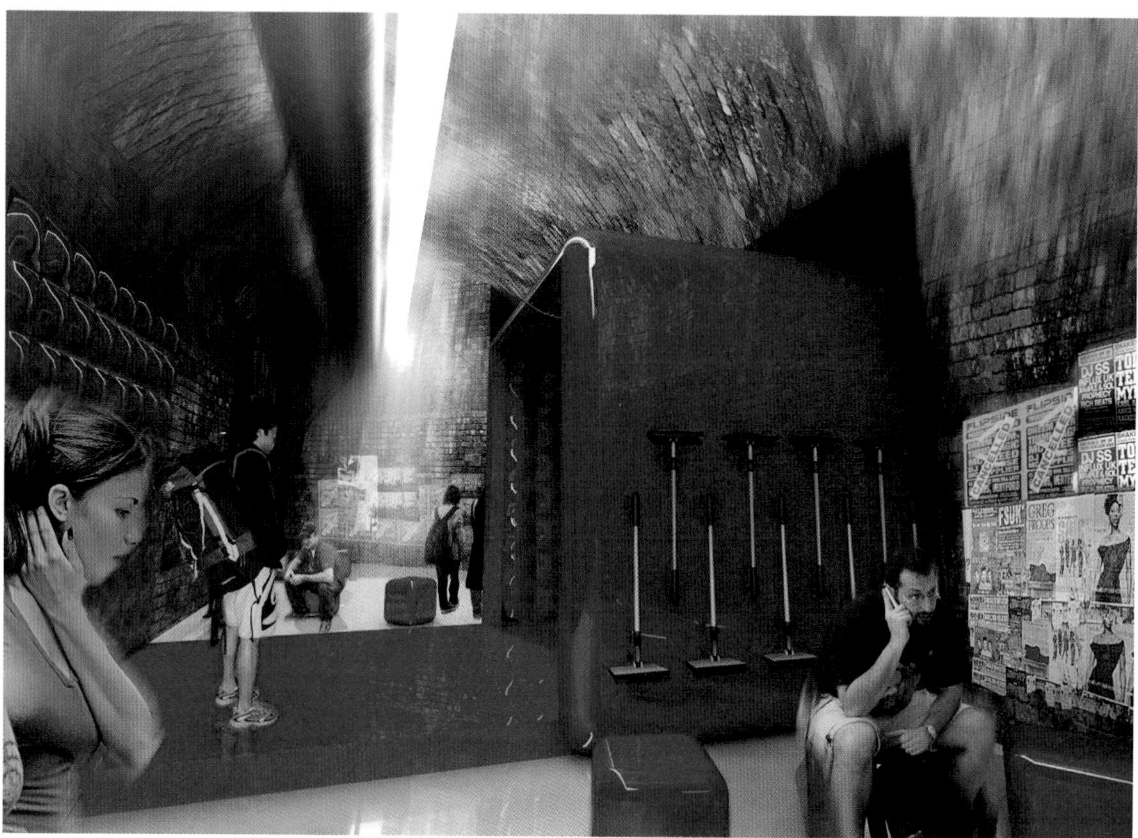

A backpacker hostel with moulded units for sleeping, washing, cleaning and socialising, designed under railway arches, Tim Miller, spatial design student

How do our homes, villages, towns and cities work as human habitats and what is the role of design and the creative industries in shaping our environment? These questions form the foundation for a massive global industry that encompasses design, planning, construction, surveying, engineering and conservation.

This particular discipline includes many different sub-disciplines, such as interior design, landscape design, spatial design (a specialist branch of architecture and interior design dealing with installations and design insertions, as well permanent projects), garden design and urban design. The following section outlines the opportunities that exist for the spatial design graduate in general and will help you to understand which possible future careers may suit your talents, skills and personality.

Today's spatial design industry

This industry is concerned with two principal issues. First is the creation of *places* – urban and rural settlements and their associated buildings and landscapes; and second the way in which people *use* these places.

The role of design within the **built environment** historically has focused upon architecture and planning. This chapter does not consider the design of buildings themselves (which is the role of architecture), it instead focuses on those design industries that work in combination with architecture, or what happens when more than one building is placed, or found, together. Buildings therefore generate a need for interiors and associated landscapes to be designed – collections of buildings require urban and rural planning and infrastructure. Buildings and landscapes take on specific uses – homes, schools, industry, retail, health care, hospitality, tourism, recreation, play, and so on. In order to be successful they must meet the needs of users, but designers also have a responsibility to consider visual

Built environment: this refers to spaces and buildings, rooms and gardens, shops and hotels; in general, all architectural forms and their surroundings.

appeal while producing spaces that must function effectively.

Designers work in teams, often collaborating with other designers, artists, manufacturers, suppliers and other specialist consultants. Design ideas and proposals are submitted for approval by clients and planners and tested against financial feasibility. Once approvals have been granted, the design process involves careful development of design details, selection of materials, testing the design against user needs and ensuring that the finished proposals match the overall design concept and requirements of environmental sustainability. Projects are normally handed over to professional construction teams, with designers supervising the quality of the work and ensuring that the projects are completed within budget.

If you want to make a positive contribution to people's daily lives, then this industry might appeal to you. You will also be leaving behind a legacy that can be enjoyed by generations to follow. Design sits within a time continuum; what you design now will not only be reviewed in later years but may also be around for many years to come. Many of us live out our daily lives within places that were created two, three, four, five generations previously and we enjoy the generosity of our ancestors. Just think of a church, theatre, public building or country house that has stood the test of time, or perhaps even more powerfully our enjoyment of Victorian parks with their huge, mature trees. The people who created the parks would never have known anything but grassland and small young trees. Having a good understanding of this is crucially important to know whether you want to work in this field.

Rho Fairground, Milan
Photo by Jane Bartholomew

The industry demands high levels of creativity; it is constantly seeking new ideas, new aesthetics, new technologies and new materials, such as in this image of the Rho Fairground in Milan. Iconic buildings can define a city; landscapes can provide a setting and a venue for recreation; interiors can change our emotions or help sell products; gardens produce food and can be an escape from the stresses of everyday life. The industry's creativity does not sit in isolation – indeed it can't work alone. There is, therefore, a necessity to develop good relationships with other design industries, such as graphic design and product design, and the most successful outcomes are typically characterised by the creative buzz of working within multi-disciplinary design teams.

There are also opportunities for you to design for specific users:

- children
- the elderly
- teenagers
- those with special needs (mental health, learning difficulties, mobility and sensory deprivation)
- communities
- shoppers
- tourists

Some of the most interesting work comes through the design of specialist places:

- schools
- hospitals
- residential homes
- playgrounds
- educational environments
- museums
- theatres
- cinemas
- shopping centres

If you are interested in using your artistic creativity, research, problem solving, communication, team working and professional skills to change the world around you, then this subject will appeal to you.

If you are driven by 'sustainable' issues, then you might want to consider the ethics concerned around

Colour section through garden, by Hugo Bugg

building new human environments, which some might say destroys natural ones through the use of the world's depleting resources, notably certain raw materials. You are the next generation of designers so you will need to respond to this as a first priority. Clearly you might also find yourself in a position to bring about a change in policy in the future.

In the future the world will start to look different. The industry needs people who care about the environment, but who are clever enough and have the foresight to respond to a growing need for new building projects while protecting future generations by developing sustainable approaches to design. There are few careers that allow such a long-term view of the world around us, so this could be a very rewarding pathway.

What courses are there?

Vocationally oriented courses combine both education and training, with a clear focus on ensuring that their graduates gain the relevant skills to be employable within associated industries. A generic design course may provide a wider approach to the subject, perhaps allowing you to take a more creative, experimental, risk-taking approach. This can appeal to students wishing to pursue a more creative journey, but it might be less obvious how this maps on to a direct career path. All courses will prepare you to be a highly employable graduate but, for some, the next move after graduation may be less obvious until you have identified your own strengths and matched these with personal interests to identify your own career path.

The courses available fall into these three categories:

Accredited courses: the content of the courses are confirmed and approved by internationally recognised, chartered professional institutes, mainly found within the Landscape Architecture courses. By being professionally accredited, students may be able to become a member and there may be other benefits, depending on the organisation.

Courses that associate themselves with recognised professional groups and meet quite stringent requirements of practice and ethics set out by the governing bodies (but which are not professionally accredited): Interior Design, Urban Design and Garden Design can fall into this category.

Courses that deal with the scope of specialist areas of built environment design, but for which no profession is widely recognised: Spatial Design, Interior Decoration, Interior Architecture and Exhibition Design.

Courses that have a vocational title will normally align or associate themselves with a particular sector of the industry and, sometimes, a 'professional' body.

Landscape architecture:

This is a professional career and is governed in the UK by the Landscape Institute. The Landscape Institute has a Royal Charter that protects the title 'landscape architecture' and 'landscape architects' and requires all those using these names to be professional members of the institute. Membership is achieved through successful completion of accredited courses and by satisfying the professional institute that you can perform to a high professional standard and work within its codes of conduct and to rigorous rules surrounding professionalism and professional ethics.

Only courses that are professionally accredited by the Landscape Institute can use the title 'Landscape Architecture'. Courses named 'Landscape Design' will often cover a similar curriculum but without professional accreditation. This is not to say that they are any less exacting in their teaching or assessment, only that they have not, or have not yet, received professional recognition.

Garden design:

Garden design has a semi-professional society (the Society of Garden Designers), which does not accredit courses but provides a framework for the profession of garden design. Interior design, similarly, has a number of semi-professional bodies, such as the British Institute of Interior Designers. However, it is not necessary to have any qualification at all to call yourself a garden designer or interior designer. Graduates can find themselves competing for work against others who therefore do not have a recognised qualification. This is not to say, obviously, that the unqualified designers are necessarily bad designers.

Urban design:

Urban design is an interesting title. Most urban designers come into the profession from an allied discipline, such as landscape architecture or architecture.* This is a popular course as a second degree or an area of postgraduate study, although specialist courses exist and lead to good employment opportunities.

While there is no professional institute, there is an Urban Design Group (www.udg.org.uk), a campaigning organisation, and an Urban Design Network (part of the Royal Town Planning Institute).

There is a potential disadvantage of a highly vocational or professionally aligned course. Many applicants to design courses know that they want to work in this area at degree level but are unsure about what they want to do for a longer-term career. The broader disciplines, such as spatial design, can be ideal choices if you are in this position as you will be working on a range of interesting projects but keeping your options open. Most graduates find themselves working in associated practices (landscape architecture, architecture, multi-disciplinary design) or pursue their emerging specialism through postgraduate study. This is also true of graduates from the less vocational courses in garden and interior design.

Universities and colleges specialising in art and design will tend to take a more creative industry approach to the subject, often combining design and art from the outset. On the other hand, courses within departments that focus on built environment or engineering courses may specialise more on building technology, materials science and traditional engineering routes.

The design 'crit' is an important part of design education and develops confidence in explaining your ideas to different audiences

General course content

The following section summarises the content that a typical generic course might contain, so it is worth reading this thoroughly even if you want to focus on landscape design, for example (more details on specific courses can be found further on in the chapter). The core taught elements may be similar, but all courses will place their own emphasis on the curriculum and this might depend on staff specialisms or even regional business needs.

As a general overview to all courses within this subject area, it is not only about *what* you will be taught, but *how* you will be taught. All courses will combine three principal areas: practice (studio-based design projects and exercises); theory (history, context, precedents, semiotics and critical appraisal); and research (site survey and analysis, user survey and analysis, academic writing and research methodologies). You will be expected to produce resolved projects, reports, dissertations, essays, seminars, discussions, presentations and exhibitions. You will be encouraged to enter national and international design competitions that will provide you with a good focus within a project, and any successes will be well publicised and therefore noted by the industry.

As a useful start point, take a look at the following list of themes and the issues arising. Consider using these as a guide when undertaking any project. This list could easily turn itself into a checklist for you to use when you are working through a project or drafting your own brief.

Spatial design – themes and issues

The design of contemporary space usually concentrates around specific issues. These are some areas that commonly inform a design approach:

1. **Activity and interaction:** How the introduction of shared activities and interactive events can animate and revive space.
2. **The old and the new:** How the 'insertion' of a contemporary design into a traditional or historical context can establish a dialogue between the old and the new.
3. **Light and colour:** How space can be transformed through the use of colour alongside natural and artificial light.
4. **Sensory/experiential:** How sound, light, texture, colour and even smell can be used to make space a more interactive and sensorial experience.
5. **Narrative, place making:** How spaces can tell or hold a story; the theatrical unveiling of a particular place and its unique qualities – a 'sense of place'.
6. **Identity and landmark:** How space can be given a unique identity through a process of community branding and ownership.
7. **Orientation/route making:** How clearer indications can be made on how space should be used and travelled through more easily and comfortably, using **zoning** and **way marking**.
8. **Design for need:** How design for specific needs can benefit us all.

The first year

The first year will cover key areas that will prepare you for more complex design projects later in the course. Typically there will be a number of themed design exercises, often characterised as shorter, skills-based, practical sessions, introducing concepts of pattern, scale, colour, texture and three-dimensional design principles. Other exercises might be based around specific types of users, famous designers, personalities, localities, materials, sustainable practice and how ideas are tested against successful examples from the past.

Early on you will be introduced to site surveying, which covers the measurement of sites (indoor and outdoor) using equipment to measure site levels (optical or digital level, tripod and staff) and techniques to plot the data onto plans. Digital surveys may be introduced. Site visits will cover the collection of information on existing structures and features, materials, views, circulation routes, climate, vegetation, overhead structures, the condition of the existing fabric of buildings and valued judgements on whether site elements should remain or can be removed. A **user survey** and predictions will often be included to help position what the project is for and to inform an understanding of the design brief.

Technical drawing will be introduced to allow you to develop drawing conventions that are understood by the construction industry. These will include an understanding of the importance of scale plans, sections, elevations and orthographic projections to allow a more three-dimensional representation of space (axonometric/isometric/perspective). Technical drawings are used to communicate discrete elements of the design to builders and manufacturers – for example, details of walls, screens, barriers, steps, furniture, lighting and water features. They need to be very accurate rather than sketchy and are usually produced using specialist computer software.

Design exercises develop the skills to tackle increasingly complex projects, investigating themes

Zoning: this portrays the way that designers plan the spatial arrangement of the proposed project into discrete areas of similar use, scale and appearance. For example, an active space (playground) may be planned separately from a quiet space (a library room). The two areas may have a close relationship (i.e. linked by a door or corridor) but may be quite different in terms of space materials, colour, scale, etc.

Way marking: is used to help people navigate their way around and learn how to use the spaces. Main paths will be wider or colour coded – perhaps brightly lit – while secondary paths may be narrower with more subtle lighting. Signs and symbols may be used, as well as key design elements such as gateways, entrances, exits and landmarks.

User survey: can be anything from an interview-based discussion with a selection of potential or actual users of an environment, up to a full investigation of the purpose and requirements of a new multimillion-pound investment such as a hospital. For more information on these research methods see Chapter 14, 'Future directions'.

Glimpse into new entrance for a circus school; note how the activity of the school is revealed to the public, Katie Low, spatial design student

Rapid group drawing exercise develops freedom of visual expression, negotiation skills and builds confidence in committing ideas to paper

such as user need, brand identity and materials selections.

Fundamental to your development as a designer will be an understanding of a 'sense of place' – what makes one area different to another. This might be to do with:

- materials: based on underlying geology, such as golden-coloured limestone in the Cotswolds or red brick in London
- building/craft traditions: shapes, forms, patterns, textures, combinations of materials
- cultural influences and personalities: Robin Hood in Nottinghamshire, Elgar in the Malvern Hills
- sensual aspects: climate, quality of light, smells, etc.

Theoretical strands will cover the history of art and design (in the main, 1900 onwards), significant cultural, social, economic and political movements and their influence on design. You will be introduced to the idea of design precedents (what has happened in the past; how others have responded to the same problem; what past examples look like; why, how and what they are made from, and so on).

Your projects will probably encourage you to focus on:

- research skills (library, web, reading, note taking)
- writing skills (language, conventions, referencing)
- presentation skills (verbal and non-verbal)
- group working (critical to your future success as a designer).

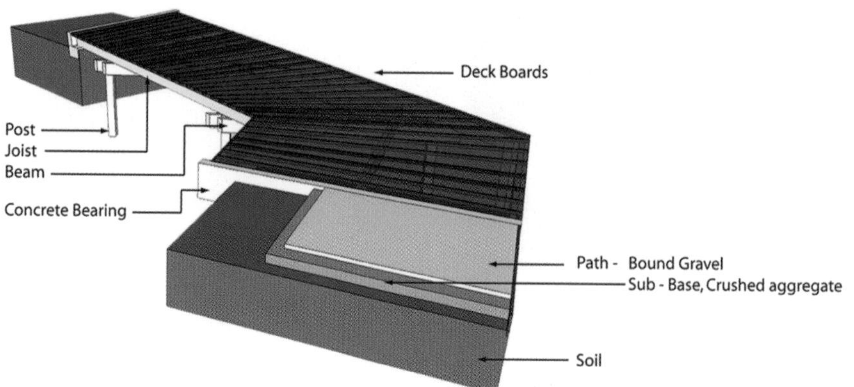

Technical drawing illustrating construction of an outdoor decked path; the finished drawing would include more information about material sizes and overall dimensions, by Hugo Bugg

Generic computer skills will be developed, to include word processing, graphic layout, presentations, photo manipulation, as well as training in specialist computer design software to focus on two-dimensional plans, sections and elevations.

3D ideas development through modelling exercises

3D concepts communicated through sketch modelling

The first year often culminates in a more substantial design project where you make connections between different parts of the first year and combine them into a single project that responds to a design brief. The project briefs will normally be provided for you in the first year.

The second year

This sees the courses move towards longer and more complex design projects. At this stage it is quite common for projects to be themed around specific areas that require new knowledge or a more contextual approach. For example, projects could be user-based:

- design for children and play
- design for dementia
- community-based design
- educational environments
- specific hospitality environments (hotels, restaurants).

Other projects commonly covered include:

- site master planning (arrangements of a number of buildings, spaces, topography, site circulation and servicing)
- branding (retail environments)
- redevelopment of historic buildings and landscapes
- working within sensitive environments (wildlife habitats, listed buildings, conservation areas).

The choice of projects will be based upon the particular interests of the course. At this stage students may be expected to start to develop an individual design brief.

Think about becoming aware of the necessity to be able to have answers to a client's specific requirements and identify the existing and potential new users for each space. For example:

- **Age-specific groups** – the elderly/young children/family groups, etc.
- **Interest-specific groups** – clubs/societies/sporting groups, etc.
- **Occupational groups** – workers/students/nurses/policemen, etc.
- **Groups with special needs** – specific disabilities/socially disadvantaged, etc.
- **Educational groups** – primary/secondary/higher education/mature students, etc.

These groups must also be aware of their needs:

- **physical needs** – size, shape, strength, endurance, ability
- **physiological needs** – food, air, water, exercise, protection, comfort
- **psychological needs** – social needs, stabilising needs, individual needs, self-expression, enrichment
- **aesthetic satisfaction** – physical arrangements that promote or reduce social interaction.

The second year will also see an increased focus on professional practice. This is where you develop skills and knowledge of the industry and its conventions. Some courses will start to introduce contract drawing conventions, specification and choice of materials, costing and professional ethics (although some may postpone this until the third year).

Temporary theatre space based around folded paper shapes by Spatial Design student Charlotte Clement

The final year

Many courses include a dissertation in the third year, along with the research methods and writing skills required to complete this task well. It is at this stage that most students start to think about post-graduation decisions and may choose to specialise in particular areas of interest if the course permits; the sooner you begin to think about this the less daunting it will seem.

Skills sessions will include more advanced computer-aided design techniques, including three-dimensional design, visualisations, photo-montage and rendering packages.

The final-year experience varies enormously. Project work falls into two areas. Firstly, those that give students independent choice of projects and secondly, those where all students work on the same projects. Independent choice works well for students who have developed a focus for their study.

In both cases it is most likely that you will complete a dissertation on a subject of your own choosing, together with a comprehensive major project. You will also have some aspect of professional practice built in to help you consider how you might bridge the gap between studying and employment. Careers advice will be important, as will marketing yourself through the development of an effective CV, digital portfolio and web presence (see Chapter 15 'So, where is this going to take you?' and Chapter 16 'Working for yourself').

If your course belongs within an art and design department it is likely that the final year will culminate in a degree show where you will exhibit your work for public viewing. Focusing on effective ways of doing this might also be a significant part of a third-year experience.

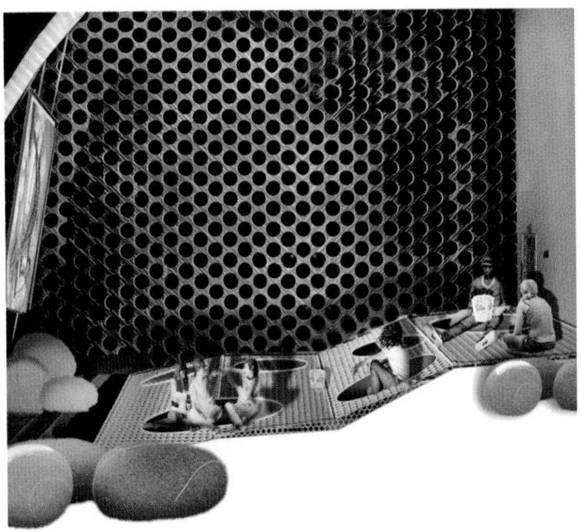

'Cave' – social space by Anna Huczek

Circus school by Katie Low, BA (Hons) Spatial Design

Two of the main course titles, Interior Design and Landscape Design, are being outlined next. Clearly there are similarities between all of the courses in the spatial design discipline. It is recommended that you read all of them to get the widest view and gain a greater understanding of the differences. The first two contain all the information any spatial design student will need. The other sections, on Spatial, Garden and Urban Design, offer further detail on how these differ from Interior/Landscape Design. Some courses do not cover business skills and self-employment skills and it will be worth considering a top-up course if this is the case.

COURSE TITLE: **INTERIOR DESIGN**

COURSE DESCRIPTION

Interior Design is a major creative discipline offered at a wide range of academic and vocational levels throughout the UK. Many people think of interior design as a largely domestic opportunity as illustrated through magazines and media via decorative and stylistic approaches to altering the interior within homes. Studying at degree level deals with the subject at a much deeper level with academic rigour and leads graduates into a significantly wider range of career opportunities. It has strong links with associated design disciplines, including architecture, fashion, graphic design, product design, textiles and even fine art.

This wide range of approaches has generated a similarly wide range of courses, and colleges and universities provide general and specialised programmes that meet the demands of students and employers. A good starting point is to consider what you hope to gain from your qualification and how to match your goals against the specific courses on offer. Interior courses fall into four key areas, with many developing skills and knowledge by combining and emphasising particular approaches:

- Domestic interiors
- Commercial interiors
- Architectural approach
- Decorative approach

Decorative approaches have their origins in domestic styling, with a focus on encouraging a well-developed understanding of colour, pattern, textures, furnishings, textiles and the way these can be used to create a specific style or atmosphere for a new interior room or space. Some courses focus on interior styling through the selection of a range of commercially available products to develop a stylistic theme, which can range from recreations of traditional interiors to cutting edge contemporary spaces. Clients (those who commission the work) are usually home owners, and budgets can sometimes be quite limited. Designers often work alone as design consultants.

Architectural approaches to interior design are offered by some universities, either within generically named awards or as separate named courses, such as Interior Architecture. These courses have an increased emphasis on the technical design of buildings and component parts of buildings and often have their origins in architecture or spatial design. This approach towards architectural interiors requires an understanding of building technology, materials, user requirements, histories and theories of architectural design, the language and semiotics of design, professional approaches to contract law, planning systems and planning control and building regulations. It also requires students to gain skills in architectural communication through competence in computer-aided design, technical orthographic drawing, specification writing, budget estimating and costing, engaging manufacturers and contractors and health and safety matters.

Domestic interior design tends to be combined with a more decorative approach. Clients tend to be home owners or housing developers and the emphasis is on creating a particular theme or style on behalf of the client.

Commercial interiors are for non-domestic clients. Most commonly this will involve a more architectural approach in the designing of high-quality spaces for shops, restaurants, hotels, transport hubs and lounges, leisure environments, health spas, offices, theatres, schools and similar spaces. In this case, interior designers will be working within a team alongside architects, landscape designers, graphic designers, branding consultants and sometimes artists and product designers. It is critical that interior designers are able to communicate with the rest of the team and so a detailed understanding of design communication and the construction industry will be important. Budgets are larger, with more design options available but with, perhaps, less individual control over the finished space.

Undergraduate group work

CAREER OPPORTUNITIES

There are huge opportunities in this area, but increasingly interior design is redefining itself as a much broader subject encompassing aspects of interior architecture, commercial branding, corporate design, retail environments, hospitality and tourism, leisure and well-being. Depending on your levels of interest, you might pursue a career in interior design that has either a commercial interior focus or a decorative one, applied within either a corporate or domestic setting. If you have an interest in technology and urban landscapes then you may wish to consider working within an architectural remit.

Employment opportunities are also quite broad, both in the UK and further afield. Most graduates will seek employment, either as an employee within a design practice or consultancy or as a sole trader/self-employed designer. The qualification can also be used as a route into building conservation, construction design, media work and specialist areas such as design for yachts, the film industry and exhibition spaces.

Interior designers will be sought by larger architectural firms. Those seeking employment within this area should develop a portfolio that shows skills in design creativity as well as technical resolution, computer skills and presentation skills. Multi-disciplinary design practices also regularly employ interior designers. Employers will be looking for willingness to work in a team, skills in group work and communication and what you can bring to their existing range of skills and knowledge.

Those looking towards self-employment as interior designers will need a portfolio that shows their ability to work independently and give confidence to clients that they can manage budgets and contracts as well as creative abilities.

WHAT WILL YOU BE TAUGHT?

In order to develop ideas, you will be taught how to understand and develop your own position as a designer. This will typically involve understanding what has been done in the past and why some solutions are more successful than others. Some of this will involve research into visual and non-visual precedents,

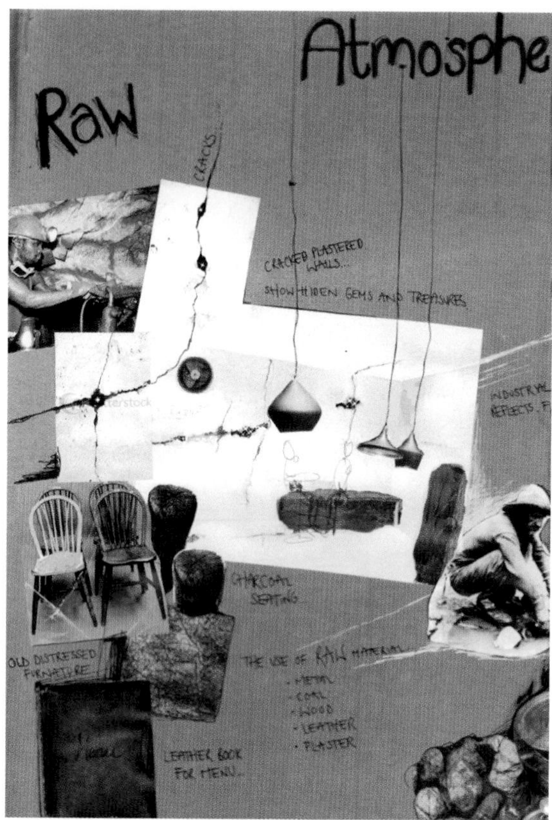

Illustrated board to show a client some emerging ideas for a new restaurant; the images and words communicate moods and emotions rather than layout and materials

High-quality rendered computer model by Anna Huczek

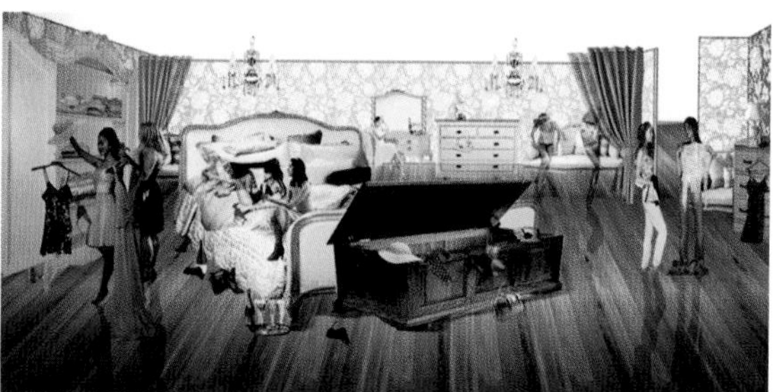

Photo collage by Giorgina Holland

looking at the work of particular designers and artists and often experimenting with designing in the style of another designer.

Perhaps the most important aspect of developing ideas will be practise in designing; in other words, 'doing' design. You will be encouraged to experiment with a wide range of approaches, techniques, styles, themes, materials, colours and textural palettes, and product applications. Some exercises will be done through sketching and drawing, others through sketch models, collages and mood boards. As ideas develop they will be interrogated and tested through discussions, design critiques, technical resolution, scaled drawings, computer models and exhibitions.

You will be introduced to a range of skills that will develop throughout the course. Visual communication skills will cover a range of techniques:

1. **Drawing skills** – sketching, drawing, and technical drawing (plans, sections, elevations, perspectives, axonometric), scale drawings, annotating drawings, collages and mood boards.
2. **Computer-aided design and graphics** – CAD software, 3D visualisation software, image manipulation software and production of exhibition boards.
3. **Three-dimensional modelling and prototyping** – scaled models, sketch models, component/detailed models, use of different modelling materials, interface with computer design.
4. **Writing and research skills** – research methods, essay writing, report writing and publishing software.
5. **Verbal communication** – presentation skills, group work skills and communicating with a variety of audiences. (See Chapter 9, 'How to succeed as a design student'.)

DESIGN FOR USERS AND AUDIENCES

Approaches to this within colleges and universities will vary widely, with some institutions developing particular reputations for expertise in design for specific user groups. Generally, visual, stylistic and aesthetic ideas will develop in combination with a user group, their needs and aspirations. Most users

Use of a photo-manipulated collage to illustrate the idea behind the project – in this case a story-telling space for children, by Charlotte Clement

A second-year interior branding project for a restaurant, specialising in chocolate, by Baiba Ciekure

have generic needs for aesthetic satisfaction, comfort, safety, security, practical needs, socialising and play, and so on, and you will be taught how to design successful spaces that meet these needs.

Some courses will encourage a more in-depth approach to user-centred design, and may include detailed lectures and projects. See Chapter 14, 'Future directions', for more information.

COMMERCIAL INTERIOR APPROACHES

Here the focus will be on the way that successful interior design can develop or enhance a brand or commercial enterprise. Courses may introduce aspects of graphic design, fashion and textiles within the context of how businesses operate and what interiors or brands say about the company or products and services being offered. Students often work with 'real world' clients to prepare and develop a design brief and make proposals.

Projects often include work with restaurants, shops, performance and arts centres, galleries, health environments, schools, hotels, community centres, play environments, offices and other opportunities through collaboration with local enterprise and business.

ENVIRONMENTAL RESPONSIBILITY

Designers within the built environment specify materials and products, many of which can come from natural environments or which are potentially damaging to the environment. Some courses teach a sustainable approach to design through an understanding of the environmental, social and health impacts arising from the use of certain materials, and how alternative choices are able to be avoided or mitigated against negative impacts. See Chapter 14; 'Future directions', for more information.

DIGITAL DESIGN

As technology advances, many more opportunities present themselves to designers. Courses may be able to offer teaching about digital design approaches, specialist software, technological products for use in interiors (projections, screens and interactive boards) and even digital strategies towards solving design problems.

A site survey

PRACTICALITIES AND TECHNICAL DESIGN SOLUTIONS

As design skills develop, you will increasingly be required to develop your designs so that they 'can be built'. This will involve accurate drawings and representations of your proposals, but also a detailed knowledge of site surveys, materials, joints, fixings, connections and finishes. Technical drawings will be produced to a more professional standard, with increased emphasis on industry conventions. Drawings will be accompanied by written statements that describe materials, quality of construction as well as cost estimates of design proposals and information required for building contracts. You will cover aspects of planning and development control, as well as health and safety and contract law.

WHAT IS THE DESIGN PROCESS LIKE?

The design of interior spaces begins with a problem that needs to be solved. This might be to do with a change in the type of user, or because the space needs renewing, or perhaps the building has undergone a significant change of use. It is often not simply a case of designing a single interior space. Many buildings require designers to explore a number of associated spaces and the connections between them, perhaps developing ideas and concepts for whole buildings, developments and complexes.

Approaching a specific design problem is both a linear and cyclical process. The formulation of a design brief outlines a list of requirements that can be checked at the end of the project and 'signed off'. However, the idea will come from somewhere else. Also, and critically, the user group's needs and aspirations need to be taken into account, and so the whole design process becomes one of developing ideas – interrogating them – revising, editing and developing them – and continuing this process until all of the requirements are met and the proposal is ready to be built.

This process often throws up further questions and new avenues of enquiry that can extend the creative vision, but can also be distracting. The idea is to set clear goals and embark upon an exciting and creative journey to meet all of the demands and requirements.

The following identifies the process of undertaking a design project:

Site survey – The normal sequence is firstly to carry out an initial site survey. This will require taking detailed measurements and plotting them onto a drawing so that it forms a foundation for any changes. The site survey will also allow designers to make judgements about the quality of what is already there, what should remain, what should go, what might influence the design proposals and to look for inspiration that might help to create an idea.

User group – Then a detailed understanding of the user, or user group, will take place at this early stage. Often this can involve interviewing existing or potential users, researching into similar projects and what made them successful, and perhaps researching into specific needs and how to design effectively for them.

Design concept – Sometimes called 'the big idea', this frames the whole project and contains both practical and aesthetic statements about

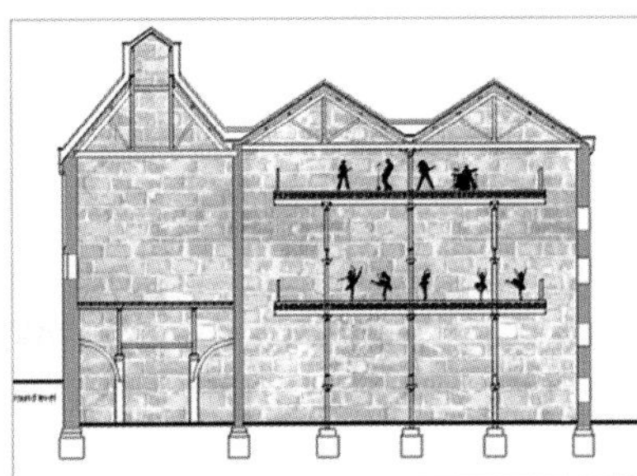

Investigations began with how I could open up the space and address the issue of the low ceiling heights.

The exploration then went onto the idea of dropping elements into the space that reflect the materiality of the existing hoist housing on the exterior facade, but also provide the requirements for the new use of the building.

Following this, these visual experiements were all investigating the idea of dropping in a series of 'pods', with each creating a separate room that could be used for rehearsal. These elements would feel very much like a new layer, however the size, shape and location would each be dictated by existing features of the building, such as the beams, doorways and floors.

These images were looking into the belief that one key insertion could be more dramatic than a series of spaces. The idea of using an organic, curved form came about as a way of reflecting the very organic barley that was brought into the original Malthouse. However the more linear shape, influenced by the exterior housing was still the stronger idea.

These model experiements were further investigations into the ideas above, regarding the use of an organic form and the final idea of the parasitic insertion, inspired by the exterior hoist housing. They allow for a clear understanding of how the object could work in the space, and its scale and relationship to both the existing building and the proposed users.

Design process communication board by Amanda Fisher

what the final space might look like, what it might contain and how it might work. This is where the designer's knowledge, experience, creativity and imagination combine with the site and users to develop a design concept. The design concept is illustrated with examples of similar projects, sketches, some key statements and possibly a mood board. There may be a concept model at this stage, which helps to show spaces and the relationship between them.

Design strategy – The design concept is often combined with a design strategy or approach that explains the particular way in which the problems will be identified and solved in the brief, and how they will work in creating the design concept and vision for the project.

Once the survey, user needs, design concept and approach (strategy) have been declared and agreed, the designer embarks upon the creative process of design development. Here ideas are mulled over, interrogated, tested, rejected, developed and generally bounced around between those involved in the project, your peers and tutors, until a more meaningful solution comes into focus that deserves to be taken forward to a next stage. The interrogation and testing of emerging ideas is critical. Appropriate and relevant ideas will be developed into detailed design solutions that can be drawn and modelled in more

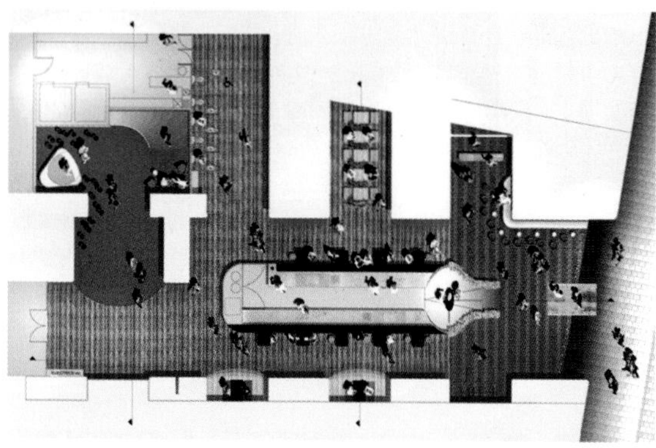

Plan view by Tamasine McNabb

Cross-section view by Tamasine McNabb

detail. You will then develop final proposals and produce them in a form suitable for costing, manufacture and construction. (Refer to Chapter 16, 'Working for yourself', to understand the fundamentals of costing products and services.)

This description assumes quite a logical and linear process. Of course the creative mind does not necessarily work like this and some of the best and most influential interiors seemingly come from an extraordinary leap of imagination. It is often a combination of a logical practical sequence and a creative exploration that combines into a single solution which can then be tested and interrogated for its feasibility and value.

WHAT WILL THE FINAL YEAR BE LIKE?

The final year of degree-level courses will most often contain a significant major design project and a substantial piece of research and writing, normally in the form of a dissertation. Both provide you with an opportunity to develop personal interests and expertise, build upon your strengths, tackle any weaknesses and produce your best work to show to potential employers and clients.

The best students will choose projects that reflect their strengths, allow them to stretch themselves, learn new skills, develop knowledge and make a contribution to the world of interior design. The more complex and interesting the problem to be solved, the more important the solution will be. If this can be combined with very high-quality communication (graphic, three-dimensional, written or verbal), then the work will be worthy of the best portfolio and will place you in a strong position when it comes to a job interview or further study.

Further study options include postgraduate diplomas and master's-level degrees, which will help to extend specific interests and knowledge. (Take a look at Chapter 15, 'So, where is this going to take you?', to find out more about further study.) Many specialist shorter courses exist, which are an ideal way to add particular strengths to your work as well as extending the variety of skills and services you can offer.

The end of the course usually culminates in an exhibition of student work, which is often open to the general public and offers you an exciting opportunity to interface with the industry and members of the public, putting your design ideas in the public domain for the first time.

COURSE TITLE: **LANDSCAPE ARCHITECTURE/ LANDSCAPE DESIGN**

16

The timber balancing beams form part of the playful elements that provide children with the play movements, which will improve a child's balance and coordination as well as make the journey through the garden a more enjoyable one.

17

The stepping stones across the stream provide an exciting and varied journey through the garden. It also gives children the permission to use the water as part of their play and get closer to it, which the existing site lacks.

Trenance Children's Garden Maren Hallenga

Proposal for a large park for children, by Maren Hallenga

COURSE DESCRIPTION

Courses are based within a range of specialist institutions and these will provide a clue to how they approach the subject. Those that are based within schools of architecture and the built environment will tend towards a more architectural approach, with a more technical and 'building' emphasis. Those within faculties of science or land-based institutions may have an increased emphasis on natural science and horticulture. Other courses exist within an art-orientated environment. The nature of professional accreditation means that courses meet industry standards and cover a similar curriculum – it is the emphasis that needs to be identified.

Landscape architecture has its origins in garden design, one of the oldest professions and, some have argued, one of the greatest contributions Britain has made to Western Art. The profession developed after WWII, when a massive demand for new homes, rebuilding of towns and cities and important developments in transport and infrastructure required specialists to plan and design new urban environments. Since then, it has developed to cover most aspects of design outside buildings and includes:

- urban design
- planning
- site-based design
- industrial landscapes
- land reclamation
- urban streetscapes
- housing developments
- schools
- public parks
- playgrounds
- sensory gardens
- private gardens
- institutional grounds
- recreational landscapes
- forests, lakes and country parks
- sculpture parks

and almost any other outdoor spaces you can think of.

Landscape architecture has sometimes been described as an ideal partnership between art and science, and this gives some clues about the type of

ARTISTIC INFLUENCES

These contrasting shapes and forms being introduced into the architecture at Part Farm are similar to those found in the art work from the cubism art movement e.g. Pablo Picasso and Georges Braque and the modern abstract work by Wassily Kandinsky.

In cubist art works, objects are broken up, analyzed and re-assembled (instead of depicting objects from one view point). This represents the subject in a greater context. This has been repeated within the design process, but instead of objects, key factors such as view points, circulation, contours and micro-climates etc. have been manipulated helping create a design representing a greater context, whilst meeting all the clients' needs.

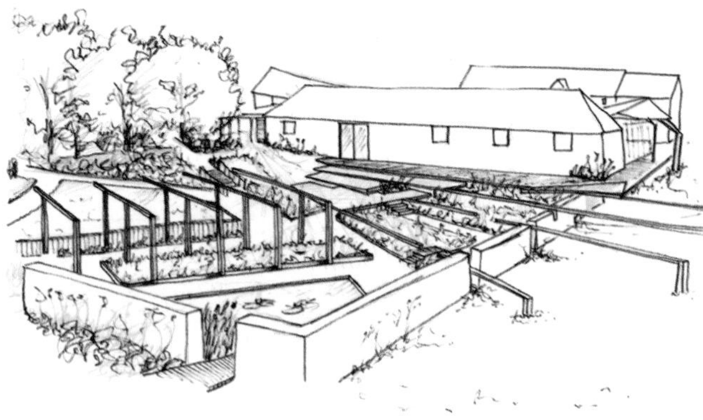

Shapes and forms similar to those in these pictures (right) by Picasso and Kandinsky are clearly visible within the design process.

Initial ideas for an organic productive vegetable garden; the inspiration for the layout comes from re-working abstract fine lines, by Hugo Bugg

people who are well suited to the profession and who will find it fulfilling as a long-term career. Designing outdoor space is fundamentally a creative process, but within the important context of a living landscape. While it shares many common areas with architecture, it is the importance of the dynamics of a living environment that separates it from all other areas of design within the built environment (with the notable exception of garden design). Designing with, and for, living things is where the science combines with art and technology, and landscape architecture will lead you into areas of ecology, horticulture and physical geography in its widest sense.

Studying landscape design will appeal to anyone who likes to work in an artistic way but with great environmental awareness and sensitivity to the detail.

These outdoor spaces are designed for people – normally communities or wider user groups. Some will have special requirements and needs, while others will have more generic requirements. A significant part of landscape architecture education will be based around user-centred design and there will be many overlaps with sociology and human psychology.

Landscape architects regularly work in design teams with architects, civil and structural engineers, planners, artists and ecologists. They cover both public and private spaces, urban and rural, inland and coastal, commercial and domestic, natural and man-made.

In addition to a creative and artistic approach to design, all courses will cover the technical design of outdoor space, which is split into 'hard landscape' (meaning built structures) and 'soft landscape' (meaning vegetation). Designing the hard landscape requires an understanding of building technology, materials, construction methods and techniques and engineering requirements in order to design landform and earthworks, drainage, paving, retaining structures, walls, barriers, structures and shelters, lakes, ponds, fountains, lighting and furnishings. You will be taught how they are constructed, and industry-standard ways of communicating built elements to contractors.

Soft landscape means using plants, and a significant part of the curriculum will focus on the role of plants in the natural and designed landscape, how they work as ecological communities and how designers can exploit their aesthetic attributes to create attractive spaces. You will learn about using vegetation to define spaces, to screen unsightly views or frame attractive ones, using colours and textures to develop particular effects, how plants live and survive and about soils, water and drainage. Some courses will cover the role of vegetation as habitats for wildlife.

As landscape architecture is a professional subject, some of the course will introduce aspects of professional practice, although most of this tends to be taught within the postgraduate year. The language and semiotics of design, professional approaches to contract law, planning systems and planning control and building regulations will all be introduced. There will also be introductions to acquiring skills in architectural communication through competence in computer-aided design, technical orthographic drawing, specification writing, budget estimating and costing, engaging manufacturers and contractors and health and safety matters.

CAREER OPPORTUNITIES

Landscape architecture is governed by the Landscape Institute, a chartered professional organisation that has exacting requirements for membership. The Landscape Institute is affiliated with the International Federation of Landscape Architects, meaning that your qualification is internationally recognised and that employment opportunities are possible worldwide.

Training to become a landscape architect is a long process. A three-year undergraduate degree is normally followed by a year in practice (a 'year out') and a further postgraduate year leading to a PGDip or MA qualification. A further two, or sometimes three, years in practice (and an examinations process) is necessary to meet the entry criteria required for professional entry into the Landscape Institute. In practice, this normally means between seven and eight years from entering an undergraduate course.

This might seem like a long time, but career prospects are extremely good and for three out of the seven years you will be fully employed. Professional membership of the Landscape Institute is also recognised worldwide and international employment opportunities are very good.

WHAT WILL YOU BE TAUGHT?

Landscape Architecture courses follow a very similar general pattern to Interior Design, especially in the first year, so make sure that you read that section first. This section therefore concentrates on those areas that are specific to Landscape Architecture and Landscape Design (as well as Garden Design).

The Interior Design section looked, in detail, at the way design is introduced as an activity; the skills that designers need and use and a variety of users and audiences. These areas are very similar, and often identical, in the work of landscape architects.

A significant and specialist role of landscape architects is to integrate new developments into the natural and built landscape. This is closely linked to the UK planning system and you will learn about planning policy and how this requires certain rules to be applied to buildings and landscape infrastructure. Some of this involves master planning large sites to inform – for example, the best locations for new developments, their size and scale and how to assess the visual and environmental impact of proposed developments. This covers industrial sites, housing developments, large-scale retail and leisure developments, as well as changes to the rural landscape. You will learn about **Environmental Impact Assessment (EIA), Zones of Visual Influence (ZVI)** and how to model proposed changes.

Environmental Impact Assessment: a systematic approach to predicting the potential repercussions of the changes proposed within a design.

Zones of Visual Influence: an appraisal of how people will view the interior and exterior of a project.

Many of the most important new landscapes, worldwide, are the result of landscape architecture and come from both the imagination of designers and through collaboration with artists. Students often work with 'real world' clients to prepare and develop a design brief and make proposals.

Again, have a look at the sections on environmental responsibility, practicalities and design solutions from the Interior Design section, which cover the same ground as Landscape Architecture.

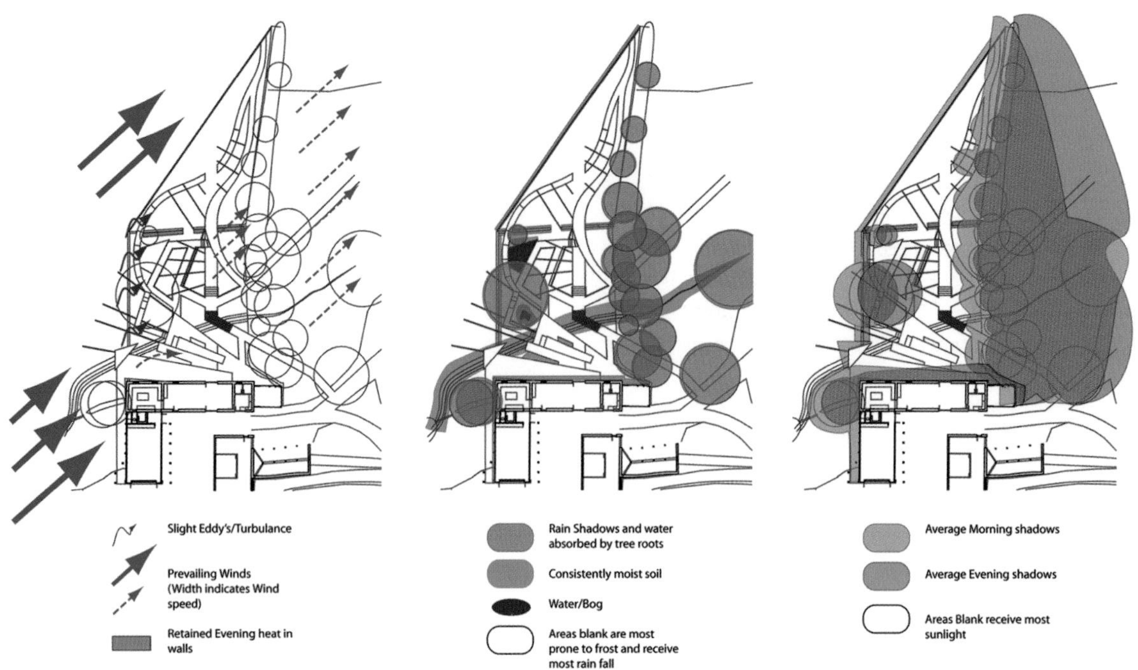

The use of simple diagrams superimposed onto the master plan shows how the design responds to varying factors – here prevailing wind, rain and shadows, by Hugo Bugg

WHAT WILL THE FINAL YEAR BE LIKE?

As with Interior Design, the final year is both the substantial way in which your award is assessed and your opportunity to produce your best work to show to potential employers and clients, and therefore acts as a bridge between your undergraduate course and any postgraduate diploma or MA. Landscape Architecture graduates typically carry on to a post-graduate year that covers, in more detail, aspects of

Emma Gullick's quayside hut illustration; the use of sketches to illustrate the proposed design being used. This type of image is critical to helping clients and potential users understand the project and how it would work in reality.

professional practice and work with more complex sites. This is often carried out at the same university, but opportunities to move to another institution to complete your education or even study abroad are also options.

Employment opportunities are very good, both at home and abroad. There are many landscape architecture practices in the UK who offer rolling 'year out' positions for graduates for a 12-month or 9-month period. Following the postgraduate year, students seek employment within a practice to take them through to professional membership of the Landscape Institute.

The end of the course, as with Interior Design and many art and design courses, usually culminates with an exhibition of student work open to the general public and offers an exciting opportunity to interface with the industry and members of the public, putting your design ideas in the public domain perhaps for the first time.

COURSE TITLE: **SPATIAL DESIGN**

USING THE SIDE ENTRANCE TO CREATE ACCESS TO THE SITE/USE WITH OTHER CONCEPT

Proposal to turn a former monumental gatehouse into a new recreational facility for local residents, by Stuart Martin

Spatial Design is offered, at degree level, by a number of specialist institutions. It can be considered as a hybrid between architecture, interior design and landscape design and has its origins in all three disciplines.

Spatial Design will appeal to those of you seeking a broader approach to design within the built environment and provides a wider foundation for further study and employment. Spatial designers also work with spaces and design opportunities that are numerous, but fall outside the remit of more specific interior or landscape-based professions. You will generally be given more freedom to explore possibilities arising from design and to experiment more broadly with outcomes. Particular focuses for Spatial Design courses can include:

Interior remodelling – working with the old and the new and inserting proposals to revive and animate space and the introduction of colour, light, sound and texture to makes spaces more interactive and sensorial.

Design for events – the creation of recreational spaces, performance space, assembly and community spaces for specific short- or long-term events.

Design for specific use – age-specific, clubs, societies, sporting groups, occupational environments, special needs, educational groups, interactive and interpretative spaces.

Architectural conservation – work with historic buildings, finding new and used roles, respecting existing architectural qualities and heritage, breathing new life into buildings.

Exhibition design – temporary installations to excite, inform and provide a short-term user experience.

Approaches to design can be divided into three areas: 'insertions' of entirely new elements into existing spaces; 'interventions' that alter parts of the site; and 'installations' where a permanent or temporary element is added, but which is independent of the original site.

In other respects, Spatial Design courses follow a very similar pattern to Interior Design and Landscape Design, and graduates tend to find employment with architectural, interior and landscape design practices.

COURSE TITLE: **GARDEN DESIGN**

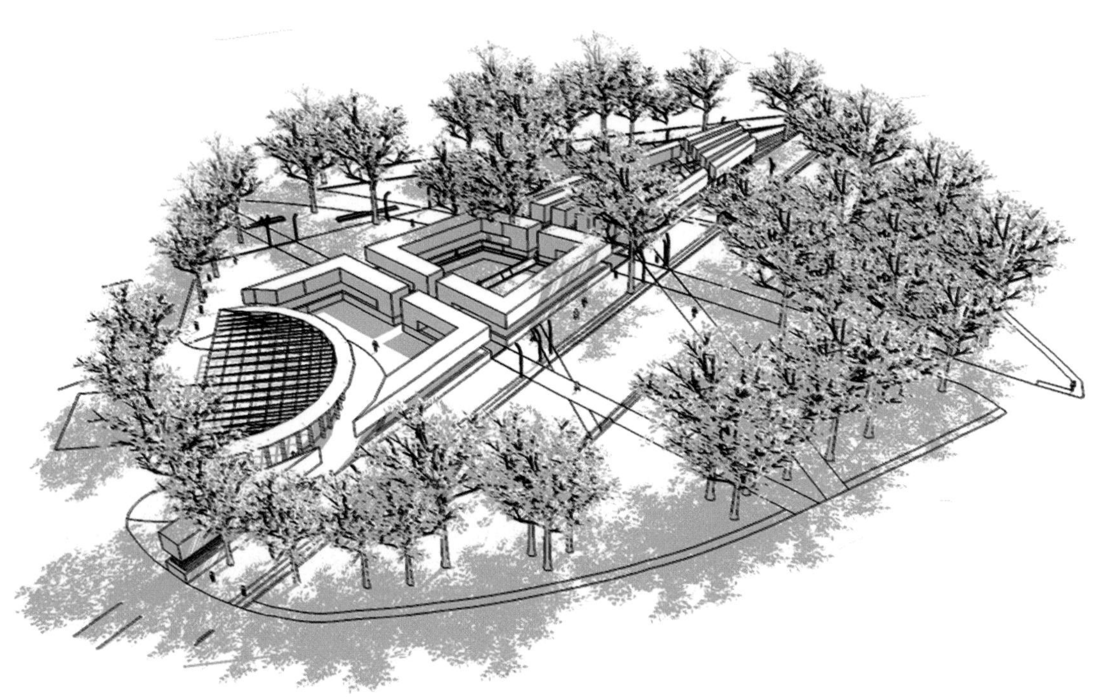

Computer-generated perspective sketch of a large public garden, by James Harris

Garden design is a very old profession, dating back at least 4,000 years. It provided the foundation for landscape architecture and is celebrated each year through important international events such as the Chelsea Flower Show in London. Gardening is the UK's most popular pastime and these courses provide an opportunity to become involved in this broad area, but with advanced qualification, knowledge, skills and creativity.

It sits between gardens as human space, gardens as an art form, gardens as a place for growing and showing plants and gardens as activity. It is distinct from 'gardening', in that garden designers rarely make the gardens they design themselves. Gardens designed in this way arise from a conscious design process, rather than evolving over time, and the description of landscape architecture in this section covers a very similar approach to the way that garden designers are taught.

Some Garden Design courses have very close associations with Landscape Architecture and often share large parts of the curriculum. Those that are offered as discrete courses all have different emphases and it will be important to understand the main focus of the course – horticulture, art, users, outdoor architecture, environment – although all courses will cover all of these areas to a greater or lesser extent.

The significant focus on garden design, as opposed to landscape design, is to do with detail. User groups can be much more accurately defined, normally by talking directly to the end user when framing the project and identifying the design brief. Because spaces tend to be smaller, plants and built elements will be seen and enjoyed from much closer proximity and the careful selection of plants, flowers, colours, textures, as well as the way in which built elements can be combined and detailed, turns garden design into, arguably, a more creative art form. 'Garden Design' is not a protected title, but is overseen by The Society of Garden Designers – a professional organisation with codes of conduct and specific requirements for membership.

Garden Design graduates will be sought by landscape design practices, and those seeking employment within this area should develop a portfolio that shows skills in design creativity as well as technical resolution, computer skills and presentation skills. Landscape contractors, garden centres, plant nurseries, grounds maintenance companies, local authorities and recreational companies will all employ garden designers. Employers will be looking for willingness to work in a team, skills in group work and communication and what you can bring to their existing range of skills and knowledge.

Hugo Bugg graduated with a first-class degree in Garden Design and won the Royal Horticultural Society Young Designer of the Year award in 2010. He now runs a successful business:

> 'I mainly work on large, bespoke, private garden design commissions but also tackle smaller projects. I supplement my design work with some teaching and work with emerging designers. The work involves both design and working with clients and contractors to supervise the project on the ground. Designers cannot be around all of the time and trusting your team, including all of the suppliers of plants and materials, is critical to the success of the scheme. The gardens have very high-quality details – sometimes using expensive materials, but always demonstrating close attention to the quality of the finished work.'

Show garden design by Hugo Bugg and Maren Hallenga

Those looking towards self-employment as garden designers will need a portfolio that shows their ability to work independently and give confidence to clients that they can manage budgets and contracts, as well as their creative abilities. As stated earlier, some courses do not cover business skills and self-employment skills, and it will be worth considering a top-up course if this is the case.

COURSE TITLE: **URBAN DESIGN**

If you are interested in a degree or career that considers the wider build environment, beyond individual buildings or landscape spaces, then Urban Design could be a discipline worth considering. Urban Design is most usually offered at degree level as an option within another core subject, such as Town Planning, Architecture or Landscape Architecture, although some universities do offer it as a separate degree subject. It is a popular option at postgraduate level.

Urban Design is a broad discipline that ranges from the development of legislation and policy through to design guidance and master planning of existing towns and cities, as well as new urban areas. It deals primarily with the design of public spaces, how these are planned and used and how a combination of factors contribute to the social, economic, aesthetic and functional success of these spaces.

Typically, courses will cover how towns and cities work – how places relate to one another, how people

After buying your ticket and discovering that your tour boat doesn't leave for another half hour you stroll back along the pontoon and up onto the Wharf. Here you buy a cup of tea from the small cafe and sit on the chunky steps leading down from the wharf to watch the activity, the sun on your back.

Proposal illustration by Fiona Walters

understand, use and navigate their way around and how towns and cities are planned and organised. This involves an understanding of social, economic and environmental sustainability, transport infrastructure, architectural design, landscape planning and the way that these areas can be used to develop new spaces and improve existing places. It involves large-scale planning, zoning and mapping, as well as more site-specific interventions and draws upon skills and knowledge of human geography, sociology, law, architecture, environmental science, design and art.

An important focus of urban design is the adoption of a sustainable approach to design and change. Urban design is a collaborative activity and professionals in this field work with other disciplines. Successful urban designers must therefore be skilled communicators with good interpersonal and management skills.

Urban design is a subject that can build upon knowledge of other subjects, such as geography, biological sciences, architecture, engineering and law.

Urban space concept sketch by Jon Penn

Preparing your portfolio

For all design disciplines there is general advice on preparing your portfolio later in Chapter 9, 'How to succeed as a design student'. Here is some extra information pertinent to all spatial design disciplines featured in this chapter.

Your portfolio needs to evidence, not only project outcomes and design solutions, but also a demonstration of your wide-ranging skills. Try and include all of the following:

Evidence of the levels of creativity in your research and idea development – in particular your ability to identify and solve a creative problem. This could be a design project or the way you have interpreted a brief. Perhaps the best way to show this is through your sketchbook work, which will show your thinking process and where your ideas have come from. If you don't normally work in sketchbooks, try to capture your thinking process and sequence.

Evidence you are thinking about what you were asked to do and how you have responded to this. What were the key decision-making stages within the project? You should certainly plan the way you explain your projects before an interview, but also think about how to express your design or creative journey visually. Jot down the key stages in your project and make sure that you have illustrated this journey. Don't be afraid to include ideas you have rejected along the way.

What was successful and what was edited out? Remember, this is an extension of your design or creative journey, but try to explain the criteria you used to judge the success of your emerging ideas. Sometimes these decisions are very difficult and might help to make the story behind your portfolio more interesting.

Your own personality – what excites you, why you have chosen a particular topic or idea. It is likely that this will form a focus for discussion at an interview, together with your view on how you are going to bring a new perspective to the company.

Group project mood board for restaurant proposal

Conclusion

Spatial design is a wide-ranging design discipline with varied modes of practice. You could work within a large consultancy, as a freelancer or self-employed or within a partnership, on a wide range of interior and/or exterior projects, from domestic interiors and small projects to large offices suites and entertainment venues.

These different spatial design disciplines can change lives. By designing buildings, landscapes, interiors, schools, hospitals, parks and gardens, shops, hotels, fun parks, theatres, festival sites, and so on, designers are providing the environments within which we live. Most graduates will work in the UK, but many will also work abroad, taking their education and experience to new parts of the world, to new cultures, and to help solve new problems.

Good spatial design professionals are highly regarded, well paid and sought after. Most importantly, it is a profession that allows you to make a serious contribution to the lives of future generations. The connection with users and the quality of their experience lie at the heart of what spatial designers must achieve. It is a discipline that mixes research, analysis and creativity in the biggest sense.

Further resources

1000x Landscape Architecture, Braun (2009)

Featuring 1,000 projects from around the world, this is a great visual feast and highly inspirational.

Alexander, R. and Sneesby, R., _The Garden Maker's Manual_, Conran Octopus (2003)

Still the best introduction to the principles of working with built and architectural elements within gardens.

Bently, I. (et al), _Responsive Environments_, Architectural Press (1985)

A beautifully illustrated book (in slightly comic-book style) that puts people at the centre of urban design. A classic text.

Brooker, G. and Stone, S., _Basics Interior Architecture 01: Form and Structure_, AVA Publishing (2007)

This takes a broader architectural approach to the subject, looking at contemporary and iconic projects.

Dodsworth, S., _The Fundamentals of Interior Design_, AVA Publishing (2009)

A comprehensive introduction to the profession of interior design, this usefully considers a broad definition of the subject, from commercial to domestic.

Kingsbury, N., *Gardens by Design*, Timber Press (2005)

Despite being a few years old, this is one of the best collections of contemporary gardens, with excellent photographs and informative text. Look for other titles by this author as well.

Littlefield, D., *Metric Handbook: Planning and Design Data*, 4th Edition, Routledge (2012)

A classic text, now in its fourth edition. There are no photographs, but it is the most comprehensive single volume of architectural design details available.

Ross, A. (et al), *Architect's Pocket Book*, Architectural Press (2011)

An abbreviated version of the above.

Waterman, T., *The Fundamentals of Landscape Architecture*, AVA Publishing (2009)

This is a good, basic introduction to the main principles and approaches to landscape architecture and design from a UK author; a useful starting point.

Urban Design Associates, *The Urban Design Handbook: Techniques and Working Methods*, W.W. Norton and Co (2003)

This book provides a generic international approach to urban design and covers basic approaches and principles; a good starting point.

Vernon, S., Tennant, R. and Gormory, N., *Landscape Architect's Pocket Book*, Architectural Press (2008)

A reference guide full of technical specifications and design guidelines, this is probably the best short reference available.

Young, C. (Ed.), *RHS Encyclopedia of Garden Design*, Dorling Kindersley (2009)

A good overall introduction covering design principles, garden styles and planting design, this is full of inspirational images.

Project story

Designer:
Anna Deery, final-year interior architecture student

Project title:
kidsPACE

kidsPACE exterior view

Tell us about the project

Being the oldest of five children within my family, I have been surrounded by and looked after children my whole life. I wanted to embark on a project that I was passionate about and that I had knowledge of. I have seen my brother and three sisters grow up and I am fascinated with how a child develops as a person and how the things they experience and the people they interact with can influence their development.

I wanted to focus on the design and planning of a built environment for young children through early to middle childhood. I created a series of spaces and zones that will enhance a child's learning and stimulate optimal child development.

I took a trans-disciplinary approach to introduce and integrate my knowledge of childhood development by translating it through stimulating interiors and landscape architecture; ultimately creating a unique, playful learning experience.

kidsPACE is a learning and supportive environment for children and parents that focuses on the design and planning of architectural spaces with regard to a child's physical awareness, allowing independent exploration through sensorial interaction and social activities. It is a centre that brings parents and children together to learn, play and bond, while offering help and support for all of the family.

Architecture for children needs to be functional, yet playful. KidsPACE offers children a sense of control, enabling them to engage in focused, self-directed play. It evokes a sense of security that is a prerequisite in the formation of a healthy identity.

Throughout a child's life, there are many influences that shape and aid their development. It is the hope of parents that their children will grow up to be independent people, confident in themselves and confident in their relationships with others. This parental desire is as natural and spontaneous as their desire that their child will grow up physically healthy. I interpreted the concept of development and the growth of child independence architecturally through a series of 'building block' structures, which represent the playful aspect of childhood.

What processes and skills were most relevant to this project?

The research process is the initial process that needs to be carried out in order to fully understand the topic that you wish to explore. During my research phase I visited children's centres to see how they dealt with design for children and to find out what staff believed were the most important activities that both parents and children should be involved with, whether it be together or separately. Through these

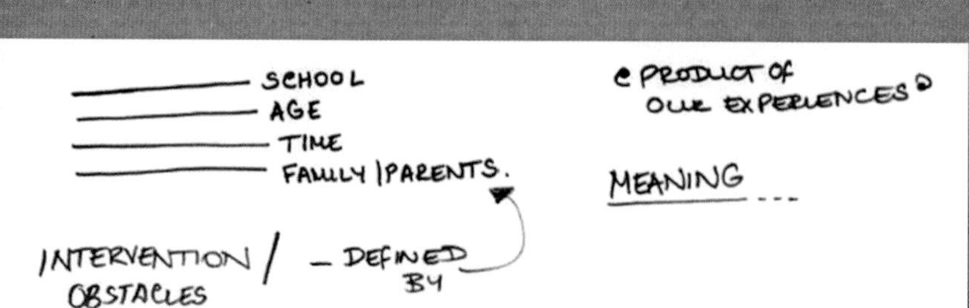

Part of Anna's thought processes sketched out

visits, I got to hear from members of staff about the pros and cons of the design and planning of their spaces, which I then took into consideration in the design for kidsPACE.

What challenges did you face?

One of the main challenges I faced when designing for parents and children was making sure I was not imposing on parents. Child centres are there as a choice for parents to use with their children; it is an option for them to explore if they feel they would like to and the support is there if they feel they need it. Having not had the experience of being a parent myself, I interviewed numerous parents to get a full understanding of their experiences and their wants and needs in a space such as this.

When it came to choosing the colour for my spaces it was difficult trying to get that right. I looked into the psychology of colour with regards to children and this informed what colours I was using and which spaces they would be used in. It gave me a better understanding of colour and how colour can influence child behaviour. During the first eight years, children are developing their visual acuity. Their perceptions of objects, movement and print are expanded as they have opportunities for experiencing interesting visual images and colour. Changes and variations of design and colour intrigue children and cause them to visually attend to the unusual. It promotes a sense of adventure and exploration.

I find that justification is of utmost importance when designing for anyone – a reasoning and a full in-depth understanding of the client allows for a more successful project.

What were the highlights of the project?

Being able to write your own brief with no budget is always going to be a challenge, but a challenge that has no boundaries. It was interesting and rewarding to see how my passion, ideas and designs came together to create a usable space for my clients. Developing a project from start to completion is quite a steep learning curve. I have really grown as a designer and I now feel that I can confidently take on challenges in the design world.

It is important always to be willing to learn throughout your time at university. I would always be teaching myself new computer programs to help me better my visuals, for example, which I found helped me immensely in communicating my final design project. Not only that, the more you know, the better the CV!

Time management is of utmost importance within this profession and the ability to work under pressure is an advantage. The best advice I can give to students embarking on a design course is to treat it like a nine-to-five job, that way you get a good balance of both work and 'you' time . . . and it prepares you well for life after university in a professional practice!

Project story

Designer:
Harriet Fleur Curtis, final-year interior architecture student

Project title:
The 'How are you today?' clinic for chronic pain management

Tell us about the project

'How are you today?' is a chronic pain management clinic and my final-year university project. It grew out of the brief that developed over my placement year, when I spent a lot of time looking after a close family member who suffers with chronic pain. This gave me a real connection with my project and a drive to design for the end-user. The concept for the project, connect/disconnect, grew out of this personal experience as well as research I did into chronic pain. The clinic is a treatment and social centre for chronic pain sufferers.

The situations I found myself in during my placement year made me very aware of how the able bodied sometimes don't see what's right in front of them. Coming from a design background, I found it frustrating that small things could make life so difficult – it is impossible to go clothes shopping in a wheelchair as the racks of clothes are just too close together. I often felt that accessibility was the smallest afterthought in the designer's mind, a place to scrimp and save rather than meet the needs of the end-user.

My project aimed to change this focus, with accessibility and the end-user at the forefront. This dictated some of my space planning, with corridors and routes around the building having to be a certain width to accommodate wheelchairs and those on crutches. I also focused on the individual rooms in order to decide where they needed to go. Model making was a huge part of my course, so whipping up a few light models (boxes with holes in) wasn't difficult. Based on these, I arranged my treatment rooms according to where they could get the most similar light.

Additionally, I wrote and sketched out a narrative, focusing on having two routes around the building – one for the good days and one for the bad – in line with my concept and brief development. A narrative really helps with developing a concept, especially one where the emotions of the user are so important. Do one early on and then, as you develop your thinking, keep editing it so that it's always up to date.

Space minimums are exactly that, minimum. I feel that if you don't have the space to give something more than the building regulation guideline minimum amount then you need to re-think your design. It's one thing to give as little space as possible to the bin storage area, but another to make a corridor only just wide enough to walk down. The design of a space is exceptionally important; all the work you do in the first year teaches you that. However, equally important is the spaces that get you there – the corridors, atriums and entrance areas. Doing something purposeful with

Visualisation of the 'How are you today?' clinic library by Harriet Curtis

Visualisation of the atrium of the 'How are you today?' clinic by Harriet Curtis

these will lift your design and make your spaces that more exciting to reach. If you make a corridor the absolute minimum you can, make sure it's because you want people to feel claustrophobic and hemmed in (perhaps just before reaching a wide open space), not just because you don't have room to give it any more space.

My design came out of personal experience and research gave me the starting point of 'connections' because of how nerves connect with each other to pass information and the effect damage to them can have on how you interact with the world. Personal experience taught me that there can be good days where the pain is well managed but then there are bad days where it is every single thought someone has. This led me on to the idea of two routes around my building, I coupled this with the idea of damaged nerves and connecting nerves – providing spaces where people could connect with each other, interact, regain confidence and develop and spaces where people could hide away, ignore the world and withdraw from life.

What processes and skills were most relevant to this project?

It's amazing how much faster you can do something the second time around, so I would definitely say 'learn to learn' from your mistakes and get on with it. It can be very hard to do this, you feel so protective towards your projects that it's difficult to give up that idea of possession. The less attached you are, however, the easier it is to scribble it out and start again. Of course, there is merit in feeling connected to your projects, you take pride in them, you take time over your models and you don't want them to get trashed. But make sure that during the development you are constantly prepared to start over, change everything and use a lot of red pen.

Time management is important and if you're not good at it, get good at it. Know when to say you've done enough and print as early as possible. Be confident in your ideas, you don't always have to do what your tutor says, but you do need a good reason for not doing it. Always have your *Architect's Pocket Book* to hand.

What challenges did you face?

The final project for my Interior Architecture and Design course is over six months long, so maintaining interest can be very difficult; choosing something you feel strongly about definitely helps. The main problem I came up against was frustration, when I felt my project should be further along, or when it wasn't going the way I wanted it to. Taking time out and looking at it with fresh eyes always helps – there's no point staring at it because you just end up digging yourself a bigger hole. Equally, don't stare at a blank piece of paper when you're trying to sketch; even if you just cover the page in pencil marks it will get you further than sitting mindlessly for hours and then feeling bad because you haven't done anything. For me, I slept or watched TV when I found myself grinding to a halt, and gave myself decent lunch breaks. Don't kill yourself over it until it's really worth it.

What were the highlights of the project?

I really enjoyed the research stage of my project; having a basic knowledge of what I was researching really helped push my understanding. It led me to consider different technologies, treatments and approaches, as well as teach my tutor a few things too. We had to find our own site as well. It turns out it's really good to have a site you can visit whenever you want and can take photos of at the last minute.

Exterior view of the 'How are you today?' clinic by Harriet Curtis

Theatre, film and television design

By Patrick Connellan Nottingham Trent University and
Jayne Harvey, Nottingham Trent University

'Designing for the screen means combining a great number of skills and processes in order to achieve the final output, which can be very satisfying. Using hand-drawn and computer-aided design skills merged together is, for me, what produces such dynamic and exciting creations.'

Laura Ellen Ward, design for film and television student

Aida by Giuseppe Verdi set designed by Paul Brown, Bregenz Festival 2009–2010

Designers who have made a career in designing for the performance industries have done so because they have a combined interest in performance (drama, dance, opera, music theatre and film, costume, event, puppetry and objects and rock gigs) and design. The combination of the two can be extremely exciting. Design in a performance context, whether it be for theatre, film or TV or whether it be for live or recorded events, brings a visceral and immediate quality to design that few other design disciplines can achieve. Design for Film & TV student Mike Jones, working on his project *3:10 to Yuma* (a spaghetti western), describes the challenges of capturing the mood of the piece:

> 'One of the challenging aspects of the hotel bar area was finding the right balance of how wealthy the establishment would be. As it was a frontier town, the décor and ornaments would not be particularly extravagant, but the script described the hotel as the biggest in town, so it would need to impress a visitor on some level. The amount of lighting was very important as the hotel was to be dark and smoky but also had to be bright enough for the viewer to take in all the elements of the set. The lobby had to be big enough to accommodate the bar, piano and gambling tables but small enough to create a dark dingy atmosphere. The exterior of the hotel also had to balance a sense of grandeur with the materials and building techniques common in frontier towns of the period, and also had to match the layout of the interior – wall dimensions, window positions, etc. It was creating fine balances such as these that was quite challenging but enjoyable and I feel I accomplished this in my design.'

Mike Jones, final-year design for film & TV student

Hotel bar illustration by Mike Jones

Today's performance industry

The industries themselves are exciting to work in: the collaborative nature of the work and meeting a lot of people with all sorts of skills is stimulating; deadlines that simply can't be missed give an extra excitement to the work; the transient quality of an intensity of work for a period of time as everyone works toward a shared goal, then finally seeing your work, whether it be on stage or screen, and sharing it with an audience.

Although the different design-for-performance disciplines work in distinct areas, they do influence each other. For example, film design often borrows from the rich visual metaphor and the glorious artifice found in theatre and costume design. Think of Baz Luhrmann's *Moulin Rouge*.

Moulin Rouge, costume design by Catherine Martin and Angus Strathie

20th Century Fox/Godfrey, Chris/ The Kobal Collection.

Equally, theatre design is highly influenced by film. The *montaging technique* developed by Sergei Eisenstein and many film directors since, where one shot is juxtaposed with another to make comment and develop the narrative (an overarching thematic story and structure), has been explored, to heightened effect, by many theatre directors and designers to create **epic theatre** stage pictures with multi-detailed moments and activity within.

Think of Maria Bjornson's designs for *Les Miserables* and the complexity of the scenes on the barricades. However, it used to be rare for designers to move between the disciplines of film and theatre, for example, but recently it is becoming more common. Es Devlin, an award-winning international stage and costume designer, moves with apparent ease between designing for the Royal Shakespeare Company, International Opera, Take That's live tours and the closing ceremony for the Olympic Games London 2012. So, in real terms, anything is possible.

When imagining what theatre design is, many will be thinking of the sets and the costumes that they have seen that create an atmosphere or an environment for a play, often referred to as the *mise en scène*. It is most likely that what is being imagined is in a conventional proscenium arch theatre. This is a framed stage, as seen in most theatres, with a curtain separating the audience from the stage. But, this essentially naturalistic approach (a perfect illusion of reality that reveals a hidden truth) providing an often-necessary visual and historical context for a production, is just one aspect of a very broad discipline. Theatre, film and TV design can, and often does, become a conceptual vehicle for interpreting and commenting on a play or any other material that you might be working with and the context that it comes from. It may also, as one element of a production, make insightful commentary on our lives and society. At its best, when interacting with these elements successfully, it can aid the profound revelation of meaning within a text or performance.

Theatre, film and television design is a collaborative art form where you may be working with an artistic team including a writer, director, lighting designer, sound designer, choreographer and musical director or conductor. Some of the best theatre has been created from a shared conceptual vision. Theatre design can be practised in many genres, including theatre, dance, opera, musical theatre and installation-based work. All of these genres can take a number of forms, such as site-specific work, verbatim and memory theatre, puppetry, carnival, immersive theatre, educational work and many more. The possibilities are almost endless. Film and TV design has a similar range to its practice, except that a blockbuster set of films, such as the *Lord of the Rings* trilogy, can require huge numbers of set and costume people and the facts and figures of what is required can be staggering – for example, 300 wigs for *The Fellowship of the Ring*.

Therefore, theatre design can be applied to many forms that have live performance as their foundation. Most performances include some kind of story or narrative at their heart, so we might define theatre design as a visual narrative for a performance context. Sometimes, you might come across theatre design being referred to as **scenography**.

Epic theatre: a political form of theatre where everything presented has an objective basis and is devoid of illusion. A form of theatre made popular by Bertolt Brecht.

Scenography: an artistic perspective concerning the visual, experiential and spatial composition of a performance. It is an holistic approach to theatre performance that embraces all of the key artistic disciplines associated with theatre production.

This is partly an international term for theatre design, as theatre design can mean theatre architecture (the design of the theatre itself) in many countries. The theatre designer usually designs for a number of forms of theatre and will need to be able to respond intellectually to a plethora of texts, ideas and materials. This can mean that a designer will engage with visual culture, literature, history, politics and philosophy to inform their design. This intellectual and artistic process, combined with a team-solving approach, can be very rewarding. Ultimately, the theatre designer is interested in everything that informs and challenges the audience through a visual and dramatic language. This can be reflected through choosing the correct period and colour of button for a jacket to making an epic visual statement.

Theatre design is practised wherever there is theatre, from small community-based theatre groups to international opera houses. Although theatre is often staged in purpose-built theatres, it can be made for site-specific spaces, outdoor environments, in tents and even for the streets.

What courses are there?

Courses in this broad discipline of performance-related design subjects include Theatre Design, Costume Design and Design for Film and Television, to mention just a few. These subjects can be studied at degree level and are available in two very different types of academic institution, broadly known as the drama school, or conservatoire, and an art school within a mainstream university. The advantage of a drama-school training for theatre designers is that you are likely to be working closely with students studying other theatre disciplines, such as Directing, Lighting, Writing, Acting and Stage Management. You may also have the opportunity to work on several drama-school productions, both as a designer and maker (set, scenic art, costume, puppetry and properties).

The art school usually offers opportunities to work with external theatre companies in the absence of being surrounded by the other disciplines. It can also be argued that the art school environment encourages greater concentration on the subject as an art form.

Some Theatre Design courses, both in drama school and art school, may have a bias towards the vocational aspects of theatre design, such as training in scenic art, costume making, set construction and puppetry. However, there are some non-degree courses that specialise in these disciplines. You can also specialise in Theatre Production and Management.

In all of these courses, there is a good balance between the deeper study of the context for the subject, the necessary practical applications and an understanding and practice of the wider choice of supporting craft skills, such as prop design, for example.

COURSE TITLE: **THEATRE DESIGN**

Costume design by Jennifer French

There are a number of courses that teach theatre design as the core subject, but may also teach applied craft skills and performance application to support the learning of the subject. Titles might include Theatre Design, Design for the Stage, Scenography, Design for Performance, Theatre Performance and Event Design.

Then, to contrast with these, there are theatre design and craft skills courses. These put craft and skill at the forefront of what you will be taught. These titles are Theatre Practice, Performance Design and Practice, Theatre Art, Technical Production Arts, Technical Arts and Special Effects and Design and Craft for the Stage and Screen.

COURSE DESCRIPTION

Studying one of these courses typically will teach you a deep knowledge and understanding of the subject through a mixture of lecture and direct teaching, combined with enquiry-based research that will inform and support your work. Most courses will provide a deep engagement with the practice of design in a wider cultural context through study and trips to the theatre. Some courses take an integrated approach to the theory and practice of theatre design so that one clearly and demonstrably supports the other, while other courses may have separate cultural studies modules to support and enhance the understanding of the subject.

CAREER OPPORTUNITIES

There are many routes and opportunities to practice theatre design as a career. Getting started as a theatre designer may mean following several routes before settling on a clear career path. Experiencing the industry is also very important. So, to begin with, you may need to work as a model-maker for an established theatre designer, use your craft skills and make or paint for an established theatre company or production company, or team up with a director you admire and design small-scale theatre. You might decide to work with like-minded contemporaries and start your own theatre company and perform pieces at festivals.

Here are some descriptions of the types of roles found in this industry and a brief description of the types of skills required, which should help you think about what might suit you.

Assistant designer – Model-making and technical drawing skills are usually required. You can contact designers or their agents via The Society of British Theatre Designers' register of designers (see further resources at the end of this section).

Heritage and exhibition designer – The National Trust and other heritage organisations can be very open to theatre designers bringing their skills into historic homes and landscapes.

Scenic artist – for theatre, film, events and TV. This includes many specialised skills and techniques, including cloth painting, set painting, set distressing, spraying, marbling, wood graining, texturing, staining, effects and working with fabrics. A highly skilled scenic artist will never be out of work.

Set construction and property maker – There are many theatre and independent set construction companies that offer work to skilled set and prop makers.

Event and festival designer – There are many festivals, from the long-established Edinburgh Fringe Festival through to Camp Bestival, that provide opportunities for self-motivated people and companies to perform their work.

Puppetry and carnival maker – This is a growing area of theatre and there are many opportunities to get involved in carnival design and making for long-established Afro-Caribbean-inspired carnivals such as Notting Hill and Nottingham, to newer carnivals such as the Godiva Carnival in Coventry.

Costume maker – If you have sewing skills working with period costume then you could approach theatre costume departments or the growing historical re-enactment industry.

Prosthetics designer – Britain has a very lively theatrical and film prosthetics industry that is often looking for people who have skills in this area.

Set and property constructor – for theatre, events, film or TV, building sets and objects in timber, steel, aluminium, plastics and other materials. Often construction drawing skills are needed using AutoCAD and other technical drawing software.

Lighting designer – for theatre, film, event and TV. A complex, highly skilled and specialised area with little crossover between theatre and film.

Lighting technician – for theatre, film and TV.

Projection designer – for theatre and live events. A fast-growing area of highly specialised work.

You may be determined to go freelance or become a self-employed theatre designer. Whether you are working on your own or as part of a group, there are a number of opportunities open to you. Look out for awards, bursaries and prizes – the biannual Linbury Prize for Stage Design is open to all graduating students. If you are determined to create a company, as you may have production ideas that only you can develop, try and find the support of peers and work collaboratively. Find out how the Arts Council can support you in funding projects, and try and raise private sponsorship to support your venture. (For further information on self-employment, see Chapter 16, 'Working for yourself'.)

Here are a number of notable examples of British theatre companies who work in different and innovative ways and in differing venues that you might like to look out for:

Forced entertainment – A Sheffield-based international touring company that is Britain's foremost experimental theatre company. Work includes: *Void Story*; *The Thrill Of It All*; and *The Coming Storm*.

Kneehigh Theatre – An international company based in Cornwall that tours to many kinds of venue. It specialises in a very visual, physical theatre, with music that often works with material adapted from a play or film. Work includes: *Brief*

Encounter; *Red Shoes*; *Cymbeline*; *Hansel and Gretel*; and *The Umbrellas of Cherbourg*.

Wildworks – A theatre company that specialises in community-based memory theatre (personal, traditional or cultural stories shared). It often works with rural, although latterly with urban, communities to create epic outdoor landscape events that combine myth and legend with everyday stories and memories. Work includes: *The Beautiful Journey*; *Enchanted Palace* at Kensington Palace; *The Passion* in Port Talbot; and *Babel* in east London.

Complicite – This company specialises in devising work from existing stories and retelling them in a very innovative and visual way. Work includes: *The Street of Crocodiles*; *Mnemonic*; *A Disappearing Number*; and *Master and Margarita*.

Punchdrunk – The pioneers of what has become known as *immersive theatre*, this company works in site-specific locations and creates events often using classic texts, where audiences can roam around and immersive themselves in a very visual and visceral experience. Work includes: *Faust*; *Tunnel 228*; and *Sleep No More*.

Dreamthinkspeak – This is a theatre company that specialises in site-specific work that often includes landscape and found interiors combined with projected material to create a beautiful relationship between audience, space and word. Work includes: *Underground*; *One Step Forward*; *Absent*; *Before I Sleep*; and *The Rest is Silence*.

Quarantine Theatre – A Manchester-based company whose starting point is real people, real events and real things that are then woven into an innovative and visual form of theatre. Work includes: *Geneva*; *Old People, Children and Animals*; *Susan and Darren*; and *Soldier's Song*.

Birmingham Opera Company – This company works on site-specific locations in Birmingham, with large community casts and professional opera singers. They work with classic texts and newly commissioned pieces. Work includes: *He Had It Coming*; *La Traviata*; *Idomeneo*; *Othello*; *Life Is A Dream*; and *Mittwoch Aus Licht*.

DV8 – Britain's premiere physical dance company. Work includes: *Can We Afford This?*; *The Cost Of Living*; *Just For Show*; *To Be Straight with You*; and *Can We Talk About This?*

Blind Summit Puppet Company – A touring company that often collaborates with other companies and is leader in the inventive use of Banraku-inspired puppetry to create a unique and challenging approach to puppet-based performance. Work includes: *Low Life*; *1984*; *The Other Seder*; *A Dog's Heart* (with ENO); and *The Table*.

Other notable companies to be recommended for researching, which produce work in more conventional theatre spaces, are: National Theatre; Royal Shakespeare Company; Royal Court Theatre and Edinburgh Traverse (new writing theatre); Donmar; Almeida Theatre; Young Vic; Chichester Festival Theatre; Sheffield Theatres; Birmingham Rep; Manchester Royal Exchange; Bolton Octagon Theatre; New Vic Theatre; Glasgow Citizen; Opera North; and English National Opera.

You might like to go into teaching, as there are many ways to become involved in theatre in education. Some graduates get teaching assistant jobs with a view to undertaking postgraduate training, to then become a teacher in a secondary school. As a practitioner or designer in the industry, you may find that you can teach in further or higher education, and this might be a useful way of supplementing your income if you are self-employed.

Network! See as much theatre as you can afford and begin to form an understanding and passion for your favourite theatre companies and the work they do. You can then, with genuine enthusiasm, target those people you would like to work for through letter writing and networking (go to the 'first nights' of their latest productions).

If you want to assist an established designer, make sure you know their work before you make contact with them.

WHAT WILL YOU BE TAUGHT?

Much of your learning in theatre design is through clearly briefed design projects. In essence there are three types of design projects that you might come across:

- a pure design project, or speculative design project, where typically a text will be set, researched, interpreted and communicated through models and drawings as a complete design response

Studio-working in groups

- a craft-based design project, where an element is designed but the emphasis is on the craft and making of an artefact, costume or set
- a design project that is applied to a performance or a realised production; often these are live projects, working with professional theatre companies.

Running alongside the core curriculum might be other modules that support your communication, craft and practical skills. The communication skills taught may include model-making, life-drawing, perspective drawing, spatial drawing, technical drawing (both by hand and using software such as AutoCAD), 3D digital design and 3D printing.

The craft skills taught will depend on the nature of the course and are often optional strands to a course. They may include costume making, set and property construction, scenic art and puppetry. Additional specialist skills may be taught, but not always to the high level that you find on specialist courses or on a postgraduate course.

THE FIRST YEAR

From the outset you will be encouraged to keep a sketchbook that may become the place for all sorts of working out of ideas in sketch form, written notes, story-board form, photographs and research images. The sketchbook should be the central document to the whole design process and therefore becomes an important tool for explaining where your ideas have come from. A sketchbook is not a scrapbook, it is an ideas container.

An undergraduate Theatre Design course typically will concentrate in the first year on the fundamentals and principles that underpin a deep understanding of the subject. This will include theatre design practice combined with contextual work and essential craft skills.

THE SECOND YEAR

The second year tends to be a more exploratory year where some specialisation may occur as particular interests are developed. This may be a craft skill or a design specialisation.

It is most likely that your tutors will have substantial experience as practising theatre designers and will therefore attempt to reflect the real-life design process in your projects. Therefore, right from the first briefing your tutor may act as 'the director' and give interpretive guidance as well as respond to your ideas. Like all good director/designer relationships, this should involve an engaging and informed dialogue.

Once briefed on the text, you will be working on analysing the text and its literary background. You might be organised into groups to read the text together and discuss it, as this often yields greater levels of creativity. Some directors and designers read the text together so that they can make immediate responses and begin a fruitful dialogue.

WHAT IS THE DESIGN PROCESS LIKE?

Sketches by Katie Grinsell for her design of *The Rise and Fall of the City of Mahagonny* by Bertolt Brecht and Kurt Weill

A 1:25 scale 'Othello' set design by theatre design student Alexander Green

You cannot design for a play unless you have a clear understanding of the play and your interpretation of the text. You probably won't have that unless you have researched the literary, historical and cultural context that surrounds the play. In doing so, you may also come across some useful reference images to inspire your work. Sometimes, it is one image or artefact that becomes the inspirational driver for the design. It may be worth visiting appropriate galleries and other places of interest from a research perspective at this stage of the project.

So, how do you start designing your set and costumes? At this point several things need to happen, virtually at the same time. You will need to have a clear understanding of the performance space you are setting the play in. You will need to visit the space and acquire a set of ground plans and elevations so that you can build a scale (usually 1:25 scale) model box of the theatre space to start designing in. You will need to have made a unitisation (sometimes called section) of each scene or moment in the play that asks fundamental questions about the scene. Who's in it? Where is it? What time of day is it? What objects are required? Most importantly, what is happening in the scene?

It is the moments of activity in each scene that will inform the space and dynamics of your design. This may be a good time to begin a storyboard using the information from your unitisation. You don't need to know what the set looks like before you sketch out each moment of activity. A good tip is for the moments to be defined from a line in the text so that there remains a clear relationship between the script and the design.

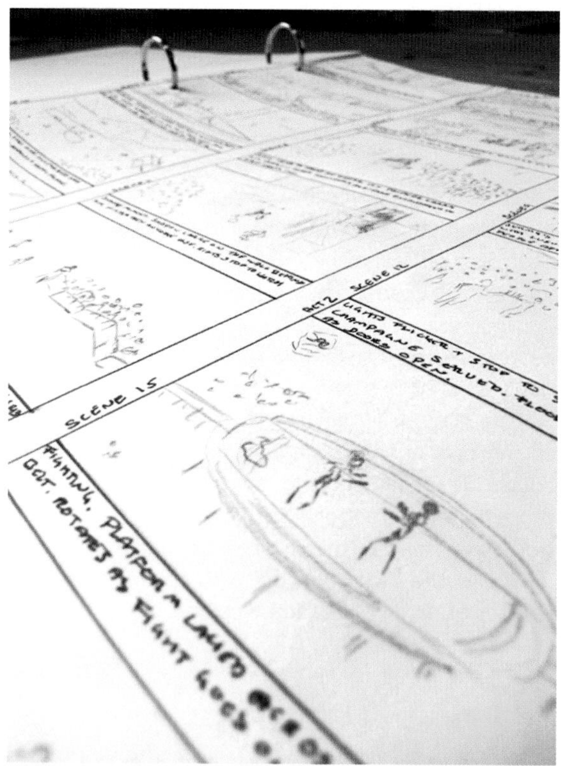

A storyboard of each moment of activity, by Katie Grinsell

The final year on a Theatre Design course is the opportunity to prove to yourself, your tutors and the theatre industry that you can work to a professional standard. Therefore, you may have the opportunity to work on several live projects and even have your work realised. With an audience!

Costume concept by Charlotte Bakewell for *The Winter's Tale* by William Shakespeare

Cottage Countess costume by Jennifer French

COSTUME DESIGN

This can be studied as part of a theatre design course or in its own right. It is surprising that most designers don't start by designing the costumes as the characters are usually central to understanding the play. Should the environment they inhabit be secondary to the characters and what they wear?

It is important to be considering the costumes from the outset. You will be required to make a costume plot that lists all the characters and what they wear in each scene.

You will need to research the historical background to what the characters wear to determine exactly what clothes and their cut, style, fabric, class and colour are appropriate.

There is another list of career opportunities that is based on costume design. Usually, theatre designers design all of the major visual aspects of the production, including the sets, costumes and properties. However, sometimes designers specialise in one aspect, such as costume design. The separation of disciplines is more common in opera and dance than theatre. Theatre design as a skill can be used in many other creative areas. Some job descriptions that may require a background in theatre or costume design include:

Costume design by Emily Ahearne

1. **Theatre designer** – Set and costume design. Required to produce scale models of the design, rendered drawings of the set, costume drawings, technical drawings and property designs.
2. **Carnival designer** – Designing floats, back-pack worn costumes and large animated objects.
3. **Costume designer** – For theatre, film or TV.

4. **Costume supervisor** – For theatre, film or TV. Supporting the designer, sourcing fabrics, sourcing costumes, coordinating the makers and costumiers and arranging costume fittings.
5. **Costume property maker** – For theatre, events, film or TV. Specialising in an area where there is a crossover between property making and costume making.
6. **Costume maker** – Often making historically accurate costumes. Requiring specialist skills in cutting costume patterns and sewing techniques.
7. **Theatrical tailor** – Making historically accurate tailored costumes.
8. **Milliner** – Bespoke hat making for theatre, film and TV.
9. **Wig maker and wig supervisor** – Knotting, styling, maintaining and supervising wigs for theatre, film and TV.
10. **Dyer and distresser** – This includes dying fabrics and 'breaking down' costumes for theatre, film and TV.
11. **Theatrical make-up and prosthetics artist** – For theatre, film and TV. Working with designers and actors to create a period look or a special effect.
12. **Opera designer** – Set and costume design.
13. **Ballet and dance designer** – Set and costume design with a specialised knowledge of dance-based costume.

Most designers design in the model box, working in three dimensions and creating sketch models to develop their idea. You will need to make a scale figure of one of the characters that can be present at all times in the model box so that you have a scale character to refer to. Some designers may also develop their designs through sketches and a more developed storyboard.

At this point the project will stop for an interim presentation point. In theatre this is called 'the white card stage'. This allows the director and production manager to consider the design before it is fully rendered.

Model box for *The Rise and Fall of the City of Mahogonny* by Jayne Riddell

The rest of the project will be devoted to the developing and rendering of finished designs. This includes a painted scale model with exquisite detailing (if you produce a very accurate model it is more likely that the finished set will be of higher quality and closer to your vision), a full set of colour rendered costume designs with fabric samples, a finished storyboard, prop and furniture designs and references and a full set of technical and working drawings drawn using AutoCAD software. This will be presented to your tutors and peers at a final presentation that mirrors the theatre production meeting.

WHAT WILL THE FINAL YEAR BE LIKE?

The focus in the final year tends to be on preparing the student for professional practice by testing their knowledge and skills by realising productions and working with professionals.

COURSE TITLE: **DESIGN FOR FILM AND TELEVISION**

JAYNE HARVEY, NOTTINGHAM TRENT UNIVERSITY

> *'One of the most satisfying aspects of my job is the research – collating visual references in order to add realism and believability to what you're designing. For example, I had to learn about horse tack (saddles, reins, etc.) for Ridley Scott's* Robin Hood, *I studied Regency architecture for* The Golden Compass *and military hardware for* The Dark Knight. *With each new job you will garner more knowledge, helping you become more visually articulate.'*
>
> Dan Walker, Concept Artist

COURSE DESCRIPTION

Designing for film and television can be very exciting! Production designers, art directors and their art departments are responsible for providing an appropriate environment (film or TV set in studio or on location) for any given production. The scope for variety is enormous: for one job you might be helping to recreate a medieval castle, the next designing a futuristic space city, the next an *Eastenders* living room or a Victorian slum, a period ballroom for *Downton Abbey*, or a children's game show – perhaps a world that, so far, only exists in your imagination. Everything that is brought to our screens, in the cinema or in our homes, has to be imagined, designed and brought into being by someone – maybe you? A degree course in Design for Film & Television will prepare you for work within an art department and set you on the path to achieving your aspirations within this diverse industry.

The course will introduce you to the skills and knowledge you will need to be successful in this field. You will develop skills such as drawing (both traditional and computer-aided drawing), life-drawing, technical drawing and storyboarding. Drawing is one of the intrinsic skills that needs to be a large part of your skill set. You will be encouraged to adopt an experimental approach to media and technique

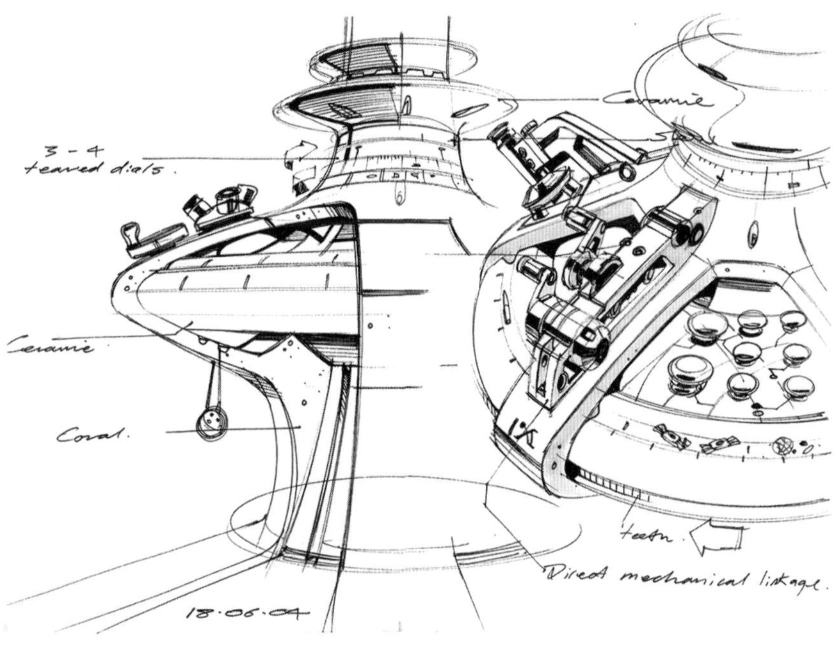

Sketch for Doctor Who's TARDIS, Dan Walker

Mike Jones' atmospheric computer rendering of the hotel from the film *3:10 to Yuma*

using pencils, pens, paints, graphics tablets, anything and everything. Drawing is important because this is how ideas and designs are communicated, both in the early development stages (drawing to design; thinking with your pencil) and in presentation visuals (illustrations of your design as it would be seen on screen – communicating your concept before it is built) and technical drawing (your instructions to construction workers so that your set design can be built). You will also develop skills in scale model making, both physically and digitally (creating 'fly through' animations of your set design). In some courses you may also try prop making and even a little set construction.

There will also be a contextual studies aspect to the course, where your skills in visual and textual research will be tested and developed. You will need to learn the relevant technical vocabulary and consider what the technical requirements are for, for example, a certain TV show's target audience. You will look at films and TV shows with a designer's head, looking at the impact of colour, atmosphere, etc.

Tutors are often practising designers and art directors and as such have strong industry links, affording exciting work experience opportunities for their students on numerous productions.

CAREER OPPORTUNITIES

Few people outside of the film and TV industries are aware of the role of a production designer (though perhaps some of you may have caught a glimpse of them on the DVD extras available these days). The art department, of which the production designer is head, is mostly concerned with 'the background', i.e. what is behind the actors' heads. If they do their job well, most people don't notice; if they do their job badly, people do notice.

Designers and art directors within the film and television industries use their skills to create a huge variety of environments for a vast array of exciting productions; everything from your favourite soaps to blockbuster films. They produce imaginative set designs through their skills with concept drawings, storyboards, technical drawings and scale models. They consider props, set dressing and location modifications. They work with teams of craftspeople to bring the imaginary to life. They make what you see on your television or cinema screen 'look right'.

If you aren't a nine-to-five sort of person you may find the scope for variety within this area very attractive indeed. Many people move from education to the industry through contacts, internships and placements and the networking you do on the way is very important.

Mary Barton- Murder Scene Page 3

Carson heres click and stops before alley way.

Carson shot dead, his body falls in slow motion.

C.
Carson continues to fall.

D.
CUT
Carson hits the floor and the stagnant water splashes all over him.
End Scene.

Final-year student Dan Blackmore's storyboard

WHAT WILL YOU BE TAUGHT?

> *'I really feel strongly about this; that when the students graduate they have covered all the basic needs. Apart from having designer heads on their shoulders they can draw what they need to, both technical and freehand. They know how to pass information on to construction, directors and crew. I need people who can draw, make models and, very importantly, know how to read drawings and plans.'*
>
> Charmian Adams – Supervising Art Director (*Downton Abbey, My Week with Marilyn, Nowhere Boy*)

THE FIRST YEAR

You will get a broad grounding in many of the required skills during the first year relating to researching and designing in 2D and 3D. You will work on real-situation-based design projects in dedicated design studios, where an 'art department' environment is created. You will benefit from visiting industry practitioners and guest lecturers, keeping you in touch with contemporary practice. In supporting subjects you will begin to understand the historical and cultural aspects of your subject, by film-viewing, listening to lectures and discussions in seminars.

Set design plan for a 2 min. 41 sec. scene by film and TV student Steven Hollywood

THE SECOND YEAR

Your second-year projects will be more varied, conceptual and challenging. You may experience team projects and fast-paced shorter projects. There will be more knowledge and skills to be gained in areas such as computer-aided design (CAD), which will allow you to enhance your work and develop a personal style. This would also be the time to arrange a summer work placement. Remember, many of the staff who teach on these courses are or were practitioners and will have great contacts in the industry. In some courses you may be able to take part in industry work experience, for short periods, at any time.

WHAT IS THE DESIGN PROCESS LIKE?

The process is not always linear. It is always a good idea, no matter which discipline of design you are in, to generate lots of ideas and push them around. Models play a large part in this.

'The Grand Conservatory' for the set of Jean Cocteau's *La Belle et la Bête* by film and TV design student Laura Ellen Ward

The industry standard is to produce these at a scale of 1:50 (TV) and at a range of scales for film. You will also get used to using a studio plan. Producing visuals in various media, both traditional and computer-aided, will communicate your design as it would be seen on screen. Here's a good, first-hand report of the issues from recent Design for Film & TV graduate Mike Jones:

> 'I usually begin designing by sketching out floor plans of sets with rough dimensions. When I have the seed of an idea I start to design in 3D space using computer modelling programs such as Google Sketchup. I prefer this computer method to sketching out ideas by hand as with a 3D space I can instantly see the size of the design in relation to a person, as well as being able to position the camera anywhere.
>
> I like to look at how the set can be shot. I can instantly see problem areas, such as where the camera may shoot off the set, and I feel it is a better way of communicating a design when you can show it from every angle. I always try to picture which shots will be most important and most interesting to the viewer and design the set accordingly, where possible. This helps ensure that the set has a nice amount of depth – the camera captures interest in the foreground right through to the background of the shot.
>
> It always helps to storyboard a scene first if possible, so you know what shots will be needed and don't waste time designing details that will not be seen on camera. That being said, set designs should always be thorough, to give the director as much freedom as possible to shoot the way he/she wants.
>
> During my final-year project I took a week out to do some work experience at Pinewood Studios in the art department on the new James Bond film, *Skyfall*. Seeing the sets first-hand made me truly appreciate the level of detail that is required to make the set come to life, and I tried to replicate this in my own project.'

The design process involves the gathering of visual inspiration, producing mood boards and using these, as well as historical or contextual research (be it a period drama or a late-night show for teens), to sell your idea. You will also get used to presenting your work and having it criticised by your peers and tutors. And praised! Mike Jones again describes this:

Stagecoach model from *3:10 to Yuma*, final year project by Mike Jones

> 'Researching the period for my project, the western film *3:10 to Yuma*, was particularly enjoyable. I watched countless westerns, from the black-and-white era to present day, and learnt a lot from these films before I had even begun the project. One of my favourite aspects of digital film and television (DFTV)/production design is that the subject/genre of the script could be absolutely anything – allowing me to delve into an interesting period in history which previously I may have known nothing about – and emerge with in-depth knowledge of architecture, technology, clothing and more. The challenge I most enjoyed was designing and drawing up the armoured stagecoach.
>
> I looked at films, such as John Wayne's *War Wagon* for inspiration with regard to textures and materials. I only watched the film *3:10 to Yuma* once before I began my project, I did not want to keep referring to it as I was conscious it may affect my design. I always enjoy designing from a script and making a story into reality – from the concept stage right through to the drafting, which will enable my designs to be constructed. Every DFTV project I complete leaves me with new skills and this one was no different – I developed my skills with computer rendering software such as Artlantis and built on my existing Photoshop skills.'

If there's a narrative, then a storyboard will be required and technical drawings so the whole project could be costed and made. Presentations are your opportunity to pitch your ideas and 'sell' them. These skills are important in real life and you will be exposed to the kind of critical appraisal you will experience in the real world.

WHAT WILL THE FINAL YEAR BE LIKE?

The final year is geared towards the individual to highlight their interests and strengths. You will need to work out where you're heading and aim your project in that direction. Your projects need to be agreed with staff in case you set off doing too much. You also need to get your head around who the characters are in your films and programmes and understand the story and performance. It is an opportunity to work out what the scope of the project needs to be in order for you to demonstrate the skills you want to be known for and to achieve a believable outcome. And get a degree.

The scope for variety is enormous. You can create a world that doesn't yet exist, or a fiction that is very close to home, such as *Coronation Street*. For drama projects the connection you make with the characters and environments, making them come alive in the viewers' minds, is very important. On the other hand, you may prefer to design a production without a narrative, such as an award show or a light entertainment programme such as *The X Factor*.

You must also ensure that, beyond the conceptual, the set design works on a practical level too:

- how will the desired shots be achieved?
- what will it be made from?
- what props will be necessary?
- where might we be able to hire them from?
- is there enough room for cameras?
- is the distance between the set and the scenic cloth or green screen correct?

Designing for film and TV can be complex and involve a great deal of problem solving as well as vision and creativity, which is why it can be so challenging and ultimately rewarding. By the end of the final year you will be equipped with all the necessary skills to begin a satisfying and exciting career (just add experience).

Final year film and TV student Hollie Cleaver produced this stunning visual to portray the Lazarus machine in her adaptation of *Casper the Friendly Ghost*

Final year student Dan Blackmore demonstrates a moody story line

Preparing your portfolio

With your realistic projects and the links with the industry that you've forged during your time as a student, your portfolio should demonstrate a depth to your work and your design process. In this day and age you can have all manner of technology to help communicate what you've done, even as far as videos on your laptop or phone. However, the traditional skills required in the industry are very much at the forefront of what potential employers are looking for from you. Bringing a project in on time and on budget will always impress.

Remember to include your research folders and notebooks in your portfolio. These demonstrate your thinking, not just how well you can draw or make.

Conclusion

Many of those attracted to a career in the theatre, film and television industry may have been excited and intrigued by a production that they have seen or become enthused by one they have been involved in making themselves. This often starts a love affair with the power and possibilities of theatre and film making. It is commonly believed that working as a theatre or film and TV designer is highly creative, immensely exciting, supports the designing of spectacle, introduces you to the camaraderie of theatre and showbiz and is even glamorous. All of this can be true, but it is also so much more!

A huge variety of people with a wide variety of backgrounds become set designers, costume designers, production designers, art directors and fill related art and design roles in the performance industries, so be encouraged by this. You don't have to necessarily 'fit the mould', be a brilliant artist or have a certain background. As long as you have potential, creativity, bags of enthusiasm, determination, a strong work ethic, dedication and, importantly, a passion for the subject, you can succeed.

Further resources

Books

Davidson Cragoe, C., *How to Read Buildings: A Crash Course in Architecture*, Herbert Press Ltd (2008)

This does what it says – a good introduction to understanding architectural styles – necessary for the designer.

Ettedgui, P., *Production Design and Art Direction*, RotoVision (1999)

An inspirational book including interviews with successful production designers.

Hart, J., *The Art of the Storyboard: Storyboarding for Film, TV, and Animation*, Focal Press (1999)

This is a good introduction to this art department skill.

Huaixiang, T., *Character Costume Figure Drawing*, Focal Press (2010)

An excellent book for any design student wanting to be able to draw the human figure.

Millerson, G. and Owens, J., *Television Production*, Focal Press (2009)
A good, all-round reference book.

Woodbridge, P., *Designer Drafting for the Entertainment World*, Focal Press (2000)
This book explains this particular type of drawing and communicating – a core skill.

Websites

www.guisemagazine.com
New online magazine for all things costume-related.

These are all theatre companies' websites:
www.birminghamopera.org.uk
www.blindsummit.com
www.complicite.org
www.dreamthinkspeak.com
www.dv8.co.uk
www.forcedentertainment.com
www.kneehigh.co.uk
www.punchdrunk.org.uk
www.qtine.com
www.wildworks.biz

Organisations

International Organisation of Scenographers, Theatre Architects and Technicians
www.oistat.org

The Linbury Prize For Stage Design
www.linburytrust.org.uk
www.nationaltheatre.org.uk/linburyprize

The RSC Assistant Designer Scheme
www.rsc.org.uk

The Society of British Theatre Designers
www.theatredesign.org.uk

Project story

Designer:
Laura Ellen Ward, design for film and television student

Project title:
BEAST

Tell us about the project

One of two major projects completed in my final year, BEAST was based on the original screenplay *La Belle et la Bête* by Jean Cocteau, focusing specifically on the film set for the 'Great Hall'. Using the script as a basis for key scenes and sequences that would need to be communicated for the feature film, I created an alternative design for the production. I designed 'The Grand Conservatory' and exterior surroundings. Many sources of inspiration contributed to the final concept. The conservatory structure was based on original Victorian architecture, which was researched accurately and thoroughly throughout the ten-week project.

As you are very much in control of your own learning and development at university, the chosen architecture for 'The Grand Conservatory' was considered carefully to ensure that I could be creative with the design, incorporating elements of fantasy, but could also strengthen my knowledge of a specific architectural period. The university provided such fantastic research facilities, which if used to their full potential can really broaden your opportunities.

I visualised 'The Grand Conservatory' to be a giant glass container, hidden deep in the forest and filled with overgrown tropical plants, in thriving condition, despite a lack of care and attention in this somewhat neglected part of the Beast's home. Broken glass panes in the structure enabled the conservatory to merge somewhat with the surrounding exterior, spilling out onto the overgrown surroundings, which also made up part of the design. The exterior was designed to look as green and lush as is possible within the boundaries of studio filming, in an attempt to look damp and moody but alive. The colour palette used worked to enhance this atmosphere, though splashes of vibrant colour were introduced to steer from the somewhat overused dark mysterious aesthetics of the film.

The design was communicated through widescreen visualisations, storyboarding, 1:50 scale model making, digital modelling, prop listing, photography and film, with all these elements allowing for significant detail to be considered. As well as these requirements, technical drawings support the build and technical aspects of a production design.

The project BEAST was a result of everything I had learnt by being part of a design course, but it was also

Visual by Laura Ellen Ward

A close-up of 'The Grand Conservatory' model, by Laura Ellen Ward

the start of new pathways I was considering after student life. The project was utilised as a core part of my portfolio that could be showcased to employers in order to gain work in the industry. As well as creating visually pleasing work for university, it is important constantly to consider effective ways in which you can demonstrate your skills and understanding of design.

In your final year, there will be a freedom to steer your projects towards your interests and strengths, but it is vital that you continue to push your skills to high levels. Never become complacent with what you have achieved. Obviously there will be methods that work well for you, but there will always be areas that you can develop further. Take risks, immerse yourself in inspiration and surround yourself with creative people. Peer learning, advice and discussion are invaluable.

What processes and skills were most relevant to this project?

Of the 40 students on our course at Nottingham Trent University, every individual had varying techniques in creating a production design. For me, designing work for the screen meant combining a great number of skills and processes in order to achieve the final output, which can be visually very satisfying. You will hopefully see a combination of both hand-drawn and technical design skills merge together to produce dynamic and exciting creations. Many students work through technical programs such as Google Sketchup, Adobe Photoshop and Vectorworks. Every student should be confident in using such programs at a basic level, but it is important to realise that although introductory tutorials are integrated into many courses, the level of skills that I and many others have learnt come from hours of individual exploration and development to gain a deeper understanding of what the programs can really offer.

Freehand drawing is still a major part of designing, however, and I relied heavily on my abilities to draw initial ideas from scratch in this way using just ink and paper. Multiple computer programs can be used to colour, enhance and add effect to a final sketch, which I found to be an efficient way of delivering visuals with high impact. You really have to explore all opportunities to find what works best for your personal style. At university you have to be highly motivated to succeed, with independent learning being a crucial factor in becoming a confident and independent creative individual.

What challenges did you face?

Discussing your work and processes within tutorials is an essential part of the university experience. It can be challenging at times to effectively express a design you are trying to communicate, however, talking to tutors in the industry is of the utmost importance in order to strengthen a design and broaden your outlook. Criticism is key, and will essentially drive you to develop elements of your designs that are far more considered and valuable.

Collaboration and communication with industry contacts and companies is very important in your final year. For me, this happened both through work experience with the BBC, but also through researching aspects of my projects and making contact with companies that are involved in production design, including prop houses, scenery specialists and lighting designers. They may not always want to talk to you as a student, and you often have to push for answers, knowledge and opportunities in a busy and competitive industry.

Project story

Designer:
Zosia Stella-Sawicka, theatre design student

Project title:
'The Resistible Rise of Arturo Ui'

During your Theatre Design course, what led you to specialise in puppetry?

I have always wanted to design as well as make, but didn't feel that Set or Costume Design alone were for me. In my first year at university our final project allowed us to go in any direction we wished, and my group worked together to create a shadow puppetry performance. It involved a lot of 'play' and experimentation with materials and it was the excitement of creating a sense of character and atmosphere from simple objects that really inspired me. From here on I jumped at the chance to do projects based around puppetry.

I have a passion for creating character, but what led me to focus my work within puppetry specifically is the sense of freedom that it allows. Puppets can be any size, shape or material and can be manipulated in a huge number of ways. In order to create the right sense of character, these elements must be carefully considered, yet, at the same time, the more you stray from the rules, the more exciting puppetry can be!

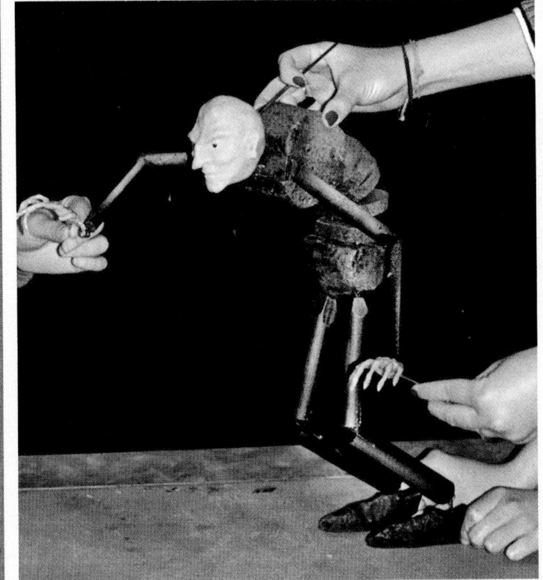

Arturo Ui puppet

Tell us about the project

The project was in association with the Blind Summit Theatre, and I was mentored by Nick Barnes who provided feedback as well as workshops on puppet-making techniques and puppet manipulation. The brief was to create a puppetry design for the play *The Resistible Rise of Arturo Ui* by Bertolt Brecht. This included a fully functioning puppet, or set of puppets, and a performance context for a studio space. The brief required technical drawings, a storyboard and individual costume and puppet designs. The play is a highly satirical allegory of Adolf Hitler and his rise to power. Each of the characters in the play has direct parallels with historical figures in real life. In this way, the play reveals the simplicity of how such a monstrous criminal can rise to power. My design response focused on the transformation of Arturo Ui (representing Hitler) and aimed to capture the grotesque and monstrous nature of his character. The main inspiration for my character development was taken from the movement and aesthetics of creatures from the insect world. I looked to combine these characteristics with those of a human in order to convey the true essence of Arturo Ui. In this way I hoped to explore the full potential of puppetry, by expressing an idea or character that an actor could not.

What processes and skills were most relevant to this project?

The design process was an important part of the project – undertaking extensive research into the characters, historical context and the narrative of the play as a whole. An important aspect of any puppetry project is experimentation with materials. To make my own Arturo Ui puppet I used a combination of wooden dowel, Plastarzote and Super Sculpey. I found that I was continuously learning about the different materials and relevant tools throughout the project, discovering what worked and what didn't as I went along.

By the end of the project I had a much clearer understanding of how to use the different machinery and tools in the workshop, as well as the technical construction of a puppet – for instance, the different types of joints that can be used within a puppet and how they can be constructed. It is vital to get the construction of the joints right, because the way that a puppet moves when manipulated is key to the creation of its character and personality.

We had to produce technical drawings for the puppets, which consisted of accurate scale drawings providing detailed information regarding measurements and the materials used. Storyboards were also one of the project's requirements, which are important in order to convey the narrative of the performance, combined with a sense of atmosphere and character.

What challenges did you face?

The greatest challenge that I faced during this project was with the construction of the puppet. I had a clear image in my mind of how I wanted the puppet to move, according to the character that I had designed. The most important aspect of the design was the spine of the puppet. I wanted it to be elongated and articulated in such a way that the puppet would at first appear hunched over, with its posture reminiscent of the curve of a beetle's shell. The puppet would also, however, have the power to transform, by unfurling its spinal cord and extending upwards to take on a powerful and intimidating stance. The process of getting this construction just right involved a lot of experimentation with materials and a lot of trial and error! I eventually succeeded in bringing my design to life through the use of a thick wire spine, which was articulated with several Plastarzote discs attached along it.

Another challenge that I came across was devising the performance and considering how the puppet worked in relation to the rest of the play. I found that one of the most successful ways to devise the performance was during the improvised manipulation sessions with the puppet itself. It is this physical 'play' that helps to develop the sense of character and allows the puppeteers to see how the puppet can move and exactly what it is capable of.

The process of creating a storyboard is another very helpful technique, allowing you to quickly sketch out different scenarios, looking at the puppet in relation to the scale of its surroundings. Storyboards are also a good way to experiment with mood and atmosphere, since they enable you to play around with things such as lighting effects, simply through the use of shading and colour.

Storyboard sketch

What were the highlights of the project?

The highlight of this project was seeing the whole design process come together for the final performance. When you're focusing on the construction of one particular puppet it is easy to forget the impact of the performance as a whole, and all of the other elements that bring the piece together. An important element within my own performance was the use of sound. I came across the work of Graeme Revell and his album *The Insect Musicians* (1986), which consists of pieces of music composed entirely from recorded insect sounds. Using the program Garage Band I edited two of the tracks ('Phobia' and 'Sleeping Sickness') and combined them to form one piece of music. This created the perfect soundtrack for my performance, being reminiscent of the insect world while also remaining abstract, mysterious and intense.

Being a keen maker at heart, I particularly favoured the practical elements of the project. Having constructed the main form of the puppet I then had to render it, using spray paints to achieve the metallic speckled effect reminiscent of a beetle's shell. Sculpting the face of the puppet was one of the most enjoyable moments for me, for which I used Super Sculpey moulded over a tin foil core (to keep the weight down). I began moulding the Sculpey with my hands, and then continued with the reverse end of a paintbrush to sculpt the finer details. I felt that this was one of the strongest aspects of the puppet, which really brought the grotesque and monstrous character to life.

In retrospect this project was one that really helped me to realise the strengths of my own design process. I found that my most successful work came about from three-dimensional sketch modelling, as opposed to 2D sketches. As you begin to play around with different materials you can get a sense of how they work and move, and how they can be best used in relation to the structure of the puppet. This allows an idea to be quickly brought to life, enabling you to judge its success, which in turn helps the design to develop further.

There is more information on all of my previous work on my website, **cargocollective.com/zosiasawicka**.

Career profile

Name:
Dan Walker

Current job:
Concept Artist

Describe what you do

Every concept artist in film works in a self-employed capacity. Some design jobs, for example in advertising, can take just a couple of days, whereas other large-budget features can contract you for many months. *The Golden Compass*, which I embarked on in 2005, employed me for almost a year. This was due to the sheer amount of design involved in creating a convincing parallel universe. I was responsible for all the vehicles and airships. The generation of 3D computer models played an important part in their design development. After the initial concept sketch and Photoshop rendering phase, all the concepts were built and refined in CAD. The models were then passed on to special and visual effects for the physical (film sets) and virtual (computer animation) builds. Finally the data was used by the toy manufacturers for all the tie-in merchandise.

I've worked on many feature films and television shows over the last few years, each being a new experience from the last. You not only have to be capable of turning your hand to designing for different periods or genres, but you also have to accommodate the particular style of the production

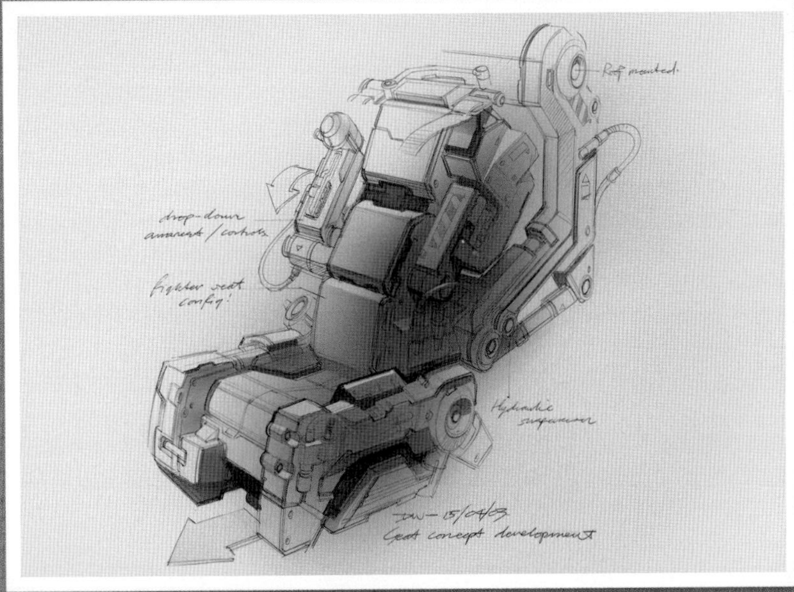

Concept seat sketch

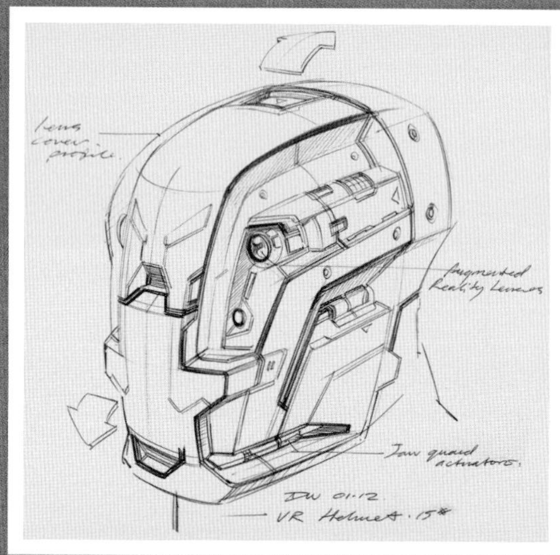

Concept virtual reality helmet sketch

designer. Much of the time you are a conduit for their ideas, realising their vision. This can result in many design iterations before the right one is generated.

Tell us about your career so far

Having always had a passion for film and a desire to work within the industry, I decided in 2003 to broaden my career in product design and become a freelance concept artist. My first employment in this capacity was on Chris Nolan's re-imagining of the Batman franchise for Warner Brothers. Usually working within an art department and under the creative guidance of the production designer, a concept artist is responsible for generating design visuals for sets, props, vehicles and costumes. The role is best described as a combination of art director and production illustrator.

On *Batman Begins* my main job was to design the interior of the Tumbler vehicle. This was an ideal scenario for me as, having a great deal of experience working in the automotive industry, my skills were easily transferable. In film I apply almost exactly the same working process as I do in product design, starting with the research and sketch phase, working up colour visuals and finally generating a full 3D data set on computer. There are some differences, most of them to my advantage! Unlike designing for the real world, most of the time it only has to look convincing in shot. The Batmobile interior was actually a set, split into three in order to film from different angles and scaled up 25 per cent in order to package the complicated working parts, which, like the outside, were built by the SFX team to actually work.

Is there a particular project you'd like to tell us about?

Doctor Who has become regarded as a British institution, so creating something new and exciting for a new generation, while respecting the show's long heritage and mythology, was a particularly difficult balancing act.

Under the guidance of the show's production designer, Ed Thomas, I was charged with the task of concept designing two of the show's mainstay trademarks – the sonic screwdriver and key elements of the TARDIS interior. Unlike larger features, television is very different with regard to scale, budget and lead-times (how quickly it needs to be finished!). Hence you have to be economic with your proposals. With this in mind, the TARDIS console, with its six organic spoke layout, was designed as an easy-to-construct 'chassis'. This provided an ideal canvas for the set decorators to dress it with an assortment of eclectic, bought-in props. The TARDIS is a very good example of the collaborative effort that goes into creating a set, with the art and construction department and set decorators all playing an important role in bringing the concept to life. Of all the jobs I've worked on to date in my career, *Doctor Who* has been the one I'm most proud to have been associated with.

What advice do you have for students considering a similar career?

Concept art is a highly rewarding job, especially when you finally see your creations come to fruition on the big and small screen.

Timing is always crucial when freelancing from one job to another. The end of my contract on *Batman* led perfectly into another re-invention of an established cultural icon – *Doctor Who* at BBC Wales. This was an entirely different working experience from working at Shepperton Studios on *Batman*. I went from being in an art department of 50 plus people to one of just 5.

Having design skills that are transferable between industries has been an important part of my career so far, as employment in film and television can be unpredictable. Having other areas to work in, such as product, automotive and games design, has proved very important, providing a steady rate of employment.

Project story

Designer:
Katie Grinsell, theatre design student

Project title:
A live design competition working with Birmingham Opera Company's Associate Artistic Director

Our brief was to design the set and costumes for *The Rise And Fall Of The City Of Mahagonny* by Bertolt Brecht and Kurt Weill. The found space or site-specific design was to be set in a vast WWII aircraft hangar on a disused airfield. The director showed us around the site. The hangar was nearly 100 metres long and had huge doors at each end to roll the aircraft out. The director emphasised that it was as important to respond imaginatively to the space as it was to the opera itself. Due to the dynamics of the vast space, we all relied heavily on the storyboard to 'map out' the moments in the opera, and had to build a 1:100 model to respond to the epic qualities of the opera's narrative.

The director had several individual meetings with the student group, as well as introducing us to the particular approach to opera that Birmingham Opera is known for. The project culminated in a competition and I won the prize of working with Birmingham Opera Company after graduation.

Stage sketch by Katie Grinsell, *The Rise and Fall of the City of Mahagonny* (TRAFOTCOM) by Bertolt Brecht and Kurt Weill

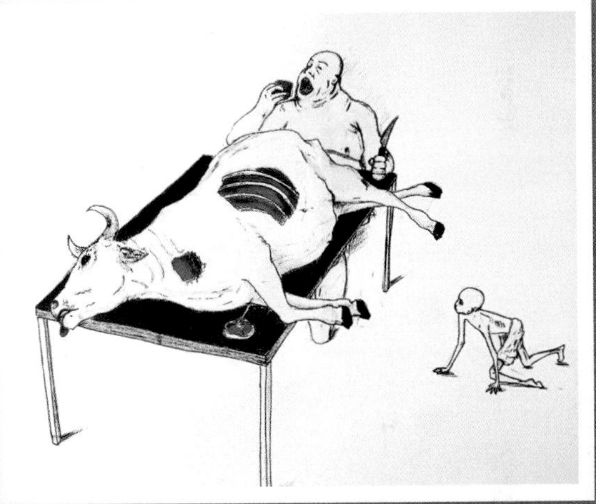

Scene sketch by Katie Grinsell, *TRAFOTCOM* by Bertolt Brecht and Kurt Weill

Close-up of model by Katie Grinsell

Project story

Designer:
Dorrie Scott, theatre design student

Project title:
Nottingham Playhouse Prize: *This Story of Yours* by John Hopkins

Act 3 'Johnson at Baxter's side', storyboard for speculative design by Dorrie Scott

Tell us about the project

This was for a competition that is run annually by a local theatre, the Nottingham Playhouse. Once the director had selected his chosen play, *This Story of Yours* by John Hopkins, you could then begin developing a speculative design for the script.

What processes and skills were most relevant to this project?

Communication was the most valuable skill to have developed during this project. Working with the director mirrored the director/designer relationship that is found in industry and allowed us to practise and test our abilities in communicating design ideas in 2D, 3D and verbal communication during the development and final design stages.

What challenges did you face?

Honestly, upon reading the script for the first time, the themes and structure of the 1960s police drama did not interest me and I really struggled to make a start with the designs. However, following further research, involvement and investigation into the script I identified areas of interest. An alternative approach to the script then began to emerge to form the underlying basis of my design. I became much more confident with these identified themes and went on to develop a strong design that I was really happy with. My storyboards were further developed by taking inspiration from photographers Henri Cartier-Bresson, Euan Duff and Weegee.

Presenting the final design to the director and members of the production team was a huge challenge. While developing the design, every decision needed a clear and rationalised reasoning behind it, knowing I might need to take these designs further into a confident and concise presentation at the end of the term. This drove me to work thoroughly for the whole duration of the project. The final presentation was essentially a time for the director to gauge who was capable of working as part of a team with him in the future. It was a very nervous time for me, as the competition amongst fellow students was very high. We were working towards the same script, yet our individual designs and approaches needed to be distinctly different from one another, and memorable.

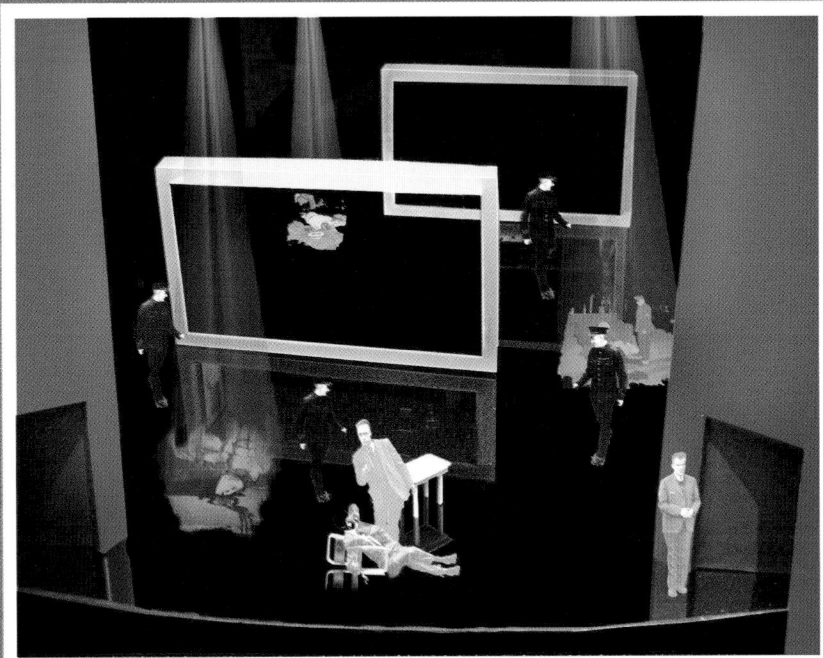

Act 3 'Police enter', storyboard for speculative design by Dorrie Scott

What were the highlights of the project?

An initial highlight for me was realising I had reached my final design concept. I remember the moment when the research, discussions with the director, development and process all suddenly fell into place and the design concept became clear.

Working so closely with the project, I became very familiar with the ideas and concept. I often doubted the strength of the design, so to present the final design and then be selected as the winner of the competition was a huge highlight of the whole project! The highlights keep on coming, as I now begin to work as a designer with the director at the Nottingham Playhouse.

Winning this prize has given me the perfect step into the industry as a new graduate in a competitive world. Already the lessons I have learnt are vital to developing as a professional designer. Working with such an established and talented team of theatre makers is nerve-racking but extremely exciting and rewarding.

Three-dimensional design

By Bill Schaaf, Buckinghamshire New University

'Challenges are what make your project exciting. If you are not challenged by your work then you are not really learning.'

Nell Bennett, industrial design student

Visual model of a digital radio by Product Design student Rachel Elizabeth Ward

Studying 3D Design can prepare you to design anything, from a pencil sharpener to a jumbo jet. The increase in new materials and manufacturing techniques, alongside the increased complexity of our designed environment (with digital technologies gaining a wider integration with the domestic environment), means that the need for a wide range of design skills is ever present.

The UK structure of education provides a focused curriculum in which virtually all of the modules of study are directly linked to the design discipline. This provides the distinct advantage of enabling a deep level of skill and understanding to be developed in a subject area. This strength is mirrored in the UK design industry, which enjoys a worldwide reputation.

The UK is well known for its design industry, which operates in many different guises, as you will see. There are also many independent designers in the industry who collaborate with others as and when necessary, depending on the nature of their projects. While operating as individuals, they do not hesitate to join forces with another practitioner or small company to investigate a potential project or produce a limited run of a particular object. This freedom of working is a function of not only a culture but of a university education system, with emphasis on both building effective teams and collaborations (often part of the second year of a course) as well as independence and self-direction in the final year of study.

Today's three-dimensional design industry

Graduates of 3D Design courses have a range of subsequent employment opportunities that could be described in three broad categories. The range of skills required and the working practices in these categories vary widely:

Self-employed designer-maker:

You will need to have a good understanding of business opportunities, a wide range of abilities and good networking and communication skills. In addition, designer-makers in particular may have a deeper understanding about batch production methods, rapid-prototyping or other means of bringing products to market in a fast, inexpensive manner.

Design consultant:

As a self-employed freelance design consultant you will be working with other companies and their teams. If you work for a design consultancy you will be on a fixed salary and working with a range of clients. Consultancies range in size from just a few people up to over a hundred. They provide services to other companies that either do not have an in-house design team or that do have a design team but require occasional specialist skills. Designers in such organisations tend to like the challenge of undertaking a wide range of products and the variety of projects they may be working on.

Designer in industry:

Working in a company as part of its design team tends to be more prevalent in larger manufacturing companies that have separate departments to undertake the different specialisms necessary for mass production, from sales and marketing to engineering. Those designers who like specialising in one area and covering similar markets may find this employment more suitable.

More information on this is available in Chapter 15, 'So, where is this going to take you?'.

Many people move around these categories as they develop areas of expertise and want to develop their careers. The UK design industry is thriving, working not only for UK manufacturers but attracting design business from around the world. International businesses also recruit designers from the UK – some of the trendiest Italian furniture and some of the most stylish German cars have been designed by UK designers.

What courses are there?

Due to the wide range of product areas and the range of manufacturing processes, from one-off to mass-produced, there is a very wide range of courses available. They can therefore be quite different but can be loosely viewed in three categories:

Broad-based course:

These courses tend to have titles such as 3D Design, and will introduce you to the design approaches and sensibilities of working in three-dimensions. Assignments will cover a range of different disciplines so that you gain experience designing everything from small objects to large spaces and mass-produced products to one-off items. In your final year of study, you will identify a major project topic and build a portfolio appropriate to your intended profession.

Specialist pathway:

Some courses may have broad titles, such as 3D Design, but then offer specialist pathways, which enable you to begin your studies with a broad investigation of the options and then choose from a fixed selection of pathways (normally in the second or third year of study) that focus on a specific specialist issue. Such courses frequently enable students to graduate with a certificate identifying this specialism in the degree title (e.g. BA (Hons) 3D Design: Product Design).

Subject-specific course:

There are many courses that offer a degree aimed at a definite discipline, such as Industrial/Product Design, Furniture Design or Transport Design. These programmes cater for students who have already identified their particular interest within three-dimensional design. Some of these degrees can be focused further upon particular aspects of the discipline, such as Sport Equipment Design, aimed at product designers who wish to focus on the world of sport, or Furniture Design and Craftsmanship, aimed at producing designer-makers who can also physically make their designs to a high standard.

A good programme will provide a student with both a breadth and depth of experience and understanding of their field of study. You should evaluate your own understanding of the territory and whether or not you wish to specialise early on in your studies. You should also get involved with industry relationships, through your course or on your own initiative. Broadly speaking, despite the large variety of 3D courses, there are widespread similarities in content and teaching style. Read on to find out more detail about the following courses – Industrial/Product Design, Furniture Design, Automotive Design, Boat Design, Lighting Design and Sports Equipment Design.

Second-year students Stuart Brown, Alexandra Chin and Laura Sutton manufacturing prototype furniture

The first year

This will develop basic skills in your area of study. Students come from a wide variety of backgrounds and educational experiences, so the first year ensures you are introduced to core skills and processes that the later years are reliant upon. Common skills you will be introduced to are creativity and design methods, drawing and making and research and communication. Projects tend to be not too complex and the teaching styles are more prescriptive than later years will be, as you are guided through the process and told what steps to follow. Some students still find the leap from school or a foundation course to be difficult, in that the 'right answers' are rarely provided. Students need to learn at an early stage to develop their own views and recognise that, in design, there is rarely only one good answer.

Computer-aided design (CAD) is integrated into most courses and you will probably be introduced to it in your first year. It is a central part of many design processes, from portraying initial ideas to producing dimensioned drawings of manufactured parts. It also ties in with rapid prototyping, the method by which machines can take a digital 3D model of an object and produce a physical example. A wide variety of software is used and you will be introduced to the specific ones that your discipline uses.

The second year

This will introduce more complex issues. It is intended to develop your ability to draw connections between different areas of study. Some of the projects are team-based and/or are centred around industry or other external projects. You may find yourself working with a team of peers on a 'live brief', with professionals from an outside organisation or company who will have determined the brief and who will review your project outcomes at the end. You will be using your skills in design, research and development and improving your skills in communication. You will learn to refine and present your work to this audience.

The end of the second year is the most common time when courses may offer work-placement opportunities. The summer between your second and third year is a great period to experience the professional environment. You will have developed enough skills at this stage to understand the field and have something to offer a company. Some institutions offer four-year 'sandwich' courses, with a year of employment normally sandwiched between the second year and final year.

The final year

This can define your own personal path within your discipline. You will demonstrate the ability to position your own work in relationship to others and the ability to present your research and insights in reasoned and critical arguments. You should also be working towards your degree show and possibly one of the national events, such as New Designers or FreeRange.

The emphasis is on honing your skills to a high level. You need to develop your own approaches to the development process, be able to synthesise your research and demonstrate your ability to propose and present creative solutions in a coherent manner. In order to do this you should have regular contact with your tutors and your peers and create time for critical reflection on your work. Many students also try to interact with industry professionals, using them for research, running ideas past them, trying to get manufacturers interested in producing their work and also scoping out which companies they might like to work for in the future.

New Designers 2012, The Business Design Centre, London

Your skills and thinking in the final year are generally developed through two important pieces of work, a major independent practical project and a significant written thesis or business proposal. These projects involve you in selecting the topics, defining the briefs and developing the entire project through advice and consultation with tutors (rather than operating on their instructions). The project need not be related to the written component of the final year, which could be research-based or a piece of critical writing. Refer to Chapter 13, 'Design history, culture and context' to find out how the written component enables you to develop a sustained body of research that critically analyses a chosen topic.

TIP

The three main 3D Design courses are being outlined next. Clearly there are similarities between the courses. It is recommended that you read all three to get the widest view and gain a greater understanding of the differences.

The first, Industrial/Product Design, contains all the information any 3D Design student will need. The following sections, covering Furniture Design and Automotive Design, offer further detail on how these differ from Industrial/Product Design.

COURSE TITLE: **INDUSTRIAL/PRODUCT DESIGN**

COURSE DESCRIPTION

Industrial Design and Product Design are interchangeable terms. There is a wide range of courses in both, and they overlap in what they all cover. There are many similarities between projects that students on Industrial and Product Design programmes study and what professional designers actually do on a daily basis.

Many Bachelor of Arts (BA) Industrial/Product Design courses run in art and design schools within universities and lean towards the artistic side. Bachelor

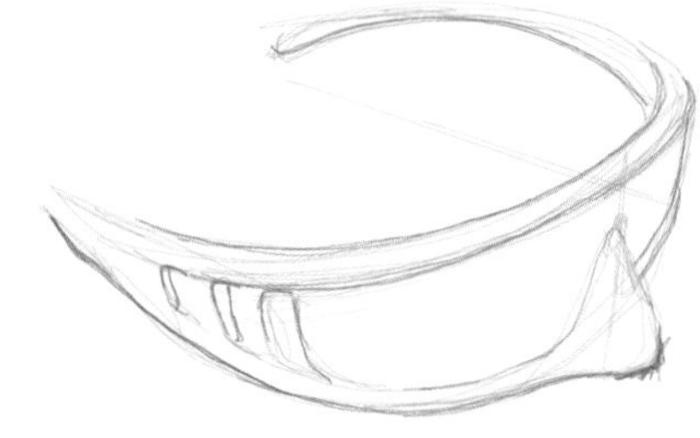

Concept sketch of sunglasses

of Science (BSc) Industrial/Product Design courses tend to involve more technical training, developing in students an understanding of industrial materials and processes, together with a certain level of mathematical and structural engineering ability as well as aesthetic appreciation. Many of these newer BSc Product Design courses have emerged in more recent times, often in the older universities.

People who study Industrial/Product Design tend to be interested in improving quality of life through the development of new products and services. Creativity, a passion for learning and a desire to bring new things into existence drive students into this field of study. To develop your skills and experience in the design process you will be addressing aspects such as technical function, **aesthetics**, **ergonomics**, costing, production processes, **branding** and marketing.

Ultimately, designers define the interfaces between people and their environment; these skills allow you to balance the constraints of people's needs and wants to those of production, manufacture and consumption. Generally, Industrial/Product Design is about bringing together various quite different strands of research with creative ways of looking that allow you to be innovative.

Aesthetics: the study of form, proportion and beauty. See the book *Beautiful Thing* by Robert Clay.

Ergonomics: the study of people's physical and psychological interaction with objects and systems: how we see, understand and use things. See the book *Bodyspace* by Stephen Pheasant and Christine M. Haslegrave.

Branding: the study of a company's or product's identity: how this can be designed in relation to the market and the consumer. See the book *Wally Olins. On B®and* by Wally Olins.

For example, a project on the use of future technology in the home would need research into people's lives, desires and needs (read about the IDEO Research Method Cards in Chapter 10, 'Being creative and innovative'), it would need to look at recent technology developments in the communications industry (hardware and software) and it would need to observe the latest practices and thinking in sustainable product design and sustainable consumption (as mentioned in Chapter 14, 'Future directions').

Practise obviously helps you understand this process, therefore courses often consist of a continual programme of projects covering a wide variety of product areas and experiences. This is the kind of education that prepares you for any experience that might come along in the future.

CAREER OPPORTUNITIES

This discipline may lead to designing and developing almost anything you might lay your hands on – handheld staplers, mobile phones, computers, handbags, surgical instruments, sofas, interiors, the list is endless. There are many examples of designers from this discipline crossing over to other disciplines such as graphic design, web design, 3D computer visualisation and simulation and film and TV design. See Chapter 7, 'Theatre, film and TV design' for Dan Walker's story, taking him from designing kettles to spaceships.

Industrial/product design covers a vast array of markets, products and services. It requires a broad and ever-increasing set of skills that are rarely covered by one person but instead are frequently deployed by a team of people from many related disciplines, inside and outside of design. The UK has a strong tradition of design and a healthy design industry. As well as opportunities in the UK, British designers (and British-trained designers) will be found around the world on every continent. Arguably Britain's most famous designer, Sir Jonathan Ive, lives in California and works for . . . well, you know who, don't you?

Their way of working varies too, as stated at the start of this chapter: you might be self-employed working on your own products or as a consultant to a company; you might be a salaried member of staff in a design consultancy working for various clients; or you might be an in-house designer for a manufacturer.

WHAT WILL YOU BE TAUGHT?

Industrial/product designers need 3D skills in drawing, model-making and computer-aided design, idea generation, research and communication skills. They also need a deep understanding of human interaction. This aspect will be taught through a variety of different approaches, perhaps under the title of 'human factors', 'usability' or 'ergonomics'. It may be that this is covered in another context – a series of tests with real users and prototype models that represent how your design might look or function.

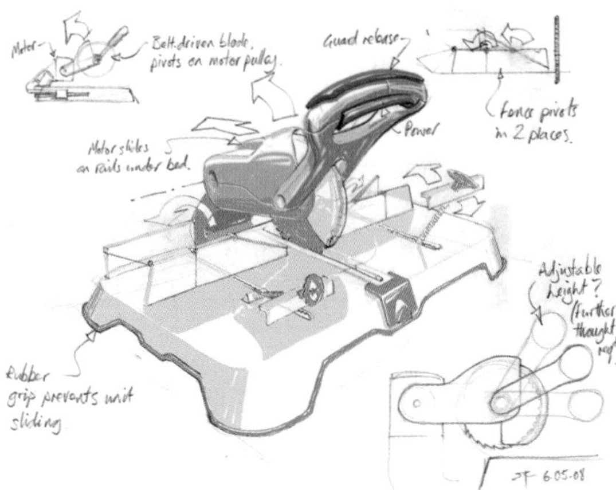

Concept sketch of a crosscut saw

Some courses may place an emphasis on computer modelling and simulation; some courses may emphasise aspects of business practice. Depending on the emphasis of a particular course, this may be focused on mass production and global markets and resources for understanding the sales potential of product proposals. Or it may be broader still and examine full business proposals for those who want to be self-employed and produce and market their own products – how to plan, cost and market your products and secure the backing of **investors**. There is also room within your workload to research and develop these skills and knowledge in your own time if they aren't explicitly covered in the syllabus.

Investors: a product might require large sums of money to gear up for production and to build a stock for sale. There might also be marketing costs. Companies and individuals can secure loans from banks or look for venture capital investors who specialise in taking risks on new products, such as those featured in *Dragon's Den*, © BBC.

Most importantly, you will gain experience at balancing the often conflicting needs of a project, your client and the users of your designs. This project management skill is one of the most useful transferable skills in demand in industry.

WHAT IS THE DESIGN PROCESS LIKE?

Design is a term that can be applied to both an object and an action. It is a process of defining a context and exploring possible solutions. While every design project varies, the general process should contain:

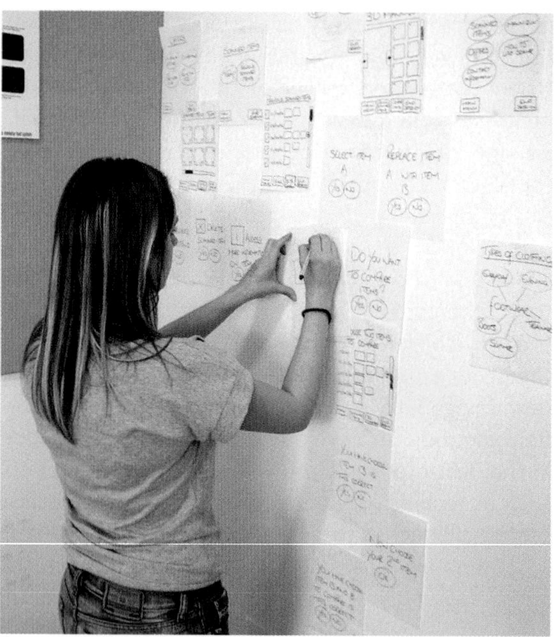

Rebecca Clewes mapping out her design process

- **definition**: what are we doing/designing?
- **research**: what do we need to know?
- **conceptualisation**: how might we solve this problem?
- **evaluation**: are any of these ideas any good? and
- **refinement**: let's make sure this is good and people will love it.

Definition involves defining what the project is about: the brief, the parameters, what the project aims to achieve. Occasionally, this information may be provided to the designer in a written brief, but is more often derived from a dialogue between the designer and other stakeholders to identify the best approach to defining the territory to be explored. Mapping out this data is an important part of defining what the problems are in a project. Many designers produce physical mind maps using Post-it notes or paper, usually filling a spare wall, see Chapter 10, 'Being creative and innovative' for more guidance on this.

Research enables a designer to learn more about the general environment of use, or specific details about the people who will use their product/system. You might find yourself learning about new materials in an effort to identify a suitable new application. Other forms of research may be more specific, such as learning how a surgeon uses an operating theatre in order to help you design medical equipment.

Vanessa De Zilva sketching quick concepts

Conceptualisation involves thinking of as many ways as possible that the problem might be solved. Creativity plays a part in this, as well as synthesising your research.

Evaluation requires a formal process of deciding which factors are important and evaluating them, to arrive at an accurate answer. Typical evaluation for an industrial designer involves deciding between various optional designs or testing with focus groups to see how end-users might take up their design.

Every project involves some aspect of evaluation. A mass-produced product may require significant testing of the market or the performance to ensure that the proposed design solution will be successful. If you set yourself a personal one-day project then evaluation may be a matter of personal judgement or experience brought to bear in deciding between various options – almost never will a final design be produced from the first idea sketch (there is more on this again in Chapter 10, 'Being creative and innovative').

Refinement, the last stage, may be considered as simple as drawing a more accurate line prior to production. More than likely, though, this will be a rigorous block of time dedicated to making modifications in response to testing or increased knowledge of the manufacturing process. For mass-produced products, there is typically a significant proportion of time dedicated to refining for production – for creating drawings or models necessary to communicate to manufacturers exactly how the product should look and function.

MOVING FROM 2D TO 3D

The jump from working in 2D to 3D can happen at various points in the design process. You may need to consider the form of the object you are designing and you might make a quick foam model. If there are technical aspects that need proving then you should build a test rig. These are the main, widely recognised forms of three-dimensional representation that are used:

1. **Scale model:** a visual representation of a design in 3D, smaller than actual size.

   Materials: card, foam, wood, plastic.

   Reason: to communicate ideas and to aid analysis of concept choices.

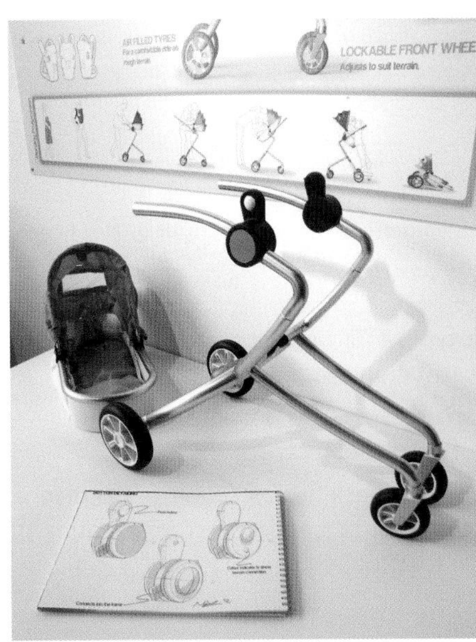

Scale model of a pram by Nicola Danks

2. **Appearance model:** a full-size accurate and well finished model that looks like the real thing.

   Materials: wood, plastic, metal, rapid prototype; high-quality paint finish with product graphics.

   Reason: to communicate exact look and image of product.

High-quality concept model by Nell Bennett

3. **Test rig:** a full-size working object that reproduces some or all of the functionality so that it can be tested.

   Materials: anything.

   Reason: to prove a concept; to analyse the use and efficiency of a mechanism.

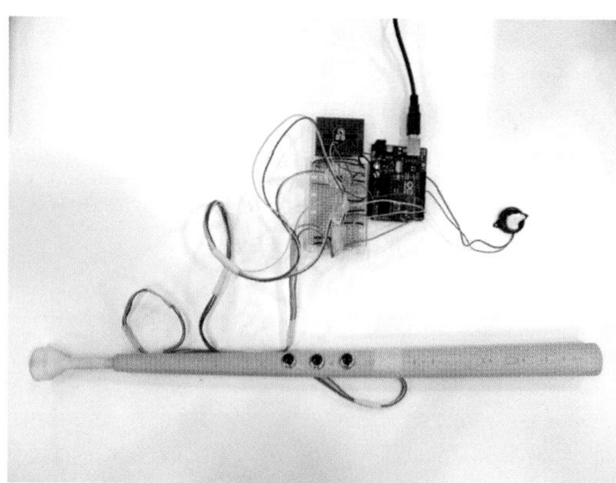

Working prototype electronic trumpet by Nell Bennet

4. **Prototype**: a fully working and visually accurate representation of the final design.

   Materials: as close as possible to the final manufactured object.

   Reason: to check final detailing, performance and customer perceptions before releasing the design for production.

Corky Stool: height adjustable stool made from 100% recycled cork granules. Designed and manufactured by Richard Collings

You must always try to pick the most appropriate 3D outcome to match the project, your resources and the point at which you find yourself in the design process. A series of models, test rigs and prototypes could be made to test various aspects, if it is appropriate to the length of project. For instance, designing something hand-held you may wish to make a series of wooden models to see how different shapes and geometries affect the beauty or functionality of the product.

Realistic models need not be beautiful or, indeed, take long to construct. A cardboard mock-up, quickly stapled or taped, may be sufficient to see how the proportions look or how pieces articulate. You still want your models to impress anyone who sees them, though, so be careful. And don't make things at half scale. There's something about it that doesn't look right; certainly furniture doesn't. Cars at half size would be almost pointless as it would involve a huge amount of work. Appropriateness is the main criteria by which 3D work should be undertaken:

- Is it the right time to make something? Do you need to sketch or research more?
- What will I find out by doing that? Should I make a visual model or something I can test?
- Can I do it with the resources I have to hand? What is available in the workshops?

WHAT WILL THE FINAL YEAR BE LIKE?

In your final year you will be consumed with developing your major project, as well as one or two minor projects, writing a thesis and refining a portfolio to get you a job or to gain entry to postgraduate study. As the jobs market is so broad, it is important that you understand what sorts of opportunities you wish to pursue upon graduation and that you focus your projects and portfolio toward those goals. While a good portfolio will gain you employment in a number of diverse companies, the most competitive ones require you to demonstrate you can tackle their particular challenges – the markets, the technologies, the materials or other factors that define their business.

You should map out what your current capabilities are, what you can reasonably develop across the final year, what your passions are, what opportunities are available and what sort of presentation materials you may need to prepare to demonstrate to others that you are capable of satisfying their requirements. This should help you select topics for final-year projects and dissertations that generate the materials you need. It is frequently the case that students use their preliminary research in their major project to help them refine their understanding about particular markets or employment needs.

Out of all the common skills that you and your three-dimensional design peers share, you will be aiming to differentiate yourself in the jobs market either through:

- a higher level of skill in a particular area, such as sketching
- a unique linking of skill to application, such as computer-aided design and rapid prototyping, or
- a well-developed, deep understanding of a particular domain that may be applied to a specific market, such as designing medical equipment.

COURSE TITLE: **FURNITURE DESIGN**

COURSE DESCRIPTION

As well as the qualities outlined above in Industrial/ Product Design, the Furniture student will have to immerse themselves into the history and tradition of the furniture field. The subtleties of form, colour, texture and proportion play an extremely important part in contemporary furniture design and these need to be explored in detail. The furniture field has been responsible for pushing the boundaries of materials and manufacture, from the early wooden construction techniques to the pioneering use of tubular steel in the early twentieth century. This interest in and understanding of materials is concentrated in the mind of a furniture designer.

Furniture projects are often likely to go as far as detailed manufacturing information and the production of full-size, working prototypes, as the feel and comfort of a piece of furniture are difficult to predict by looking at a representation of it on paper or screen.

Factors such as ergonomics, textiles and market research play an integral part in these projects, as well as trend analysis – what colours and surface finishes will people be buying next year? See Chapter 12, 'Working with colour'.

'Finn' chair by Oliver Hrubiak, winner of the John Lewis Award at New Designers 2012
Image courtesy of Oliver Hrubiak

CAREER OPPORTUNITIES

Furniture design offers many diverse career opportunities for those with skills in specifying, designing, hand-crafting, manufacturing, promoting and selling furniture. Although the markets are highly competitive, furniture is everywhere. The challenge for furniture designers seeking employment is differentiation – to offer a new response to an existing market or a new product taking advantage of recent advances in materials. Graduates from Furniture Design courses can find employment in various forms:

- in-house with furniture companies
- as freelance designers for furniture manufacturers or design consultancies, or
- working for companies that provide full-service interior design, specifying all the furnishings and furniture that make a home, as a self-employed designer or as an employee.

Many graduates set up their own companies, frequently working in pairs or small teams of colleagues in order to sell their own designs to order. Such designs tend to be developed for **low-volume batch production** or for **one-off production**.

Low-volume batch production: a certain batch of products will be produced at the same time for distribution to a range of different clients or shops. There are economies of scale in manufacturing a certain number at the same time.

One-off production: a designer might produce a 'one-off' piece for a particular client, to their own brief. They will pay more for the 'exclusive' right to the design and appreciate the fact that no one else owns the same piece of furniture.

Furniture designers also move into other discipline areas. Many practice interior and/or lighting design at some point, and the discipline can evolve into a craft-type activity where pieces are designed and made by hand for individuals to their own specification.

There is a sizeable UK furniture industry and associated companies, producing textiles, forecasting trends and making furniture for the office and for the home. There are also possibilities for working abroad. As well as furniture design and manufacturing companies throughout the world, many UK companies manufacture their products abroad and the development process often requires designers to visit the production facilities to oversee detail changes and sign off final production specifications.

WHAT WILL YOU BE TAUGHT?

Above and beyond the core skills that all 3D design students study, as outlined at the start of this section, furniture students also develop skills particular to their field.

There can be a great difference between courses majoring on making and those with more emphasis on designing. The making-based courses dedicate a larger amount of teaching and learning time to building the craft skills necessary to control the material and form highly-sought-after bespoke furniture. These students often go on to set up their own designer-maker businesses. Design-focused courses teach students about an awareness of materials and manufacturing details but don't always expect them to produce the final artefacts. These students will more normally find employment in the contract and domestic parts of the industry.

You may also be taught specific histories of furniture, as well as contemporary market studies. There is a history, in the Furniture discipline, of its links with materials and manufacture, society and consumerism that needs to be understood, and it takes time to immerse yourself in the Furniture discipline and understand its subtleties.

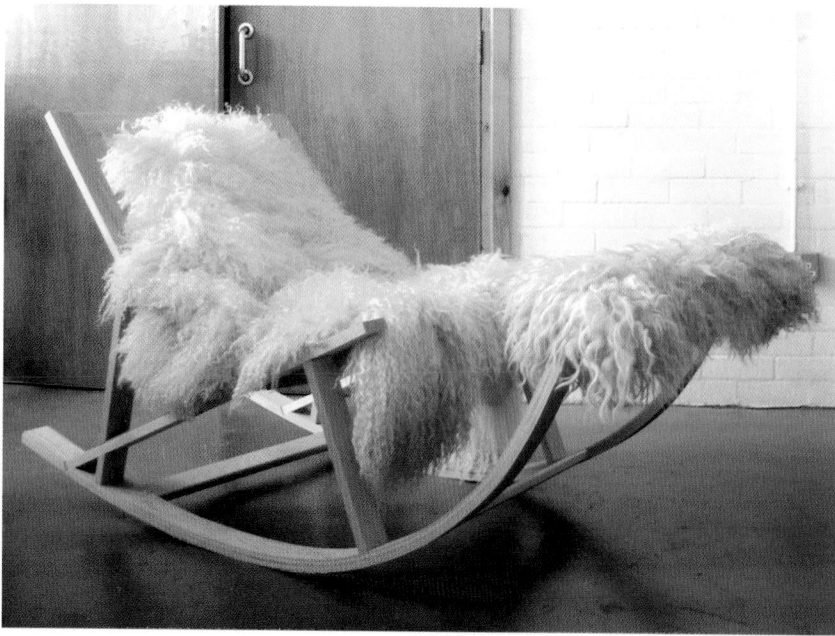

Full prototype sheepskin rocking chair by Hannah Welsh

Material and process study is vital but may be handled in a number of different ways depending on the course intentions. You may find yourself learning different ways of joining furniture components and producing examples in the workshops; similarly, you may be learning how to research, identify and use new materials that have only been developed in the past year or so, applying them in novel ways to create thought-provoking and creative furniture – for example, testing lightweight composite concrete structures.

The furniture designer is, of necessity, more interested in consumers, trends, lifestyles, textiles and how people live their lives. Projects will at times concentrate on one or more of these themes. Furniture is a specialism that is not easy for any product designer to dip in to, and the range of experience you will get in your projects will help you understand the requirements of this traditional yet, at times, very contemporary field.

WHAT IS THE DESIGN PROCESS LIKE?

Furniture is a very tactile product; we sit on it, touch it, lean and lie on it, as well as admire it. If it is not comfortable or breaks too easily, then it is not going to do well in the marketplace. The design process, therefore, tends to involve a fair amount of working in 3D, making mock-up models and full-scale test rigs (see Industrial/Product Design, earlier in the text, for a description of these) that can be tested, both physically and aesthetically. A full-scale chair model, even if it is constructed out of lightweight materials and not the final materials, is invaluable for providing insight into structural weaknesses or proportional flaws, even before it is robustly manufactured in the final materials.

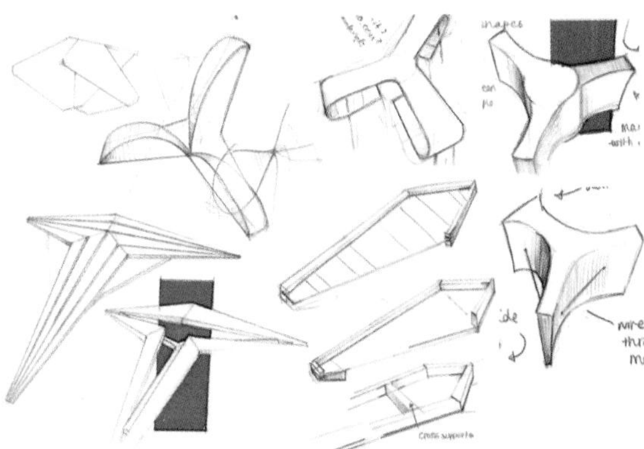

Stool concept sketches by Sarah Hoyle

Construction drawings (front, side and plan views, to scale and fully dimensioned) are made either to inform the maker or, when created as a computer-aided drawing file, to drive a computer-driven manufacturing machine, such as a mill or lathe. Very complex wooden parts can be produced at very high speed in this way, working to a level of complexity that a skilled technician may struggle to match.

WHAT WILL THE FINAL YEAR BE LIKE?

The furniture industry pays a lot of attention to year-end degree shows and industry exhibitions. As a graduate, your university may pick you to exhibit at the national graduate show, New Designers, in London. As a working professional you might be lucky enough to exhibit at the International Contemporary Furniture Fair in New York or the Milan International Furniture Fair. Both exhibitions attract graduates and professionals to see the latest furniture concepts. As a final-year Furniture student aiming to be chosen for New Designers, you would therefore focus on developing

Collaborative project – prototype stool by Sarah Hoyle, fabric by Holly Atkinson

a series of final pieces that can either be sold or licensed to interested manufacturers or, at least, capture the interest of a potential employer who is looking to bring in new talent.

So the focus of your final year will be on presenting, in most cases, fully functional artefacts that can be manufactured either singly or in low numbers, enabling manufacturers and consumers to understand exactly what they will get. Behind every project will be a series of explanatory material, such as presentation boards that show the process of developing the ideas, the details of how it is or will be manufactured and the costs of one or multiple pieces. The exhibition experience is very important, particularly to the Furniture student, because of the way it drives the industry and facilitates networking, trend analysis and the buying and selling of design and designers.

COURSE TITLE: **AUTOMOTIVE DESIGN**

COURSE DESCRIPTION

An Automotive course will have an exciting range of design projects, some small, some big, that will build your skills. You will probably, at times, work on futuristic, conceptual projects, with other projects tied in to today's market and consumers. You probably won't work on an endless series of car projects. There will be skills acquisition projects that might concentrate on sculptural form; there will be others with an emphasis on research – consumer trends, ergonomics, aerodynamics, etc.

Automotive Design courses thrive on their links with the industry professionals and you will be exposed to their critical appraisal of your work. Industry-linked projects are sometimes carried out in teams, where you can experience the challenges and opportunities of working closely with your peers.

You will also get a sense of the enormity of this discipline in your final year when you will have to pull together a multi-faceted project combining all you have learnt. Your project management skills will be tested to the limit. Automotive courses often culminate in a large, well-finished scale model, plus photo-quality computer renderings and often a substantial project report, all presented to an industry standard.

All of this means that the competition for employment in the automotive design industry is very high, as are the rewards (see 'Land Rover: Discovery' at the end of this chapter). There are not many courses in this subject and they are very popular!

Concept by Frederico Zanjacomo

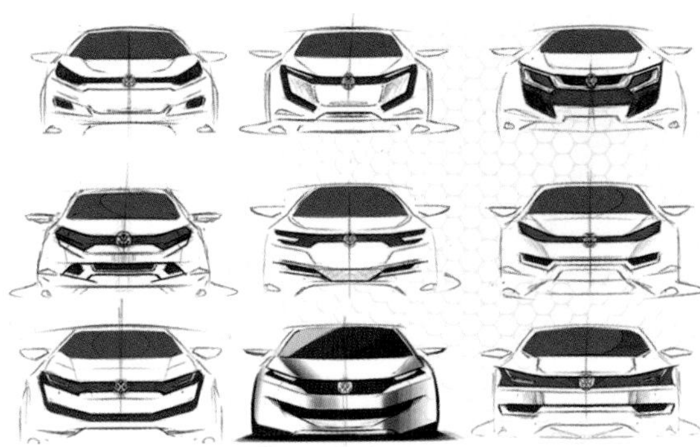

Vehicle identity concepts by Frederico Zanjacomo
© Frederico Zanjacomo

CAREER OPPORTUNITIES

The design and manufacture of cars involves, quite obviously, numerous disciplines working together in a development team. Automotive design may involve designing the external appearance or interior layouts and components. Typically, these areas involve many designers, each taking up different roles – defining the overall appearance, the design detailing of every component and the colour and material scheme. Development teams would normally contain a wide range of professionals – automotive designers (interior, exterior, controls, seating, displays, etc.), powertrain, chassis and suspension engineers, IT specialists, psychologists and trend analysts – all working together to create the emotional experience of buying the biggest, most expensive product we ever buy – a new car. They also have to work ahead of the trends – cars can take from two to five or more years to go from sketch to showroom.

Upon completion of an Automotive Design course, a successful graduate may expect to be involved in a junior supporting role and will typically be allocated the role based upon their demonstrated specialist knowledge and skills in drawing, modelling, trend forecasting, material properties, usability, etc.

Employment with car manufacturers can be highly competitive, and there are distinct advantages to having studied on a course involving some placement or internship with an automotive company. In addition to designing cars, graduates may seek employment in a variety of fields directly related to this, or building on their core skills. Other transportation areas, such as bus, lorry and construction vehicles, employ automotive designers, and their often highly developed aesthetics and detailing skills are also sought by large equipment manufacturers and other industrial design consultancies.

WHAT WILL YOU BE TAUGHT?

The generic requirements of Automotive Design courses mirror those of Industrial/Product and Furniture Design courses. The elements of marketing, ergonomics, consumers, trends and materials and production processes are all shared.

The automotive industry pays particular attention to the ability of a designer to communicate the complex surfaces of a car body accurately and portray the important factors that it must embody – elegance, power, stability, strength and beauty. The techniques that the industry has developed over the last one

Interior concept by Frederico Zanjacomo
© Frederico Zanjacomo

Rear light cluster proposals by Frederico Zanjacomo

hundred years are the most sophisticated of any of the design disciplines. Students of Automotive Design therefore need to develop a high level of skill at drawing in order to succeed. The industry uses many media to portray its products, both traditional and technological. Students will be required to use traditional pens, pencils and markers and also graphics tablets, digitisers and 2D and 3D software to a very high standard.

As well as an emphasis on photo-realistic, sophisticated computer drawings there is, traditionally, the use of clay as a sculptural medium to develop and refine the final shape of a vehicle.

Even with the amazing power of computer-generated imagery available today, to see an accurate, physical representation of a proposed car body is an absolute necessity. Most real-world car design projects therefore have a full-size clay model of the design, which develops in parallel to the drawing/ CAD process. This is used to work on developments and to check changes. The special clay is heated up in an oven and can easily be added to the model when it is warmed up and soft. When it cools down it is quite hard and can be carved very accurately. This means that there is still a requirement for Automotive Design students to be skilled in the manipulation of clay on scale or full-size models as demonstrated in the Land Rover Discovery project story at the end of this chapter.

A materials and manufacturing module will likely be explicit in the curriculum. There are also requirements for market and trend analysis and a sophisticated understanding of the modern consumer.

WHAT IS THE DESIGN PROCESS LIKE?

Buying a car is an emotive experience, and designers are always looking at what kind of emotions and connections people make when they look at a vehicle. Many people think we buy cars to portray to the outside world a version of ourselves that we want people to see. Part of the process is, therefore, to understand consumers, their motives and their desires, to inspire new visions.

Due to the long development process of complex products such as cars there is even more of a requirement to predict future trends and make sure your designs are heading in the same direction as consumers. You will experience trying to connect these disparate themes and merge them into a coherent body of work that explores the future of personal transportation.

The real-world automotive design process involves large teams of multi-disciplinary designers and a lot of digital work. Automotive design projects are necessarily more complex than most product design projects. After all, a car is the largest, most complex and expensive product we might ever buy. There will be an emphasis on the market and branding in the

Early sketch concepts by Frederico Zanjacomo
© Frederico Zanjacomo

initial phase of a big project. There is also the 'packaging' to consider. This is the term used for the engineering and ergonomic factors that define how a car works, how big it is and its features. Automotive designers are often presented with this engineered package before their part in the design process begins.

This is not to say that designers don't engage with engineering occasionally and try to suggest new ways of offering a vehicle's functionality. There is also a point in a real project where the designers must make sure that, as the vehicle is readied for production, the overall design stays true to the original vision and is not compromised by production issues.

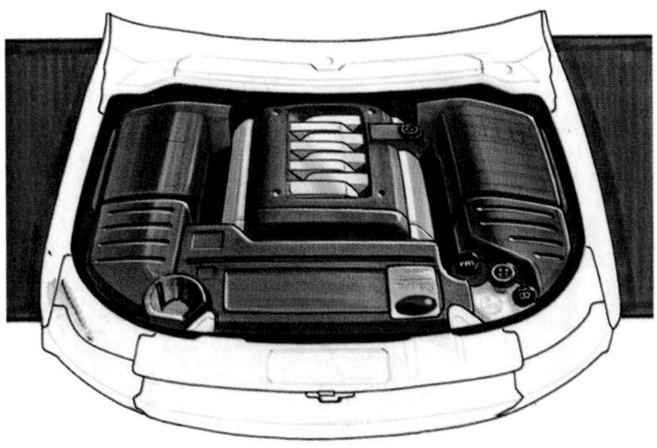

Under-bonnet architecture concept
Picture courtesy of Land Rover

WHAT WILL THE FINAL YEAR BE LIKE?

Due to the specialised nature of the industry, Automotive Design students will be highly focused in their final year, typically working on one or two designs. It is quite common to find pairs of students working together on one vehicle so they can each focus on one particular aspect in greater detail. You may find, therefore, that portfolios on graduation include interior concepts or novel, engineered solutions to a specific sub-market, such as a motorbike, as well as a car. Many students take the view that a large, scale model of their biggest project is the thing that will get them noticed by the industry and they take a lot of pride in producing them. Managing a design project that culminates in such a large outcome requires excellent planning and time-management skills.

As with Industrial/Product Design, the final year is an opportunity to examine who you are as a designer and target your work towards acquiring skills and experience that will help you achieve your aims. There may be a requirement for a written component in the final year, normally a 5–10,000-word research-based essay, report or dissertation, possibly related to your major project.

Some courses have significant links with industry, and final-year projects can be sponsored or in some way supported, such that the student is engaging in a very real-world context.

COURSE TITLE: **BOAT DESIGN**

Computer-rendering of a yacht for Totalmar by Albert Montserrat

There are few courses of study dedicated exclusively to boat design; those that are, tend to focus on the engineering design of hulls. In the industry, this creates a division of labour separated by the water-line: those who design the shape and material properties of the hull and engineer the powertrains, and those who design everything above the water-line – interiors, furniture, communications (the areas that people interact with).

Below the water-line is the preserve of BSc/MSc Naval Architecture and Marine Engineering courses. For designers it is quite common, therefore, to enter the field of boat design having completed a degree in a related field, such as automotive, transport, product, furniture or interior design. Speed-boat manufacturers may seek out an automotive designer for their styling capabilities. Luxury yacht manufacturers benefit from the skills of furniture, product and interior designers, and also demand design sensitivities typically found in craft or textiles graduates, with their ability to place colour and texture.

COURSE TITLE: **LIGHTING DESIGN**

Lighting design tends to be treated as a sub-category of Furniture or Product courses, yet the market demand remains reasonably high for those graduates who can demonstrate the particular understandings of lighting design.

With increasing environmental agendas and governmental initiatives to increase the use of low-wattage, energy-saving bulbs, the variety of lighting products is increasing and with it the demand for creative lighting designers. While there are few

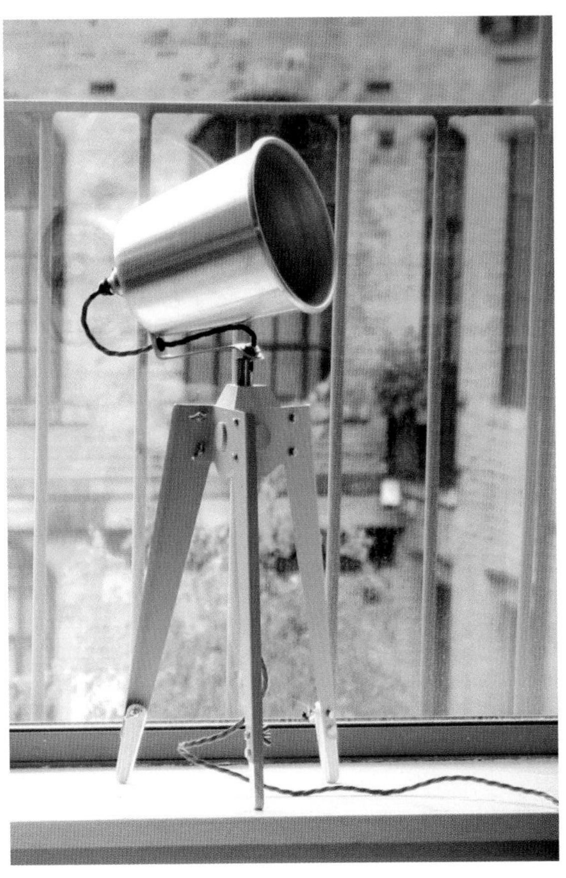

Light by Oliver Hrubiak, final-year furniture & product design student
Image courtesy of Oliver Hrubiak

courses dedicated exclusively to lighting design, many students on Product or Furniture courses build up a portfolio of lighting concepts to enable them to focus on this in the workplace. The most successful of these are highly attuned to material development, as well as human nature, and generate lights that use innovative materials in novel and engaging ways. Even industrial materials, such as light-permeable concrete, are becoming part of the lighting designer's palette.

Later in the chapter you can read about Sarah Turner's lighting company. She uses recycled drinks bottles to produce a wide range of light shades. Because she uses lots of Coca-Cola bottles she called her first light the 'Cola 30', and this brought her to the attention of Coca-Cola. They commissioned her to supply lights for their hospitality centre at the 2012 Olympic Park. She explains:

> 'When I first started making lights from waste plastic bottles I always hoped that Coca-Cola would notice me and my work and get in touch. So being commissioned by them for the Olympics really is a dream come true for me. It feels like a great honour to be working with Coca-Cola for this once-in-a-lifetime event.'

The lights make quite a statement, being 2 metres wide, and are made using 190 plastic Coca-Cola bottles. There are five of the large lights in total; each is made up of rings of the plastic bottles and a globe in the middle. The globe in the centre of each one

Sarah Turner's lights in the Coca-Cola Hospitality Centre, Olympic Park, London
Sarah Turner

is Sarah's 'Cola 30' design, which is made from, of course, 30 Coca-Cola bottles hand-cut and sculpted into decorative forms. This was the design Coca-Cola first noticed and expressed interest in; after all, it is made using their bottles and is named after them!

> 'I wanted the lights to have an Olympic look to them, which is why I chose to make them as circular disks with rings of the plastic bottles, reminiscent of the Olympic rings. I also liked the idea of having the classic-looking light bulbs visible, it reminded me of the famous Coca-Cola Christmas truck. I really wanted to include my 'Cola 30' light. I thought the contrast between the whole plastic bottles and the transformation they go through with the "Cola 30" was a fantastic thing to show. The project was a huge success and hopefully the relationship with Coca-Cola will continue.'

Coca-Cola did commission more work – Sarah created table centre-pieces for the Langham Hotel in central London. Coca-Cola took over this prestigious hotel for their guests to stay in for the duration of the Olympics.

COURSE TITLE: **SPORTS EQUIPMENT DESIGN**

'Dax Scooter' – a folding scooter by Chris Lamerton

As with Boat or Automotive Design, most of the students on a course in Sports Equipment Design will have that extra, specialist passion for their subject. It may or may not align with the sports equipment they end up designing, but a love of some form of sport or a love of taking part in sport tends to drive this student. Sports Equipment Design also covers a wide territory, from graphic design to engineering.

Sports companies hire graduates with graphics and fashion skills, textiles designers with surface, pattern and construction skills, product designers and engineers, so there is no general requirement to study Sports Equipment. You may also find that some courses offer Sports Equipment as an optional module.

You may enter this discipline with your own set of skills, but you will also need a personal understanding of what motivates people. While some companies may offer specialist products developed with batch-production manufacturing, this market tends to be dominated by larger manufacturers, so you will get the opportunity to be working as part of a team, bringing to bear your skills, with the support of others, to create robust, stylish products that support sporting activity.

Preparing your portfolio

Your coursework needs to be very professional and communicative for presentation and assessment purposes. You will also need to keep it together in portfolio form for the purposes of placement interviews and possibly end-of-year portfolio reviews.

In addition to the general portfolio suggestions to be found later in the text, those studying 3D design need to ensure they focus on material particularly relevant to their discipline, especially a variety of evidence demonstrating their ability to work in three dimensions.

Three-dimensional design requires an ability to communicate 3D concepts in lots of different ways and you should show that you understand when to choose which method. A fully illustrated process with a variety of different imagery can demonstrate this. Freehand sketches can show first ideas and quickly communicate aspects of proportion, scale and colour. You should then develop the ideas through other means – by doing dimensioned drawings, for instance, in plan and elevation, and models.

Seeing those drawings realised in scale models provides a viewer with an insight into your skill level with materials and your ability to construct something accurately to scale. Optimally, you will show several variations that illustrate how you experiment with colour, texture or material and that you don't just select the first thing that comes to mind.

Translating 2D into 3D in the workshops

Conclusion

3D design is a vast and exciting discipline that offers stimulating and varied careers to people with a wide variety of interests. Designers in this area create, literally, the three-dimensional world around us, providing tangible ways of interacting with, understanding, controlling and enjoying our environment. This discipline of design provides you with a large number of opportunities for exploring your love of people, spaces and products. The pace is fast, enabling you to have an idea and see it realised in a matter of months. Indeed, through the mutual support of universities and industry and the placement experience, some students these days see their designs on the shop shelves before they graduate.

On the way, you will foster relationships with peers and professionals that will enable you to find your place in the industry and hit the ground running in your chosen profession.

To demonstrate the range of practice within 3D design, the following case studies each describe the designer's involvement in two very different areas, in their own words. The first is a personal, self-initiated project by a young, self-employed designer; the second outlines a designer's role in a large team project for a world-famous car manufacturer.

Further resources

Books

Byars, M., *The Best Tables, Chairs, Lights: Innovation and Invention in Design Products for the Home*, RotoVision (2005)

With excellent information and inspiring objects, this book is especially important for its information on manufacturing.

Hallgrimsson, B., *Prototyping and Modelmaking for Product Design*, Laurence King (2012)

The follow-up book to *Drawing for Product Designers* below. This is just as good.

Henry, K., *Drawing for Product Designers*, Laurence King (2012)

The latest and most sophisticated sketching and rendering guide for product designers, this takes this genre of book to a new level and surpasses everything that has come before.

Parsons, T., *Thinking: Objects – Contemporary Approaches to Product Design*, AVA Publishing (2009)

The latest academic viewpoint in the subject, this book is very important for students approaching their final year and considering their role in design in the future. It maps out the subject and will aid the reader in his/her career planning.

Pheasant, S. and Haslegrave, C.M., *BodySpace: Anthropometry, Ergonomics and the Design of Work*, CRC Press (2005)

The 3D designer's ergonomics book, this contains advice, anthropometric data and information on how to use it. You will use this book for the duration of your career. It contains wonderful insight, information and hard and fast data on the variation in the size of the human frame, complete with guidance on how to use it.

Websites

core77.com

The definitive product/industrial design website: news; articles; images. Also host to one of the best portfolio sections in the business for this discipline area.

www.designcouncil.org.uk

The UK's design ambassadors. This site contains great information and advice to the business world on how to use design. It also champions the socially responsible and inclusive aspects of the discipline and leads by example by facilitating lots of great projects. These are written up in great case-study style, with videos. Brilliant.

Project story

Designer:
Alexander Taylor, Furniture & Lighting Designer

Project title:
'Fold' Lamp

Tell us about the project

This was a self-written brief, and from the beginning of the project was driven by budget and material constraints and limited by the materials for modelling and development that were easily available to me in my studio. The 'Fold' lamp was the first project I worked on when I opened my studio, so I needed to keep costs down. I wanted to design something that was simple to make, clever, functional and aesthetically pleasing – a signature design. A small, domestic table lamp was the ideal product. I also had the aim of making the entire lamp – base, stem, shade and lamp-holder – from one sheet of material, which would be cleverly shaped and then simply bent up into a three-dimensional object, as can be seen in the diagram. While working on earlier projects in my previous job, I had made contact with an acrylic fabricator, so I was able to realise a working prototype very quickly and at very low cost.

I now licence the product to Established & Sons, a British furniture and lighting manufacturer, and receive a financial return in the form of a **royalty**. I retain the **licence** and they have the rights to produce.

What processes and skills were most relevant to this project?

It was important for me to reduce the processes involved in manufacturing a light to the minimum – aiming for just one process with a single material to produce a 3D form that would function as a light. This was so that I could have complete control over the process.

The form developed very quickly and easily over only a few days by cutting and folding paper to a scale of 1:1 before moving on to working with acrylic, a more rigid material. This helped me understand more about the structure and strength of the design.

This prototype formed the basis of the project and provided me with the confidence to move forward with it and invest time and a small amount of money into producing the design in a small batch. The aim was to self-produce a small number for both private and retail sales. Acrylic was not the material with which to develop this light for production due to the cost of the raw material and the forming and finishing costs. Therefore, I wanted to develop the design using metal – either brass, aluminium or steel sheet.

All three materials had to be tried and tested. I produced samples from different gauges (thicknesses) of material to gain an understanding of structural qualities.

At this point I also had an understanding of costs and was forming an idea of how much I would like the light to retail for. This is a very important part when developing any project. Most products will have an ideal **price point** where it will sell, and it is important

Royalty: a payment from the manufacturer to the designer for every item sold in the shops, normally expressed as a small percentage of the retail price.

Licence: this denotes my ownership of the idea and allows me to decide who will produce it.

Price point: the price that consumers will pay in sufficient numbers to justify the production run and generate a profit.

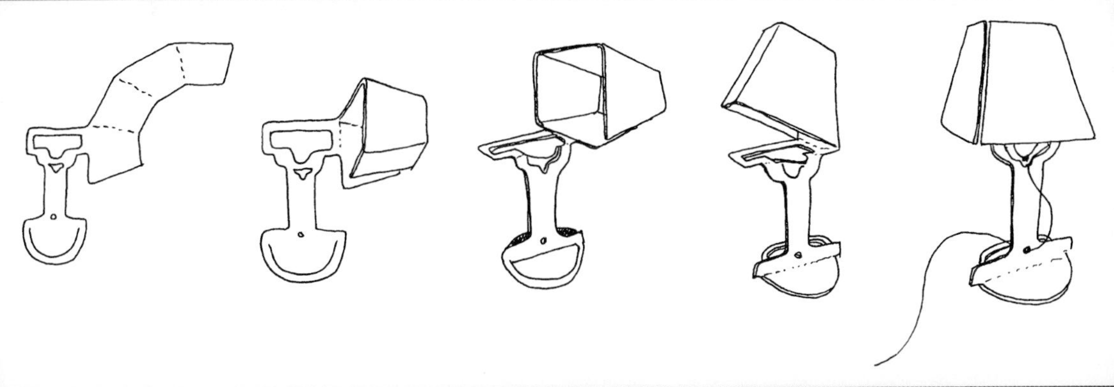

The 'Fold' lamp is created by folding one sheet of material

The finished product

The 'Fold' lamp in various colours
Photograph: Alexander Taylor Studio

to have an understanding of pricing structure whenever designing any product.

Small details on such an outwardly simple project are important. Simply by specifying an old-fashioned electricity cable made from a woven, coloured fabric, I was able to add character and identity to the piece and also incorporate a familiar, traditional element that works well in contrast to the rest of the lamp.

What challenges did you face?

Understanding time-scale is very important in all projects. It took approximately 18 months from the first prototype to getting confirmation that Established & Sons were on board. It then took another year before the lamp was finished. As a self-employed designer-maker, it is important when you are establishing your own business to try to have a number of projects running at the same time, as it is always difficult to anticipate just how long it will take to develop and launch a product. You also don't know which ones will be successful and which ones will not. Perseverance is a quality that helps in these circumstances.

I was delighted when the company decided to invest in the project and develop a family of 'Fold' lights for its first collection, to be launched at the Milan International Furniture Fair. Although developed to a certain level, the light needed investment to move it forward and develop it into a range of lights. We decided to design a larger table lamp and a standard lamp. The design was altered slightly to incorporate a separate base and shade for production and material reasons. A lighter-gauge material was also used for the shade and thicker, heavier material for the base.

Once the product line was complete, promotion, branding, distribution, packaging, etc. needed to be designed and completed to get it to market. The 'Fold' light was finally presented in Milan 2005 and received a good response and lots of publicity.

What were the highlights of the project?

The first was getting the product into serious production. Initially I decided to invest my own money and produce 200 'Fold' lights to sell and present during the London Design Festival in 2004. However, my goal was always to design products for industry-leading manufacturers to put into production. The best way to get your designs into production is to make them and exhibit them. If the work is good enough the right companies will get in touch. If you do something that you really believe in then the self-belief will help give you confidence and strength.

The 'Fold' light generated a lot of great publicity but the highlight for me was when I received the news that it had been selected to be part of the permanent collection of the Museum of Modern Art in New York.

See more of Alexander's work here:
www.alexandertaylor.com

Project story

Designer:
Andy Wheel, Designer for Jaguar Land Rover

Project title:
Land Rover Discovery

Tell us about the project

The Land Rover 'Discovery 3' project was formally started on July 1st 2000, and the first customer-ready car was manufactured in November 2004. My part in this project was as lead designer for the exterior design, reporting to Chief Designer George Thomson, Studio Director David Saddington and, ultimately, to Geoff Upex, Design Director for Land Rover.

When the project started there was already a great deal of planning in place regarding timing of the design phase and the start of production, the investment budget, the general concept for the vehicle, where it would sit in the Land Rover portfolio of vehicles and in the marketplace in general.

The design brief had three distinct elements:

1 **'Discovery-ness'** – We had already established that this product was a replacement for Discovery 2 and not a new product proposal, so we had to ensure that we communicated this heritage through certain key aesthetic elements.

2 **'Package Efficiency'** – The new vehicle had to offer the best seven-seater package in the marketplace, with a flexible and optimised cabin space.

3 **'21st-Century Land Rover'** – This would be the first all-new Land Rover-badged vehicle of the 21st century, and had to embody the lineage of Land Rover in a contemporary way. (This last element of the brief was easy to write, but was the hardest element to define!)

A small team set to work on several concepts, developing ideas in 2D (sketches, renderings) and in 3D (computer models). Each design proposal was presented to senior staff as 2D images accompanied by a verbal presentation.

I was extremely fortunate to have my design theme proposal chosen as the primary route for the exterior design. This occurred at a milestone called 'theme selection', or 'go for one', where several proposals are considered and a single theme agreed on for further development.

Based on these drawings, we created a set of elevation drawings over key hard points, such as the interior space and the drivetrain and suspension components, and went straight into a full-size clay model of the body.

Early sketch of the Discovery concept
Land Rover

Design proposal
Land Rover

Full-size clay model
Land Rover

This master clay model was placed on a measuring plate so that we could regularly check overall proportions, feature lines and surfaces. We could also move it out of the studio into an outdoor 'viewing garden'. Here, in a more real-world lighting scenario, we could see it more realistically and from a greater distance than the space available in the studio.

The major turning point in the design and development process was a project milestone known as 'Stop Clay', which fixes the exterior design. 'Stop Clay' is the end of the creative phase and the start of finalising the design for production. Control of any design changes that might be needed transferred at that point from the design team to the product development team, whose responsibility it is to then make it ready for mass manufacture. From this point onward, any changes to the exterior design could only happen as a result of discussion between design and product development, and a detailed analysis to resolve a feasibility or cost issue.

What processes and skills were most relevant to this project?

Research. At the very start of the project my hypothesis was that Land Rover had a proven track record of creating vehicles that stood the test of time: that their aesthetic was set apart from other vehicles as a result of their functionality, and as a consequence they stood apart from other vehicles where aesthetics and style played a part. This meant that my research and communication skills were paramount in realising my ideas and getting them across to the design managers.

Sketch proposal for lights
Land Rover

Final resolution to the headlights
Land Rover

I collated evidence to indicate that a minimal aesthetic approach, where there is no superfluous decoration and everything has 'a reason to exist', was the key contributor to customer relevance, product longevity and 'timelessness'. This evidence came from all disciplines of design and indicated that a minimal 'product design' approach would create a modern Land Rover.

Along with the evidence to support my hypothesis, I also had illustrations of how this philosophy translated into a product proposal. I spent a great deal of time critiquing all components, establishing whether they were necessary and, if so, could they be integrated in a seamless way with the rest of the vehicle – for example, designing flush surfaces.

The 2D images at the 'theme selection' stage consisted of both 'influence' or 'mood' boards and 2D renderings of the vehicle from many angles. Realising my vision for the shape in a way that communicated what I meant to the decision makers was extremely important. It was necessary to communicate the shape in a way that would produce the right emotional response from the viewers.

Detail sketch produced late in the project
Land Rover

Later on, in the development and detailing stage, sketch work was still an important communication method. There were many refinements to the design, some driven by aesthetics but most driven by production and engineering feasibility requirements. We produced a great deal of sketch work that went into more and more detail but always acted as a fundamental and nimble communication tool. This is a very important skill for any designer – the ability to engage your audience and get them to see, understand and support your vision is fundamental for success. This applies to all aspects of the design, from the smallest detail to the overall theme.

The sketching supported the development of the clay model, which was continually updated and changed. This was regularly scanned in 3D to generate the latest 3D computer data model of the design.

What challenges did you face?

I knew how big the investment budgets for new vehicles were, and had an awareness of how the sales of 'fashion'-led vehicles, such as the VW Beetle, tend to be high initially and then fall off very quickly. I was also considering that this product would start production at the end of 2004 and would expect to stay in production for a minimum of eight or nine years, a long time in the car world. Therefore the exterior look of the car was even more crucial to its success.

This meant that the vehicle needed to stay relevant to customer needs throughout that timescale and beyond. Discovery 3 would be around until the mid-2020s! Translating all this into a shape that embodied 'timelessness' was a complex process involving drawing, computer modelling and the hands-on shaping and re-shaping of a full-size clay model.

Later in the project this clay model became a very important tool to focus all the other, associated design modules that would deliver the vehicle to production – 'bumper systems', 'glazing', exterior lighting', etc. Each module has a set of targets for their system of parts. These targets are often associated with performance (e.g. no damage to bumpers on impacts up to 8kph) or technical sizes and requirements, and are usually driven by changing customer needs or legal requirements.

At this point, the role of the design team changed fundamentally. We moved from being 'designers' to 'design police'! We had designed the exterior and needed to make sure that it wasn't compromised by any of the modules clashing with it. During the development of the exterior design there were several situations where different module targets were mutually incompatible – we could not achieve one requirement without compromising another. The master clay development brought these issues into the open and brought the module teams together with exterior and interior design to review options and agree on a balanced way forward.

What were the highlights of the project?

As well as being very fortunate to have my theme concept chosen for the final exterior design of Discovery 3, the high point for me was to be in a car in New York being followed by a Discovery 3 on the day after it had been officially unveiled to the world. We were taking some publicity pictures and I took the opportunity to tag along and see the final product 'in context' among other vehicles.

When I was able to see the modernity of the exterior style set against the vehicles and backdrop of New York I felt honoured to have been a part of the huge development group that took those initial ideas and worked as a team to deliver an incredibly complex and innovative product, wrapped in an equally innovative exterior design.

See and hear me talking about the design of Discovery 3 here: **www.youtube.com** (search for LR3 exterior design).

The final production model
Land Rover

What do you need to know?

Part two

How to succeed as a design student

By Jane Bartholomew, Nottingham Trent University

'The more open you are to change, the more surprises will come your way, whether it is involved with your studies or life as a student. The whole experience has been very valuable to me. Getting involved in lots of aspects of university life has taught me a lot about people and team work and how to accept difficulties as a challenge to work through and learn from. The mantra that has stuck with me this year is that a designer thrives on limitations; limitations produce an opportunity to think outside of your comfort zone.'

Sophie Alidina, postgraduate student studying Smart Design

A graphic design studio

You are at the point where you recognise that you are motivated by your own creativity, inspired by designers and perhaps beginning to appreciate that you could contribute to the design industry in the future. You may also have noticed that, over the last few years, you have been identified by teachers and friends as an 'ideas' person, or someone who pushes a project further by asking questions or by looking at it from a number of different perspectives. This is all evidence that your lateral thinking and questioning skills are beginning to take shape. Whether you decide to become a designer or not, you will need these core skills and attributes to prepare yourself for a challenging and rewarding career in the creative industries. In design, there are many varied subjects to study and they all focus on building your theoretical, practical and research skills. They will also provide you with opportunities to broaden your awareness of the industry. Design is an excellent discipline within which you can gain many transferable skills, from communication and team working, to improving your organisational and managerial abilities.

This section covers what it's like to be a student studying a design discipline and contains many practical tips on how to get the most out of your course. It will encourage you to reflect on who you are and how you operate, and provide you with ways of thinking that will help maximise the opportunities that might present themselves while studying. There is also information to help you prepare for a career in the industry, including practical guidance on what to include in your curriculum vitae (CV) and how to prepare a portfolio, together with advice on being interviewed.

What motivates you?

> *'Can you prioritise and commit? Quite simply, you need to ask yourself if you're prepared to do what it takes to achieve the qualification you want. Are you willing to put in the effort and time to achieve your goal? A strong and focused mind-set needs to be nurtured well in advance, instead of during the last weeks or days before your project deadlines.'*
>
> Lincoln Fan, Graphic Designer

It takes many students the majority of their course to become aware of the skills they are developing and to understand what excites and motivates them. Picturing yourself the year after you complete your degree in a particular job, or perhaps earning a certain amount of money, might give you a clear focus that could then become your motivation while you study. Spending time asking yourself what you want from your course and devising a plan to help you achieve your goal can increase your motivation levels and help you get the most out of your studies.

Career development is a bit like preparing to write an essay; you can't possibly just sit down and write it, you have to go out and undertake the relevant research first! The sooner you start this process the better, as this then allows time for some important reflection about the different career routes open to you and gives you time to develop a plan of action that will help you move toward your goal. Think about career planning as a project in itself that will need managing. Work out what steps you need to take to move it forward and then protect some time on a regular basis to progress it.

Many courses now have self-awareness and personal development planning built in to the curriculum. This will help you evaluate your strengths and weaknesses. You can then consciously focus on working to your strengths and improve some of your weaknesses.

Developing other essential personal attributes along the way, such as good visual, verbal and written communication skills, also plays a part in determining whether you have the confidence and motivation to go it alone or whether you would prefer to work on specific tasks as part of a team within a larger organisation.

Remember that the resources available to you at college or university are second to none, so make the most of the workshops, the library, the online provisions and the staff! Be aware of how you are spending your time and making use of these precious resources to develop your skills and knowledge. These resources are often not replicated in the workplace to the same scale and breadth.

How confident are you?

Coming up with a fantastic idea, resolving a project to the best of your ability and receiving positive feedback about your work are all ways in which you can begin to build your confidence as a designer. Not all projects run smoothly, but being receptive enough to absorb critical feedback from your peers and lecturers will help to improve your work next time. Another way to build your confidence is to do as much as you can to understand the context in which your chosen design discipline sits.

For example, as a fashion designer you have to have an understanding of the type of customer for whom you are designing and the different market levels within the fashion industry. You would also need an awareness of the type of fabrics that would be best suited to the silhouettes that you are creating and be clear in your understanding of how much companies can charge for any given item for a particular consumer group. It would also be important to have an awareness of the supply chain and production costs, e.g. price per metre of fabric, garment 'making-up' and distribution costs. There is also much to learn from how the other design disciplines operate, so be sure to read about all the disciplines covered in this book, and not just the one that you are naturally drawn to.

During your final year you might have to ask yourself whether you are going to step forward as a 'leader'. There are usually a few people in each year group who take on the organising of fund-raising events to raise money for the degree show's brochure, or coordinate the publicity and photography, etc. As an organiser, you can really benefit from the experience that this can bring as you will have further developed your skills and enjoyed seeing your hard work come to fruition.

Making the most of being a student

So, how can you ensure that before you complete your studies you have exposed yourself to as many sources of information as possible?

Think of your three- or four-year degree course as a period of time in which you can gain as much knowledge about the industry as possible. This 'period of time' is full of opportunities! The way you decide to approach your course will have a direct impact on how prepared you feel for your first job following graduation. Whether you choose to go on to be a designer or not, you will have developed a clear understanding of your design personality during your studies. The taught curriculum will provide you with a wide variety of projects, some set by the industry, some short, some long, but all involving opportunities to gain more skills and understanding about the way you work and how you might fit into the industry.

There will be national and international trips to take advantage of that may involve attending **trade fairs** and exhibitions relevant to your subject, which will enhance your cultural awareness.

In addition to this, your lecturers will mention other events that could be of interest to you, such as conferences and symposiums . . . these are organised, themed events where specialists from particular disciplines come together to share ideas about new work and new approaches. You may also be told about companies that you could perhaps visit that may support the development of your project. It might be up to you to pursue these and it is a good idea to consider doing your own research and identifying how else you might spend your time to support and enhance your awareness and increase your understanding of the subject. You will be expected to take responsibility for your own development at this level of education.

> **Trade fairs**: these are large-scale, organised exhibitions, where companies from a specific industry sector come together to launch their new products and services. This makes it easier for trade customers to compare the products available from different suppliers. The design sector has many of these events across the world and throughout the year. It is worth finding out which ones might be important for you to visit.

Look out for local, national and international events and talk about them with your friends and lecturers and work out how you can fit them in alongside your studies. You could get together with a few other students (preferably those with a car so you can share the petrol costs) and arrange your own visits to companies.

Being proactive and putting some time aside to undertake these extracurricular activities will stand you in good stead, and may even result in you having other things to talk about at a placement or job interview.

The majority of students on a design course in the United Kingdom have come from studying GCSEs at school or college and perhaps have completed an Art and Design foundation or pre-degree course. Having said that, students are also coming from an increasingly broad set of backgrounds and it is this dynamic mix of cultures, age and experience in the student cohort that provides a rich and diverse learning experience for all. Sharing the learning experience with your peers and embracing everyone's different backgrounds will enhance your studies and ability to work with diverse teams of people. The design industry spans the world and many companies work across the continents to manufacture products and sell goods and services. Therefore, the more awareness and empathy you have for the different approaches to design that other cultures deploy, the more comfortable you will be in the workplace, interfacing with an ever-increasing international agenda.

Being assessed and progressing through your course

The majority of you will have experienced a percentage mark and grade-obsessed approach to measuring your own levels of achievement. In higher education, however, it is just as important, if not more so, to obtain a breadth and depth of the subject. The general advice is, wherever possible, to avoid 'chasing' marks. You can easily become over-worried about what it is exactly that your lecturers are expecting from you in terms of the way you work and your level of output, but if you are confident that you know what you are trying to achieve, then it's your job to convey this vision accurately to your lecturers. It is certainly very important that you understand how your work will be assessed and against what criteria, but it is just as important to understand who you are as an individual and become more familiar with the way you naturally research and undertake design

A fashion design studio

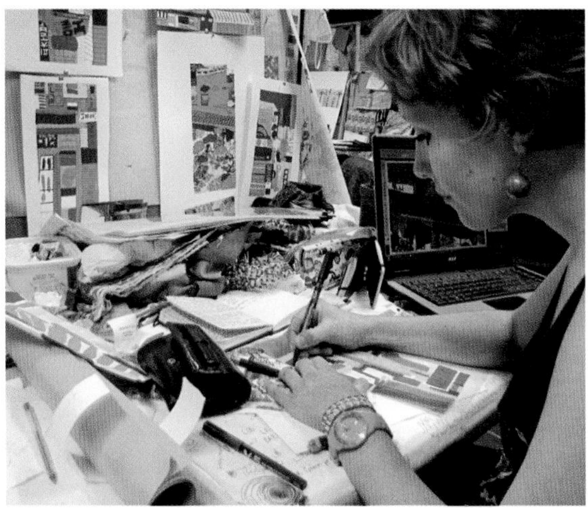

Decorative Arts student Jade Crighton developing her designs

'As a student, I found one of the ways to get the best result was, prior to submitting my research projects for marking, to seek highly critical peer review to assess whether my work was up to standard and to identify where improvements could be made. This, I feel, is a very important process that ought to be incorporated into any type of study projects. As one can be quite oblivious to his or her mistakes, critical **peer review** *can make the difference between an average or distinction outcome.'*

Lincoln Fan, graphic designer

challenges. Once you have a clear understanding of the assessment process, it is best if you put it to one side and focus on your design work and allow your creativity to flow!

If you are true to yourself, know what motivates you and are inspired by your projects, your subsequent final mark should accurately reflect your level of achievement.

In design, achievement is measured by ingenuity, innovation, the design journey and whether the design outcomes are fit for purpose and appropriate for the intended end-user or customer. Occasionally you might find that the design outcomes for a project don't work as well as expected, or aren't particularly effective in resolving the brief, but as long as there is evidence that you have been working systematically throughout the project and have documented the development process, this will usually be taken into account at the final assessment point.

It is also worth noting that the design development journey is of paramount importance. So, if all that you have done is come up with a couple of polished presentation boards right at the end of, for example, a six-week project, even if the ideas are brilliant, you might be surprised to see that top marks aren't awarded. Without the supporting development work evidencing the way you research and generate ideas, you may find that you do not meet all of the learning outcomes and may end up having to re-sit the module.

Communicating your ideas effectively, both visually and verbally, are often fundamental to the assessment process (see 'Communicating design' later on in this chapter).

Peer review: this is when fellow students give you their opinion about your work in a formal situation, probably as part of a taught session. It might be carried out as a group or individual exercise and could form part of the assessment process. It is a very useful way of finding out what others think about your work and can help develop your projects enormously. You do need to be receptive to others' opinions though.

'In my first year at university I have learned to become much more independent, especially in relation to time management and producing work for specific deadlines. I have learned the importance of taking part in all the competitions that we were encouraged to enter, as, although I wasn't successful, going through the process helped me to learn new skills and tackle new software programs. The most satisfying time was getting really positive feedback during a final assessment – it was great to hear such encouragement and praise after a few stressful months. My advice to those about to start a degree course is to embrace and absorb all the information that is given to you, don't wait to be "spoon-fed" – really think for yourself and make your own decisions.'

Sarika Patel, a final-year student

Whether you are writing an essay or report, or engaging with the design process through sketching or research, the ongoing levels of investigation are being monitored by tutors during tutorials. Find ways to 'think on paper' and document all your research and process thoughts as you progress through a project. Remember that everything you do from day one of the project counts.

Planning your time

The student lifestyle is fun, scary, intriguing and often a very new experience, particularly if you haven't lived away from home before. Creating a balance between your studies, your social time and any part-time job can be a real challenge. It can take half the degree to begin to manage your time effectively and to stop feeling that you are being pulled in too many different directions!

Some of your friends might be studying other subjects that allow them to have a lot of fun in the first few weeks of the year and then switch to studying hard for exams toward the end of the academic year. This pattern doesn't fit with studying design! It demands that you give each project your full attention from the outset, as it is hard to catch up at the end of a project. Your progress is monitored through seminars and tutorials (read on through the chapter for descriptions of these different types of taught sessions) by the tutors who will probably be involved in the assessment. Some students only find this out the hard way – they leave aspects of the project to the last minute and run out of time to complete it to a satisfactory standard in order to pass.

The first two years of the degree are crucial in gaining the knowledge and skills necessary to be able to tackle projects effectively and be able to approach

'Procrastination is a devil, especially between the Christmas and Easter breaks, four months of eating, sleeping and crying over your project can make it very difficult to concentrate. The type of procrastination where you don't even realise you're doing it is the worst; there are so many everyday things that need doing – the shopping, the washing, the dishes, the tidying and making breakfast . . . lunch . . . dinner.

Now, I'm not saying only do your washing up once a week, but it can be helpful to take one day a week to do all these things. If you can't go a week without washing your clothes, then buy some more (especially for the end of the project!). Do your food shopping once a week, put the washing on before you go out, vacuum and tidy your room when you get back and so on. These things shouldn't have a daily place because you'll just find time slipping away – especially if you tend not to get out of bed until 10am!'

Harriet Fleur Curtis, final-year interior architecture design student

Tom Atkins, designing for film and TV

'My first year has made me realise how important time management is. At the beginning of the year, my control of this was not great, but then in one lesson we planned a weekly study timetable and I've stuck to that ever since. This way I had a good balance of work and social time. The rewards come at the end of the year, when the extra effort and time invested are rewarded with good grades – so the late nights, the energy drinks and chocolate pay off in the end!'

Emma van Blommestein, fashion design student

the final year with any true design confidence. This accumulation of skill begins in year one, on day one, and every opportunity you have to improve your abilities will stand you in good stead as you progress through your design degree. In higher education the focus is on encouraging you to develop a greater awareness of what it means to be in control of all that you do and develop greater self-awareness skills in relation to what it takes to progress. This is often referred to as 'becoming more autonomous'.

'The final year is intense for someone who's set out to achieve the best grade possible. Sleeping and eating comes second to creating 3D CAD models and queuing at the print shop (a little too close to the hand-in time). Social lives become a distant memory and, at the busiest times, friends and family will text to ask if you're still alive due to the lack of contact. Surprisingly, it all becomes worth it in the end, and being picked to exhibit at New Designers was an added bonus.'

Craig Foster, product design student

How will you be taught?

Depending on the design subject that you choose to study, there could be different amounts of practical and written work expected. Practical work will consist of plenty of drawing, lots of research, working through design problems by developing prototypes and making final design outcomes. Understanding who your ideas are for and the reasons why you are doing a particular project are a very important part of studying design. Therefore, analysing the marketplace and understanding the types of people that might use or interface with your creations are also crucial to the success of a project. It is probable that you will be required to write reports or deliver presentations that include analysis of your research that supports your practical work.

Historical and cultural studies

In all design courses you will find content that covers 'contextual studies', and this might include learning about art and design history, cultural awareness and subject-specific context. This aspect of your course will demand a high level of analytical and critical thinking and you may be asked to produce essays, verbal presentations and, in the final year, a **dissertation** or similar output. This will allow you to demonstrate a deep understanding of a subject that you have become inspired by.

Dissertation: this piece of work is completed in the final year and is often illustrated and about 6–10,000 words long. A specific interest in a concept or subject is at the heart of the piece. This is a critical and theoretical piece of writing, offering personal insights into your chosen theme, supported by plenty of research.

Other subjects you might study

Alongside what are commonly known as the two core subjects, design and theory, you will possibly be offered other smaller modules that cover a diverse range of topics to broaden and deepen your awareness and further enhance other subject interests. You may have opportunities to study languages, business studies or subjects with a vocational focus, such as buying or marketing.

Appreciating the relevance of all of this and making connections between the theoretical and practical subjects that you study will enhance your creativity and make you a better designer. So, be assured that the curriculum will have been designed to offer you the broadest possible scope for you to explore and experiment fully, so that you graduate with confidence and are ready to enter the creative industries.

Size of year groups

Depending on the size of the course and the year groups involved, your learning experience will differ from one institution to another. For example, on a course with more than 100 students in each year group it is going to be difficult to meet and work with all of them. You will therefore need to build your confidence to be able to interface with students you don't know very well in group tutorials, seminars and presentations. Smaller courses may offer the opportunity for groups to work more closely together on a regular basis, which can build the team dynamic. All of these experiences will certainly enhance your communication skills – a 'must have' for the creative industries.

Participating in discussions

Receiving an effective education is about your ability to participate actively in all sessions and to feel confident in contributing to the learning experience of others. Both of these skills will develop your communication and analytical skills. Your course is bound to consist of many, if not all, of the following types of teaching activities.

What will the taught sessions be like?

Lectures – Styles of lectures can vary hugely, from formal delivery-based to a more informal activity- or discussion-based approach. The lecture might be specific to the subject you are studying or be an 'open lecture', made available to students and staff on a broader 'creative' subject, for example 'Designing Sustainably'. Lectures could be delivered by the staff assigned to the course that you are on, by visiting specialists giving a one-off lecture or by other staff from within the department. If modules are shared across courses then you can expect to meet more lecturers and specialists. Student numbers in lectures could vary from 30 to 200.

Lecture titles may not always appear to be closely linked with the subject but you can guarantee that they will always provide the necessary richness and the all-important broader context that studying a degree in design should contain. The content of lectures is often followed up and discussed in subsequent taught sessions, such as seminars or tutorials, so make a note of any particular issues that you might want to clarify or discuss.

It is a good idea to find out about other scheduled lectures happening across the institution. For example, a visiting speaker for graphics students might also appeal to product students. It is also worth noting that galleries, organisations that support creative businesses, trade fairs and conferences often have their own lecture programme where designers, makers and theorists will present their work and ideas.

First-year students Alice Hood, Stephanie Oddy, Sam Hawkins and Mike Fielding in a group

Seminars – These are often discussion-based events where you are asked to prepare beforehand your thoughts on a subject that might be linked to either your theoretical or practical work. They could also include practical exercises. The content and delivery style will vary, but they are always excellent opportunities to engage intellectually with fellow students and build your ability to reflect on your progress in relation to the development of others. As a guide, student numbers might be between 8 and 30.

Lecturer Huw Feather and Student Sam Hall in a personal tutorial

Group tutorials – These are often in relation to your design work and are very successful if you participate fully by bringing all that you have been asked to bring and are ready to contribute to others' learning and be receptive to feedback about your own work. The size of a group might usually be between 5 and 10. These sessions can take a bit of adjusting to if you are coming straight from school where group numbers studying A level art and design subjects can be small, or if you have only been accustomed to one-to-one discussions about your work.

Group tutorials are very good opportunities to raise questions about the direction of your project and to receive ideas from your tutors and fellow students. It is a good idea to keep a record of your project's progress by taking notes.

Personal tutorials – These tutorials might be about your work or about the progress you are making on the course. Many courses build in 'one-to-one' tutorials when they feel they are appropriate to the learning needs of the student.

Workshops – Acquiring practical skills is certainly high on the agenda of all design students and making sure that you attend the workshops that are timetabled for you is essential. In many institutions there is a rota system for undergraduates to access workshops, as there may be other

Brittany Delany lathe-turning plaster to make a mould

courses, for example foundation, postgraduate or short courses for industry, that require access to the same resources at different times of the week. Make the most of the time you are allocated to resources and be prompt, if not early!

Online learning – Many institutions will provide a supportive online resource that allows all students to access teaching material and other information that will support them through their studies. This has become commonplace across the higher education sector in recent years. The level and focus of the provision will vary depending on your institution. Some institutions have virtual learning environments (VLE), which are used to part-deliver modules and may even have built-in assessment tools. It is commonly recognised that, whether you are studying full-time or part-time, you may need to have a job, therefore being able to access teaching material and information about the course at a time to suit you is clearly beneficial.

Live projects – Your course may include projects that are being set, delivered or judged by a company; these are often called 'live projects'. Interfacing with the industry and testing your abilities to design appropriately for a customer type or market level is one of the most important skills you will need to develop. This will also help you to understand what it is really like to be a designer, as you will be working to tight deadlines and having discussions with industry specialists about a specific market sector or client. Whether the work is to be completed in a team or as an individual, ensure that you plan your time appropriately and that the design outcomes are clearly understood and visually or verbally well-communicated. Remember to include these projects on your CV.

Live project being presented to Autofil Worldwide Ltd (suppliers to the car upholstery industry) at their headquarters

Teamwork – One of the most important skills to develop while studying is your ability to increase your awareness of what it takes to work well within a team. Being put in a team and expected to produce effective outcomes is no easy task, but most of the time it is a very rewarding experience. Teamwork is always a subject that comes up at interview, whether you are applying for work experience or a job. The interviewer will want to gain an understanding of your personality and the role you might play within a team to understand how you might fit in with their existing team. It is also generally true that a large proportion of design is carried out in teams.

When you are put in a team at college, spend time understanding and building a clear picture of the individuals within your team and identify their strengths and weaknesses so that tasks can be given to the right individuals to maximise their motivation levels and participation. This will help to increase the team's output. If all team members have agreed a common goal and are focused on it, then the project is sure to be more successful. Students can become really agitated if some members of the group are not working as hard as others, and this can hold back

Group meeting to determine a project concept

project development. It is a good idea to tackle these issues as a team if they arise. Generally speaking, a fast-approaching deadline usually brings a team together.

Working collaboratively – You might think this is the same as teamworking, but there are fundamental differences. In this context, a collaborative project is where two or more people, with perhaps very different skills and specialist knowledge, come together to work on a single outcome that might not have been possible if only one of them had taken the project on. Teamworking is more about a group of people, with similar skill base, pooling their ideas together to arrive at a comprehensive design solution. Designer Philippa Hill will explore what it means to work collaboratively by referring to her own experiences in a story about a project at the end of this chapter.

Competitions – One of the most satisfying and rewarding ways to build your confidence and interface with the industry is to enter competitions.

Many national and international competitions will be supported by your institution, and might even be built into the curriculum to maximise the number of students producing the standard of work required, so that they have a good chance of some successes. This is clearly newsworthy for the institution and the company, and highly beneficial for you as you are receiving external recognition for your ability to design. Prizes often involve financial rewards, placement opportunities or having your work put into production, all of which is fantastic for your CV.

'I studied BA Graphic Design and was awarded the prestigious "Yellow Pencil" award at D&AD (a graduate exhibition in London). The exposure my work has received after winning this year's branding brief has been invaluable and really kick-started my career. I am enjoying the new knowledge and varied experiences while I undertake various placements.'

Catherine Perrott, BA Graphic Design, graduated 2011

In addition to these there will be many other competitions available to you that are perhaps worth entering that your institution might not be focusing on. This should not deter you from entering if you feel you have the time and the motivation to put in a good submission. Just read the guidelines carefully and submit the work in the exact format that they have specified, paying particular attention to the judge's criteria. Always photograph or scan your work before you send it so you have a record of it, particularly if it is to be assessed as part of your coursework.

Final-year student Craig Foster won the Business Design Centre New Designer of the Year award at New Designers, London 2012:

> 'Winning the BDC New Designer of the Year came as a bit of a shock. Just having the opportunity to show my work at New Designers was a goal I had set myself at the beginning of the year, but I never thought I would win a prize! I was caught completely off guard by winning the overall award and the experience was a bit surreal. Throughout the day, before the awards are announced, the judges speak to students individually and it's difficult to know who you are talking to. My advice to anyone at the show would be to talk to anyone and

Judging of a live project

everyone who shows interest in your work. Don't be too keen or pushy, but make sure you don't miss an opportunity. Anyone could be a judge or potential employer.

The award came with a cash prize and free legal, intellectual property rights and press advice. My plan is to take advantage of the situation and use the award to put my work into production. The opportunities that come with winning the award are great and it's a massive help to self-employment start-up costs, which can be a deterrent to many people looking to start a small business or consultancy.'

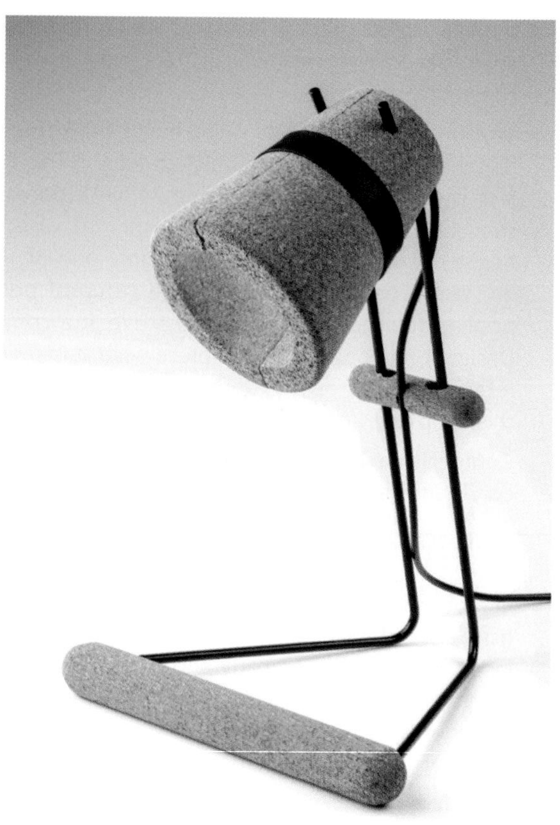

Craig Foster's light, 'Kurk' – cork and steel rod, 2012

In the event of being awarded a prize, find out a bit more about the company or organisation and who the panel members were. These individuals liked your work so what harm would it do to make contact with them a couple of weeks after receiving the prize to thank them and update them as to how you have benefited from winning? Perhaps there are other opportunities on the back of the prize; they might be happy to look at your portfolio in the future. You should certainly invite them to your degree show!

Liaising with the industry

Networking

You are going to be meeting people in the industry who may be able to help you in the future to get where you want to be. Remember to be professional at all times and always follow through by doing what you say you are going to do. Companies talk and people change jobs – everyone knows or is aware of everyone else in this industry, so remember to make a good impression, as first impressions really do count!

Design a business card that communicates basic contact details, to make the process of exchanging contact details an easy task when meeting someone for the first time. You might need some cards if you

'The whole of my class exhibited at New Designers, which was an experience. It was a little bit daunting seeing all the other graduates and their work but it was great to talk to other people with a similar research process or design method. I have got some student contacts and aim to keep in touch with them over the coming year so we can share our post-uni experience! A few companies did talk to me about my work and now my time needs to be spent following up on these leads!'

Philippa Hill, textile design student

are going to a trade fair or visiting a company. If you only need a few at a time then it is easier to design and print your own; however, there are plenty of printing companies willing to do small runs of business cards. Before you spend any serious money, check that your email address is 'professional' and that you aren't going to change your mobile number anytime soon. You might not need your address on the card, as these days email and mobile contact details are enough for the first contact.

A designer, Sarah Dryden, shows work to a student group

TIP

Start an address book that is specifically for contacts you make in relation to the industry you are entering. In addition to the people you meet when you're away from the university, remember also to include any speakers who visit the university, your peers and even your tutors, as you never know when you might need to contact them in the future!

Placements and work experience

As part of your course there may be opportunities to undertake a placement or work experience. If the whole of the third year is spent on placement, then this is often referred to as a 'sandwich degree'. Students are usually paid, and these placements are often part-organised by the institution.

A work experience might last a week or two and might be paid or unpaid, often dependent on the size of the organisation. If there is no formal opportunity built in to the curriculum to undertake a work experience, then your tutors may advise that you gain experience of the industry during the vacation periods. This will give you the opportunity to appreciate the different roles within the industry and help you evaluate which ones might suit you in the future.

'I don't think you can underestimate the importance of getting work experience. Employers look for relevant experience of working within the design industry as it shows your level of passion and commitment to do this alongside your studies. Without having various placements on your CV, it may prove difficult to find paid employment in this industry straight out of university.'

Adele Parsons, print designer

Francesca Muston is a senior editor for trend prediction company WGSN (Worth Global Style Network), responsible for the 'What's in Store' fashion retail directory. Francesca studied a Fashion Promotion degree and graduated in 2002. Her career took off through completing a successful internship with WGSN. The internship was longer than most, and money was very tight, but eventually her tenacity and determination paid off. Seven years on, she reflects back:

> 'The experience and contacts I gained as an intern helped me enormously to obtain my current position with WGSN. I did a full placement year at uni and began my fashion career with a mammoth stint of work experience lasting almost two years, working as everything from graphic designer, to catwalk stylist and even working for a photographer in New York. In February 2003 I began a two-week work placement with the graphics team

at WGSN. My placement in fact lasted six months and I had an opportunity to work across all the departments of WGSN. This really developed my writing, photography and trend analysis skills and then I was offered a job. I would like students to know that an employer is far more likely to recruit someone based on their personal experience of the individual and a proven track record than they are to take a chance on a faceless CV, so my advice is to get as much work experience as possible.'

It is a good idea to vary your work experiences and try working in a large organisation and then a small one and compare the differences and your personal responses to the different situations. It is a good idea to complete at least two work experiences before you graduate, so that you have current and relevant experiences on your CV, and something to talk about at interview!

It has also become common practice for students to consider taking a year out while studying on a full-time three-year course. Many students use this year to gain work experience, to travel and to earn money. These are all useful experiences to include on your CV and will certainly help you to prepare for your final year, not to mention your future employment. It is a relatively easy process to defer the next year of study – just communicate with your college or university and your local education authority.

To organise your own placement, think about where in the country (or world) you might be able to stay cheaply and then begin your search for subject-related companies. Identify who you might like to work for and what particular role within the company you would like to undertake or shadow. Write to them, clearly outlining your intentions and dates of availability. Include a CV and some images of recent work, if appropriate, and then follow it up! Think about how many enquiries a company such as Paul Smith or the BBC might get a year – so be proactive and have a plan to call them a few days after they receive your information and be one step ahead of the rest!

Whether your course is organising a placement for you or you are doing it yourself, here are some practical ideas to help prepare:

- Develop a full CV that captures all of your knowledge, skills and experiences. (There is more information on how to develop a CV next.)
- Scan or photograph, in advance, some finished work and development pages from your sketchbook so that you can send good quality images to companies to accompany your CV when required.
- Find out as much as you can about the company before you apply so that you are sure that you can base yourself nearby for the duration of the placement, if it is not in easy travelling distance.
- Research the size of the company and understand its place within the market. Find out who buys the goods or services, as this knowledge will be useful if you are called for an interview.
- Put a 'portfolio' of work together in readiness for an interview. Neatly mounted work to include examples of all of the types of work that you have done, together with a sketchbook, will suffice.
- Have a strategy for the different companies that you are going to apply for, i.e., have an 'A' list of a handful of the companies that you really want to work for and approach them first. If you have no success, then move to your 'B' list (and then your 'C' list) but basically keep on trying.

- Don't be put off by a placement that is unpaid. It is not an ideal situation to find yourself in, but having particular names on your CV can be priceless. Also, some companies are just too small to be able to afford it, but you could be right in the centre of the whole business and learn a huge amount.

'I would strongly recommend a year in industry to anyone who gets a chance. Without a doubt, my year as an intern at design studios in Milan and London changed my outlook on design and was a massive influence on my final year. The knowledge I gained of different parts of the process, from conception and manufacturing to costing and distribution, helped me understand that designing is more than nice drawings and making a sketch page look good. An internship or placement gives you valuable experience, contacts and also adds to your employability when you leave university. It gave me a more professional and organised approach towards my final year and definitely helped me achieve my goals.'

Craig Foster, furniture and product design graduate

Developing a curriculum vitae (CV)

A company's first contact with you might be via your CV arriving in their office in response to a job or placement vacancy, so first impressions count. Your CV should always be accompanied by a covering letter, which should clearly state the purpose of the correspondence and include details of any specific dates that may help the company make decisions about opportunities for you within their company. The letter is also an opportunity for you to tell them why you have chosen to apply to them. Take care not to duplicate information that they will find on your CV. Covering letters should always be written targeted to the specific company. Research the company and refer to them in some way to demonstrate that you have an interest in them and are familiar with what they do.

A CV needs to contain up-to-date, accurate information and should include the following:

- **contact information** – full name, email, mobile number, address (students often supply both term-time and home addresses if appropriate)
- **personal information** – date of birth
- **personal statement** – in a few sentences, capture your subject interest and future plans and include skills and attributes
- **work experience** – list them in the order of most recent first and include date, company and brief details of work or project undertaken; minimise the inclusion of unrelated, casual part-time work in pubs, etc. (a catch-all phrase that includes some examples of where you have worked and skills obtained might be enough)
- **relevant industry-related experiences/attainments** – include attended trade fairs, conferences, exhibited work, competition successes, etc.
- **skills and attributes** – be objective and list your transferable skills and abilities, including some detail where necessary; consider presentation, organisation, communication, numeracy, computer literacy (information technology), people skills and team working
- **qualifications** – list most recent first and remember that your degree is one of the most relevant experiences on your CV, so, in brief, include module titles and any specific course content that you feel would be beneficial; do include a list of qualifications post-GCSE but use a catch-all statement in relation to GCSEs themselves – for example, 'seven GCSEs including English and Maths'

- **voluntary work** – many companies look favourably on those applicants who have contributed to society in some way. If you haven't, could you?
- **interests and achievements** – other attainments, e.g. sports, music, etc.
- **other information** – include languages, driving licence, etc.
- **referees** – full contact details for two referees (one from education and one from a place of work); include their name, job title, address, email and phone number and get their permission.

As a designer you will need to spend time on the layout, content and design of such a document, as this will be introducing you to those within the 'creative' industries, all of whom have a keen eye for detail and what constitutes a good design. Think about positioning the most relevant information about a third of the way down the page as this is where the eye naturally falls first.

Many students develop two types of CVs – a traditional CV with perhaps some image interest, often in A4 format and on good-quality paper, and an illustrated CV that might take the form of a small booklet or a well-designed, full-colour document that contains images of work. Remember that not everyone has perfect vision, so avoid over-complicating the design and be wary of small text size and coloured backgrounds.

Remember to keep your CV designs simple. A CV that comes out of the envelope and transforms itself into a three-dimensional object is referred to in the industry as a 'comedy' CV and is often filed in the bin. This is not to say that you can't produce a wonderful, simple, elegant object that will get you noticed.

Here are a few examples from Amy, Victoria, Alexandra and Stuart, illustrating the way that some of the sections within a CV can be used effectively, with the information clearly communicated.

Including your skill base on the CV is very important, and these examples pictured are particularly successful as they not only list the skills but take the CV as an opportunity to demonstrate other design skills: Amy has included a visual of a mock-up of an article about a young product designer, mimicking the layout in a magazine.

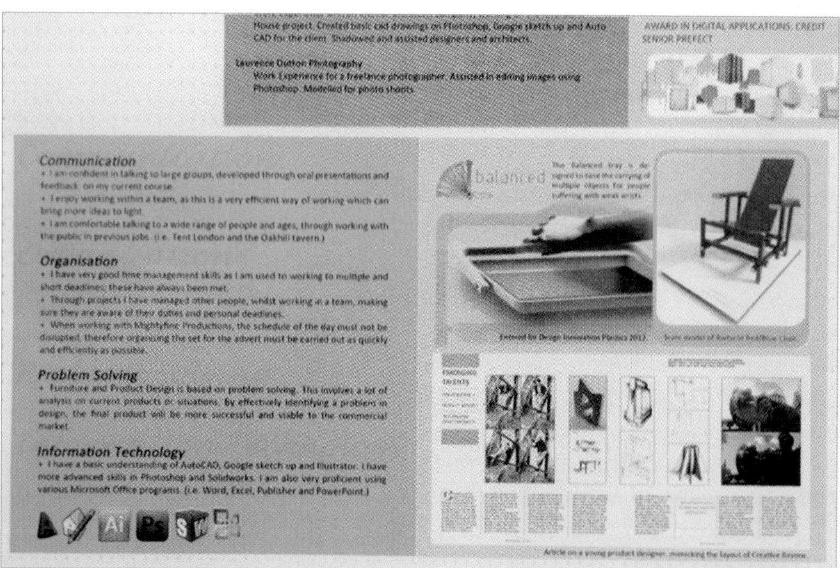

House project. Created basic cad drawings on Photoshop, Google sketch up and Auto CAD for the client. Shadowed and assisted designers and architects.

Laurence Dutton Photography
Work Experience for a freelance photographer. Assisted in editing images using Photoshop. Modelled for photo shoots

AWARD IN DIGITAL APPLICATIONS: CREDIT
SENIOR PREFECT

Communication
- I am confident in talking to large groups, developed through oral presentations and feedback on my current course.
- I enjoy working within a team, as this is a very efficient way of working which can bring more ideas to light.
- I am comfortable talking to a wide range of people and ages, through working with the public in previous jobs. (i.e. Tent London and the Oakhill tavern.)

Organisation
- I have very good time management skills as I am used to working to multiple and short deadlines; these have always been met.
- Through projects I have managed other people, whilst working in a team, making sure they are aware of their duties and personal deadlines.
- When working with Mightyfine Productions, the schedule of the day must not be disrupted, therefore organising the set for the advert must be carried out as quickly and efficiently as possible.

Problem Solving
- Furniture and Product Design is based on problem solving. This involves a lot of analysis on current products or situations. By effectively identifying a problem in design, the final product will be more successful and viable to the commercial market.

Information Technology
- I have a basic understanding of AutoCAD, Google sketch up and Illustrator. I have more advanced skills in Photoshop and Solidworks. I am also very proficient using various Microsoft Office programs. (i.e. Word, Excel, Publisher and PowerPoint.)

balanced

The Balanced tray is designed to ease the carrying of multiple objects for people suffering with weak wrists.

Entered for Design Innovation Plastics 2012.

EMERGING TALENTS

Article on a young product designer, mimicking the layout of Creative Review

CV of Amy Bicknell, furniture and product design student

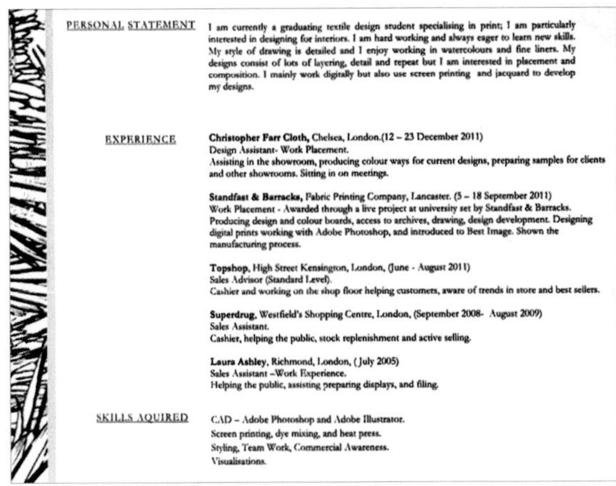

PERSONAL STATEMENT

I am currently a graduating textile design student specialising in print; I am particularly interested in designing for interiors. I am hard working and always eager to learn new skills. My style of drawing is detailed and I enjoy working in watercolours and fine liners. My designs consist of lots of layering, detail and repeat but I am interested in placement and composition. I mainly work digitally but also use screen printing and jacquard to develop my designs.

EXPERIENCE

Christopher Farr Cloth, Chelsea, London.(12 – 23 December 2011)
Design Assistant- Work Placement.
Assisting in the showroom, producing colour ways for current designs, preparing samples for clients and other showrooms. Sitting in on meetings.

Standfast & Barracks, Fabric Printing Company, Lancaster. (5 – 18 September 2011)
Work Placement - Awarded through a live project at university set by Standfast & Barracks. Producing design and colour boards, access to archives, drawing, design development. Designing digital prints working with Adobe Photoshop, and introduced to Best Image. Shown the manufacturing process.

Topshop, High Street Kensington, London, (June - August 2011)
Sales Advisor (Standard Level).
Cashier and working on the shop floor helping customers, aware of trends in store and best sellers.

Superdrug, Westfield's Shopping Centre, London, (September 2008- August 2009)
Sales Assistant.
Cashier, helping the public, stock replenishment and active selling.

Laura Ashley, Richmond, London, (July 2005)
Sales Assistant –Work Experience.
Helping the public, assisting preparing displays, and filing.

SKILLS AQUIRED

CAD – Adobe Photoshop and Adobe Illustrator.
Screen printing, dye mixing, and heat press.
Styling, Team Work, Commercial Awareness.
Visualisations.

CV of Victoria Robinson, textile design student

This is an excerpt from a more traditionally styled CV that clearly states, at a glance and on one side of A4 paper, the candidate's skills, experiences, education and intentions to become a textile designer for interior fabrics, as set out in her personal statement. The style is uncomplicated and easy to navigate. The section about Victoria's relevant experiences is positioned in such a way on the page that it is the first bit of information that your eye naturally falls upon, which is about a third of the way down the page. This has been designed purposefully, and some of the more basic sections, such as 'education' and 'referees', for example, have been placed toward the bottom of the page.

Alexandra's illustrated CV provides a snapshot of information about particular projects that she has chosen to highlight as good examples of her design abilities. She has also chosen to capture her develop-

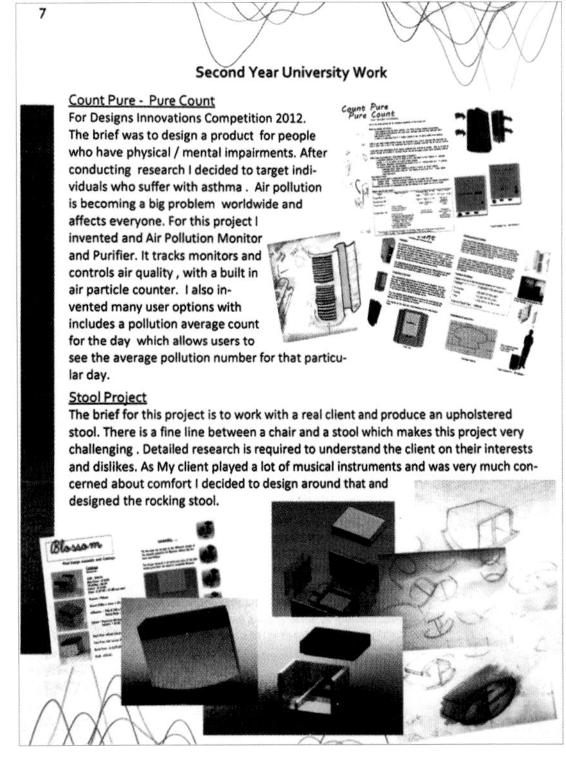

7

Second Year University Work

Count Pure - Pure Count
For Designs Innovations Competition 2012.
The brief was to design a product for people who have physical / mental impairments. After conducting research I decided to target individuals who suffer with asthma . Air pollution is becoming a big problem worldwide and affects everyone. For this project I invented and Air Pollution Monitor and Purifier. It tracks monitors and controls air quality , with a built in air particle counter. I also invented many user options with includes a pollution average count for the day which allows users to see the average pollution number for that particular day.

Stool Project
The brief for this project is to work with a real client and produce an upholstered stool. There is a fine line between a chair and a stool which makes this project very challenging . Detailed research is required to understand the client on their interests and dislikes. As My client played a lot of musical instruments and was very much concerned about comfort I decided to design around that and designed the rocking stool.

CV of Alexandra Chin, furniture and product design student

mental thoughts behind her projects by using visuals that depict initial ideas through to prototyping.

Stuart chose to return to studying and retrain as a furniture designer, after many years in business. These two pages, from an eight-page illustrated CV, clearly identify how he has developed during his studies and how he intends to move on into the design industry.

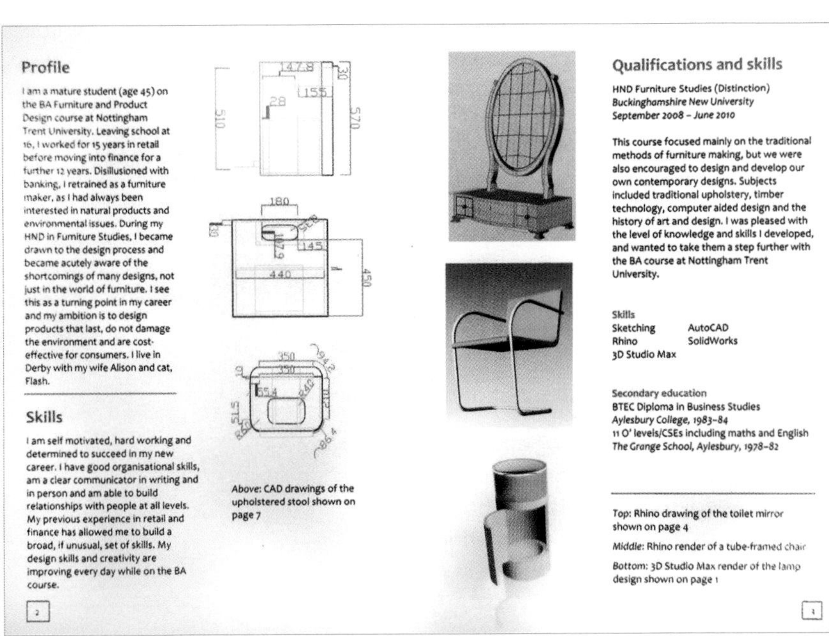

Profile

I am a mature student (age 45) on the BA Furniture and Product Design course at Nottingham Trent University. Leaving school at 16, I worked for 15 years in retail before moving into finance for a further 12 years. Disillusioned with banking, I retrained as a furniture maker, as I had always been interested in natural products and environmental issues. During my HND in Furniture Studies, I became drawn to the design process and became acutely aware of the shortcomings of many designs, not just in the world of furniture. I see this as a turning point in my career and my ambition is to design products that last, do not damage the environment and are cost-effective for consumers. I live in Derby with my wife Alison and cat, Flash.

Skills

I am self motivated, hard working and determined to succeed in my new career. I have good organisational skills, am a clear communicator in writing and in person and am able to build relationships with people at all levels. My previous experience in retail and finance has allowed me to build a broad, if unusual, set of skills. My design skills and creativity are improving every day while on the BA course.

2

Above: CAD drawings of the upholstered stool shown on page 7

Qualifications and skills

HND Furniture Studies (Distinction)
Buckinghamshire New University
September 2008 – June 2010

This course focused mainly on the traditional methods of furniture making, but we were also encouraged to design and develop our own contemporary designs. Subjects included traditional upholstery, timber technology, computer aided design and the history of art and design. I was pleased with the level of knowledge and skills I developed, and wanted to take them a step further with the BA course at Nottingham Trent University.

Skills
Sketching AutoCAD
Rhino SolidWorks
3D Studio Max

Secondary education
BTEC Diploma in Business Studies
Aylesbury College, 1983–84
11 O' levels/CSEs including maths and English
The Grange School, Aylesbury, 1978–82

Top: Rhino drawing of the toilet mirror shown on page 4

Middle: Rhino render of a tube-framed chair

Bottom: 3D Studio Max render of the lamp design shown on page 1

1

CV of Stuart Brown, furniture and product design student
stuartbrowndesign.com

The technical drawings in his CV illustrate his design ability, as do the images of his final products.

The order of the sections is up to you and will probably be determined by the amount you have to say in each and the level of importance each section has in sending clear messages about your skills and experiences to the reader.

The layout is of utmost importance, as is the choice of font and size of type. If you choose to incorporate images, then select these carefully and consider the placement of them. If you are determined to attempt to place an image behind or with the text in some way, then be mindful not to obliterate the text as it is this information that the companies want to access. Print the CV out and check the layout before you send it, as they often look very different when they are printed out.

Every time you are to send your CV out, it is worth considering who the recipient will be and tweaking the content of the CV to make it even more appropriate to them. Make sure that you utilise the personal statement section in this instance and imagine what they might want to read here. By doing this you can tailor each CV to suit each type of enquiry.

Test the content of your CV on tutors, fellow students, careers advisers, friends and family to gauge reactions, and ask them directly if it best reflects who you are and whether there are any improvements to be made to the design or the content. Then, keep it up to date so it's ready to send in response to the fast-paced enquiries from the industry!

Communicating design

By Steve Rutherford, Nottingham Trent University

To design is to create. At first these creations exist as sketchy ideas in the designer's head, in computers or in sketchbooks. At every stage of your education and career you will always need to communicate these ideas effectively to other people:

- as a student you need to understand what you have designed and who it is for
- new designers in industry need to impress their new managers
- famous designers need to excite magazine editors with fantastic images to keep their names on the front pages.

All designers need to develop the skill of being one step ahead and having new ideas daily, so communication will always be very important to you, now and in the future.

In a design context, communication has two important meanings:

1. **The act of communication** is about talking, writing and presenting clearly. We deal with this by looking at peoples' perceptions of our communication and how we might use this knowledge to create the right image of us and our work, during the design process and during presentations.

2. **The physical or virtual creations** that communicate our ideas are about developing pictures, models, artefacts, animations, presentations and reports. We look at the approaches that we can take to create two- and three-dimensional communications of our work.

Verbal presentations

When we experience any kind of communication our senses are overwhelmed with sight, sound or even smell. The brain evaluates this information and works out what we should think of this and what our perception of this should be. We use our previous experience of life to decipher what we're seeing and make sense of it. This is what happens when we take information in. But what if we are the providers of that information? What happens when we have designed something and it doesn't exist yet but we have to convince someone of its worth?

In order to communicate well as a designer you need to think widely about what is important and how you portray it. You also need to be thinking about how it might be perceived by the audience. You therefore need to know who you are addressing. Part of your communication skill set is to be confident and knowledgeable about the project, understand the audience and choose a delivery and method that is appropriate for the work and your audience.

Know your audience

Good communication is not possible unless you have investigated your audience, and knowing your audience is crucial. Communication must be appropriate, so find out who your audience is:

- if you are presenting as a competitor in a competition, find out who the judges are
- if you are presenting to a company at the end of a live project, find out if the company representative is a director, accountant or the head of design and tailor your presentation to suit them.

Delivering a presentation

Nervous? Of course you are! The majority of students find the experience of presenting their work really daunting, particularly when they don't know the audience. A bit of adrenalin goes a long way when you have to stand up in front of a group, as it helps keep your mind on the job, and if you have prepared and practised your piece, it should allow you to enjoy it a little. Clearly, if nerves and an extreme adrenalin rush take over to the point where you just can't cope, then you must seek advice about this as there will be ways that you can ease the pressure on yourself in these situations.

Students often deliver their presentations in a chronological format, where they start at the beginning of a project and 'tell the story' of how they worked through the project and arrived at the conclusion. This might not always be the best format to choose, as your audience can often guess what's coming next and you might lose their attention. Try imagining you are delivering to someone who

Joe McAlonan presenting work to fellow students and tutor Kathryn Pashley

knows nothing about your work, then try and think of what they might like to hear about first. Then refer back to your tutor's guidance and plan a presentation that keeps your audience awake. Here are some useful pointers to think about:

- Don't overrun your time; this is unprofessional and generally unforgiveable. It means you won't get to the end if you are stopped, and the end is generally the whole point of the presentation. If you have people presenting after you it is very unfair on them. The only way to keep to time is to practise!
- Deliver the presentation in a positive manner.
- Draw attention to the strengths in the work.
- Reflect on the areas of the project that could have been improved.
- Explain who the designs are for and why – e.g., market, customer, etc.
- Involve the audience by passing pieces of work around or posing questions.
- Avoid using slang or turns of phrase that are casual, such as 'you know what I mean' and 'like'.
- Use the correct terminology to describe technique and process.
- Make eye contact with the audience, even if your presentation is being projected behind you.
- Don't fiddle with clothing, hair, jewellery, etc. or move around unnecessarily, as this can be very distracting.
- Pitch your voice so that it can be clearly heard by the people at the back.
- Speak more slowly than you think you need to and pronounce your words clearly.
- Verbal presentations are not reading exercises, so avoid bringing sheets of paper with you! It is your project that you are delivering. If you really need something, then a few key words on a piece of card should be enough to prompt you.

- Have back-up presentation plans. If you have produced an electronic presentation, put it on a memory stick and have it available online too. Have a print-out of your presentation so that you can carry on if the projector or computer dies on you. Check the resources before you start and make sure that what you want to do is compatible with what is available.

It is a nice touch to end with thanking the audience for listening. You can also invite them to ask any questions and you may receive feedback on your work at this point. If you do get asked any questions then give honest answers; it is the act of audience participation and overall engagement on both sides that the tutor will be looking for, and not just a well-rehearsed and delivered verbal presentation.

Here are two accounts of verbal presentation experiences – a good one and a not so good one.

Example of a good presentation

Amy March, a second-year student, produced a very good presentation that went down well with her tutor and her fellow students. Here's her story:

> 'When I stand up to give a presentation, I have no idea what I am about to say. All I know is what I'm meant to be talking about and I know exactly how I want my audience to react. I want my audience to want to hear what else I have to say. I want my audience to feel part of my presentation, and react as if I am having a personal conversation with them.
>
> I feel nervous throughout the whole experience, and it seems to happen so fast that I'm always surprised that I've finished my presentation already and especially surprised if I've captured everyone's attention. It's important for the audience to feel involved in your discussion and that they're not just being talked at! I always strive to obtain eye contact with every person at least three times, making it feel more personal to them.
>
> At the beginning of a presentation, I automatically assume that nobody is paying attention and that they're already bored with me talking. It's my responsibility to gain their attention, and my way to do this is to ask a question and create a "system" so that each and every person has to respond physically at some point during the presentation, by standing up, moving around, talking or raising a hand.
>
> If you have ever had to sit through a presentation, you know how frustrating it is if you can't hear what the person is saying and how boring it is when they're talking as if you are a flock of sheep. If you are giving a presentation, you should deliver it how you would want to be receiving it and speak loud and clear.
>
> All information in a presentation should be delivered concisely and tell the audience what they need to know in just a handful of sentences. The most important rule in presentations is to have fun: if you are not enjoying your presentation, how can you possibly expect anyone else to?'

Example of a not-so-good presentation

Jeremy McGuinness (not his real name), a second-year student, tells the story of a bad experience giving a presentation:

'It is a curious situation standing in front of your peers presenting your work. I myself am confident in my abilities and would say I'm articulate, although this is not the impression that I project while standing up there. I use the phrase "standing up there" because this is not a relaxing situation for me with all those faces judging, comparing, analysing you.

I put a vast amount of pressure on myself and the seeds of doubt start initially in the "waiting period", hoping it will be over soon, but then, at the same time, not being quite ready to go up on "stage".

The introduction to the work can make or break the scenario and determine the success of my overall oration experience. If this first bit goes wrong, a whole domino of symptoms occurs:

- the dryness of the throat makes me more self-conscious of the words that I am speaking and doubts come into my mind about the validity of these statements
- fidgeting and playing with objects is a particularly bad trait of mine and I have been known to hit presentation boards against tables without realising this in the middle of a nervous panic
- a hot burning sensation occurs in my face as my speech crumbles; I then become paranoid about my attire and mannerisms.

I know this is a psychological issue about delivering presentations and I also know that I am entering an industry where contracts can be won and lost on the success of these, so I continue to practise and hope they will get easier.'

Presentations are just as important in the real world. Design is an investment in the future and everyone involved needs to be as certain as they can be that the investment will be a profitable one. You might find yourself presenting to senior people in the business, from areas other than design – accounts, sales and marketing, for example.

If your work is presented in a graphically professional manner, with well-produced physical 2D and 3D artefacts and the content and language are pitched appropriately, then your audience will be delighted. However, if there are gaps in your work that you were hoping no one would notice, or even if you used a really blunt scalpel to cut a ragged line on some mounting card, then that is what your audience might remember of your presentation. That's why you have to be very, very professional all of the time.

It is worth becoming very familiar, in advance of a presentation, with the CAD software (e.g. Photoshop and Illustrator) that can help you scan, manipulate and collate great images for your presentations.

Interviews

Prepare! Prepare! Prepare!

Whatever the interview, and whatever you perceive the opportunity to be, always do the necessary research so that you feel knowledgeable about the company, the people and the project. This early preparation will go a long way to keeping your nerves at bay. Students are always more nervous in interview situations when they haven't done the background research. Here are some suggestions for preparation:

- find out how big the company is
- try and identify how many staff are employed and in what capacity (the business section in county libraries will be able to help you do this if the website doesn't reveal this information)
- find out about the company's vision and look out for press releases that identify the company ethos
- find out about their product ranges and who their current (and perhaps even future) customers are
- build an awareness of the trade fairs that cover the sector; for example, if the interview is with a manufacturer selling to retail, then ensure you visit the stores to gain a greater appreciation of the products and responses to them.

Then, ask yourself what they might want to know from you during the interview. A good way of tackling this is to imagine being on the interview panel and thinking about the purpose of the meeting. Every interview may have a different set of questions, but, in general terms, it is people-to-people communication that comes high on the agenda every time. It won't be appropriate to say very little and hope that a brilliant portfolio will prevail, and it often isn't relevant to focus solely on your work and talk through every project in detail. The company will want to get a feel for your personality, together with your skills, and will want to assess you as an individual and determine whether you will fit in to their existing team.

Write down the questions that you think they might ask you and then spend some time thinking about your answers. Always read through your CV, application form and any covering letter just before the interview, as this is the paperwork that will be in front of them when they interview you.

Don't forget that interviews are a two-way process – you are also there to interview them, so have questions prepared for them too. Advice has always been that it is better to accept the job if offered and get into the industry, and then move around once you have worked out what you really want to be doing. Clearly you will be focused on getting your first job or succeeding at an interview for a placement, and this may cloud your decision-making. However, sometimes students hold out for a particular job with a particular company, which is clearly a risk yet occasionally pays off.

Visual communication

As a student your work will be assessed on your ability to meet the various criteria for assessment. These vary in detail from institution to institution, but cover research, design process, synthesis of ideas and development and execution of the final idea. Your tutors need to see evidence of you meeting these criteria.

'I found that one of the most crucial elements or skills within the course was to be able to clearly communicate your designs. I feel it is important to take time to consider, when planning your intended output for the project, what are the best and most efficient ways to get across your design, as without clear communication a perfectly good design can be overlooked or misinterpreted.'

Hollie Cleaver, design for film and television student

Two-dimensional communication

This covers pictures, reports, animations and presentation boards. The range of two-dimensional (2D) communication methods available to designers is huge. Traditional practices, such as drawing, have in some instances been superseded by technological solutions. The range of design practices also means that some designers don't rely on traditional drawing – they might use CAD or produce collages of imagery/objects. As with many aspects of design, there is much to be learned from looking at how other design disciplines practice, and the earlier chapters on the various disciplines will give you a wide range of examples.

Emma Storr, a second-year textile design student, explores the importance of effectively presenting her travel and luggage accessory project for men:

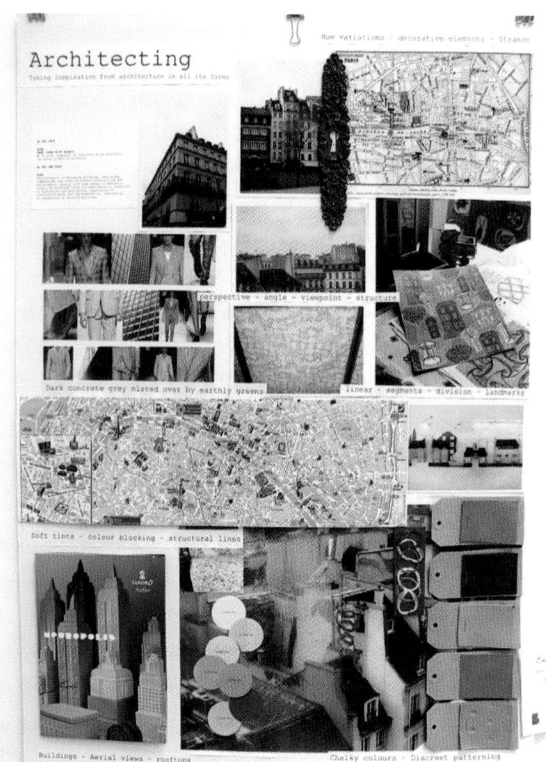

Mood board by Emma-Louise Storr

> 'The initial research for this project began on a study visit to Paris. The theme had been easy to come by, as the architectural presence and eclectic mix of old and new became evident very quickly. The product wasn't as straightforward and began life as leather satchels before arriving at travel luggage and accessories for the menswear market. After collecting many photos, visiting the Maison et Objet exhibition and many galleries and museums, I began to write down the approaches I wished to explore. Using textile exploration as a sketchbook, I began to sample and it morphed from a traditional view of architecture to a more aerial approach, removing elements to leave shapes and lines.
>
> I was really inspired by the architects' ideas and the plans and structures that are created. I wanted to explore many avenues, including building fronts, angles, rooftops, maps and aerial views.
>
> A mood board was created at the beginning of the project as an inspiration tool to help pull all the conceptual and visual ideas together. The board is made up from photographs, maps, and current trend images for fashion and interiors, colour swatch leather samples as my main fabric choice, catalogues from trade fairs and key buzz words and sentences to inspire my ideas. Having visited Paris in January the colour palette was muted. I have always been draw to the more muted tones within my work. The tones chosen worked well for menswear, and hints and accent colours were added where necessary.
>
> I developed a full range of samples and then I had a clear idea as to how I would like to portray the work. I decided to develop it into a book. Developing a digital book was daunting but soon became exciting. On realising the roots of my ideas were based on Paris, a city first seen from a folding map, I had the idea that the layout of the book could be concertinaed. This was very ambitious and difficult to execute. Having struggled with printing problems, layout issues and too many pages, I had to rethink and decided to create the effect by only printing on one side.

The pages were all punched and threaded together to represent the interlacing of a city, and during the CAD process, five strands of ideas were designed, which flowed throughout the pages, to represent my final product's outcome for "Architecting". The five themes were buildings, people, road infrastructure, ecology and green space.

The book was printed onto card, and additional smaller pages contained actual textile samples – creating a tactile approach to a digitally printed book.

Concertinaed idea development book – research
Emma-Louise Storr

These particular pages of my book are the introduction to the strands that run throughout the book. The strands represent my found understanding of the architecture of Paris, and appear in the book as linear colour blocks. The large images were selected from my own collection taken in Paris during a study visit, and these are overlaid with images taken from my samples. Sampling for this project included: embroidery, laser cutting, print and jacquard weaving.

This is where my thoughts and ideas began. The pages have samples and photographic images over laid with buzz words and ideas, continuing across to a scaled-down image of the mood board for the project to support the initial ideas positioned adjacent to the finalised colour palette.'

Emma-Louise Storr, second-year textile design student

Developing a colour palette
Emma-Louise Storr

Wherever you study, there will be a range of different design disciplines. Find an excuse to get to know about them. Is there a design society? If not, start one. Are there multi-disciplinary projects where you get to work with other disciplines? If not, suggest it. Are there end-of-project presentations on the other courses with work pinned up for assessment? See if you can get in to have a look before it's taken down. Ask your tutor to okay this.

Many projects start with 2D as it is quick and accessible. Some projects might start with a 3D investigation if the discipline and the object are best portrayed that way. For instance, a ceramicist might want to build in clay rather than do a sketch of the piece. Some disciplines, such as graphics have projects that are purely 2D and others have projects that are four-dimensional. Time is the fourth dimension – animations or interactions with products or systems have this added complexity.

Two-dimensional communication encompasses any flat work produced during a project. There are plenty of visual examples of these in the different design discipline sections and project stories. These include:

- sketches – worksheets, sketch pads, doodles
- colour-rendered drawings – colour sketches, finished high-quality visuals
- computer outputs – production drawings, templates, patterns, 3D renderings, fashion simulation boards
- presentation boards – concept designs, technical descriptions, final designs
- mood boards – trend and market analysis, customer profiles.

Your course will introduce you to the traditional and new methods for producing two-dimensional work in your particular discipline. As ever, your interest in how other designers communicate is crucial. You will be working with them in the future in a team and you might, like many designers, work across disciplines.

Designers should be good at drawing. You want everyone you come into contact with to understand what you are communicating in your drawing. In some areas of design there is an increasing accent on the teaching of drawing now because a larger proportion of the design process takes place on screen. This has

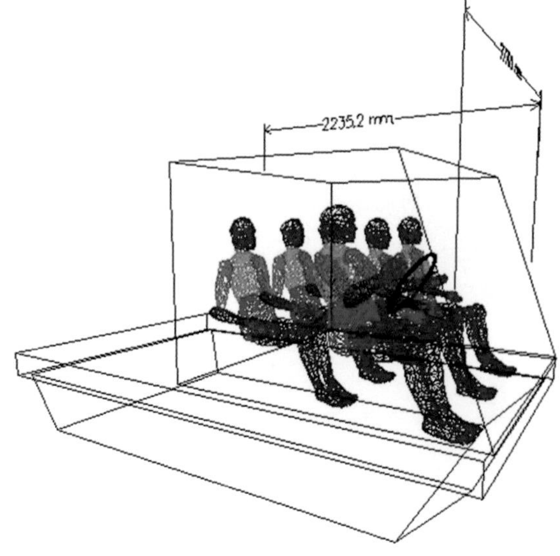

Computer calculation of vehicle interior

A Photoshop sketch

reduced the role of drawing in the design process and designers' skill levels in this particular area have suffered. However, it is still a fundamental part of the act of design and the industry realises the importance of it. The above examples demonstrate the need for this skill.

You need always to select the best tool for the job. There are some things a computer is good at – working out dimensions (as depicted in the first image, for example), and some things a sketch is good for – a quick impression of a finished design (as shown in the second). If you meet an important potential client on a plane at 35,000 feet you need to be able to sketch convincingly on a napkin with their pen. This requires a certain amount of confidence. If you're in a meeting with a client and you're talking about modifying their design, you can't always have your laptop and computer-aided design software running next to you, but you should always have a pencil.

How you organise and present this two-dimensional communication at the end of a project needs serious consideration. It is safe to assume that keeping your work sheets in some kind of numbered, chronological order will have benefits for you, both during and at the end of a project. Also, your tutors will want to see your design process and understand how you have managed it. A pile of A2 sheets of paper strewn across a table can inhibit communication. You need to maintain some order to a potentially chaotic pile of work.

Some courses encourage the use of large sketchbooks, where sketches, images, samples and various other media can all be located in the same volume. Other disciplines might require ideas sheets that are numbered and later bound together to hand in. You will develop the ability to work in various ways, depending on the type of project, however, the key (whatever your method) is communication. The work should be accessible and understandable.

Three-dimensional communication

This includes models, test rigs, prototypes, garments, artefacts and samples. As stated earlier, there is a need at times to communicate your designs to people who may not be able to read or understand drawings. Three-dimensional artefacts always make an impression in presentations.

A picture might be worth a thousand words, but a model/prototype that looks and performs like the final concept can be much more useful in communicating your ideas.

This model of a stool developed by Scarlett San Martin, a second-year student, depicts the final concept perfectly.

Stool by Scarlett San Martin, made out of wood, aluminium and silk

There is much more variation in three-dimensional communication in the design process across all the disciplines. Much of it, again, is discipline-specific, however, as designers you never know what you might end up designing, so a wide knowledge of the possibilities will be useful in the future, as these examples demonstrate.

Fairline Boats Ltd make full-size mock-ups of their powerboats from plywood, medium density fibreboard, cardboard, nails, screws and sticky tape. This gives the engineers, stylists, interior and furniture designers a chance to discuss and develop the new

Full-size model powerboat that you can climb into, by Fairline Boats Ltd

Toile development – Lucy Clark, fashion and knitwear student

designs in great detail. The models are so solid that it's possible to get inside them and explore ideas for room layouts, the size of the shower, stairs, dining areas, etc.

These designs also exist as detailed 3D computer models, but there is no substitute for being inside a physical space when you are trying to ensure that everything is designed well. The designers, directors, clients and engineers need to be able to climb into the boat to understand the issues that the designers have had to respond to.

Here also, Lucy visualises the three-dimensional aspect of a garment by creating a toile. This will let her explore the construction and consequent drape of the cloth in her garment.

Glass designer-maker, Lharne, produces a wide range of artefacts, but the making is a large part of the design process. Exploring what the material can do is often more appropriate than sketching ideas out beforehand. Pushing the material to its limit is only possible by working through the ideas in three dimensions using the right materials.

Developing a prototype – Lharne Shaw, glass designer
The Harley Gallery and Foundation

Computer-aided design (CAD)

This is a subject that crosses the disciplines. The discipline-specific software you will use and how it fits into your discipline you can only get from your course. However, the issues are exactly the same as we've been discussing here for 2D and 3D work – good communication is the key.

CAD is a tool that will be in your future toolkit. Like all tools, you have to use it at the right time, when it is the right tool for the job. As stated earlier, drawing is important because it's quick and accessible, like when you're on an aircraft and you haven't got your laptop with you. However, there are other times when CAD can deal with things – scale, sizes, etc. – that a pencil is less good at.

This is true of 3D too – sometimes cardboard and gaffer tape gives the control and feedback you need to map something out in 3D. At other times, CAD will give you a photo-realistic 3D view that will really sell your ideas to a client.

As with two-dimensional communication, the various design disciplines have their own traditions of three-dimensional making involving specific materials and processes and your own course will introduce these to you. However, the creative industries are always looking for a broader viewpoint and, you will need to cultivate a wider appreciation of how the design disciplines operate. As is often stated, the multi-disciplinary nature of design is a crucial part of your future and your development, so refer back to the design discipline chapters for more information on the range of ways designers communicate.

Developing a portfolio

The word 'portfolio' is a term used regularly in the creative industries and can often mean different things to different people across the spectrum of disciplines and at different stages throughout a career. Most people think of a portfolio as a flat black case containing design sheets ranging in size from A4 to A1, but you might need to develop a system for carrying three-dimensional prototypes, so think carefully before you invest in the wrong sort of carrying system.

At the end of your second year of study it would be a good idea to spend some time really thinking about the sorts of projects that you would like to see in your portfolio by the time you finish your degree. The final year of the majority of design degrees is organised in such a way as to allow you to be highly creative and exploratory and to undertake the projects that you want to do. Think about the potential career choices that you are beginning to focus on. Ensure that the projects in the portfolio demonstrate a wide range of skills and abilities that the industry will want to see, and at the same time maintain a focus towards the type of work you think you want to be doing in the future.

Aim to show a variety of different projects that demonstrate as wide a range as possible. The contents of your portfolio will change to suit each interview scenario, so build it to allow flexibility to change the content. It is fine to have a portfolio that

CAD presentation boards by fabric designer, Susan Hall

Product samples and research reports by Oliver Hammond

2D and 3D designs for film and TV by Hollie Cleaver – degree show

contains unique and creative art works that might make a fantastic design statement at the end-of-year show, but fundamental design skills, such as drawing and computer-aided design skills, will also need to be reflected in the portfolio. You need to demonstrate a clear understanding of what the industry requires and expects from you in terms of subject-specific knowledge.

For example, if you are developing two-dimensional imagery for a graphics or textiles application then ensure that the designs have been developed using the standard dimensions that the industry use. Companies will also want to see examples of how you think as a designer, so take sketchbooks or work sheets that detail your thought process.

Your portfolio may be required to sell itself without you present, as in this online example of Craig Foster's website. Some companies might look at your portfolio first, then interview you. The contents therefore will need to be self-explanatory. Clear and effective visual communication is a skill that all employers in the creative industries are looking for. A good tip is to place your strongest projects at either end of the portfolio. Also include any projects that have won competitions, together with any press releases.

To summarise communicating design, being a good designer is all about doing this effectively. Think about the signals, information, concepts and image of yourself that you send out through the various presentation modes. As designers we need to be aware of and control what we produce. There is a business ethos that we also have to create, and as practitioners you will be interacting with managers, accountants, entrepreneurs, bank managers and all manner of specialists.

Everything you produce, from your first big freelance job down to your business card, will be scrutinised. To paraphrase many top designers – 'It is all in the detail'.

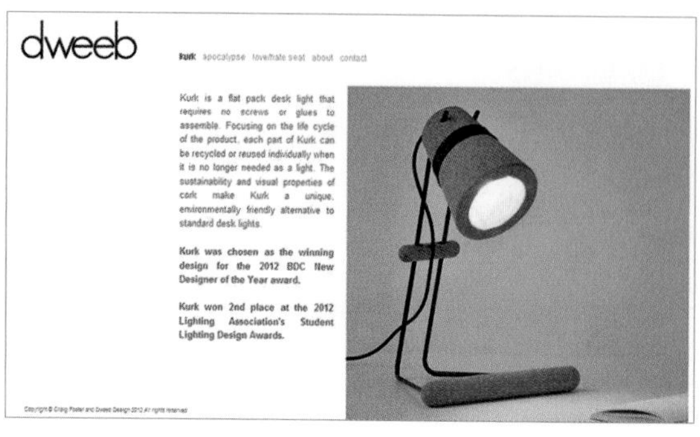

Craig Foster's website – www.dweebdesign.co.uk

Conclusion

This section has walked you through what it is like to study design to degree level. It has pointed out, along the way, the skills you need to develop and the approaches you need to take to succeed in this industry. Keep a focus on your reasons for wanting to study design and remember your enjoyment when being creative. The industry and the specialisms within are broad and varied in their use of design ability, yet it is a dynamic and ever-responsive industry to people and their needs and desires. It is certainly going to take a lot of hard work and determination to succeed, but keep a clear vision of what it is that you want and work hard towards your goal.

Further resources

Books

Cottrell, S., *The Study Skills Handbook*, Palgrave Macmillan (2008)

This is a good all-rounder full of tips and checklists.

Evans, C., *Time Management for Dummies* (UK edition), John Wiley & Sons (2008)

One-stop guide to improving your strategy for getting on with what needs doing. Contains plenty of tips and tools to support you.

Peck, J. and Coyle, M., *The Student's Guide to Writing: Spelling, Punctuation and Grammar*, 3rd edition, Palgrave Macmillan (2005)

A classic guide book, favoured by lecturers; use this and they'll love your essays.

Pink, D.H., *Drive: The Surprising Truth About What Motivates Us*, Canongate Books Ltd (2011)

Explores creatively ways in which you can control your own approaches to becoming motivated.

Visocky O'Grady, J. and Visocky O'Grady, K., *A Designer's Research Manual: Succeed in Design by Knowing Your Clients and What They Really Need*, Rockport Publishers (2009)

A useful research manual, written for graphic designers but of use to many.

Websites

www.studentdesigners.com

A social networking site for students studying design.

Project story

Designer:
Philippa Hill, textile design student

Project Title:
Working collaboratively

Tell us about the project

As part of my degree I worked collaboratively with a fellow knitwear student, Marie Leiknes, contributing to her fashion collection, which she called 'Heim'. My own project, 'I am the Jumper', was inspired by traditional fishermen's jumpers and heritage textiles. My project was dedicated to my Granny and was a very personal project, which made me more determined for it to be successful. (Refer to Chapter 3, 'Fashion and textile design' to read more about this project.)

How did you find out about working collaboratively?

Collaborations are encouraged at Edinburgh College of Art, where I studied, but students are not forced to work together in their final year. It is up to the student to pursue collaborations and make them work. As part of the final year, ECA held an event called 'Revel and Engage', which enabled each design student to display their work and interact with each other to identify possible collaborative opportunities. This is where my two collaborations began. I collaborated with a Performance Costume student, Cicely Giles, and with a Fashion student, Marie Leiknes. I had noticed Marie's work in the print room previously and, although I had never spoken to her, I could imagine working with her, so I made it my priority to seek her out at the event. On talking to her I discovered that her inspiration was similar to mine, she was concentrating on looking at coastal knit from Norway. We then had several meetings between ourselves and sometimes with tutors.

My second collaboration, with Cicely, came about because she was looking at a fisher folk-tale – 'The Whitby Witches' by Robin Jarvis – in which some of the characters wear gansey jumpers. This collaboration was more self-initiated and rarely involved tutor contact as I was now more familiar with the process. After development of prints over two months, it was completed by printing seven metres of fabric for Cicely in which I used five different print processes. When asked about this collaboration and whether I had had enough time to do all my work, I always answered that to work collaboratively was always something that I wanted to do and saw it as an opportunity to apply my work in a different way. I thoroughly enjoyed working with Cicely as a designer – the imagery and techniques were very different from the development in my own sketchbooks.

How did the collaborative process work?

My collaboration with Marie was much longer and involved a lot of teamwork and development. Both our individual concepts were developing, which allowed us to explore a variety of ideas together.

In January we started to finalise the collection, but it was only about eight weeks before completion that the making really started. My collaboration was one of just a few in my year group and the success of collaboration is down to the individuals involved and their mind-set.

As a textile designer working with a fashion designer, I sometimes felt forgotten about. Sometimes large amounts of work were expected of me without

Ideas board

any awareness or knowledge of how long an individual process would take to develop a textile. I don't think other students realised the amount of experimentation that was involved and the lengthy trial-and-error process. My collaboration with Marie lasted throughout the majority of my final year and it was difficult to carve up my time efficiently to produce my own collection simultaneously. While my classmates focused solely on their own collections, I had to think about Marie's deadlines too.

Marie Leiknes and Philippa Hill – collaborative fashion collection 2012

Were there any specific skills you needed to work effectively?

I learnt some great skills collaborating – for example, transferable skills such as teamwork, cooperation, how to share ideas and how to avoid being overprotective about your own work. I also realised that working collaboratively mirrors the real world much more, as textile designers are rarely expected, or able, to work as individuals.

What were the benefits of working collaboratively?

The collaboration allowed my print designs to be shown in context, applied to garments, which for me was a key thing to demonstrate in my portfolio! I received positive feedback from my tutors and fellow students about the work. Yes, it was harder near our final deadline to complete both the collaborative and personal projects, but the press and publicity I gained from being part of the Edinburgh College of Art Fashion Show and Graduate Fashion Week was well worth it! (It is normally only fashion students that show work at an end-of-year fashion show and at Graduate Fashion Week). My own project, 'I am the Jumper', was influenced by my collaboration and led me to work on a variety of different fabric surfaces that I would not have considered otherwise.

What advice do you have for students?

During my final year it was quite hard to balance my studies with my part-time job and other commitments. My recommendation for your final year would be to prepare yourself over the summer so you have a running start at it, and when it gets really tough and demanding remember you're only human, but make sure you have your priorities sorted!

I loved exploring and doing various projects in the first years of my course. Sometimes I felt lost or uninspired, but you've just got to push through, take risks and try things out and be prepared for good and bad criticism. In the end you can only try your best and you should always do what you want to do to enjoy it. Learn from your classmates and different year groups and share ideas and techniques. Try not to be competitive with each other! Everybody's work is different.

www.philippahill.co.uk

Project story

Designer:

Alice Walsh, furniture and product designer

How to stay motivated

Tell us about what motivates you

Throughout my life I have always been fairly motivated. However, my motivation comes from an amalgamation of other character traits: stubborn; competitive; conscientious; diligent; impatient; inquisitive; and resourceful, to name a few. These character traits have meant that I have always sought to achieve whatever I have set out to do.

Motivation, for me, has always stemmed from the desire to do something you enjoy. Likewise, lack of motivation has always stemmed from something I do not enjoy, do not care about or in some way am afraid of.

My motivation for art and design was quite natural. For my first 15 years of life it was a hobby. It was only when I chose to pursue it as part of my A levels that it became significant to my career path. I have to thank my school for a certain amount of my motivation. They were flexible enough to allow me to do both Art and CDT (Craft, Design and Technology), a timetable clash that had never been requested by a previous student. I was given keys to the art and design departments and the flexibility to ensure I could do both subjects using my extracurricular hours, on the condition that I achieve good grades and make this a viable solution for future students. I felt honoured that they trusted me to maintain this, and motivated to achieve the results and allow other students to do this in the future. This was an example of positive feedback encouraging my motivation.

On the flip side, a bit of negativity also motivated me – my art teacher had warned me about applying to what we perceive to be the very best institutions, saying that I was not good enough! Whether this was true or whether this was a psychological bluff I'm not sure, but either way I went on a mission to prove him wrong and built up a portfolio that could not be refused.

Art college and university put a different spin on motivation. At school I was motivated to achieve results and pass exams, whereas at university I was setting myself up for my career in art and design. University was a less spoon-fed environment, one that still required you to jump through hoops but here I had the freedom to choose how I went about this. I soon learnt to balance work and play and spent four busy, sociable and hard-working years studying.

What other things were important to you during your studies?

What was apparent through my time studying was how important friends were. Having a great bunch of friends that were on the same course as me allowed work and socialising to go hand in hand. We all motivated each other along the way, encouraging, supporting and competing with each other. Maintaining a busy lifestyle and being motivated out of work-hours meant being motivated in working hours.

My industrial placement in my third year at university probably had the largest impact on my chosen career path and taught me invaluable skills that I could take back into my final year. I worked at Habitat and spent eight months in the design team doing everything and anything, from making tea, to creating exhibitions, making models and designing products. It taught me the importance of working to deadlines (even if they are self-made), client relationships, attention to detail and time efficiency. I returned for my final year where I proceeded to work 9am–6pm daily to achieve a first-class degree, without the stress factor, and still retained a good social life. The placement also taught me that it is not wrong to ask for help. I built relationships with factories and encouraged them to give me some free technical advice, support and skills to help with my projects, something that we all do every day in our design jobs.

Have there been any benefits to remaining motivated?

Upon completion of my degree, I was diligent in standing by my work at the degree shows and waiting patiently for any enquiry that might come my way. It was boring, as nothing would happen for hours on end, but it paid off as I received a placement offer from Mathmos and Philips Lighting, and Habitat offered me a job.

My career continued from Habitat to Tom Dixon's Design Studios, then on to the company Forpeople and then Conran. I have also been self-employed and worked for a variety of clients, including the companies just mentioned. Working for myself allowed me to use all of the skills that I had learnt in my previous positions and be as diverse as I wanted with them. Motivation is key to succeeding as a self-employed designer, as clients don't just roll up – you have to go and get them.

I have been lucky to have never 'burnt any bridges' throughout my career. Your personality and contacts are as important as your skills, and in the design industry this is even more important as everyone seems to know each other. I feel very lucky to be able to do a job that I initially deemed a hobby.

Cufflinks by 'Alice Made This'
alicemadethis.com

My motivations lie in seeing my initial sketches through to production, meeting deadlines, answering briefs, travelling to factories, exploring materials, realising pragmatic solutions, building relationships, learning about new technologies and making a living from something I really enjoy.

Lucky? Yes! But someone once told me that you make your own luck and this, I believe, is true.

alicemadethis.com

Project story

Designer:
Anna Deery, interior architecture and design student

Project title:
Placement at Atelier, New York

Tell us about the placement

In the third year of my four-year course at university, I was lucky enough to get a job in an architectural firm, Atelier NY (New York). The office functions in a dynamic studio atmosphere situated in Queens, New York City. This is a small, diverse office with a team of fifteen individuals coming from different levels within the architecture and design field and different parts of the world. Each member of the team takes part in experimentation and creative problem solving on a daily basis. Atelier NY offers opportunities for people from around the world. With such a multicultural work environment, it stimulates a unique design dialogue and invites investigations on design trends and philosophies. The employees at Atelier NY have vast experience with the New York region's local state and building authorities and approval processes. The capabilities in this area are demonstrated by numerous successfully completed projects throughout New York.

I was given the opportunity to get hands-on experience from residential, private residential, commercial and institutional (healthcare) projects. With that, responsibility followed and I was able to help in the coordination of projects, construction of budgets and purchase orders and updating them accordingly.

What processes and skills were most relevant to this project?

My reasoning behind applying for an architectural practice, rather than a design firm, was to get a more technical background, which I believe was a side of architecture that was not explored enough during my first two years at university. I wanted to build upon my knowledge, skills and attention to detail as well as building upon my conceptual design processes.

Project presentations at university enabled me to develop my presenting skills, which I found gave me the confidence to put my design ideas, views and comments across.

My design skills have been developed through the design process using Atelier NY's approach – the communication of information, idea sketches and computer-aided design – therefore enabling me to discover, first-hand, the translation of theory and how it develops into a three-dimensional reality.

My main responsibilities included:

- AutoCAD drawing sets
- coordinating and attending meetings
- taking clients on a 'walk through' of selected furniture
- creating budgets and purchase orders
- coordinating sales and deliveries
- undertaking product selection
- undertaking cost comparisons
- producing schedules for fixtures and fittings
- coordinating site surveying
- making models for clients.

I learnt a lot on the job and was able to put what I had learnt at university into practice. Atelier NY has enabled me to venture into numerous aspects of architectural practice. I am now a confident AutoCAD user, which I wasn't before the placement opportunity, having only had a brief introduction to it at university. I was trained in the development of technical drawings and I now understand the importance of accuracy and attention to detail. My technical drawing ability has improved through the construction of:

- ceiling plans
- furniture detailing
- drawing sets
- window and door schedules
- sections and elevations.

What challenges did you face?

I have gained copious amounts of experience in surveying and site visits, which has taught me the importance of detailing and time keeping. During the site visits, I attended meetings with clients, contractors, project managers, landlords and furniture bidders, which has built up my confidence, not just as an architect, but as an individual.

My role within Atelier NY was multi-dimensional, and exceeded my expectations as an intern. After completing a 36-week placement at the company, I was able to venture into four main areas of architecture – commercial, institutional healthcare, residential and high-end residential.

Anna Deery's kidSPACE design

What were the highlights of this project?

The experience in a professional architectural practice has enabled me to grow and develop as a designer. I have demonstrated the professional skills and interests needed to proceed not only onto a further academic year at Nottingham Trent University, but to a professional career in the architecture world. The experience also helped me with my major project in my final year, the kidsPACE project.

The experience was not just an educational one, but a cultural one, and I would highly recommend a placement abroad to future student applicants.

Project story

Designer:
Hollie Cleaver, design for film and television student

Project title:
Casper

Tell us about the project

'Casper' was one of two major projects I completed in my final year studying Design for Film and Television. For this project I produced designs for an adaptation of the original 1995 film *Casper the Friendly Ghost*. I used the original films storyline as a basic structure but then adjusted the concept by changing the period it was set in and the target audience to create a completely different take on the design of the original.

I chose 'Casper' for my final major project because it was different from my previous project and allowed me to produce designs that would demonstrate more challenging model making and technical drawing skills. I felt it was important when choosing my final-year projects to consider how I would demonstrate a variety of skills and strengths that would result in a strong portfolio.

This project involved a lot of research, as my design was set in a Victorian water pumping factory, which was an area I had little knowledge of. After collecting a variety of research and imagery using the Internet and books, I felt the only way to have a better understanding of such a complex building was to visit one first-hand and talk to someone who understood them. I visited two pumping stations, one of which I was shown around by a volunteer who had a vast knowledge of Victorian pumping stations and was able to answer my questions and demonstrate how the mechanism worked. I feel this was my most beneficial method of research because, when designing a 3D space, although images from books and the Internet are all important references and inspiration, they will only show a 3D space as a flat image.

CASPER
SCENE 28/ "UNDERGROUND PASSAGE TO PUMPING STATION"
CASPER LEADING KAT INTO AN UNDERGROUND PASSAGE INTO THE PUMPING STATION
INT. LOCATION PAPPLEWICK RESERVOIR TUNNELS & STAGE: THE PUMPING STATION
DAY. LATE AFTERNOON
1. LONG SHOT AS CASPER LEADS KAT THROUGH THE TUNNLE LIGHT BY KATS LANTERN
2A. LOW SHOT LOOKING UP AT KAT AS SHE HOLDS LANTERN OUT INFORNT OF HER.
2B. CONT. LOW ANGLE SHOT OF KATS FEET AS SHEEP STEPS OVER CAMERA.....
2C. CUT TO. LOW ANGLE SHOT AS KAT STEPS INTO PUDDLE
4. LONG SHOT AS CASPER LEADS KAT DOWN NARROW PASSAGE.
PAN SHOT AS CHARACTERS WALK AROUND CORNER TOWARDS LADDER
5. LOW SHOT AS CASPER CLIMBS LADDER TOWARDS LIGHT SHINING THROUGH THE METAL GRILLS OF THE PUMPING STATION FLOOR ABOVE.
6. MID SHOT AS KAT BEGINS TO FOLLOW CASPER UP LADDER
7. CLOSE UP AS CASPER REACHES THE TRAP DOOR STARING INTO THE PUMPING STATION
8. LONG PAN SHOT FOLLOWING KAT AND CASPER LOOK AROUND THE PUMPING STATION IN A AMAZMENT

'Casper' storyboard

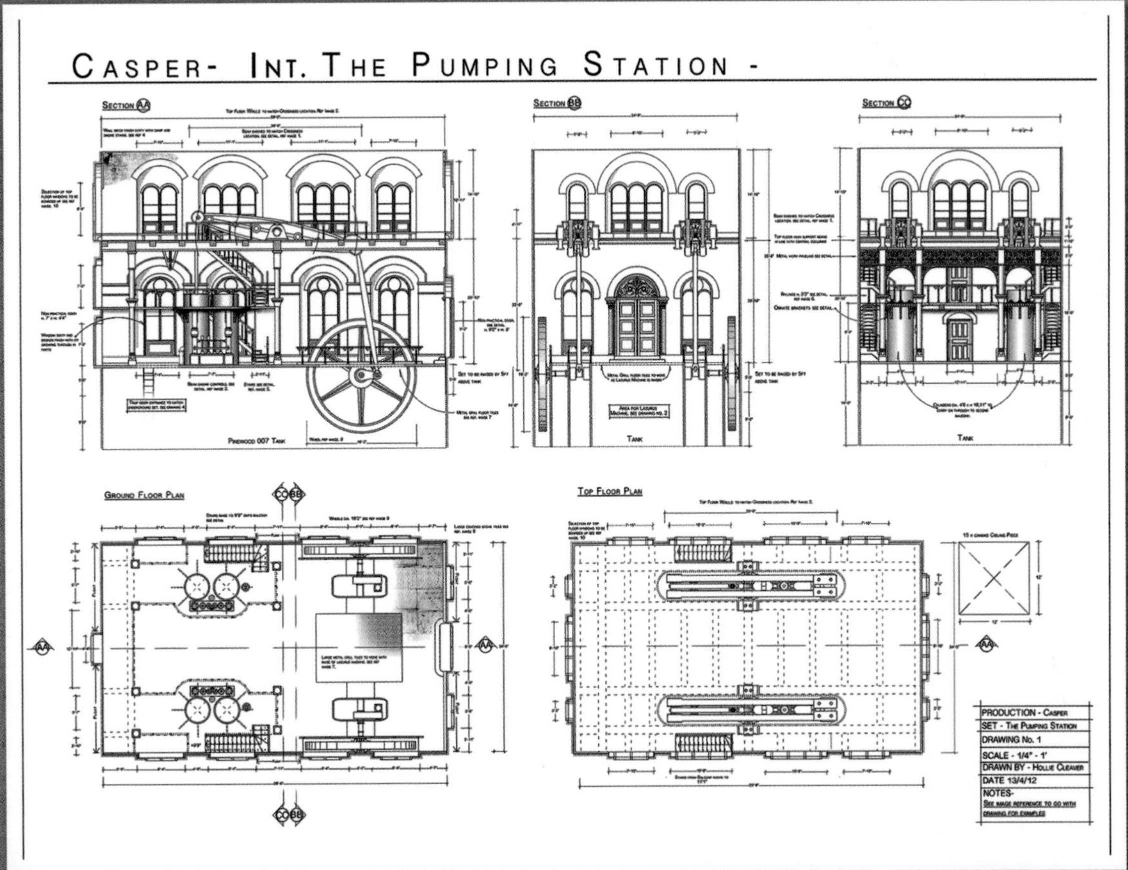

Technical drawings of pumping station

To understand a complex space, as I had to, I was only able to understand fully by visiting one. I also made sure I took my camera and tape measure with me to take down measurements and take pictures of features and parts, which I felt would help me when I got back home. My visit also meant that I was able to get hold of some original architectural drawings, which were particularly useful for me, but also to communicate the development of my design in presentations. I found researching out in the field really rewarding, as sometimes sitting at a computer or studying books can become tedious.

What processes and skills were most relevant to this project?

I had to think carefully about the best ways to communicate my design. For my project I chose to do a white-card model rather than a rendered one because I wanted to focus on refining the complex structure of the film set and then attempt to display in detailed visuals the atmosphere and use of space, so when model and visuals are seen together they portray my design and idea comprehensively. I also spent a fair bit of time using a Google SketchUp model of my design to choose the viewing angles that would be best to convey and sell my design, trying to include the important features. I produced a set of technical drawings detailing my design, which I then referenced to my research with examples of textures and finishes, enabling a more in-depth communication of my design. I think it is important to not just rely on the minimum requirements assigned by the tutors as a way of presenting your design, but to use your own initiative and find ways that make your work easier for others to understand.

What challenges did you face?

For me, model making was one of the more challenging aspects of my final-year projects, but for my second project, 'Casper', I wanted to make sure I produced a model I was happy with. I didn't feel my model for my first project was at the same standard as some of my other work. By learning from my first project I realised that I needed to allow myself more

White-card model of Victorian pumping station

time to make my model. I found the best way of model making was to work from technical drawings, but also that patience is fundamental. Before starting I spent a fair bit of time planning how everything would need to fit together and made my own assembly instructions to work from. I then made the pieces bit by bit and, although it is so tempting to start gluing things together and building the model up straight away, it wasn't until the very end, when I had all my pieces and walls laid out, that I actually started to fix things together and the model began to take shape and look like a set.

What were the highlights of the project?

I found my final year was a lot of work and pressure at times, but I did find a lot of the work very rewarding. I also found that because you have a lot of freedom to play around with your own brief it allows you to cater it to your own interests. If you are working on projects for weeks it is important to keep yourself interested, as it is easy to lose motivation. I particularly enjoyed researching for my 'Casper' project and getting the chance to look around some of the pumping stations first-hand. I also found a lot of mechanical designing, such as 'The Lazarus Machine' for my Casper, really interesting to research and be creative with. I was lucky to gain a variety of different work experience throughout my three years' studying, including working on Tim Burton's 2012 *Frankenweenie*. I feel that work experience within the industry is invaluable, especially with such a vocational course as Design for Film and TV. Work experience within the industry was not only inspiring for future prospects, but also really beneficial during my studies.

10

Being creative and innovative

By Mark Jones, Southampton Solent University and
Steve Rutherford, Nottingham Trent University

'I like to blur the boundaries between fine art drawings and textile art, flat and 3D work, illustration and embroidery, as I feel exciting things happen when one pushes the limits of a discipline or material. If I can keep exciting myself by what I do, I hope I can continue to excite others.'

Debbie Smyth, textile artist

Ferris wheel, 50 × 50cm, pin, thread & acetate by Debbie Smyth
WWW.DEBBIE-SMYTH.COM

What exactly are creativity and innovation?

- creativity is the origination of new ideas
- innovation is what we do with creativity to make something of our great ideas and bring them to the market.

Together they make up a large part of the design process. How do we recognise and manage them? How do you make sure you're creative? How do you ensure you are one of the innovators of the future? These are the questions this section will answer. By the end of it you should be able to see how and why the design process is different if you are designing a bracelet, a coat, a kettle, a website, a book cover or a hotel bedroom.

A successful project needs creativity and innovation and they need to be well managed. As we have seen in the design discipline sections, the way we design can vary widely, depending on the object that is being designed and the designer themselves. You have seen examples of design process theory and many examples of projects. To build on this, this section will highlight the importance of the process and demonstrate the roles that research, creativity and innovation play in the successful running of a design project. It will also increase your awareness of your own process and help you manage it. To do this the section will outline:

1. **Background theories and methods of creativity**: This will help you to be more creative. These methods are the backbone of design. You will see ways of working, ways of thinking and read theories for you to understand.
2. **The nature of innovation**: This is the management and synthesis of creativity and research. It's what you will do when you start to take your ideas to market.
3. **The design process**: You will see that the activities required in different projects vary across disciplines. You will also be prompted to analyse your own design process.
4. **Research methods**: This will give you knowledge and insight into how to research, validate and prove the worth of your work.

What is creativity?

Can it be taught? Can we measure it? These questions are difficult to answer. We know it when we see it, as in this 'Mango Tree' sculpture in Taipei, Taiwan (www.tsai-yoshikawa.com).

As a designer, at various times you will stare at a blank sheet of paper and think 'Why can't I start sketching something?' The pen is poised to start sketching but your mind has gone 'hazy' and even plays tricks, wanting you to believe that you're not very creative after all. Don't worry, we've all been there and these are common feelings!

One thing that definitely helps if you want to be creative is to be a bit of a polymath. This is the name for someone who has a wide range of knowledge and experience. However, the wonderful thing about being a designer is that you don't have to be an expert in all things, just interested enough in lots of things to want to know more.

> *'It's a myth that creativity cannot be learned, and that you are either born creative or you are not. Creativity is not genetically determined. Creativity means looking at the same information as everyone else and seeing something different. To get new ideas, you need to rethink the way you see things, to look at the world in a different way and to consider different ways to see problems.'*
>
> Michael Michalko, one of the world's leading creativity experts

That's great news for you. Even if you're not feeling particularly creative right now, you've just been reassured that you can take steps to improve your creative-thinking ability.

Creativity is a term that describes how the human brain makes connections between all of the information contained within it. Think of your brain as a big room, with all of the walls covered in lots of little cupboards. The cupboards are filled with ideas, concepts, knowledge, wisdom and other useful information. Creativity occurs when you have two or more cupboards open at the same time, that have never been opened together before. Suddenly you see the connection and the possibilities that result from combining the contents of the cupboards.

This is the insight that drives creativity and leads us to new ideas, ideas that are what we call 'outside

Group creativity session

'The Mango Tree' Public Art Installation by Hsiao-Chi Tsai & Kimiya Yoshikawa 2011

the box' (not easily discovered but obvious when they're pointed out). In the image above we can see students taking part in a creative thinking exercise that allows them to share all their thoughts and to think both inside and outside the box.

So how does our brain process information?

Our brain processes information in two ways: the left side looks at information in an analytical way, looking to process information logically and step by step. To take the 'cupboard' analogy a little further, the cupboards are opened one at a time, or one row at a time, or in some other logical format. In contrast, the right side of the brain is more intuitive and processes information more holistically, looking for patterns and connections, e.g. cupboards are opened in random order and then the contents are compared, allowing a free flow of ideas to bounce around the room.

For most of us, creative ideas are the result of using both these methods – i.e. the brain processes information using both the left and right side. However, for others, information will be processed with a dominant left or dominant right. There are online tests you can do to find out whether you are naturally left, right or what we call left/right double-brain dominant. The test is called 'Creativity Quotient' (CQ). There is no correlation with the 'Intelligence Quotient' (IQ) test that you may be more familiar with, and it isn't true that the more intelligent you are the more creative you are likely to be.

Creativity: what exactly is it?

Creativity is the discovery of new ideas and concepts. **Albert Einstein**, although associated with brilliance in the field of mathematics and science, was a creative thinker. Perhaps less well known is that he was also an accomplished poet. Science united with creativity is a powerful combination. The same qualities are also true of many people throughout the ages, from **Leonardo da Vinci**, who was a prodigious inventor and discovered many scientific and medical principles, to **Thomas Edison**, the inventor of the humble light bulb. Einstein put it succinctly:

> 'The intuitive mind is a sacred gift and the rational mind is a faithful servant. We have created a society that honours the servant and has forgotten the gift.'
>
> Albert Einstein

Edward de Bono is a world authority on creative thinking and Professor of Thinking at the New University of Advanced Technology in Phoenix, Arizona. His work on creative thinking techniques, including that of 'six thinking hats' (more on those later), focuses on improving the construct of perception and the frameworks that exist to enhance creative activity. He comments:

> 'It is better to have enough ideas and for some of them to be wrong, than to be always right by having no ideas at all.'
>
> Edward de Bono

Being creative

A creative person can be described as someone who has both fluency and flexibility in their thinking. As a design student you are in a very unique and privileged position. You will be surrounded by many other creative individuals who share an equal passion for 'being creative'. Your course tutors want you to succeed and will encourage you to take risks when you're designing, in a environment that is safe, supportive and fosters creative thinking.

You will be encouraged to challenge traditional thinking that governs and shapes the world in which we live and to explore what lies 'outside the box'. This comes from you taking more responsibility for what you do and becoming comfortable in your approaches to design, even if your ideas don't always work. Creative people are often expected to be creative all the time and the pressure this can place on you can in itself be inhibiting. You also need to be aware that you may find it difficult to relate to people who are less creative than yourself. This can lead to frustration, particularly if you are presenting what you believe to be a fantastic concept to a client, boss or even your tutor and they don't seem to share the same passion or level of enthusiasm for an idea as you do!

At university you will inevitably make mistakes and things won't work. You will experience unsuccessful starts to projects and design something only to find out later you can't make it, but that's ok as it is part of the learning process. In fact, being creative and trying new things is risky and quite a few of your ideas won't succeed – this is a fact.

Having a great idea is only the start of the journey that will take you through many twists and turns before knowing whether the road you are on is likely to lead to success. With that in mind, what better time is there to start this journey than now? Search for inspiration, be inquisitive, be the one who always asks questions in class. If you push and explore creativity, your ability to produce challenging solutions will become second nature by the time you graduate.

Elissa Bleakley, a Textile Design graduate, explains how inspiration leads to creativity:

> 'While flying over Europe and Egypt I captured aerial images of the landscape and cloudscapes below. I was instantly inspired by the starkness of the open sky and the ways this contrasted with the structured landscape. There was a strong atmosphere and mood to these images, with an ethereality that I could strongly envisage working on the body. This concept of contrast, looking upwards and downwards, allowed me to engage unconventional ideas and push this imagery into a new and exciting direction.
>
> I kept my ideas evolving with fluidity, analysing my visual research to create processes that reflected my concept. Thinking about the structured landscape, I began to explore creating three-dimensional fabrics and developed a process of ink-less printing, creating laser cut templates to emboss structured

Early colour work based on aerial photography

Test fabric by Elissa Bleakley

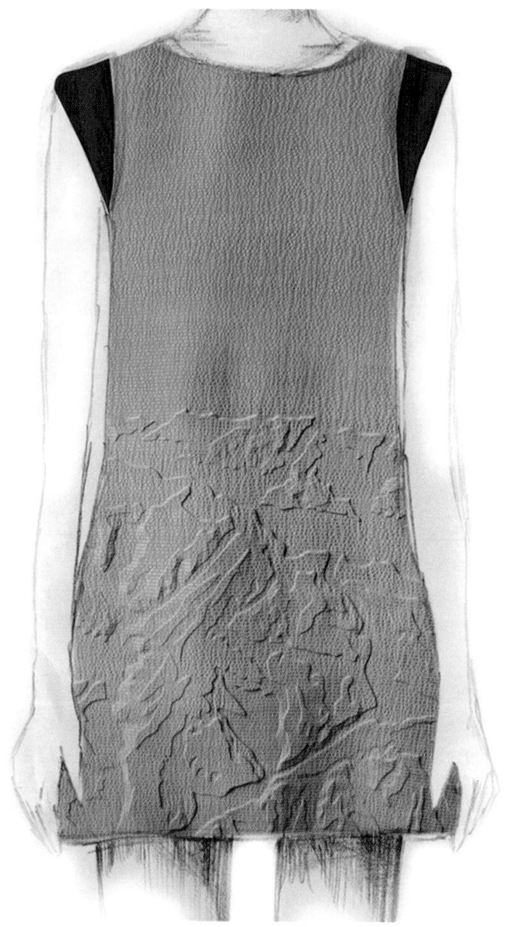

Final piece by Elissa Bleakley

patterns onto fabric. This minimal technique was extremely effective in creating two-dimensional relief fabrics with layered patterns.

My final colour palette and colour gradients were extremely successful in abstractly capturing skyscapes and landscapes from my visual research; this success was achieved through delicate observation.'

Creative behaviour

Remember the brain lined with lots of little cupboards? You can fill them, much like Elissa above, by being interested in what you see around you. But how do you know which cupboards to open if you're starting a project? If you're looking for something and you've got all those cupboards to choose from, how do you start, especially as this is something we do subconsciously (often while sitting, staring at a blank sheet of paper with a pencil in our hand!)?

The answer is that there isn't any one single combination that will guarantee success or will be right every time. This is because there are potentially lots of different combinations that will prove to be fruitful. That might sound confusing, but opening a combination of cupboards in the hope that you will find some creative connection is a bit of an 'art form'. This is why it can be difficult to teach creativity and why some people struggle to understand how to use it.

Let's start by saying 'creativity' is a behaviour and 'creative behaviour' can be taught. It's really about being different, a willingness to try out new and different things, some of which will seem unfamiliar and therefore can feel uncomfortable both for you and others. Creative behaviour challenges us in the way we normally think, respond and act. It is often a subconscious process, with ideas coming out of nowhere when your attention is diverted. Anna Glasbrook, who produces art installations for interiors and gardens, has her own methods for developing creative approaches to her projects ('filling the cupboards'):

> 'I have always taken endless photographs of things around me and collected stuff – bits of coloured plastic I find on the beach, scraps of fabrics, interesting yarns and ropes. Often an idea begins by stitching through clear plastic packaging – ready-made poppadom packets are great, but I have to eat a lot of them. I like to work big and I like to work small. Obsessed with transparency, I collect plastic packaging destined for the recycling bin. These ideas act as my sketchbook, encouraging free experimentation and the development of fresh new ideas. I now have a large collection waiting to be scaled up and developed into large pieces of work.'
>
> (Read the full story about her business in Chapter 16, 'Working for yourself'.)

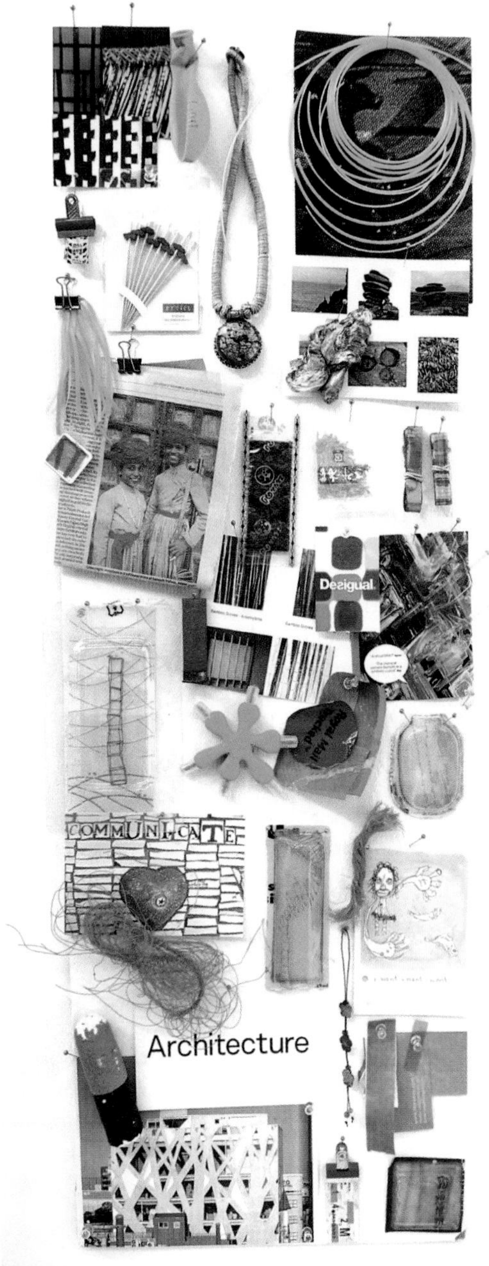

Visual inspiration board by Anna Glasbrook

Techniques to develop creative behaviour

By definition, the 'creativity' aspect of design should start at the beginning of the design process from the moment you engage with the project brief. There are a number of tried and tested techniques and exercises to help you navigate around the right side of your brain. Try some of them out. You may not find them all helpful but some you will, and it's those ones you need to practise and become fluent in using. You can even adapt them to suit your own 'learning style'. Using these techniques regularly helps exercise the brain. One thing we know about exercise is that the more often you do it, the easier you will find it and the more natural it will seem, and the more efficient you will become.

A summary of the key excercises that are commonly used in design are listed below, with an overview of the benefits in using them. Links to other books, useful resources, online tests, videos and podcasts about these specific exercises are contained within 'Further resources' at the end of this chapter.

Idea-generating techniques and exercises

Positive thinking:

A technique that places emphasis on the positive aspects of an idea. Set within a short time frame of one or two minutes, the results are instant. Thoughts are captured within a list of words, with ideas ranging from crazy thoughts to more obvious ones. This exercise can be done individually or in a group. Basically, everyone must only speak or write positive statements. No negativity is allowed.

Mind mapping:

A systematic approach to recording thoughts, involving creating 'branches' that stem from a central idea. Words or visual images are recorded and form related sub-ideas. This exercise is often done individually but can also be done in small groups.

Problem definition:

A technique used to generate solutions by redefining the problem in order to enable a different approach to be taken. This exercise can be done individually or in a group.

Brain storming:

An exercise usually conducted in small groups (6–12 people) in an environment that is supportive and non-judgemental. A systematic approach, with quick-fire thoughts, exploring themes (analogy, biological, transitional, juxtaposition) and links within headings that relate to the brief.

Lateral thinking:

Involves looking at a problem from different angles rather than tackling it straight on. Solutions can often be derived from looking at alternative man-made or natural resources and solutions outside the immediate field. This exercise can be done individually or in a group.

Role-playing:

This exercise works on the premise that there is a strong correlation between role play and enhancing your creative thinking. All team members need to be involved. As a team you look at the problem or issue and respond in the style of (something we are all familiar with), such as: a chat show, a children's TV show (such as *Blue Peter*) or a programme such as *Top Gear* – how would the responses and solutions differ in each scenario?

Random word association:

A technique used whereby two or three words are randomly selected and then you allow your mind to visualise the words separately and analyse what they mean and represent. The fun starts when you link or combine the words and your thoughts. This will spark a range of ideas that should be recorded as a description, or a sketch drawing. (See the iTune App in the resources section.)

Six thinking hats:

The premise of the method is that the human brain thinks in a number of distinct ways. The hats – six different colours – represent the different ways of thinking. This is a time-framed exercise and as the coloured hat is worn by one member of the group, the others are instructed to think of certain aspects that relate to the colour being represented. This is a group activity and can be very effective for both generating and evaluating ideas.

Mind mapping

One of the more common techniques you will have used already is mind mapping, which is where the project title, subject area or problem statement is written down in the middle of a piece of paper and then a series of thoughts and related words or images are also recorded. This technique is also referred to as a 'spider diagram' due to the visual look of the map, which, having circled and linked a number of related ideas, will very much look like 'spider legs' stretching out from the original word (spider body).

Here are two examples below of mind maps. The first, by Product Design student Emma Alderman, uses just words that relate to the theme, and you can see a number of related thoughts starting to emerge. One important aspect that is often missed is that they are working documents; when you've finished producing it you then have to analyse and discuss it. You can also revisit it later in the project.

The second illustration, by Interior Design student Hayley Collins, demonstrates the power of communication through a visual mind map, where thoughts are represented as small sketch drawings. Hayley's 'draw the problem' method is very similar to the **IDEO** 'Draw the Experience' Research Method Card described later in the text. This fact blurs the boundary between creativity and research; in this instance they are similar activities.

Sketch drawing will always be important for artists and designers to communicate ideas and thoughts. Yes, you can read books about creative thinking and source information on the Internet to get a better understanding of how to be creative, but there is nothing more powerful and effective than using a piece of paper and a pencil! With practice you will become more fluent at transferring thoughts directly into sketch drawings, which will allow others to better understand your ideas. Refer to Chapter 9, 'How to succeed as a design student', for more information on this.

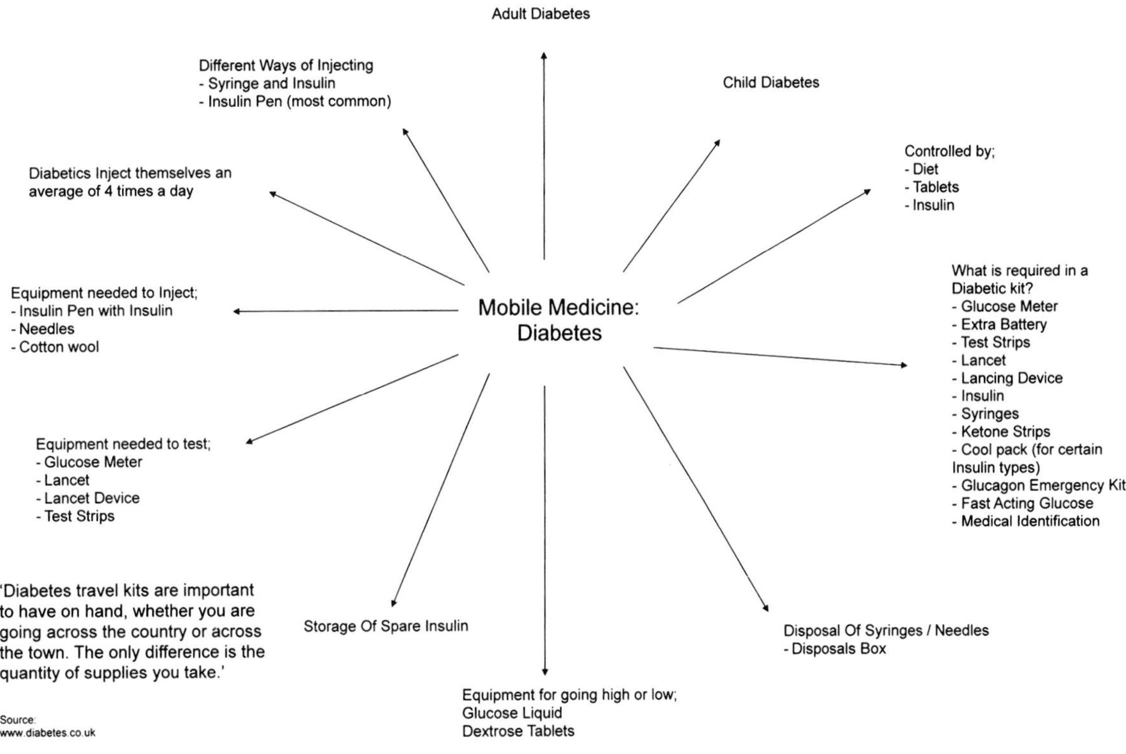

Emma Alderman's textual mind map

Visual mind map by Hayley Collins

Interestingly, even though the very nature of these exercises will get you exploring the far depths of your brain, challenging your understanding of the world you live in and questioning how you've been conditioned, in themselves they often work best under a controlled environment using specific time frames. Some are designed to be used working alone, while others are team-based and more effective if the process is interactive, with everyone getting involved and making a contribution.

Having explored some of these exercises, if you reach a point where you or the team are struggling to be productive then that's probably a good time to take a break and do something completely different. When you come back to what you were doing before you will discover how amazing a fresh set of eyes can be, even looking at the same information. If you still feel you have exhausted all your ideas then try a different idea-generating technique and see what can be achieved.

Creativity exercise

Try using at least two of the above techniques on the following scenario:

Think about a common problem – *what happens as we get older*. This could be related to any design project from any of the disciplines: you could be working on a play, a garden, a website or an object and looking for inspiration. In order to do well you need to understand people, because many of them are very different to you.

After you've completed the exercise above, don't do anything else. Put the results away. You will find that for a few days afterwards you'll keep thinking about the problem and thinking about ideas. This is one of the keys to creativity.

We often have no control over exactly when we have that creative moment, and often it's when our attention is distracted by something other than work – in the middle of the night or in the bath – we call this 'insight thinking'. This is why many designers are never without a little notebook and pencil – they know they might be inspired at any time of the day or night.

How can you control creativity?

Creativity can usually be assessed in design terms by evaluating your ability to explore a wide variety of ideas and approaches to a problem from a range of information sources, without losing sight of the overall goal. From the earlier definitions, creativity was defined as original ideas in the pursuit to create something new. You can demonstrate your creative flair through your sketchbooks, transferring your imaginative thoughts into 2D drawings.

Once you have produced a range of initial ideas and sketch drawings you may find it helpful to review which aspects of your ideas offer something new, original and challenging. This will help highlight what aspect of 'creativity' is evident in your ideas. Having this focus and direction is necessary when further evolving the idea into a more established concept. A framework for this review could include:

- how original is this idea – i.e. new thinking, adaption or evolution?
- to what extent does the concept address the brief?
- to what extent does the idea rely on or exploit new technologies?
- is the idea likely to be familiar or unfamiliar to your audience?
- what are the specific **USP**s of the idea?
- what is the likely potential commercial value of your idea?
- what does your client think?

USP – unique selling point: This is the one thing that will mark your idea out from the rest of the market.

In this image, second-year Furniture & Product Design student Claire Kirby is talking through her concept designs with her client and making notes on the drawings.

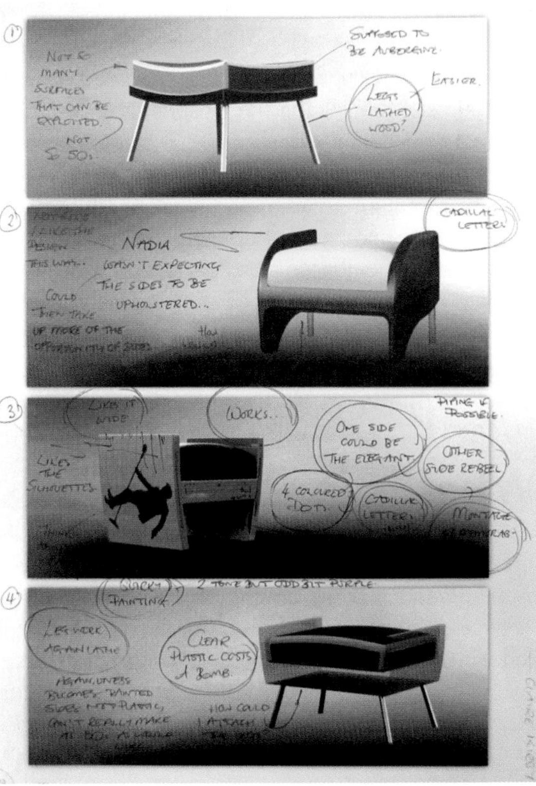

Concepts board by Clare Kirby

Your journey from brief to concept should be evidenced, no matter whether this is random words, sentences, doodle sketches, images, photographs, drawings or other visual imagery. It's about making connections and exploring lines of inquiry from this resource and information. It helps to demonstrate to your tutors that you have used some of the tried and tested techniques and exercises, or even created your own methods.

Beyond this point, the design process is concerned with development, marketing and production of your idea, i.e. making it real. This is where innovation comes in.

To summarise creativity

As a student you have a fantastic opportunity to be experimental with your ideas and the methodology you use as a designer and the freedom that comes with this. In the real world we are far too often faced with project briefs that have much tighter specifications or restrictions that you won't suffer from yet! This is your chance to explore, push the boundaries and challenge the way we have all been conditioned to think.

It's all about your approach, attitude and willingness to practise. Develop these skills now while you are in education and you are investing in your future as a creative designer. Learn to think outside the box without losing track of the overall goal and you will possess a highly valued asset in industry. Having the confidence to stand in front of other students to present and defend your ideas in a supportive environment is good preparation for the harder challenges that lie ahead in industry.

All of the best designers are 100 per cent passionate about pursuing creativity and you need to feel and nurture this too. It takes time to develop strategies to deal with creativity, steer and direct it and benefit from it. You need to be aware of your capabilities and your progress in this area and be patient. It does come with time. After ten years' experience, designers will have filled many cupboards in their brain and are well practised at searching them for inspiration and seeing connections. Younger designers can still compete – well-structured research methods will fill your brain up with useful information and wisdom. This knowledge is key to making your creativity more productive.

And remember to always explore a wide range of ideas and possibilities. The more ideas you produce during a project, the more chance there is of finding a really good one.

What is innovation?

Innovation is what you do with creativity to market products and make a difference to peoples' lives. Deryn Relph transforms furniture with her colourful knitted upholstery and matching lampshades. Producing these wonderful creations is one thing; turning them into a business is the tricky bit.

This is true of many projects, big and small. There are wonderful examples of innovation across the design disciplines:

Knitted creations by Deryn Relph
© Alick Cotterill

- Paper clips are very useful and part of their attraction is that they cost approximately 0.003 of a UK penny each to buy. If they were hand-made from stainless steel and cost 85p each they wouldn't be as successful. If they were multi-coloured, heat-treated, lightweight titanium and indestructible they would cost about £4.99 each. They probably would still sell but not in huge numbers. The paper clips we know and love were brought to the market very effectively, appropriately and economically and are a wonderful innovation.
- The notion of a reversible jacket is straightforward, even if technically it's not as straightforward as we think. There are all sorts of problems that fashion and textiles designers need to solve to turn a simple concept into something that works without compromises. How do we deal with the fact that the fabric curls in the opposite direction around the body when reversed? How do we deal with seams normally hidden on the inside?
- The revolving stage has allowed theatre productions to communicate extra dimensions to a performance – movement, time, speed. First invented in the eighteenth century in Japan, they still play a major part in West End productions, such as *Les Miserables*.

However, companies can be very creative and sometimes bring products to market that are not fully developed, or have no useful function, or are ahead of their time. The first mobile phones to include text messaging left some people confused about why you would send a text – 'Surely it's better to speak?' they cried. Eventually, we realised that texting is a useful means of communication because it can hang around and wait to be picked up. We hadn't realised it would be useful because we'd never had it before.

Successful innovation can lead to business success, but it can be risky. It is important to fully understand the value of your idea, who your target market is and the needs of the market you are entering.

Can we learn to be innovators?

Innovation is, in the real world, a strategy and management process that we can create in organisations. That doesn't mean that innovation, and indeed creativity, are widespread activities in industry. As with many research and development (R&D) activities in business, innovation is sometimes seen as an expense by senior managers and not necessarily an investment. Designers often have to fight their bosses for more investment in the design process: research, design, innovation. Some companies even see all of this as an option, not a necessity; they are simply much more interested in a shorter-term payback from investments.

Innovation requires investment in time and effort to make the most of your creativity. Looking at your ideas and making sense of their worth is a crucial innovation skill. The final-year project from Product Design student Nicholas Camilleri, coming up, demonstrates a more modern approach to new product development: using the people who will use your idea to help define and refine it. This kind of research reduces the risks inherent in launching new products. The powered wheelchair adjusts to cater for people who can stand on it or use it at times as a walking aid. Nicholas has produced a very convincing working prototype and is well on his way to delivering innovation to the market.

In the real world, as with Nicholas, the distinction between creativity and innovation is fairly specific. They are, in fact, separate aspects of the design process. Start with creativity. If you manage it and bring your products successfully to market you are an innovator.

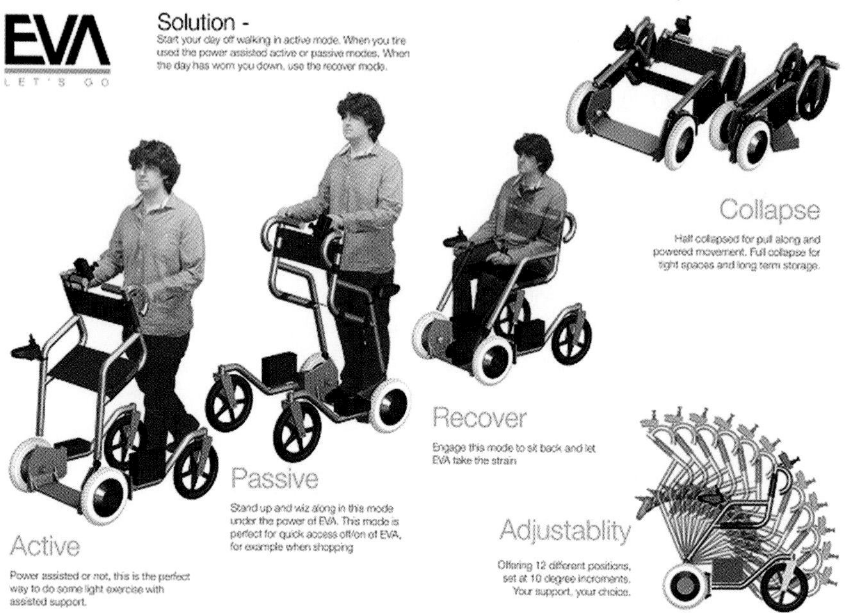

EVA personal transportation prototype by Nicholas Camilleri

How do we do it then?

Successful innovation can be described as managing creativity in an atmosphere of positive investment in the future. Some companies, or designers, or even designer-makers, simply watch the market and react to whatever is working at any point in time by producing something similar to someone else's idea. There is a limit to how successful these people can be if they are completely reactive. They will never be seen as market leaders, for instance. On the other hand, compared to this reactive view, forward-thinking innovation management is more risky. Sometimes ideas fail, as we said earlier – your research might be comprehensive but wrong in the predictions it made. This is the most difficult aspect of innovation for people to buy into – that all the investment in money, time and effort could go down the drain.

So what we must do is control the whole process, know where we are, what the options are and still be creative – churn those ideas out in big numbers. Once we've got the ideas we need to evaluate them and make good decisions. We'll now look at how important it is to analyse your ideas in a rational way and how you might do this.

Controlling innovation: objective and subjective evaluation

Somewhere in the middle of a project you will have a range of concepts, some more creative than others – hopefully ones that represent extremes of what might be considered acceptable or not – we saw earlier in the creativity section Clare Kirby's range of seat concepts and her notes on discussing them with her client.

It's also only natural that you may feel more passionate about one particular idea and share a greater affinity to it than to other concepts. This close relationship is very subjective and is to do with your feelings for this concept, having invested both time and effort. It's your 'gut feeling' that this could be a real winner, so why do you need to present all the others?

To answer, we need look no further than Edward de Bono's statement:

> 'It is better to have enough ideas and for some of them to be wrong, than to be always right by having no ideas at all.'

How might you best identify which direction to take or which concepts to take forward in development? The answer is really about the balance between subjective and objective evaluation. Subjective evaluation might be your own opinions or the opinions of people you have chosen to speak to about your project.

A well-used **objective** approach to evaluating ideas is the 'numerical selection' process. The consumer report 'Which' uses a similar method and would be a good starting point to refer to. Essentially, you identify a number of criteria from your project brief that relate to the problem or objective that your concepts can then be evaluated against, using a numerical value from one to ten.

> **Objective**: a statement, more of fact than opinion, which is more difficult to argue against.

Here, MA Fashion & Textiles student Lucy Davison is making up her own matrix to evaluate possibilities within her project. It has materials and production processes running up the left-hand side and product areas in the top row, and Lucy is using it to look for product opportunities – ways in which she might make certain objects with certain materials. She's trying to cover as many possibilities as she can.

> **Subjective**: a personal opinion that could be argued with.

From a **subjective** point of view, discussions with colleagues and other stakeholders in a project will elicit many viewpoints and you may spot correlations between what these people think of the concepts. It is possible to collect these subjective assessments and pull out the strong views that you feel accurately represent what people's subjective views might be of your concepts. In some discipline areas this is a simpler process: if you design a hat, there's the people who buy it and use it to consider; if you design a hospital, there are lots of different people you might need to talk to in order to discover what's important.

One strategy for dealing with this would be to have some background information and context in which to set your idea. Firstly, consider who your client and/or target audience is and what they know about the subject area that relates to your idea:

- what are they already familiar with and what does that look like?
- what do you think they expect from you?

There is a wider, cultural context to understand too:

- what does your client or target market understand about your ideas and how does it relate to what has happened in the past?

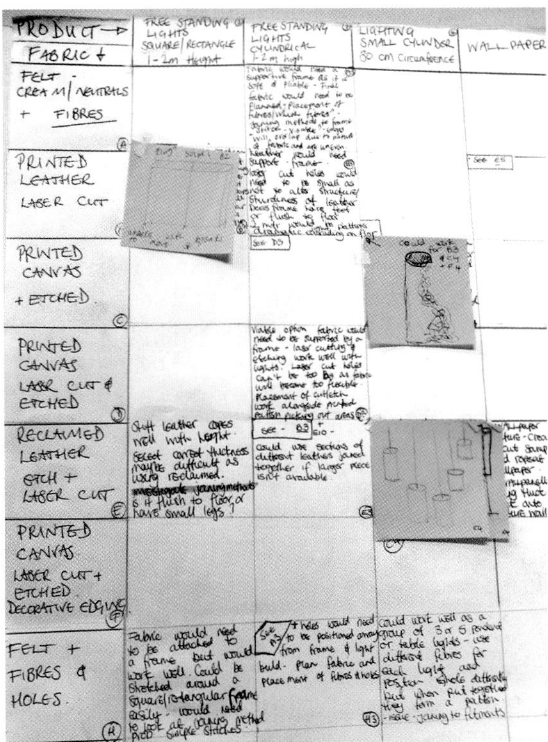

Product opportunity matrix by Lucy Davison

- what is the current thinking around the issue you are addressing and does it have other influencing external factors, such as political, economical, social, or technical ones?
- will culture play a part in your idea, or are there ethical issues involved?

To summarise, successful innovation management is often conducted at a high level in companies. It is the successful controlling of research, creativity, marketing, design process and development, therefore it's a management responsibility – the responsibility for it doesn't lie solely with designers. It's a big, corporate responsibility in large companies and, even in small companies, designers wouldn't often be responsible for it. But designers have a professional responsibility to engage with it and a very important part to play in it.

Protecting your ideas

When is it safe to share ideas? How do I protect my ideas? Sharing ideas, evaluating ideas, comparing ideas, presenting and defending ideas are not only a necessary part of the learning process but also a valued part of the design process. Protecting ideas (Intellectual Property Rights: www.ipo.gov.uk) is covered in Chapter 16, 'Working for yourself', later in the book. This outlines the legislation and the process and tells you what is covered and how much it might cost. It also points out the pros and cons of the system. Some people swear by it, others reject it. All cases have to be looked at on their merit.

The design process

To put creativity and innovation in context we need to look more widely at the design process. Some people think that there are good, reliable, well-publicised ways of designing, which should be followed at all times. Others think that design is a flexible process and that there may be as many different processes as there are designers. One thing everyone agrees on is that we should have an awareness of our own process, how we manage it and our strengths and weaknesses. We should also bear in mind, as demonstrated by Decorative Arts student Jessica Steel here, that the mess and excitement of it all should be fun.

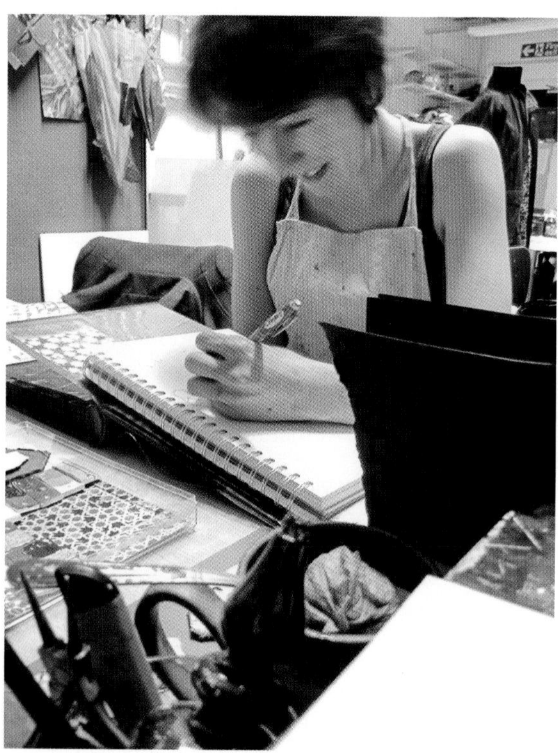

Happy in her work, Jessica Steel, decorative arts student

There are many different design processes in use and good reasons why there are many different ways of designing. Often you have to design in a certain way because of the nature of the project you are working on. As a designer you have to monitor your process and will probably have to modify and adapt your process, depending on the projects you are working on in the future. A graphic designer may be working on a website one day and a financial report for a firm of accountants the next. Those two things can't be designed in the same way because they're such different objects. This is why there are so many different possible design processes.

Why isn't it a simple process?

When we look at the various design disciplines, as we did earlier in the text, we find large differences in the way designers work – what inspires them, what kind of research they do and how they arrive at final solutions. This range is very wide. Here are two extremes:

Team-based, commercial, mass-produced	Individual, personal, small-scale
In an area such as automotive or high-street fashion design, we find big industries with huge teams of designers following what are often traditional and very highly controlled processes. We may wonder how these complex procedures can allow creativity to shine through, but looking at a fashion show or an Italian sports car demonstrates just how creative big industries can be.	In craft design we find personal journeys of exploration and creativity that are sometimes difficult to analyse and understand. The question can be asked – are they doing it to satisfy their personal creative impulse or are they designing with a client or a particular market in mind? These processes don't need to be justified like they do in larger organisations.

Even as a student, you will experience these extremes. Here are two students who are tackling very different types of project.

Anna Piper is a Textile Design student and is interested in how fabric is made and then applied to clothing. This project is an example of zooming in to the detail:

> 'For my final exhibition statement I really wanted to bring together all the areas that interest me and the skills that I had developed so far, to push them further, challenge myself and have fun. I wanted it to be a statement that represented me, and the things that are important to me as a designer. A key part of this was to work towards a sustainable outcome. This shaped the project and sparked the idea – to produce engineered, waste-free woven garments. Having made clothes in the past, it seemed feasible to combine dressmaking principles with the weave techniques I had explored in my second year to create a simple garment.'

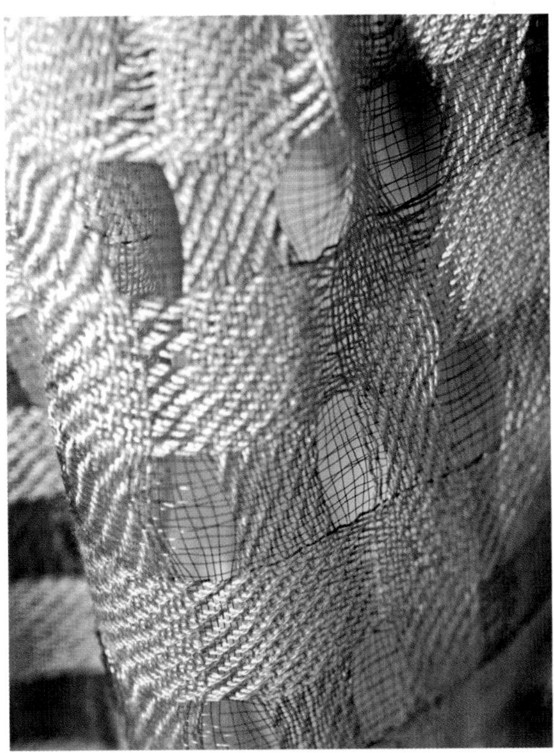

Part of Anna Piper's waste-free fabric project

Suzana Basic is a Graphic Design student and her final-year project is an example of zooming out and tackling something quite large, literally.

> 'The decision to rebrand the New York Water Taxi service (NYWT) was influenced by my experience there. Having taken a boat trip I was able to familiarise myself with the brand by obtaining a first-hand account of the service.
>
> A number of visual brainstorms drew out the key elements to take forward and adapt within the new design. Iconic imagery, such as the distinctive yellow exterior and black-and-white chequered pattern associated with

Part of Suzana's search for a striking identity

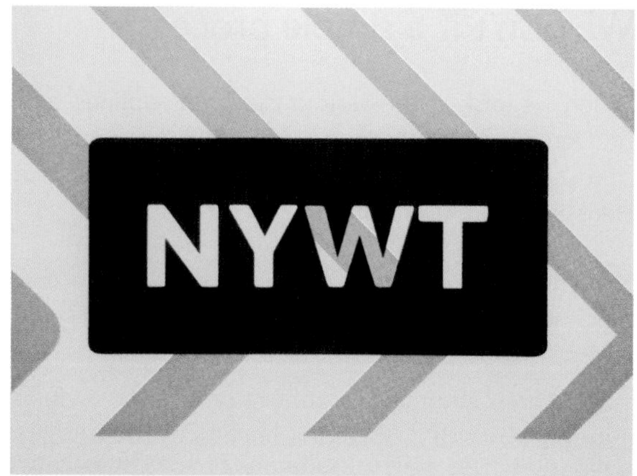

Corporate identity for New York Water Taxis

> New York taxicabs is followed through in the water taxis. I wanted to maintain this link in some way, however with greater emphasis on "water" to give it a more distinctive look and set it apart from the New York cabs. I produced numerous initial logo designs that continued to change and improve while implementing them onto the brand touch-points, such as tickets and posters.'

You can imagine how different these two design processes were; however, there are conceptual links between these two very different projects. To explain this, we're going to examine how projects are different and then we'll get you thinking about visualising them in diagrammatic form.

Let's look at three different disciplines:

- theatre set design
- textile design
- automotive design

Let's ask some basic questions:

- why?
- where?
- how?

and see what the answers tell us . . .

Theatre set design

Why . . . are sets designed?

- to captivate the audience
- because of constantly changing productions

Where . . . is the work done?

- on-site
- freelancing from home

How . . . do they design?

- by working closely with the story of the play
- with a blend of designing and making
- by developing models and ideas for producers

Textile design

Why . . . do we need more textiles?

- for new fashions and for our homes
- for new markets

Where . . . do the designers work?

- within a large production company
- as freelancers from home

How . . . do they design?

- by drawing and manipulating imagery
- by analysing trend predictions
- by responding to customer and market needs

Automotive design

Why . . . does the world need more cars?

- huge need in our everyday lives for transport and mobility products
- cars are an expression of personal image

Where . . . do the designers work?

- in large corporations
- in design consultancies

How . . . do they design?

- by understanding markets
- by understanding peoples' emotional responses
- by sketching sophisticated shapes

In the simple appraisal above, it's possible to see some similarities, lots of differences and the fact that the designers are all working in very different ways. The detail of what they do could be very different too. Let's look at them again from three different perspectives that are important to all designers – longevity, consumer trends and materials.

Longevity:

Theatre designers are often making something that will probably only last for a few weeks, so they need to be very efficient in their use of materials. Things should be just strong enough to work and be safe. However textile and automotive designers could be making things that last for much longer. They need to take a long-term view of their work. As well as being strong, their products must be durable.

Consumer Trends:

Textile designers need to be very aware of trends in the market place as fabrics are 'of the moment' and can date quickly. However, automotive designers do not have to worry to the same extent – trends change more slowly in this field – and only some companies are trend-setters, the rest follow. Theatre designers can look back at trends – a new theatre set may depict fifteenth-century Italy, which entails a very different kind of research.

Materials choices:

> Automotive designers need to make sure their designs are safe and comfortable to use and this requires making prototypes and testing them extensively. Theatre designers are often using similar materials and construction techniques from one job to the next and gain experience in what will and won't work. Textile designers are often working with materials and processes they know very well and can accurately predict how they will perform. But sometimes they too will explore new materials and processes!

We can see that there seems to be, sometimes, very little similarity in what the designers do or the expertise they require. They are all doing quite different jobs. This is the essence of design process – that there might be as many different design processes as there are projects! Suddenly, being able to visualise and control your own process might be useful. There are many diagrams of design process out there already – pyramids, funnels, spirals and even a double-diamond produced by the Design Council (see Chapter 14, 'Future directions').

You are now ready to analyse your own design process. How do you make sense of it? Think of your last project and try to generate a diagram that depicts your real experience of that project. It can start at the top or bottom of the page or the left side, and it should finish at the other end of the page with your final idea. You could use simple generic geometric shapes like this:

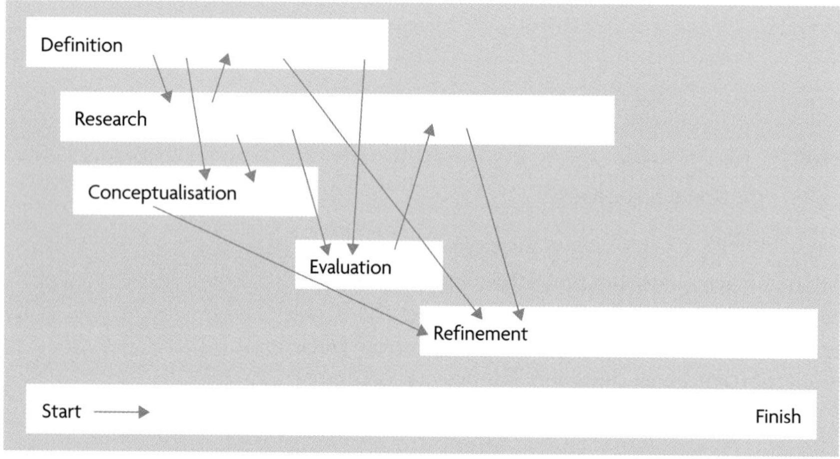

One of many design processes

You might also use circles to denote individual ideas, or 'speech bubbles' to comment on what's going on.

In reality, the different parts of the process overlap in various ways and communicate with each other (as shown by the arrows), dependent on the project, the designer, and so on. Also, at times, certain projects might only be exploratory and include just the first two or three activities.

Once you've tried to draw your process diagram for your last project you might discover that your next project needs a different diagram. There's no right way to visualise a design process so don't feel your diagrams have to look like the one above. All they have to do is make some sense to you. Try drawing the process for your next project before you start it. It will increase your awareness of what the issues are that you have to address.

This much is true about your design process

- **Be inspired**: The student project stories throughout the text will do this.
- **Produce lots of ideas**: All design processes should result in the production of various possible design solutions. We rarely just go with the first idea that springs into our head. Many clients require a range of options to choose from, so quantity of ideas is always important.
- **Do some research**: The design process is also about talking to people, looking at markets, finding out about materials, etc. Most design projects need research and it is often easy at first glance to work out what types of research are required. However, in this complex new century we need even greater awareness and ability in research.
- **Be in control**: Nurture your creativity by exploring and being inquisitive, and you can also control your analysis of your ideas and make sure they're good ones. This is good design. It leads to innovative solutions.

Is there more to it?

Your course will arm you with the discipline-specific knowledge and experience to search out what you need and use it in an intelligent way. This discipline-specific information is crucial to your development as a designer. To make the most of all this you will also need to reflect on the fact that at university the final responsibility for your education is yours. You're in charge. This is something you have to adjust to. Real life is full of variety, and university will be preparing you for that. Many of you will embark on live projects and placements with industry while at university, and these will form an integral part of your learning (see Chapter 9, 'How to succeed as a design student').

In the real world, design is very often a team effort and these teams are often multi-disciplinary; that is, they consist of various types of designers and people with different functions – sales and marketing, engineering and even accountancy. If you think of designing a hotel interior you can imagine the range of designers required – furniture, textile, interior and graphic, as well as the engineers who deliver the services to it – heat, light and warmth. Complicated! There is much to learn about how other designers from different disciplines work and there are always opportunities for designers to cross over into these disciplines. This is part of what makes design an exciting career – there are so many places to go and so many twists and turns on the way.

Even during your studies you will experience team projects. These will introduce you to the different pressures and responsibilities of being in a team. The three students of Design for Film & TV, Marjolaine Lebrasseur, Lauren Bacon and Sarah Press, pictured are collaborating and learning together. Many designers work with other designers, and even cross the boundaries of the disciplines throughout their career. Make the most of these projects, especially if you encounter any multi-disciplinary ones.

Design for Film & TV students Marjolaine Lebrasseur, Lauren Bacon and Sarah Press

Design briefs

> *'The best piece of advice I can give would be what I received while studying myself. This was to break down the brief. Being able to understand the focal point of any brief really helps me throughout a project. The key lesson I have learnt is that simplicity can produce good design.'*
>
> Graphic Design graduate, Catherine Perrott

A design brief will outline what you need to do in a project. Some are very comprehensive and quite tightly describe what you have to do. Often this is presented to you in a well-structured way, from your tutors or employers or the organisers of a design competition. Some are very broad, to the point where you might wonder how you start. Or they have a less defined period at the start of the project, when you might be developing the brief yourself or with a client.

Occasionally the brief might be very brief, and possibly verbal, in which case you might have to write it yourself. To ensure you understand all of the components and where it is directing you, here is a useful generic list of things you need to consider. It is by no means comprehensive or all-encompassing; you might want to add your own statements, or delete some of the ones here:

- the purpose of the project – what it seeks to achieve, why you are doing it and who is it for?
- a statement about the 'real world' context – why the world needs this project
- design influences and themes to explore (including likes and dislikes of the client)
- materials and processes
- statements about use and function, physically and psychologically
- general descriptions about how much it will cost to produce
- statements about the environmental cost of the project

Thinking about a brief in this way, and compiling your list of tasks, will ensure your design process is appropriate and thorough. Tutors will be looking to see that you have an understanding of the brief, the objectives or problem and a willingness to question information and push the boundaries in the search for generating something new, fresh and appealing.

There will also be, often, a personal aspect to your creativity. There will be projects that you just have to do and your personality will shine through. Don't be afraid of this, although to do it properly you need to be thoughtful about the brief you're working to. Yes, you do need to think about it carefully and write one. This attention to detail allowed final-year Jewellery Design student Amanda Trimmer to complete a project using one of her interests, 'Superhero' comic books, as inspiration.

'Superhero' brooch/badges by Amanda Trimmer admit you to her Superhero Society

Your project brief should provide a framework from which creativity can flow. It needs to be broad enough for your imagination to wander and for you to be able to explore various lines of thought while staying focused, but not so much that it channels you down one particular path. The brief should achieve the difficult balance of providing sufficient information to guide, but not so much that it restricts. As a designer you need to look carefully at this balance and may need to redefine the project brief in your own words (if your tutor or client allows it!) in order to achieve this.

This is demonstrated brilliantly by Industrial Design student Nell Bennett's final-year project, an electric trumpet:

Prototype of 'Electronic Trumpet' by Nell Bennett

> 'For my final-year major project I had to decide on a specific problem and develop a solution over the year. What you choose for your final major project will very much define what area of design you are likely to work in. It will be a strong element of your portfolio and therefore it needs to highlight your strengths.
>
> I eventually focused on this problem: all established western musical instruments have an electronic counterpart, except for brass instruments! As a trumpet player myself, it frustrates me that there is no electronic version of my instrument. Therefore I decided to develop an electronic trumpet to expand the musical capabilities for trumpeters and to push brass instruments into new music genres.
>
> I felt it was important to retain the craft heritage value found in traditional instruments. The forms, textures and materials needed to look and feel bespoke and valuable. As I was creating a "new" instrument, I wanted to explore new materials and processes that had not been used in traditional instruments, to create an object that would excite the senses. I decided to create a glass-blown bell.
>
> I found creating the glass-blown bell an enjoyable challenge. I had not worked with glass before and was unprepared for how imprecise it was. Learning how to blow glass was brilliant fun, and caught people's attention as it was out of the ordinary.'

This project required creativity and a very open outlook at the start. In order to deliver the project to the brief, Nell had to produce a working test-rig of the electronics. To visualise the final idea meant learning a new skill – glass blowing. This project is an example of exactly what is required to produce inspiring work in the design world.

What research methods are there?

Research is becoming ever more important to designers. We live in more complicated times, and the progress of technology, communications and even society set us new challenges every year. We also suffer from information overload. In a world where we can find out about and compare everything online, designers need to be aiming high – there is more competition, and consumers are much more informed than they were 20 years ago. Designers need to be cleverer than ever in a wider range of activities – at talking to consumers, researching new materials, forecasting trends, business practice, etc., as well as the more traditional ways of looking at the world.

There are two basic categories of research, primary and secondary. When you discover something for yourself and directly experience it, it is known as primary research: you watch people, investigate materials or analyse and paint an inspiring landscape. If your research takes you to a text book or magazine, or if you are inspired by a film, this is known as secondary research: you are experiencing it second-hand.

Primary research – get out there and find out about things:

- **Think**: be interested in the people you are designing for. They are often different to you.
- **Look around you**: designers are never switched off, they're always on standby.
- **Watch people**: draw, take notes, but don't photograph them without their permission – doing so is unethical and will get you into trouble.
- **Talk to people**: they're often very interested in what you're doing.
- **Do experiments**: play around with materials and processes.
- **Test your ideas**: build things, get your hands dirty.

Secondary research – get yourself into the library or on the Internet and discover what other people have found out about things:

- **Literature research**: get to know the library.
- **Published papers in academic journals**: maybe someone's done the work already.
- **Research marketing and consumer reports**: useful information may already exist about your intended market and competition.
- **Text books**: they can be useful (you knew we'd say that!).
- **Interview the experts (users, tutors, etc.)**: they've already done research – they might be flattered that you think they're experts and be eager to help you.
- **Google it! (note: this is not first on the list)**: be careful – do not just copy and paste chunks of text into your workbooks.

People and society are always changing, but the pace of that change is accelerating. For instance, in the developed regions of the world the population is ageing, as medicine and technology advance. Designers from many of the disciplines need to be much more aware of what it is like to be middle-aged with failing eye sight, or very old with reduced mobility. No matter what you are designing, you need to understand the people you are designing for, as in this project – part of Chloe Muir's final-year work on her Interior Architecture course. She worked on

Chloe Muir's 'Bereavement Centre Memorial Garden'

a bereavement centre for armed forces families, and this is her take on the people who might use this type of centre.

> 'What you think a client would need, and what they actually want are often two completely conflicting opinions. It is important to take the time to talk to the users and find out what they hope to gain from your project and what facilities they would want to help them, in my case, to rehabilitate themselves and learn to cope with grief and bereavement. Without this first-hand primary research, it is inevitable that you will design an intervention that will not be as successful.'

Connecting to people

A good starting point, if you want to involve people in your research, is observational research. If you are designing a bus stop, go on a bus ride. It's that simple (see the IDEO Research Method Cards later in the text for information on how to do this). You might find that, by the time you've done some observation, informal discussion work and interviews with people, you will have learnt a lot about the area you're designing in.

If your discussions and interviews are comprehensively done, then be aware that there is no rule that says design projects need questionnaires. You might not need a questionnaire! However, they are a very tempting method of collecting information. At this point in time, technology in the form of online surveys has made them even more achievable; they are easy to produce, look professional and will help you collate your results. But you need to be very careful about using them, as a good questionnaire needs careful design and development before you can trust what it tells you, like anything else (see further resources at the end of this chapter for a good book by A.N. Oppenheim on this subject).

This is Interior Architecture student Harriet Curtis talking about her research and the insight it can provide:

> 'You often hear people talking about "people watching" and it sounds a bit weird, but for me it is possibly the most helpful tool in creating a good design. I'm not talking about sitting with a coffee in Starbucks staring at people, just being aware of others and how they're interacting with the space.'

Working sketch of an interior space by Harriet Curtis

There will be a direct connection between you observing people in a real space and your work, like in this sketch of Harriet's, where she's exploring how people might use a space in a building. This shows her thinking about the journey down the stairs in the basement and across to the WC in the top left.

You need to be aware of the ethical issues surrounding research. Your university will have a policy on the ethical aspects of research, which you need to be aware of. Watching people at a bus stop is not harmful. However, if you take notes, or especially if you take photographs, you will probably be in breach of your university's ethical code. If you stop people in the street, you will probably get into trouble too. It's still possible to do lots of research without breaching the code and it's possible to do it without going to the university's ethical code committee for the necessary clearance, but you must first consult your tutors on this subject.

Where does research fit in to the design process?

There is a well-practised argument in the design world about which task you tackle first in a project – design or research. Surely all designing needs to be informed by the research you do first? But what is the harm in imagining some possible solutions as soon as you get a project brief, before you've had time to research?

If the first thing you do in a project is look at existing, commercially available solutions to the same problem that you're working on, you run the risk of closing your mind to crazy, off-the-wall ideas that might actually be your best ones. You might only be capable of reproducing the average of everything you looked at at the start of the research. This is a definite possibility.

This might be why some designers, the minute they find out about a project, can't help but start sketching wildly, or making something, or searching for creative inspiration. Sometimes the results are funny and all sorts of rash assumptions and mistakes are made as the enthusiasm kicks in. There's very little to lose at this stage, though, and occasionally a lovely, original, simple and beautiful idea will leap off the page. Ruth Wood, a Graphic Design student, discovered a catchy name for an arts centre she was re-branding when she started playing round with their postcode:

> 'I was working on re-branding an arts centre. In a tutorial, I said that I was having difficulty coming up with a better, more exciting name for the centre. One of my peers asked what the centre's postcode was and I discovered it was NE1. This seemed perfect. The NE1 instantly identifies that the centre is for the local area, and by sounding like "anyone" also relates to their target market, "everyone".'

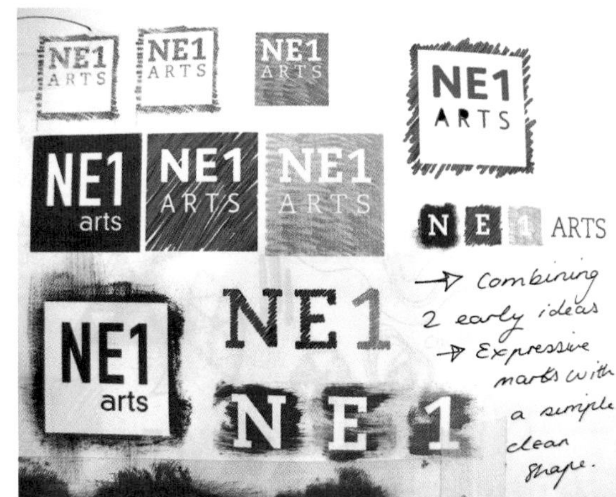

Logo ideas by Ruth Wood

There are many ways to research, and you can be quite creative about it. Try to develop new ways of finding things out. Really clever research has made a big change to design over the last 20 years. There are quite a few examples scattered throughout this book in the project stories. The future will be an interesting one in which technologies and cultures converge. Designers need to stop making assumptions about the world they're designing for and find out what's really going on and what is emerging on the horizon. Embrace research. It will inform your work and design direction and make you a better designer.

One resource that does tie into the research and design process across many of the disciplines is a set of cards that map out an array of research possibilities in an easily accessible way.

IDEO Research Method Cards

These provide 51 research methods (now that's going to keep you busy!) across four categories of research – learning about something, looking at something, asking about it or trying it. Developed by the renowned design and innovation consultancy IDEO, they can be purchased both as physical cards or as an iPhone app. These are applicable to many disciplines, not just design, and to many different aspects of the design process. Many of them come under the heading of ethnographic research. This is a qualitative method of observing cultural phenomena, which can then be related to your project. Each card states the name of the research method, how it is done, why it is used and gives an example of how IDEO have used the method.

IDEO Research Method Cards
Courtesy of IDEO

For example:

- When designing an online bank, IDEO used the 'Draw the Experience' card. This asked people to visualise an experience with money through drawing. This reveals insight into how people think about their activities with money.
- When trying to understand issues around weight loss, the researchers used the '5 Whys' card with interviewees, saying 'why' in response to five consecutive answers. This forces people to consider the reasons for their attitude and behaviour.

Conclusion

We've now seen that creativity, innovation and research go to make up the design process and are important ingredients in commercial success. It's difficult to find a success story that doesn't feature all of them. They also have to be managed well, and for this to be the case the company has to understand their importance and invest in them.

Even your university projects, especially in your final year, will require extensive management of these processes. Learn from your early years at university – your tutors will be more sympathetic then and will offer advice, but they'll be looking for you to take responsibility in your final year for creativity, project management and innovation.

The bird house by Chandni Kumari, pictured, is an excellent example of a comprehensive process delivering a very believable final outcome. She was determined to produce an environmentally friendly bird house with waste wood shavings from a furniture factory. An impressive research process involving lots of experimentation and testing led to developing a natural binding agent without any chemicals or toxins. Read more about her project in Chapter 14, 'Future directions'.

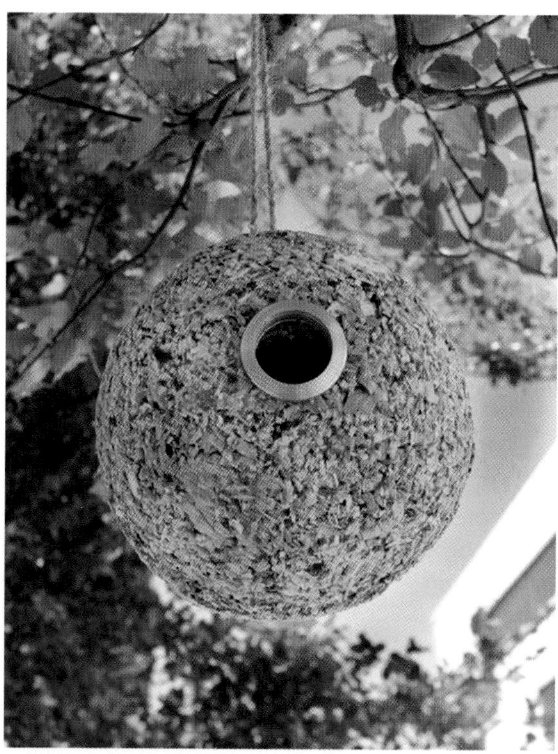

Bird box by Chandni Kumari

How do you control this?

Time management is the key to succeeding here. Students often cite it as a reason why they don't do well in their coursework:

- 'I didn't really get into it.'
- 'I had time-management issues.'
- 'I didn't know where to start.'

This is not exactly true always. In many of these cases students will simply not spend enough time on their projects – they will start it late, two weeks into a five-week project, or never get out of bed before 1pm. This is not poor time management, it is a lack of time and a lack of application to your studies.

Materials research for the bird box, Chandni Kumari

Use a time plan to plot all of the various tasks and stages of your project. Work back from the final deadline so that you know when you have to start doing things. This is known as the 'critical path'. Your plan will tell you how far behind your schedule you are and how hard you have to work.

Your plan will rarely tell you how far ahead of your schedule you are, unless you're very organised, and will never tell you, no matter how hard you look, that you can fit in a week of skiing in St Moritz before your final deadline!

We've seen so far that design processes can range from the personal exploration of the natural world by a textile designer (Chapter 3 'Fashion and textile design') to the internationally successful efforts of Land Rover (Chapter 8, Three-dimensional design'). However, the concepts we have outlined here that make up creativity, innovation and the design process relate to both of these extremes. Designers in both cases want to produce work that is creative, affordable and at the same time creates a profit for them and makes a difference to people's lives. Management of the design process is therefore a necessary tool in any designer's portfolio of skills.

At the start of a project you need to plan effectively and allow time for creativity and research, time to reflect on those aspects and the synthesis of ideas that will follow. As you develop your range of ideas, your research will move on to the technical and manufacturing phases. This is the start of the innovation phase.

Remember that there should be a pattern to your design process. Your projects may have an associated diagram, like that at the start of the design process section earlier, which will help you understand what you are doing.

Sian O'Doherty, a recent Textile Design graduate, sums this up:

> 'With any project, my fear is coming to the end and standing back and not being happy with its conclusion and knowing that I could have done more and could have pushed myself further. Therefore my highlight was coming to the end of my final-year project and not feeling like this!!'

Sian O'Doherty's final-year woven fabrics
Photograph: www.danstaveley.co.uk

Using your target market and users for subjective assessment and your research and design skills for more objective analysis, you will examine the options and hopefully arrive at a reasoned, justifiable and original final solution.

Remember that much of what is required to manage the process and deliver a good final result becomes easier with practice. Good research can help identify and define the factors that are pertinent to a specific problem and this emphasis allows young, creative people like you to shape the future.

The project stories at the end of this chapter demonstrate how these joined-up processes and in-depth thinking and research can lead to surprising and valuable results in all disciplines of design. Look for people making connections.

Further resources

Books

Buzan, T., *Use Both Sides of Your Brain*, Atlantic Books (1991)

This book provides a good understanding of how the brain works, the way we think and ways to develop right-side brain thinking. Tony Buzan is another world-leading authority in creative thinking and mind-mapping techniques.

De Bono, E., *How to Have Creative Ideas: 62 Exercises to Develop the Mind*, Vermilion (2007)

A great book from the master of creative thinking. Concepts are easy to understand, with lots of useful material to help you develop your creative thinking and approach to problem solving.

Kelley, T., *The Art of Innovation*, Profile Books (2002)

This is a description of innovation management, as practised by the design consultancy IDEO. It divulges all the secrets of this very successful, international design consultancy. It engages the reader with its honesty and wonderful case studies.
www.theartofinnovation.com

Kula, D., Ternaux, E. and Hirsinger, Q., *Materiology: The Creative Industry's Guide to Materials and Technologies*, Frame Publishers (2011)

Written in an easy to understand style with many instructive diagrams, this book about materials and technologies is available in Dutch, Czech, Portuguese, German, French and English.

Mumaw, S. and Lee Oldfield, W., *Caffeine for the Creative Mind: 250 Exercises to Wake Up Your Brain*, How Design Books (2006)

A great book, crammed full of exercises to help you exercise your brain. It challenges you to try one technique per day, so within half a year you'll become fluent in idea generating.

Oppenheim, A. N., *Questionnaire Design, Interviewing and Attitude Measurement*, Continuum International Publishing (2000)

Everything you might need to involve people in your research for your projects, with great detail and advice – a book you will keep coming back to throughout your career.

Raudsepp, E., ***How Creative Are You?***, TBS (The Book Service) (1988)

This book provides a basic understanding of the things that may prevent you from being creative, or feeling creative, and equips you with some of the tools required to become more effective in your creative thinking. The book also includes a self-assessment Creativity Quotient test. It might be an old book now, but the material is still very relevant.

Ternaux, E., *Industry of Nature: Another Approach to Ecology*, Frame Publishers (2012)

Learn from nature's smart, simple and sustainable behaviours in this book aimed at creative professionals. It will inspire and expand your imagination.

Ternaux, E., *Innovative Materials for Architecture and Design*, Frame Publishers (2011)

Contains 100 fascinating materials and their application to interior design and architecture selected by the organisation matériO. Extensive product information and contact index induded.

Websites

www.creativethinkingwith.com

Online resource for teachers and students – a great resource with lots of links to specific techniques and creative exercises you can try out.

creativity-online.com/section/on-design

Online resource – cross-discipline, featuring videos, multimedia, articles and podcasts to show how ideas can impact our world and society.

everything2.com/title/How+Creative+Are+You

Eugene Rausepp's online test to see how creative you are.

www.materio.com

This is run by the organisation matériO and is a materials library with a selection of innovative materials profiled on a daily basis. There are student rates to access the full library. There are also several showrooms across Europe.

iTunes applications

Idea Generator – a useful software gadget, it works a bit like random word association. It has the familiarity of a 'fruit machine' and randomly selects three different words. You can also input some of your own fixed words and then allow it to select others.

Project story

Designer:
Lana Crabb, jewellery design student

Project Title:
Personal Development 3

'Climber Brooch'
Lana Crabb Contemporary Jeweller

Tell us about your project

This project was the final project of my degree in Jewellery and Silversmithing at Birmingham City University. It was the collection of work that would launch my career, as well as being a resolution to the last three years of study (no pressure there then!).

The roots of this body of work began at the end of my second year, when I started to become interested in theory about 'preciousness' in jewellery – i.e. what makes jewellery precious? Is it the material worth or something more? Why do some people sometimes find things that cost very little, precious and important? In every case during my research I found that monetary worth didn't affect the 'preciousness' of an item, that usually the object or jewellery had sentimental value to the owner.

Although this research was fascinating to me, it didn't lead me to make anything, so when I returned to the workshop for my final year I knew I needed to translate my own theories about preciousness into physical work. This began with experimenting with found materials and old broken jewellery found on eBay, which I wanted to make valued and 'precious' again. During this time I was a little unconstrained and unfocused, just making for the sake of it, with no end goal or real clue as to what I was doing, so I felt frustrated because I couldn't figure it out.

Lucky for me, a happy accident happened when figuring out an attachment for a necklace, and I started to use rubber bands. This was a turning point because I realised that I could focus on re-purposing common everyday objects. I loved the idea of turning something normal and mundane into a quirky and appealing piece of jewellery.

Luckily this revelation came at the end of the first project of the final year, so over the Christmas period I could focus on my dissertation (which was about the meaning behind making art jewellery with ready-made or found objects). Choosing an area of research that will aid your understanding and develop your own ideas about your design practice is such a good idea; I'm sure I wouldn't have been able to get as far as I have in my work if I had chosen something different.

Paper models exploring shape
Lana Crabb Contemporary Jeweller

'It's a Wrap' brooch
Lana Crabb Contemporary Jeweller

This all led to my current body of work – jewellery that is precious but not made from traditionally precious materials. Instead of silver, gold and diamonds, I use base metals such as guilding and brass, with threads, balloons, rubber bands and broken jewellery. The metal bodies of my brooches are powder-coated, which is an industrial paint, to further remove their appearance from traditional jewellery. I aim to make fun, colourful and appealing jewellery that is usually small in scale.

What processes and skills were most relevant to this project?

To make the work I had to become very accurate in scoring and measuring the metal, as well as filing and soldering, and finishing the surface so it was smooth and clean looking, to then be painted. As part of my design process I experimented with different materials and shapes, and doodled in my sketchbook. Often I would make piece after piece, just trying something new and seeing how it turned out – learning and adapting and designing as I went. You have to become good at evaluating your work and being honest if it isn't up to scratch!

What challenges did you face?

The biggest challenge I faced during this project, as well as the whole three years of my degree, was to learn to trust myself and have confidence. Many times before this project I felt I didn't know what I was doing or why I was doing it, so naturally when I looked back I didn't like my own work. The final year of my degree has taught me that, certainly in jewellery, you have to make what *you* like. Don't make work for other people to like it, because if you don't, then you will never be satisfied. I learnt to trust my instincts when something wasn't quite right – sometimes you just have to think about it (or let it bubble away while doing something else) and finally you will realise what is wrong. I also learnt that some things take time. Sometimes you can't reach a stage where you know what you're doing until you've had a few weeks of going through the motions with your work; however, if you keep working hard (and I mean really hard!) you will get there.

What were the highlights of the project?

The highlight of the project, other than learning to trust my instincts and abilities, was to make a collection of work that I really like and believe in. I feel like I have finally found something I am really interested in, not just for uni, but for my designing and making practice, something that I can continue with under my own steam. My other highlights were being nominated and winning an award to travel to an important art jewellery fair in Munich next year, and winning another award for my work to be shown in a prestigious gallery. They were both totally unexpected but it goes to show that if you work hard and are enthusiastic, you will succeed. The biggest highlight of all is that two of my tutors are commissioning me to make them some jewellery – it is the ultimate form of flattery from people I admire and aspire to be like.

Next, I will be working on a project called 'Design Space' in Birmingham, a business start-up scheme for contemporary jewellers, which gives free business support and workshop space for a year. I am also working to support myself while I try to become self-employed. I aim to carry on making, broaden my horizons and hopefully get my work into some more galleries.

www.lanacrabb.com

Project story

Designers:
Sally Halls, Grace Davey

Companies:
Department of Health/Design Council/
Helen Hamlyn Centre for Design

Project title:
Design Bugs Out

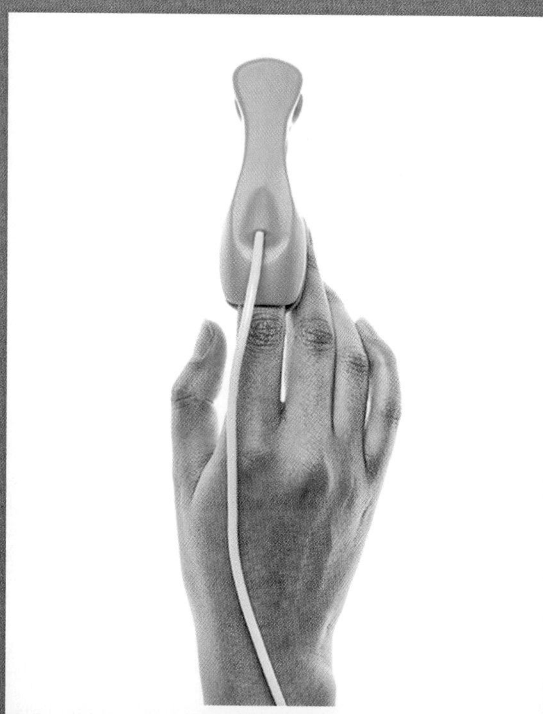

Pulse oximeter

Tell us about the project

When a number of outbreaks of infectious bacteria, such as MRSA and C Difficile, occurred in different hospitals, the Department of Health took notice and started to consider how they might be prevented. They collaborated with the Design Council to create a design challenge called 'Design Bugs Out', which challenged designers to explore ways in which cross-infection might be reduced through design. This resulted in a selection of equipment and furniture being designed to be much easier to clean. This pulse oximeter, which measures the amount of oxygen in your blood, has been redesigned to be a wipe-clean piece of flexible plastic rather than a collection of plastic and metal parts and springs, which are impossible to clean.

What processes and skills were most relevant to this project?

The project kicked off with ethnographic and design research to identify areas where design could have an impact, and then launched the briefs to the design industry, inviting design agencies to pair up with manufacturers to deliver better solutions. Initial **ethnographic research** led to the team working on six of the briefs. These were, the blood pressure cuff, the pulse oximeter, the cannula, the bedside curtains, the mattress and the issue of patient hygiene.

Ethnographic research: the scientific study of human culture. This involves contact with and in-depth study of people, society, working relationships, etc. There is a wide range of research methods that use ethnographic principles, many of which are featured in the IDEO Method Cards mentioned earlier and the sections on inclusive design and socially responsible design later in the text.

The initial research team consisted of designers, ergonomists and design strategists, and we worked with three partner hospital trusts that had been selected from across the country. We spent many days at each site, observing staff at work, understanding how the departments operated, and interviewing key stakeholders. This meant talking to doctors, nurses, cleaners and hospital porters – all the people involved in using and cleaning the equipment in the ward.

Initially, working in these new environments can be overwhelming as there is so much information to absorb. But as you learn more, you begin to see through to identify the areas where design can make a difference. Having a multi-disciplinary research team also led to some healthy discussions around what was witnessed and the different interpretations of it. To use the analogy from earlier in the creativity section, this 'filled the cupboards'.

The main issue found was that modern cleaning practices had developed into efficient, time-effective processes but the furniture and equipment hadn't been updated to accommodate these cleaning methods. Eventually our observations and insights were condensed down to 13 design briefs, of which five were tackled by different design agencies and six were tackled by our team at the Helen Hamlyn Centre for Design.

What challenges did you face?

The client liked to be kept very well informed as well as there being many different interested stakeholders. This meant there were meetings and presentations on an almost weekly basis, leaving less time for design work.

We also went back into the hospitals to continue engaging with the users, this time focusing in on the six project briefs. Healthcare staff are very busy and under pressure, so finding the right people to work with, both in terms of time and open mind-set, took some time.

Some of the proposals were also very dependent on using new and different material technologies, so it took many iterative prototypes to understand what was possible. The blood pressure cuff, for example, where we were replacing the difficult-to-clean Velcro, needed to have magnets that were strong enough to withstand the sheer forces being applied by the increasing air pressure, while still being weak enough to be opened by the users. Many different combinations of magnets and plastics were tried before eventually finding the right configuration.

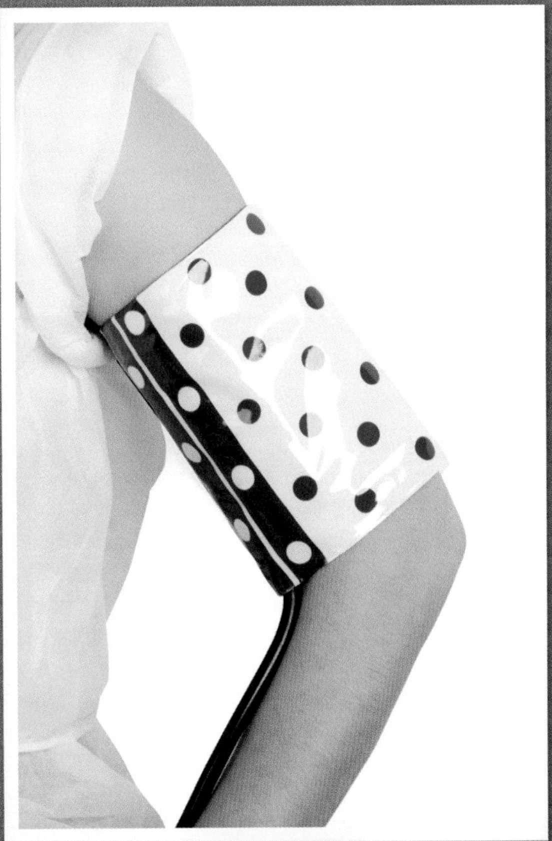

Redesigned blood pressure cuff
© Helen Hamlyn Centre for Design, Royal College of Art

What were the highlights of the project?

The project was unveiled at the Design Council with a press launch and a touring exhibition, showcasing the designs at seven hospitals around the country. It received a lot of interest, and helped both healthcare providers and manufacturers to understand the difference that design could make to the issue. Many of the products have been taken on by manufacturers and are now being developed for the market – a sign of the project's success. Take a look at the website to see what was involved and what it led to:

www.designcouncil.org.uk/our-work/challenges/Health/Design-Bugs-Out/

Project story

Designer:
Amanda Trimmer, jewellery design student

Project title:
Secret Superhero

Tell us about the project

It took me two and a half years to get to the point of creating the 'Secret Superhero' collection, a culmination of everything I had learnt in my degree with snippets of different projects being brought together.

I had always enjoyed making pieces with an element of repetition in them. This was simply because it gave my work order and gave me more control over how the pieces would turn out. However, because I was always working to someone else's brief, I would choose to look at objects that surrounded me in my everyday life, such as watches and tools in the workshop, for inspiration – I never looked at things that I was really interested in! Until one day, when I sat down to have a chat with my tutor about my work, just before my final project at university, and realised that I could make work about things that I loved, for me instead of for someone else. So I sat down in my room and thought about the work I had done up to this point, not only at university but at A level, Art Foundation, and even GCSE, and realised there were common elements that made up my work. These were bright block colour, clean-cut lines and an element of interaction.

'Secret Superhero' brooch by Amanda Trimmer

The clean-cut finish and colours had always been inspired by comic books but never actually related directly to the comics themselves. Over the years, my design process has become very ordered, almost mathematical. I started by exploring the shapes and colours in the comic books – shapes that were anything from the outline of a building to the sole of a shoe.

Once I knew which shapes I wanted to work with, I prepared them for laser cutting on Coreldraw and had the shapes laser cut and vector engraved (this is the process where lines are burnt into the surface of the paper to add detail). While this was being done, I had the metal engraved so that I could pierce out the shapes ready to solder together. The final process was to have them powder-coated (a type of coating that is applied as a free-flowing, dry powder). When I finally had all the components together, I began assembling the pieces! This took quite a long time, after which my eyes needed several minutes to readjust.

What processes and skills were most relevant to this project?

Within my own work there have been times when pieces have failed dramatically; for instance, in my first project I attempted to make a pin that would float when placed on water. It sank! It does make you despair a little bit when things don't go right, but every mistake really is a lesson learned. In the first two years of my course I learnt not only about making jewellery and how to develop my own skills, but how to design. Before, I did not think through every aspect of a design, but once I got to university I was taught how to work through the design process in a way that suited me. I was shown everything, from how to do primary research to making models.

How to run the business side of things was another aspect of the course. I was shown how to create a professional portfolio and approach galleries in a professional way, so that I could be seen as a business person and not just a jeweller. This was done by going through the process of creating a company. All 32 students in the second year were involved. We were taught all aspects of how to run a business, from the fun side (setting up venues and making contemporary and commercially viable jewellery) to the not so fun side, such as doing the accounts! This was a really good learning experience. It was stressful and there were tears, because it taught me that you can only really rely on yourself to succeed. It will be a struggle and you will have to dedicate half of your time to marketing yourself and networking.

What challenges did you face?

Before I started my course, I went to an open day and as I walked around the School of Jewellery in Birmingham, I had no idea what any of the equipment was or who any of the jewellers were who the students showing me around were talking about. I felt completely out of my depth. However, my tutors on my foundation had recommended this course, and I was confident they knew my work well and they had sparked my interest in jewellery design.

When I started my degree I felt overwhelmed by all the techniques and machinery I was learning about, but I soon realised that all the other students on my course felt the same way, especially the ones who appeared to know it all. We were all new to metalwork and it soon became clear that I was expected to mess up and ask for help – that's what the tutors and technicians are there for, after all!

When I started my studies, I thought jewellery was fine silver and gold pieces of metal that you would wear as a pendant or a ring. However, I soon learned that this wasn't the case. Jewellery can be anything that adorns the body and helps to show a little bit of your personality off to the world! It can be placed anywhere on the body and be made out of anything, from card to bath bombs! All you have to do is think 'outside the box' and create something unique and interesting that people would want to interact with. It's as simple as that . . . all it takes is time and making sure you stick to your gut instinct, making sure you create pieces that you enjoy making and are personally interested in!

What were the highlights of the project?

This was the moment when I decided that I would let out my 'inner comic-book geek' and go through my collections of comics in order to create pieces that allowed the wearer to become a 'Secret Superhero'. I have set up a forum so that people from around the world can become a member of the Secret Superhero Society!

trimmerjewellery.co.uk/secretsuperhero

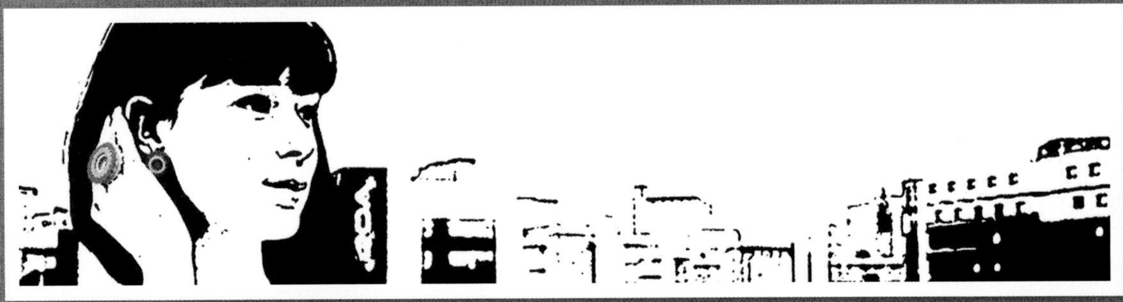

'Secret Superhero' earring and ring by Amanda Trimmer

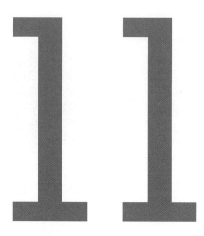

Appreciating aesthetics

By Robert Clay, author of *Beautiful Thing*
Formerly Principal Lecturer, Teesside University

'My current work is based around ideas of sacred geometry and proportion; I try and incorporate very pure and simple materials that are ecological and sustainable. I plan to make pieces that will work within and enhance the architectural space that I am designing for.'

Gill Wilson, paper-maker and gallery director

The application of new materials to once familiar objects causes us to re-appraise our notions of what is beautiful, as in this bicycle by Rafael Hoffleit (http:/vonrafael.com)
Rafael Hoffleit

Many people choose a career in design because they have a desire to create beautiful things, whether in two or three dimensions, as well as solving problems or inventing new products. Designers usually start with the end-users of their products and try to make their lives easier (or more interesting) – for example, by distorting reality to create the classic London Underground map, or by making the user interface of a ticket machine easier to operate. Of course, all designers hope that their design solutions will be regarded as beautiful by others, whether in a physical or abstract sense. Human beings are intensely visual and tactile creatures who make visual and emotional choices when choosing which product to buy or use, and a beautiful and sensually satisfying design can make us feel better and actually improve our performance. But, *what is* beauty, and do notions of beauty change over time and place? Also, what about 'good taste' in design, what do we mean when we say someone has good taste in his or her choice of furnishings or tableware, for example?

So, what is 'good taste'?

The particular culture an individual is brought up in has inescapable influences on that person's preferences and tastes, and therein lies a problem for every aspiring designer – the culture trap. The beliefs, symbols and ways of doing things in a particular culture are deeply engrained in its people, and a particular culture's art, architecture and designs can sometimes seem very strange to outsiders. Taste, fashion and design are inextricably linked together, and it is very difficult to discuss one of these in isolation. Current fashions, whether in clothes or cars, change over time, which complicates things even further. A person's taste in furnishings and clothing can also provide clues about that individual's status in society (the particular socio-economic group that he or she belongs to), and can even provide an indication of that person's set of personal values.

In late-nineteenth-century England the Victorians' appetite for elaborate pattern and ornamentation in interior and furniture design was driven to some extent by peoples' desire to show off their status and wealth in society. Because hand-crafted items were very expensive to produce, the more finely detailed, hand-crafted objects you surrounded yourself with, the more you could impress your friends and neighbours. In his book *Taste: The Secret Meaning of Things*, the first director of London's Design Museum, **Stephen Bayley**, said that:

> 'Taste is not so much about the appearance of things, but more about the ideas that give rise to them.'

Bayley states that good taste should not be confused with excess or luxury (as with the Victorians' ideas of excessive decoration) – it is better to design excellent everyday products than to conjure up some ghastly gold-plated and expensive confection.

As the art historian **Kenneth Clark** has said:

> 'Splendour is dehumanising, and a certain sense of limitation seems to be a condition of what we call good taste.'
>
> *Civilisation*, 2005

The Greek philosopher **Plato** certainly regarded luxury as a disease – the ostentatious display of luxury by leaders (and aristocrats or other high-status individuals) can be offensive to the ordinary mass of people and was therefore regarded by Plato as dangerous to society. It was simply bad manners and could breed revolution once out of control. A sense of austerity is also a valued theme in traditional Japanese culture.

Different cultures have different notions of what constitutes good taste in design. But how do these ideas and preferences relate to beauty? Is there such a thing as universal beauty, or is beauty an entirely subjective concept experienced by every individual on his or her own terms? You may feel that these ideas about aesthetics don't have much relevance to designing everyday things today. But they do, as this work by weave student Beth Snowden shows. In her words:

> 'The inspiration for my work, "Smoke and Mirrors", came from the origins of show business, from old smoky theatres and the glitz of the circus, even through to early filmography. The idea that all was not as it seems, some sort of illusion, was intriguing. I began to look at how light is affected by reflection and refraction, the way colour changes and shadows are cast. This particular piece was made to emulate the curves of light refracted through smoke.'

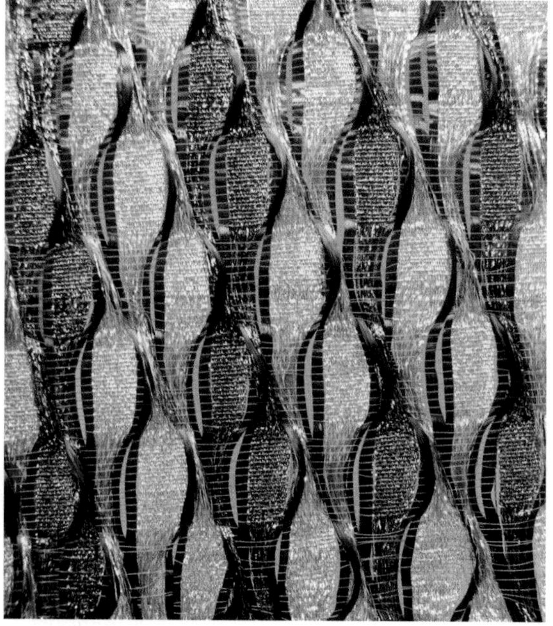

'Smoke and Mirrors' weave by Beth Snowden

Others will judge the work that your generation produces, both today and tomorrow. This is particularly relevant when you are designing something that is going to be sold in a different country and culture from that of your own. Ignorance of cultural and aesthetic factors may result in the failure of your design proposal – notice how clothes designs appear different in different countries.

Remember that you aren't always designing things for yourself; you must start by investigating the tastes and preferences of the person or group that the product is aimed at. You can also make huge mistakes designing for people, even in your own culture and time. For example, most young designers have near-perfect vision, hearing and other senses; however, a large proportion of any population includes those people whose senses are not so good or have deteriorated with advancing years. This also makes economic sense – the more people who will understand your designs and find them a delight to use or own, the more money you will make for your company, so everyone wins.

An 'inclusive design' wristwatch

The 'inclusive design' wristwatch pictured is a design that enables the short-sighted user to read the time and date without having to fumble for reading glasses. This design is just as elegant as a traditional wristwatch and is attractive to both young and old alike. This is sometimes called 'inclusive', 'universal', or 'trans-generational' design (see the section about inclusive design in Chapter 14, 'Future directions').

Western concepts of beauty

When we begin to design something, whether a new building, a new kettle or a typeface, how do we decide what it should look like? How do we decide upon an object's proportions, shape, form, colour, pattern or texture? Although there are usually many constraints on the designer in any design project (for example, the constraints of cost, materials, performance requirements and available space), there is often room for producing a variety of imaginative design proposals in answer to any one particular brief. So, all constraints being considered, how does a designer begin to create a beautiful new design?

Aesthetics is a branch of philosophy concerned with the perception and description of beauty and ugliness (and issues of taste) through language. Some philosophers say that this is impossible – the experience of beauty cannot be translated into words, nor can it be formally taught. Nevertheless, many individuals have attempted to do just that and find answers to such questions as 'what is beauty?' and 'is beauty a quality of the object perceived or does it exist only in the mind of the observer?'. The Greek philosophers **Plato** (about 427 to 347 BC) and **Aristotle** (384 to 322 BC) believed that beauty was a distinct property of the physical world and that a beautiful object or place remained beautiful even if there was no one there to observe it. In other words, the Parthenon in Athens or a mountain peak would remain beautiful even if human beings suddenly disappeared from the planet.

Opinions about aesthetics have changed considerably since the Classical Greece era. In more recent times, the German philosopher **Immanuel Kant** (1724–1804) believed that beauty was not a property of an object and that it existed only in the mind of the observer. Thus the sensation or experience of beauty is generated in response to an object – a view that is widely accepted today. Kant therefore believed that the experience of beauty was personal, and individuals made their own judgements about what was and was not beautiful or tasteful. However, Kant also proposed that, in addition to this personal experience, there was also the possibility of universal beauty – a formal beauty that has nothing to do with a particular individual's subjective tastes. He proposed that universal beauty could be established if an individual considered an object or work of art (for example a portrait painting) from a disinterested point of view. That is, view the work as an end in itself and ignore any extraneous notions of who painted it, the identity of the person in the picture, or its monetary value.

Kant admired the Irish philosopher **Edmund Burke's** treatise on empirical aesthetics, *A Philosophical Enquiry into the Origin of our Ideas of the Sublime and Beautiful* (published in 1757), which maintained that standards of beauty and good taste are universal, having been judged so by many observers over a long period of time.

This apparent contradiction of personal and universal beauty is also a theme taken up by contemporary philosophers as well as scientists. The neuroscientist **Professor Vilayanur S. Ramachandran**, Director of the Centre for Brain and Cognition at the University of California, suggests that universal aesthetic principles do exist – that perhaps 90 per cent of the variance in art is driven by cultural diversity but the remaining 10 per cent is governed by universal laws that are common to all brains. This is analogous to humans being 'wired' for language at birth, but the particular language an individual learns being dependent on the culture he or she is born into.

Unlike Kant's formalist approach to universal beauty, **George Wilhelm Hegel** (1770–1831) believed that to properly appreciate a beautiful object the observer must take into account the context in which the object was created or observed. That is, the beliefs, customs and technologies of the particular society that an artist belongs to strongly influence the creation of a particular work of art and these cultural notions must be taken into account by someone from a different society or time when judging that work. Hegel maintained, therefore, that in addition to expressing the spirit of the artist who created a work of art, objects of art also had the power to express the spirit of the particular culture that the artist belonged to.

Eastern concepts of beauty

Traditional Eastern aesthetics approach beauty from a different point of view from those of the West. Classical Japanese philosophy proposes that reality is in a constant state of change – or in the **Buddhist** sense, one of fundamental impermanence. Living things lose their beauty and die, and therefore beauty is a transitory, fleeting thing. Even inanimate rocks and landscapes change or disappear over time. Unlike Greek philosophy, there are no permanent ideals of beauty in the Platonic sense – perfect models that exist beyond human reach to which all earthly forms aspire.

The Eastern practice of Buddhism tells us, however, that an awareness of this fleeting nature of existence and beauty should not lead us to despair; on the contrary, it should make us grateful for any remaining time left – time in which to value each precious hour and fill it with useful and vigorous life. In Japan this philosophy of valuing existence and leading a useful life also goes hand-in-hand with the arts and crafts. Traditional Japanese arts have always been closely linked with **Confucian** ideas of self-cultivation and ethical 'ways of living' – for example, the way of tea making and the way of writing (calligraphy). As in China, the cultivated person was expected to be skilled in many of the arts – for example, ceremonial ritual, music, poetry and even self-defence or archery. Thus the arts were more closely woven into everyday life, compared with Western traditions.

The more fleeting the existence of a beautiful object, the greater is its value and therefore sense of loss when it is gone (for which we use the word 'pathos'). The short duration of cherry blossom in the spring is cause for national celebration in Japan, where crowds of people turn out to view the beautiful blossom and picnic under its branches. Although pear and apple blossom may be just as beautiful a sight as the cherry tree in full bloom, the cherry is more highly valued because of the shorter duration of its blossom – which can be as little as a week. This pathos, or empathy, with living things can also be expressed through inanimate objects – for example, the significant rock carefully placed in the garden can represent a friend who has died, or the beautiful vase in a room can be a symbol or reminder of the daughter who has left home.

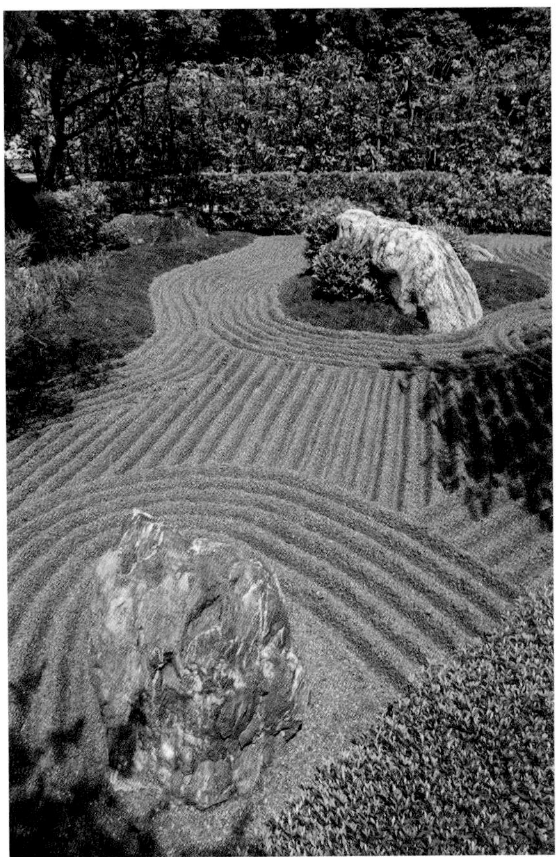

Japanese garden
Sergii Rudiuk/Shutterstock.com

The traditional Japanese garden often employs metaphors – a rock can be used to represent a person, animal or spirit, a 'stream' of pebbles can represent flowing water, occasionally crossed by 'stepping stones' of larger, flat-topped rocks. Sometimes garden designs are based on old fables and stories – the 'story' unfolds as you walk through the garden. **Shintoism**, the original religion of Japan, is grounded in the worship of nature, spirits and ancestors, and says that mankind must honour and live in harmony with nature – thus the Japanese garden provides an oasis of tranquillity in which to contemplate the universe. To the uninitiated Western eye, none of these meanings will be immediately apparent, but this does not stop us from enjoying the peaceful beauty of a Japanese garden.

Simplicity, or an understated austere beauty (*wabi*), is a Japanese ideal, perhaps the simple, raked pebble garden being the most obvious example to Western eyes – being 'cut off' (*kire*) from its natural and chaotic surroundings. This austere beauty was first admired in poetry but is also expressed through the apparent simplicity of everyday things – for example, cultivating the garden or pruning trees in such a way that removes superfluous elements to better encourage and express a tree's essential beauty. This is simplicity in a sophisticated sense – an elegant resolution of competing or confusing elements – and a philosophy that also applies to the design of the traditional Japanese house and everyday objects.

The tea making ceremony especially expresses this simplicity through ritual and the austerity of the teahouse design and its utensils. This austerity even applies to damaged utensils, providing they have been well repaired, and these imperfect utensils are often valued more highly than brand new ones, being symbols of moderation. Objects that have acquired a patina over time through use (*sabi*) – the cracks, stains and rusting of objects that have aged well – are also highly valued in traditional Japanese culture. Even in times of insufficiency and hardship, getting by with very little in an elegant way is considered a beautiful and noble aim. The term *sabi* also carries meanings of tranquillity and deep solitude.

Thus beauty does not lie in opulence and extravagance but is to be found in an elegant simplicity – an ideal not unnoticed by twentieth-century modernists in the West, with their designs for architecture, furniture and product designs – see

Mies van der Rohe's 1920s design for a tubular steel chair. The introduction of tubular steel furniture caused an outcry at the time; many people, used to 'proper' solid timber furniture, severely criticised these new designs. However, many of these 'simple', understated designs have since become classics and are still in production today.

Mies van der Rohe chair, 1920s
Bridgeman Art Library/© DACS 2012

Design evolution

For tens of thousands of years, humans have experimented continually with new materials and methods for the making of better tools and shelters. The continuing developments in technology have always led to new ways of designing things – and consequently the changing (and sometimes 'shocking') appearance of objects over time. There have been several noticeable 'spurts' of such activity throughout human history that have led to particular periods of excellence in the field of art and design. The beautiful temples and sculptures of Classical Greece and the multitude of innovations by the **Romans**, such as concrete, piped water supplies and, although the Romans did not invent the stone arch, their spectacular exploitation of the arch's benefits in huge bridges and aqueducts.

Greek temple, c. 450BC
Olga Drabovich/Shutterstock.com

The Italian Renaissance and Leonardo da Vinci

The great flowering of works of art, design and architecture of the Italian Renaissance during the fifteenth and sixteenth centuries continues to be much admired worldwide. Artists and architects such as **Leonardo da Vinci**, **Michelangelo** and **Bernini** believed that the human form was the basis of beauty, and therefore buildings and structures should be designed with human proportions in mind; the phrase 'man is the measure of all things' (attributed to the Greek philosopher **Protagoras of Abdera** c.490–420BC) became central to their thinking.

Writings on architecture by the first-century Roman engineer **Marcus Pollio Vitruvius** were very much admired in Renaissance Italy. Vitruvius had been a soldier in Julius Caesar's army and later served as a military engineer and architect under the Emperor Augustus, to whom he dedicated his *Ten Books of Architecture* – a massive work covering almost every aspect of ancient architecture, types of buildings, materials and even town planning. Vitruvius said that: 'geometry is the very footprint of man' and that towns and villages should be laid out on a squared grid. He maintained that symmetry was essential to beauty and that, 'no temple can be put together coherently without symmetry and proportion, unless it conforms exactly to the principle relating to the members of a well-shaped man . . . only a man well-shaped by symmetry can be made to produce the circle and the square'. See Leonardo's famous 'Vitruvian Man'. Vitruvius believed it is because a man's body is symmetrical that architects should follow ancient precepts about symmetrical building.

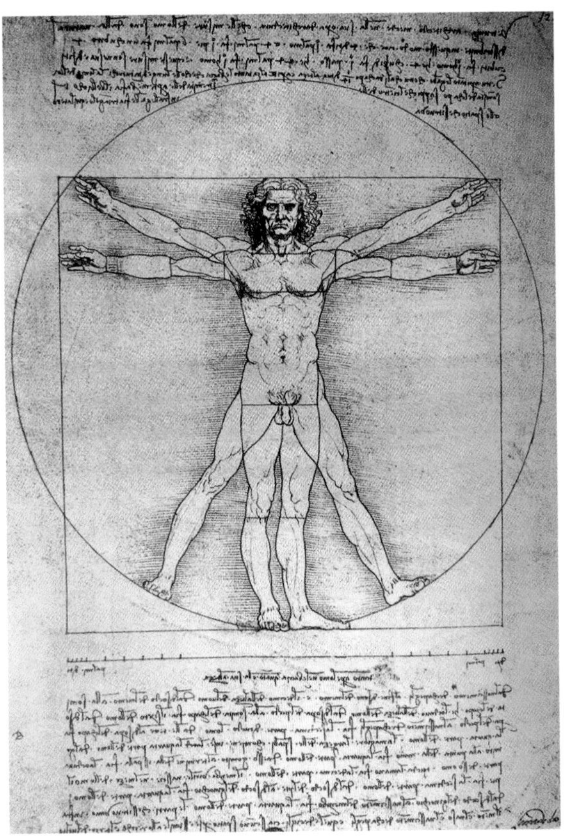

'Vitruvian Man' by Leonardo da Vinci, c.1450
Janaka Dharmasena/Shutterstock.com

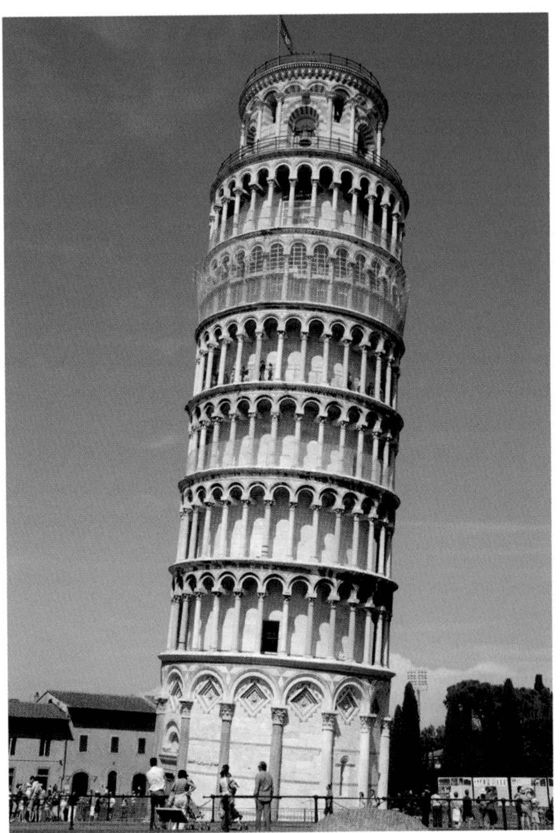

Leaning Tower of Pisa, completed 1350
nmiskovic/Shutterstock.com

Although we can still admire and apply Vitruvius' ideas today, the importance of symmetry has diminished in recent times – too much symmetry can lead to predictable and even boring designs and perhaps we have grown weary of it – resulting in today's fashion for asymmetrical design in buildings, products and graphic designs. However, we still value symmetry, not least because humans prefer balance to imbalance. We are essentially vertical creatures in a horizontal landscape and we can see at a glance if someone, or something, is in danger of toppling over – which is why the extreme angle of Pisa's leaning tower is so startling. Our visual sense alerts us to this fact immediately, and a good sense of balance is of course vital in the survival stakes for all animals.

We sense, then, a balance in the design of paintings, graphics and objects. The more educated we become in the process of looking, the better we become at discerning balance between two (or among several) elements in a particular composition, whether in two or three dimensions. Often artists and designers deliberately use 'off balance' elements in their work to stimulate the attention of the observer, and a sloping or leaning element can imply movement and direction – Pisa's leaning tower gives us the impression that it might start moving at any moment! This is also the reason why *italicised letters and words seem to speed up on the page.*

The proportions of architecture and the everyday products we use are, by necessity, strongly related to human proportions (as expounded by Vitruvius); the heights and widths of doors allow our passage through and the dimensions and layout of a mobile telephone fit our mouths, ears and fingers. However, just because an object's proportions might be strongly related to function, it does not necessarily follow that this will automatically produce an elegant design – for we still get 'ugly' buildings and products.

Porsche 911
www.carphoto.co.uk

Sloping elements can also suggest the potential speed of man-made objects; for example the sloping, 'fast' lines of the car, pictured, suggest high velocity when compared to the 'slow' shape of a delivery van. This is only a cultural notion, however, as we have learned through experience that streamlined objects have less wind resistance, which allows them to go faster – none of these notions about vehicles would be understood by someone who had never seen a car. Perhaps once they understood that vehicles are actually fast-moving objects, a new observer might intuitively make the link with the 'fast' shapes of birds and fish – streamlined by nature to assist their passage through air and water.

The Golden Section

Other notions of elegant proportions in nature have been proposed in the past; perhaps the most famous ideal proportion is the 'Golden Section' – a mathematical ratio of the length of one line compared to another, or the height of a rectangle compared to its width. Examples of the Golden Section can be seen in Classical Greek temples as well as paintings from the Italian Renaissance, and also in some more recent buildings – for example, the United Nations Building in New York. The ratio goes like this: take a line of any length and cut it at a point somewhere along its length, so that the ratio of the shorter length to the longer length is the same ratio as the longer length is to the overall length.

The Golden Section, or ratio, can be applied to a line of any length. It turns out that this ratio is 1 to 1.618 (see diagram). Therefore the Golden Rectangle would have a width of 1.0 and a height of 1.618, or vice versa. You can construct a near-enough Golden Rectangle starting with a square and, finding the centre of one of its sides, draw an arc from the corner, as shown here:

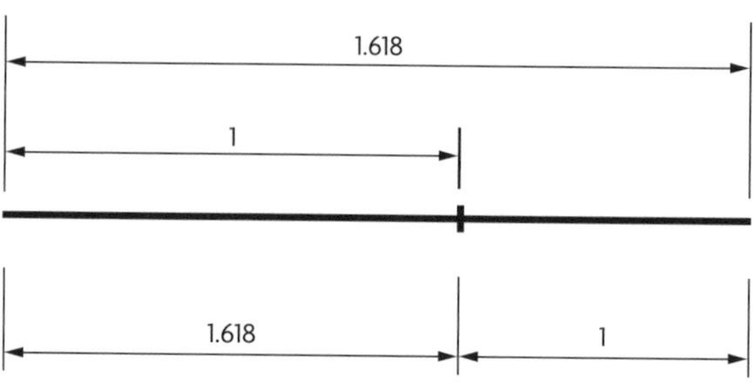

Golden Section ratio

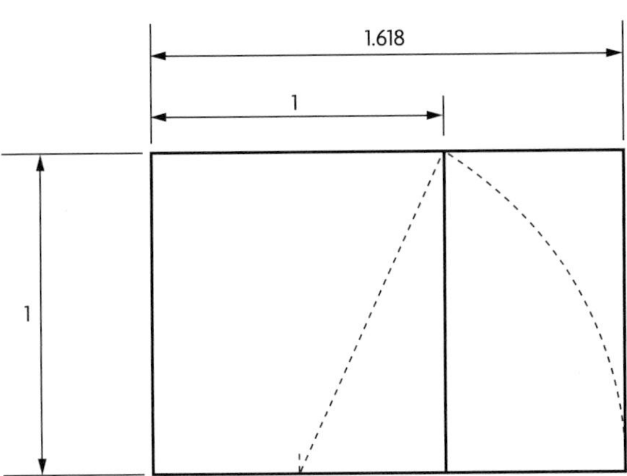

Golden rectangle

Geometry has always fascinated us and has, of course, been of huge significance in mathematics and physics, as well as art. It has been suggested that the Golden Section is also apparent in many life forms, from the way branches of a tree are spaced out up and around the trunk, to the spiral of a snail's shell (as demonstrated in the diagram).

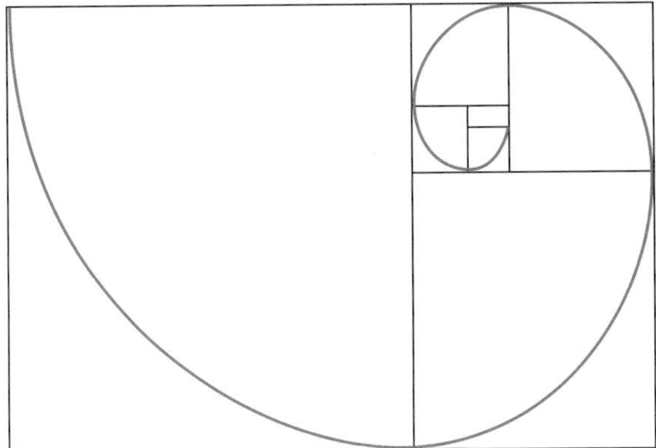

Relationship of the Golden rectangle to a spiral

Aesthetic consideration in design practice

Practicing designers and artists have their education and training to rely upon when it comes to designing a building, a set of cutlery, composing a painting or defining sculptural forms.

Designers today also have, of course, access to an almost limitless history of a variety of artists' and designers' work to admire and provide them with inspiration. No one can avoid being influenced by others, past and present, and this of course is to be celebrated, but sometimes we can inadvertently allow ourselves to be trapped by tradition and the usual ways of doing things. There exists a wide variety of established design conventions, which are often very subtle, that need to be learned and borne in mind when composing in two or three dimensions. Like most rules, however, they are to be used as a guide or intelligently

> *'Observing its beautiful linear qualities, I have combined simplistic curves into a minimal open structure that offers a complimentary contrast in the design and also accommodates the three-dimensional hollow container for fresh or dried flowers.'*
>
> Chris Castillo, designer-maker in precious metals

ignored. Experienced designers have a highly developed visual acuity that can detect very small variations between design proposals that can make the difference between the ordinary and the excellent. The 'Flower Vase' by Chris Castillo, made from sterling silver and granite, is an inspired example of closely observed nature being celebrated in a man-made, three-dimensional form. It can take years to become a competent graphic designer or typographer, for example, and to get a feel for the subtleties of letter design and layout.

'Flower Vase' by Chris Castillo, jewellery and product designer

To a layman, two similar fonts might look identical, but experienced designers can infer such abstract notions as: colder, friendlier, aggressive, aloof, and endless other terms to describe the 'emotional' qualities of a particular typeface when compared with others. Thus different designs have different 'visual personalities' that can influence an observer's emotional response, and typographers design or choose a particular typeface to reflect certain values. For example, the three fonts used for 'university', below, would be chosen to communicate very different universities. We can possibly imagine universities that would fit those fonts. Only one or two of them might be appropriate to use for the name of a bank, however – a serious business requiring a formal design of typeface, for example Goudy Old Style, as pictured below. Therefore graphic designers exploit the different personalities of fonts to better express the required message. The German typographer Erik Spiekermann said, 'choosing a typeface to set a word in is part of manipulating the meaning of that word' (Spiekermann, E. and Ginger, E. (2003) *Stop Stealing Sheep and Find Out How Type Works).*

In addition to the conventions of 'visual personality' or 'balance', mentioned earlier (whether in a piece of graphic design, a building or a colour composition),

University

University

University

Three contrasting fonts – these would be used to help communicate quite different places

SMITHS BANK

Goudy Old Style typeface – a sober, 'traditional' design, more suitable for a bank

there are other very subtle and intriguing conventions that can make a huge difference to the success or failure of a design (though these conventions are often culturally determined). Examples of these include notions of deliberate distortion to make a thing 'look right'. This is about achieving a balance between the positive and negative elements of a design: for example, the art of spacing letter forms; creating resting places (focal points) for the eye when 'taken for a walk' through text; the sensitive massing of elements; how to reduce the apparent mass of an object; a sensitivity to visual grammar and the skill in applying it; an understanding of the effects of colour on the observer's mood; and choosing an appropriate colour for a particular design, whether in two or three dimensions.

Conclusion

Design, then, is a combination of the objective and the subjective, whether dealing with shape, form and detail, or colour. In addition to learning about established conventions through training and education, every individual has to develop their own attitudes and skills towards appreciating and successfully creating beautiful pieces of work, whether in two or three dimensions. It is useful to study good examples. There is no one right answer to a particular design problem – there are potentially as many bad solutions as there are good ones and notions of beauty are not static – we constantly have to reappraise our judgements in the light of cultural and technological evolution.

You will need to build up a variety of intellectual tools and skills that will help you to solve new and unique problems as they arise – for example, dealing with a new technology, or squeezing a building into an awkward space while maintaining an elegant design. For further reading on these and other topics see the recommended book titles below. For inspiration in the field of aesthetics, take a look at the following career profile and project story.

Further resources

Books

Bayley, S., *Taste: The Secret Meaning of Things*, Faber and Faber (1991)

Bayley discusses issues of taste, primarily from a historical point of view. He argues that taste is 'not so much about what things look like, as about the ideas that give rise to them'. The book's illustrated chapters deal with notions of taste in the fields of architecture, interior design and fashion. Bayley compares the subjective values of taste with the more objective values of design, though he maintains that even twentieth-century modernists, who claimed that consideration of function alone without any need for decoration would lead to elegance, were the product of a particular culture in a particular time.

Bayley, S. and Conran, T., *The A–Z of Design*, Conran Octopus (2010)

Beginning with a series of essays about the role of design in modern cultural history, the book goes on to describe an A to Z of important people, products and processes from the late nineteenth century to today. Copiously illustrated in full colour, topics include products, transport, furniture, graphics and fashion design, with occasional references to important architects and architecture. A visual delight, easy for the reader to dip in and out of.

Clay, R., *Beautiful Thing: An Introduction to Design*, BERG (2009)

The historical, cultural, philosophical, technical, visual and practical approaches to design are often presented separately, but each approach impacts on the others and together they are critical to a rounded understanding of design. This book provides a broad introduction to design theory and practice, ranging from graphics, products and vehicles to architecture. Chapters include taste, design evolution, composition, colour, drawing, communication and expression; the book also includes a range of design case studies and recommendations for further reading, as well as a glossary of design terms.

Dudley, E. and Mealing, S. (editors), *Becoming Designers: Education and Influence*, Intellect Books (2000)

A 'reader' type book with essays by different authors (teachers, designers and writers) on various issues that the student will need to consider. For example: new technologies, history, gender, ethics, globalisation, as well as topics related to the teaching of design theory and practice.

Sparke, P., *An Introduction to Design and Culture (1900 to the Present)*, Routledge (2004)

Sparke deals with the impact of new materials and production processes on design, in the context of the changing nature of society. The book provides a history of the development of designs and includes many examples and illustrations of mass-produced products, buildings, interior designs and images. A useful glossary of designers and design movements is also provided.

Websites

aestheticsofjoy.com

This is a blog by Ingrid Fetell that would make a wonderful coffee-table book. It is full of uplifting images across a broad spectrum of life and design.

latemag.com/curated-aesthetics-driven-websites

A website about websites: it lists the curated/blog/eye candy type sites, like the one above, that are so loved by researching students. It is possible to spend days roaming around it. Be careful!

Career profile

Name:
Chris Castillo

Current job:
Designer-maker in precious metals

Describe what you do

I have a strong and never-ending passion for designing and making precious metal objects, products and items. My main area of specialisation is jewellery. The amalgamation of linear and curved movement, balanced with hollow spaces and solid surfaces, are central themes of my creative process. I love ornamental and functional forms and my main source of inspiration is contemporary architecture.

The emphasis of my designs is upon form and the three-dimensional qualities of the pieces I create. Careful consideration is laid upon the shape, such that when a piece of jewellery is viewed from different angles it instils curiosity, pleasure and elegance within the projects, for both me and for my clients. This awareness helps me to keep redefining myself as I continue to develop small-scale work and silversmith pieces made in precious metals and mixed materials.

The gilding metal 'Nibbles Bowl' was inspired by cityscape views, observing a composition of simple lines that create diverse forms and shapes when seen from aerial views. The outer bowl's pierced section allows the polished and textured surfaces to be seen, adding innovation, curiosity and contrast. The other side is etched with an organic pattern.

An area of interest is the use of computer-aided design software in combination with rapid prototyping technology, balanced with a range of hand-making techniques. This technology enables me to search, to create innovative pieces and to push the boundaries of my work.

Is there a particular project you'd like to tell us about?

This sushi set consists of three pieces, which are the dish, plate and sauce bowl. The design concept is based on the linear movement and quality I have observed in minimal contemporary architecture. Keeping in mind that the functionality of this object and the design of the food are celebrated in a minimal fashion, I focused on geometrical shapes.

'Nibbles Bowl' by Chris Castillo, jewellery and product designer

Sushi set by Chris Castillo, jewellery and product designer

In combination with the simplistic shape I introduced folds and pierced sections in the bottom layer to add more form and contrast. The set was polished for the colourful sushi to reflect on, and together these will create interesting reflections, capturing the essence of minimalism in Japanese culture.

What advice do you have for students considering a similar career?

In today's industry the technology is developing so rapidly that it enables designers to have finished pieces without using as much time or as many tools as in the past. Therefore, the way forward is to invest a lot of design time to create exceptional designs and concepts.

Project story

Designer:
Stephanie Walton, jewellery and silversmithing student

Project title:
'Metamorfos' – experimentation with nylon sintering

Tell us about the project

'Metamorfos' was my final major project of my degree year. As we were able to use any process that we desired, I decided to focus on new technology and processes such as sintering. I had previously taken part in a brand new live project, using the process of sintering with titanium, with a company in Germany named EOS. They collaborated with us as an experiment to expand their business by creating jewellery, as well as the many large-scale machinery parts they already make.

What processes and skills were most relevant to this project?

I could not have completed my project without my extensive use of CAD. My pieces required a lot of design development to ensure they stack together, are an interesting form on their own and are able to be worn comfortably. Taking inspiration from contemporary furniture design really helped me achieve the futuristic look of the product, ensuring that I did not subconsciously copy an existing jewellery design, which can happen very easily.

What challenges did you face?

I wanted my pieces to be very bold and colourful, and hand dying seemed the most appropriate method to achieve this. I used several brands of dye, and samples of each plastic to see which gave me the best results. I also had to test the durability of the material to ensure it was strong enough to wear as jewellery.

What were the highlights of the project?

Using a method that is unfamiliar to me was exciting from the very start. Researching what was possible before even designing gave me limits to design by and helped me stay on track. But I felt my style really came out when designing these pieces, and I can now continue to develop my style with stretching the limits of how we make jewellery with the available technology, and creating futuristic jewellery that is very tactile and will create curiosity.

Sintered nylon 'Metamorfos' bracelets by Stephanie Walton

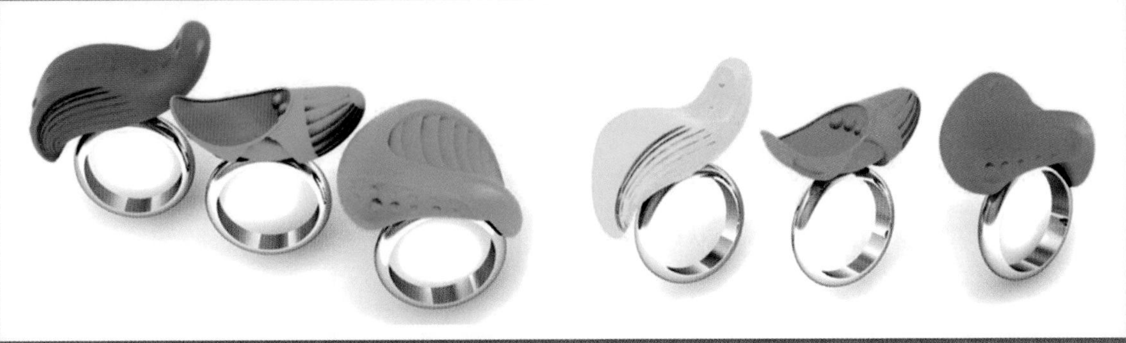

Sintered nylon 'Metamorfos' rings by Stephanie Walton

12

Working with colour

By Dr Sara Moorhouse, ceramicist and colour practitioner and Hugh Miller, Nottingham Trent University

'Colour is my starting point – it is an instant attraction and can make people love or hate what you do. It can have a massive impact on the success of a design!'

Katy Aston, MA Fashion and Textiles

'You're either a colour person or you're not. Personally, I always love to stay working in the monochrome!'

Amber Thomas, MA Fashion and Textiles

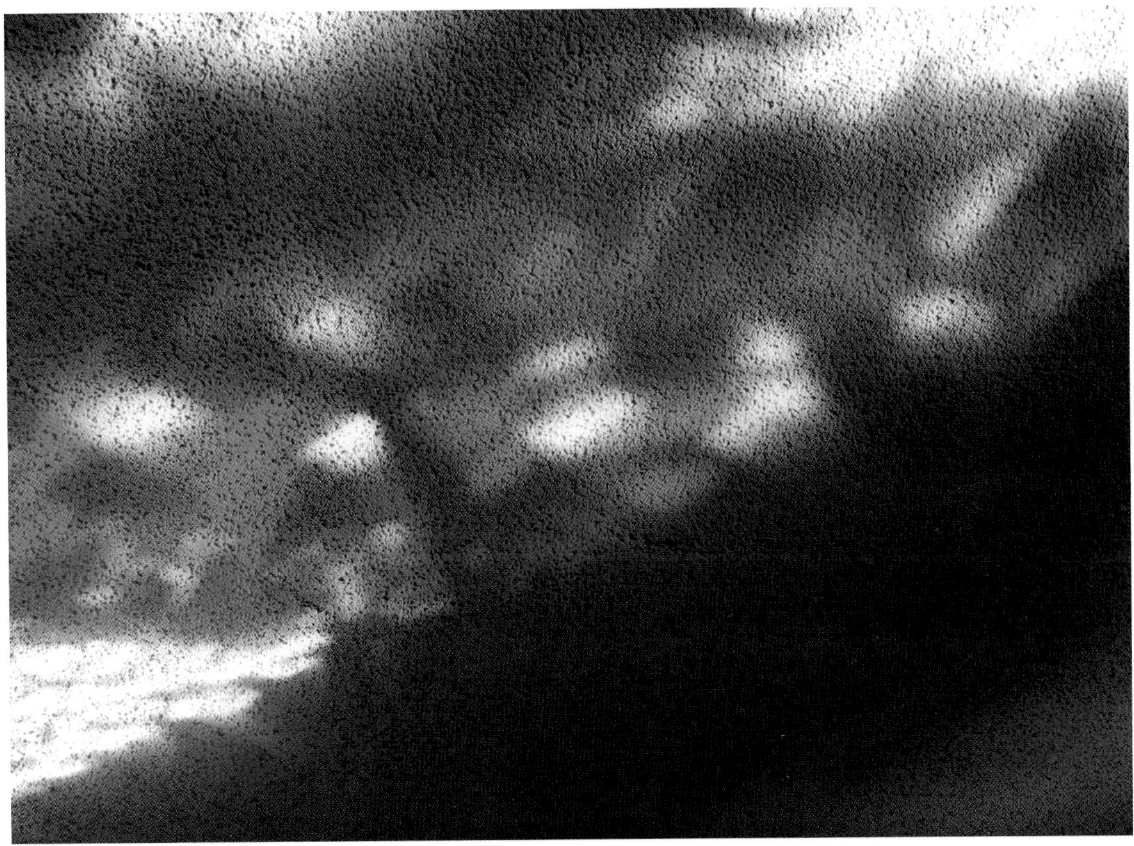

Colour inspiration
Photograph by Susan Hall

This section introduces you to:

- the terminology involved in the theories of colour
- the ways you might want to explore and experiment with colour
- the world of colour and science
- the types of jobs that involve a passion for colour

Colour is everywhere and affects us all physiologically and psychologically every day. The choices we make about what clothes to wear, how to decorate our living rooms and the tea we drink, all relate to colour. These choices are usually subconscious and our actual colour knowledge is limited. We know, for example, that the sky is blue, trees are green and poppies are red, etc. The diversity of language makes it easy to differentiate between different shades and types of the same colour, such as *lime* green and *navy* blue; and throughout school the colour wheel is taught, including primary, secondary and tertiary colour. So why investigate further? Unless, of course, you, as the designer, are wanting the public to buy your clothes or your brand of tea?

There are three key aspects when considering colour in design:

- what looks good (aesthetics)
- how colour makes people feel (emotion)
- what colour an object appears to be when we look at it (perception)

The information in this section goes a long way beyond the colour knowledge gleaned at a younger age, divulging facts and phenomena that highlight it as a

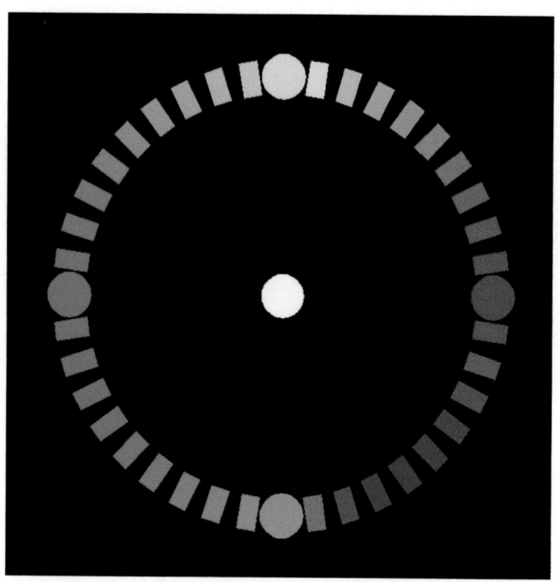

The NCS Colour Circle

NCS – Natural Colour System® © property of and used with permission from NCS Colour AB Stockholm 2013 www.ncscolour.co.uk

complex, elusive and fascinating subject. While drawing attention to colour's strangely difficult nature, the section will guide you – explaining colour science and how, with a few logical tips, colour can work its apparent magic in all aspects of visual design. You need to look for it for yourselves too. In your drawing and observation you need to spend time really looking at and analysing objects and environments, using colour media to examine colour science and colour psychology. Colour is a key component of design. We have such a wide range of dyes, printing and manufacturing techniques available to us that almost every object can be produced in almost any colour. Furthermore, the way we perceive the sensation of colour is complicated. Therefore, it is helpful to understand the perception of colour in order to predict what colours will 'look like' in use, in design.

In order to understand this better we need to first understand some of the science of colour before discussing how we perceive colour and how designers use colour. This section therefore has two parts:

- The science of colour by Hugh Miller, psychologist
- Colour: understanding and applying it by Sara Moorhouse, ceramicist

The science of colour

By Hugh Miller, Nottingham Trent University

Perception through wavelengths

The human eye has a simple and crude system for distinguishing colours through different light wavelengths. We don't have a different detector for each colour of light. We just have three kinds of bright-light sensitive cells (called cones) in the retina, the light-sensitive surface at the back of the eye. There are some cells that are most sensitive to long-wavelength light (red), some to medium-wavelength (green) and some to short-wavelength light (blue). Our impression of colour depends on how much these different sets of cells are stimulated by the light falling on the retina. This is why your computer or TV screen can produce the impression of thousands, or even millions, of colours by just varying the intensity of three lights.

Making colour

Look closely with a magnifying glass at a colour TV screen: all you'll see will be simple red, green and blue blocks, with the three colours varying in intensity depending on what colour is seen when you look at it from a distance.

Since all the colours we see are the result of the balance of activities in our three sets of colour-detecting cells, all the colours we see can be produced by different balances of intensities of three colours of light. The clearest example of this is how a screen with only red, green and blue on it can look yellow. Yellow light has a wavelength roughly halfway between that of red light and green light, and so it stimulates both the red-sensitive cells and the green-sensitive cells in the retina to about the same extent. When both our red-sensitive cells and green-sensitive cells are firing we *see* yellow – but this balance of activity in the cells can be produced equally well by shining equal amounts of red and green light into the eye. So the 'yellow' you see on the screen is actually lots of little red dots and lots of little green dots, not yellow light at all. You can see that in the two images below, where you can see a close-up of the screen showing bright yellow. All the other colours we see can be produced by mixing the right amounts of red, green and blue light, in the same way.

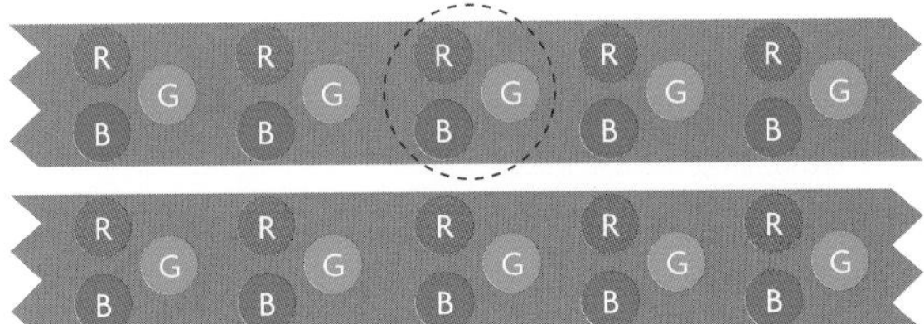

Groups of red, green and blue colour dots, as used on a television screen

Yellow object

Close-up of a TV screen

Red and green dots

Red and green dots magnified

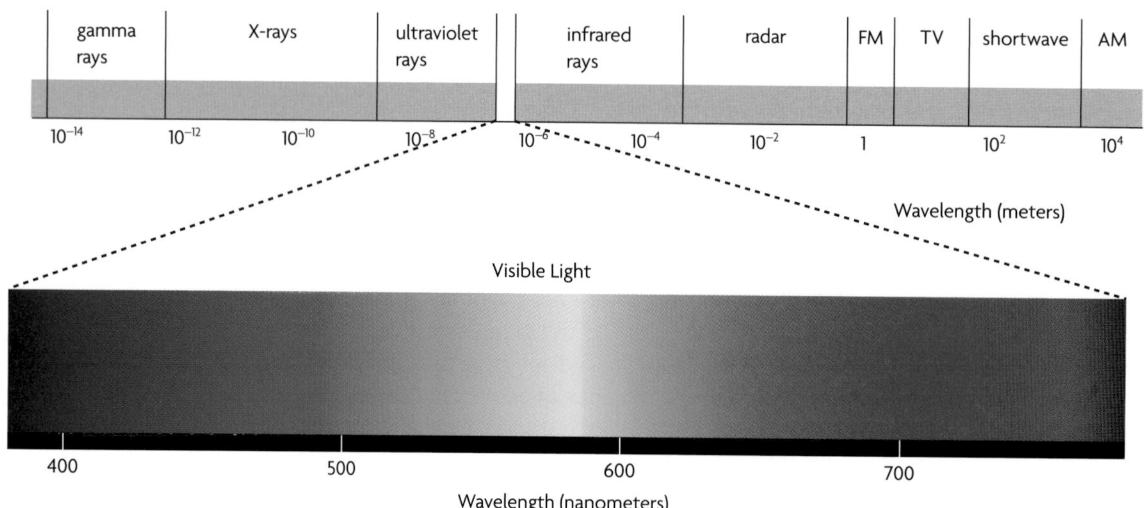

Wavelengths of visible light

Generally we see things by reflected light (daylight or room light bouncing off the object into our eye). Sunlight is a mixture of all the wavelengths of electromagnetic radiation that our eyes are sensitive to. This includes what we call 'light', as well as containing other wavelengths that we don't see, but can warm us (infra-red) or burn our skin (ultra-violet).

We see this mixture of visible light wavelengths as 'white light'. What we experience as 'colour' is separate wavelengths reflected within that mixture. The wavelengths of white light can be split into the infinite spectrum of colour that has been simplified in its description as red, orange, yellow, green, blue and violet.

The colour that objects appear to us depends on the colour of light falling on them and the nature of their surface, which might absorb or reflect different wavelengths of light differently. We only see the reflected light. Coloured surfaces look the colour they do because they absorb much of the light that falls on them and just reflect a few wavelengths.

However this 'mix-and-balance' colour detection system is easily thrown off by the different mixes and balances of the light that reaches our eyes, so different combinations of light colour and pigment can produce inconsistent impressions of colour.

Daylight

Generally, daylight is an equal mix of light wavelengths and looks 'white'. Actually, of course, the mixture of wavelengths in daylight varies due to factors such as the time of day, cloud cover and ground cover. As the colour of daylight varies, so the colour of objects viewed by that daylight vary also. Claude Monet had fun trying to record the impact of different effects of natural light, as objectively as possible, in his numerous paintings of haystacks.

Wavelengths reflected off a surface

Claude Monet *Haystacks* painting 1
Bridgeman Art Library Ltd

Claude Monet *Haystacks* painting 2
Mondadori Portfolio/UIG/Getty Images

Claude Monet *Haystacks* painting 3 (detail)
Photograph © 2013 Museum of Fine Arts, Boston

Artificial light

Things are even more variable under artificial light. Different kinds of lights produce different balances of light wavelengths. Light from old-fashioned tungsten incandescent light bulbs has more long-wavelength light than short-wavelength. It is quite yellow compared with daylight, as you can see when looking into lighted rooms from outside on a light evening. Low-energy lights produce light that is closer to the balance of daylight, so the effect is less marked in more environmentally-conscious homes.

Natural light

But if you step into one of those yellow-lit rooms, the light looks normal and a white shirt still looks white. This is because we're very good at adapting to different light conditions. We're good at perceiving relative differences in colour, but poor at making absolute colour judgements. So, unless it's very different from daylight, we take whatever happens to be the current light mix and treat it as though it was 'white', and interpret all the colours we see accordingly (see description of colour constancy later in this chapter). These examples of tomatoes and a banana are photographed, uncorrected, under different colour light.

You will probably have noticed that mechanical systems don't do this. Your digital camera needs to be adjusted (using the 'white balance' control) otherwise it will record artificially-lit scenes as very yellow, and if it has been adjusted like this, then daylight or flash scenes will look very blue. Humans have an 'auto white balance' setting built in. However, our colour vision changes as we get older. The light reaching the retina is yellower, so yellow-white contrast declines markedly in older people, which has implications for legibility of coloured displays and general vision in low-light situations.

The human vision system and the way it works with the brain is fascinating and complex. When we examine our perceptions of the world around us we find that the contribution of the brain to what we think we see is immense. To examine this further we need to look at the relationships between the colours and what factors designers need to be aware of to use colour effectively.

Tungsten light

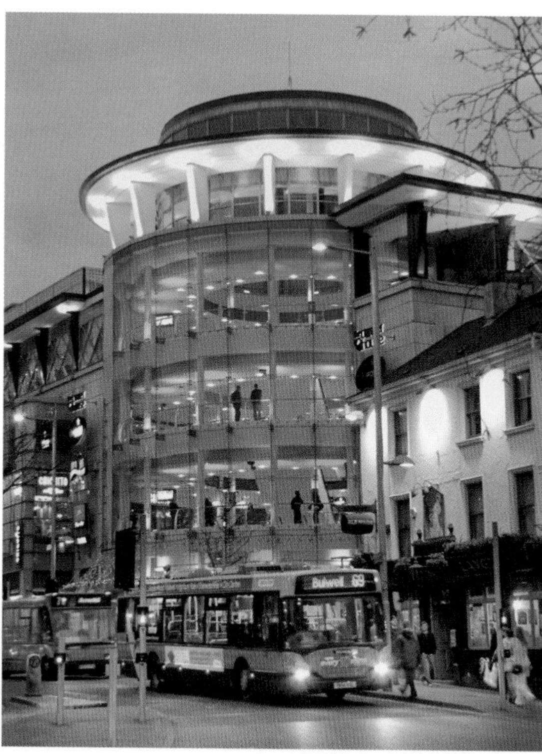

Artificial light

Colour: understanding and applying it

By Dr Sara Moorhouse, ceramicist and colour practitioner

Designers use colour all the time. Although a designer's colour palette may appear to have occurred easily and naturally, there will be a thought process and a rationale behind the choice of colours. This section will help you in those decisions. It will also be a starting point for those of you who want to explore the possibilities of the medium more deeply.

As seen in the last section, there's a certain amount of technical understanding required to use colour. However, your understanding of the power of colour from the viewer's perspective is just as, if not more, crucial. This is because when people view colours they have a response. This can be an emotive and psychological response – for example, the calming effect of pale blue or the exciting effect of bright red. It could be the relationship between colour and distance – as a green field recedes, the green becomes paler and paler, giving us a sense of vast space. There can also be a subconscious reaction – we may recognise that certain colours 'match', for example two blues, one of which is paler, without ever having worked out why.

These days, colour can be easily explored by computers, as a vast range of colours can be displayed and then changed in seconds. Many of the concepts outlined in this section can be explored easily at a computer with a 'paint' package open. Make some simple shapes, fill them with colour and explore the meaning of hue, saturation and value and how they differ from RGB values. After a while you will be doing simple sketches of your designs and filling them with one of the 16,777,216 colours a computer can generate. Just remember that the human eye can only discern up to 10 million. Unfortunately, this ability to look at them easily doesn't necessarily translate into easier decisions – we still need to be aware of how people respond to colour. Some of this is traditional, much of it is cultural, and it takes experience to learn the nuances of the field of colour.

Colour and atmosphere

Depicting an atmosphere is hugely reliant on choosing the right colours. This might be explained using impressionism, perhaps, as the artists truly captured all of the colours effectively by concentrating on mixing the colours they actually saw, even in dappled light, when one simple green tree has all of the colours imaginable in it. Look back at the painting by Monet to appreciate this.

Mike Jones, a final-year student, is just about to commence his career as a designer for film and television. Here in his renderings of a classic Western he captures an incredible atmosphere through his intense, specific use of colour, to the point where you can smell the tobacco and taste the sawdust.

Yet here, in the 'bridal suite', there is a different set of intuitive skills apparent that tackles the differences that various light sources have on the objects in the room. The floorboards, 'flooded' with intense sunlight, are handled particularly sensitively, creating a wonderfully dynamic atmosphere.

Hotel interior

Bridal suite

Anna Deery, an Interior Architect student, designed and planned an environment for young children through early to middle childhood. She planned her interior around the use of colour to distinguish the use of different parts of the building. The project was called 'kidsPACE'.

'kidsPACE' project, by Anna Deery

> 'During the first eight years, children are developing their visual acuity. Their perceptions of objects, movement, and print are expanded as they have opportunities for experiencing interesting visual images and colour. Changes and variations of design and colour intrigue children and cause them to visually attend to the unusual. It promotes a sense of adventure and exploration. I wanted each colour to represent the space in which it was.
>
> Orange promotes a sense of creativity in children and the idea of self-expression, so I coloured my creative spaces and "messy corners" in orange. I used red to highlight spaces that are for use with the help of parent and staff supervision only (kitchen/reception area, etc.) as red promotes a sense of danger and attracts children's attention as being a no-go area. The colour blue is perceived as a relaxing and calming colour, hence the use of blue in the children's sensory room. Green for children represents growth and safety. I used this colour in the spaces for mother and baby, which accentuated a sense of security for both. Yellow is the brightest colour to our eyes; it represents happiness, sunshine and fun. My storytelling and music making space is represented in yellow, promoting imagination and fun!
>
> I felt that my choice of colour had to have meaning and needed justification, as I believe colour is an important aspect in a child's life. It is the fun aspect of being a child . . . exploring and discovering new colours, new spaces and new experiences. This, in turn, ultimately shapes a child's future.'

Understanding the principles of using colour

Primary, secondary, tertiary

Although you may remember these terms from learning about colour in school, it is important to recap every now and then. Primary colour is colour that cannot be made from any other mixture of colours. The three primary colours are red, yellow and blue. Secondary colour is colour that is made by mixing two primary colours together. The three secondary colours are orange, purple and green. Tertiary colour is colour that is made by mixing three primary colours together. The most obvious tertiary colour is brown. Different tertiary colour can be achieved by combining different quantities of the three colours.

Hue, tone and saturation

Hue is a variety, tint or quality of colour. It is generally understood to have a similar meaning to 'colour'. Hue is colour in its purest form and refers more directly to a colour's wavelength. It can be used instead of the word colour as it is loosely the same thing. Tone refers to the darkness or lightness of a colour, for example if the tonal value of red is dark, then this is a dark red. Saturation describes how strong or how deeply pigmented the colour is – for example, a weakly saturated green would appear pale or even translucent and a highly saturated green would appear bright or deep in colour and opaque.

Psychology and physiology of colour

From a psychological and physiological perspective, colour is often associated with mood, temperature and vicinity. One side of the colour wheel, red, orange and yellow, is understood as warm and appearing to advance while the other side, purple, blue and green, appears cool and recessive. More specifically, yellow is expansive and associated with light and happiness, while red appears hot, alludes to anger and is the most advancing colour. Blue, on the other hand, is the coldest and most recessive hue. Green, while possessing spatial qualities similar to blue, is understood to be calming. Purple, being partly derived from red, is the warmer of the cool colours and has spiritual connotations. That said, all the colours may become more or less recessive, cooler or warmer, depending on their saturation and specific mixture – for example, warm blues can be achieved by adding a little yellow and even a touch of red.

Harmonious colour

Harmonious colour is generally understood as colour combinations that are pleasing or work well together, creating a harmonious effect. A good way to achieve this is using analogous hues. This refers to colours that are close together on the colour wheel, such as red and yellow – as seen in John Moore's bracelet.

Although the colours are bright the effect is soft. As Moore states: 'the red and yellow of the bracelet blend smoothly to give a warming glow that is soothing to the eye and the colours become muted by each other as colour is reflected between the discs. For example, yellow appears golden and warm due to the reflected red.' The effect on Moore's bracelet is quite complex due to the reflected colour. Generally speaking, on a two- or three-dimensional surface, such highly saturated reds and yellows appear vibrant while achieving a harmonious effect. By lowering the saturation, or by adding white, the effect will be calmer. Harmonious colour can also be achieved by using complementary pairs, for example orange and blue. Try lowering the tone and saturation of one or both of the complementary pair if you want a softer effect. Triadic colour schemes are those that utilise three colours, evenly spaced around the colour wheel, which also generates a harmonious effect.

Bracelet by John Moore

Complementary colours

Complementary colours are those situated opposite each other on the colour wheel. In art and design such pairings can be used to great effect. If one is arranged next to another (for example, red against green), red will appear redder and green greener due to the simultaneous appearance of the complementary from each colour, onto the other. When this occurs the colours appear so vivid that the two may seem to oscillate, move or vibrate in each other's company. This is a very good way of getting the best, or the most colour, out of colour and is often used by designers, as in Jane Moore's 'Fan Brooch with Daisy'.

Silver and enamel Fan Brooch with Daisy by Jane Moore

Colour constancy

My own fascination with colour began one brilliant afternoon in a painting studio at West Notts College of Further Education, when I was 18, and the tutor pointed out a rather unusual colour phenomenon. The white-painted, high studio walls appeared to be flooded with subtle shades of pink, blue and lilac. To my untrained eye these walls would have remained white, but once the colour was revealed, seemingly in its absence, the way I saw colour was permanently changed.

Although white surfaces change due to external factors such as coloured light and reflected colour, the reason we do not 'see' the changes is due to a phenomenon known as colour constancy. Our brain makes sure that white and other colours appear constant so that our perception of the world remains stable and uncomplicated; it is only when we really look that we see the lack of colour constancy in the world around us and notice that coloured surfaces change dramatically due to external factors. Have you ever noticed that a green field appears a deeper and richer shade of green when lit by the red light of a setting sun? Changes to a coloured surface are generally more difficult to see than those to a white surface. As an experiment, you could photograph the same scene at different times of day – but remember to use the 'white-balance' control.

Such changes to colours in the natural environment illustrates another, and perhaps overlooked, point that nature itself is a natural colourist and can always be relied upon as a source of inspiration. It also suggests that a big part of understanding colour comes from trusting your perception to notice what your eyes are actually seeing, as opposed to what is in your imagination or what you 'think the colour of something is'. By learning to see the real colour and recording it, fascinating palettes of colour can be formed. For example, in textile design take a set of colours from nature and explore how they work as a palette by manipulating them and investigating which proportions work well together.

'Blue shadows'

Coloured shadows

The coloured light mentioned above can be seen more clearly within the snow-filled landscape. Instead of the sunlit and shaded areas appearing as a range of white to grey tones, if you look carefully the snow-covered areas appear to be awash with pale tints of the sunlight colours yellow, orange or red, depending on the time of day. Not only that (and this is where the colour wheel comes in!) the shadows are the opposite, or complementary, colour to that of the light. For example, when an orange sun begins to set, the resulting shadows are blue. The occurrence of coloured shadows in relation to coloured light is well established and was first recognised by Johann Wolfgang von Goethe in 1810.

After-images, successive contrast and simultaneous contrast

External influences such as light are not the only factors that affect the way colours appear; their arrangement in relation to each other also has a huge impact. This is due to phenomena known as after-images or successive contrast, and simultaneous contrast. You will understand these more easily if you do an experiment: stare at the red circle pictured for 30 seconds, and then at a plain white surface. What do you see? A green circle should appear.

Colour circle test

This is an after-image and an example of successive contrast, whereby the complementary colour appears *after* looking at the stimulus colour. If you experiment further with coloured circles, you will see that yellow incites a purple after-image, and orange incites a blue after-image.

Complementary contrasts also appear simultaneously, or at the *same time* as looking at the stimulus colour, hence the term simultaneous contrast. In these circumstances the apparent colour appears next to, or around the stimulus colour. Imagine, then, that these sorts of contrasts and apparent colours occur all the time, simultaneously and successively, overlapping other colours in our visual field. We then begin to understand the diverse and temporary nature of colour appearance, or as Joseph Albers termed it, the 'relativity of colour', as this diagram illustrates.

Relativity of colour

In this example notice the change in appearance across the central green: the green panel appears brighter next to blue than yellow. This is because orange (the complementary of blue) appears next to the blue, lightening the left green edge, yet purple (the complementary of yellow) appears next to the yellow, darkening the right green edge. The green also affects the respective blue and yellow by projecting red, the complementary of green, onto them both. Thus, each colour simultaneously affects the other, particularly at the edges where the colours meet. For more examples of colour illusion, Albers' book *Interaction of Color* is a groundbreaking and key text.

Movement

Notice the vibration where the two colours meet in 'Bird Design' by Laura Thomas. The illusion of movement can be achieved by placing two colours side by side that are of the same tonal value, or as Margaret

'Bird Design' by Laura Thomas (original design in alternative colourway created for Inch Blue)

Livingstone suggests, 'equiluminant'. The juxtaposition of non-complementary colour for example pink and blue, of the same tonal value will create the appearance of movement as shapes seem to vibrate or even hover.

The appearance of movement can also be achieved by arranging small dots or narrow lines next to each other. This effect will be stronger if the colours contrast tonally with each other. This can be seen here in more textiles from Laura Thomas, where dark and pale lines of the same colour are juxtaposed in each of the coloured areas of 'Chromoscope'.

'Chromoscope' detail by Laura Thomas
Photo by Toril Brancher for Llantarnam Grange Arts Centre

Colour proportion

Whether you are inspired by tropical colours or harmonious neutrals, as in this image of an Italian archway, there is much to understand about what makes this image worth studying from a colour perspective. Think about the amount of each colour you can actually see, and appreciate the subtleties created by the balance. This kind of information can then be translated into a colour scheme for a textile design or an entire room, but the key factor is to retain the balance in the proportions of colour used.

Italian archway informing colour proportion

Powdered glazes

Charlotte Dredge and Kim Green preparing recipe for coloured glazes

Using colour in different design disciplines

'Will this give me a bluey green or greeny blue?'

Charlotte Dredge and Kim Green, pictured here, are choosing glazes for their ceramics pieces. The powdered glazes that they pick from the jars are hardly ever the same shade of colour when fired in the kiln, so there's an element of experimentation and surprise involved as temperature, types of clay, as well as glazes are all contributory factors that can affect the overall colour of the finished piece. One of them enjoys this process as it is unpredictable; the other doesn't because she wants a result that matches with her intended design.

'Which red? An orangey red or a pinky red?'

Fashion and textile students often need to dye their own yarns, fabrics or garments. The knowledge of the types of fibres and which dye stuff to use is paramount to the success of the outcome. There are companies that specialise in dyeing for the industry, with some focusing on dyeing man-made fibres, such as nylon and polyester, and others that focus on natural fibres such as cotton, wool and silk. In the studio pictured they have an infinite colour palette to choose from and create any colour by mixing their own dyes. As the human eye and brain can discern up to 10 million different colours, this is quite a choice.

Embroidery threads

'A colour system for all – Pantone'

Pantone is world-renowned for its system for identifying, matching, communicating and specifying colours for designers.

The PANTONE® MATCHING SYSTEM® is a book containing thousands of standardised colours. There are also design discipline-specific books for printing, publishing and packaging, fashion and home sectors and the plastics industry. There is also a range of 3,000 Pantone paints available to customers. The PANTONE VIEW Colour Planner is a two-year colour forecasting tool for companies and designers working for the menswear, womenswear, cosmetics and industrial design industries.

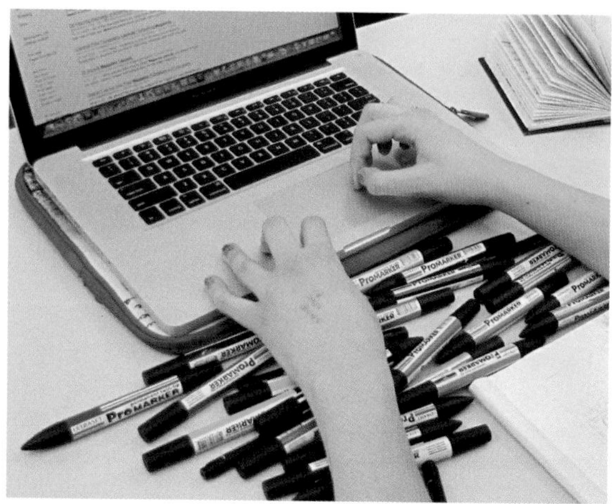

Letraset ProMarker colour system

Tips for designers

Control the number of colours you use: To ensure an effective colour outcome, try limiting your palette to two or three colours. Alternatively, select two colours and try creating several different tones of each colour and use them together across a design. Or, select four colours, keeping the tone and saturation of each colour the same, and keep changing the surface area of each colour across four colour studies. Although the colours are the same each time, you should find very different results!

Experiment with complementary colours: Try this to achieve a vibrant effect – for example, use red and green together. Or, try keeping the tonal value of the colours the same. You can check if the tonal values of two or more colours are the same, for example, yellow and blue, by making a black-and-white photocopy of them or looking at them in greyscale on an art package on your computer. If the colours are indistinguishable in greyscale then they are the same tone and you will find they have an exciting and vibrant effect in colour.

To create the illusion of movement: Try placing small dots or lines of varying colours, tones and saturations together.

Look for colour inspiration: A new colour palette can come from anywhere, whether it's a book, an advertisement, the television, nature, existing fashion or interior designs; you could try taking colours you see in fashion and transposing them into an interior design. Remember, try to see the colours that are really there, such as the colours that appear on top of another colour due to coloured light or reflected colour.

Colour saturation: Try lowering the saturation of one or more colours in the arrangement to soften the overall effect. For example, if your palette consists of red, blue and orange, you could weaken the saturation on the red and orange, making blue the dominant colour.

Creating depth through colour use: To create the appearance of depth, use low-saturated blue or green. Low-saturated pale blue is typically used to depict sky in paintings. See the painting by Ken Bushe as an example of this.

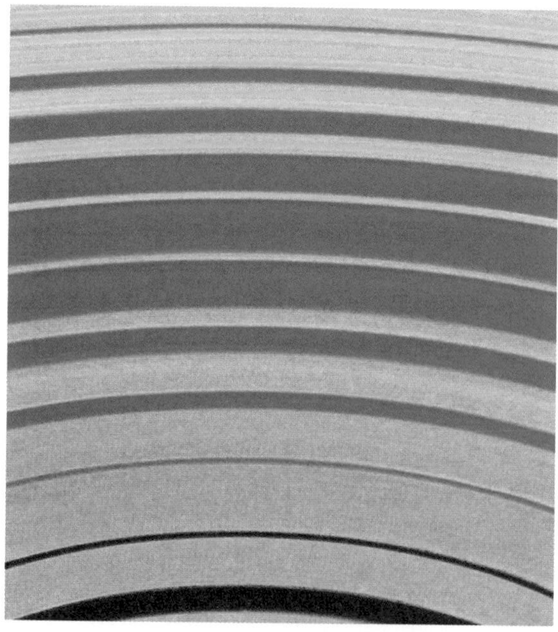

Control of colour proportion to illustrate movement, by Sara Moorhouse

Use highly saturated warm colours: To bring a part of a design to the foreground, try using colour in this way, as also illustrated in Ken's painting.

These tips are merely suggestions. The relative nature of colour and its ability to change depending on position and usage can make it unpredictable, even when rules are followed. So, although the tips above are helpful, it is really up to you to explore colour with an open mind and don't be afraid to abandon the rules in favour of your own intuitive colour preferences.

Ken Bushe's oil painting, *Landscape over Brunton*
www.kènbushe.co.uk

Roles within the industry

In all areas of design, colour plays a vital role. Colour is important to all of the industry sectors, and significantly contributes to the success of their designs. A colour specialist working in any discipline would need to use colour that is appropriate to the design and sensitive to its concerns. The use of bold and bright colour, for example, is not always necessary to make a design stand out. An interesting colour palette used in the right proportions can be equally as eye-catching. The psychological effects of colour always have an impact and are especially important in areas such as product and interior design.

Aleisha Simpson's developmental work in her sketchbook demonstrates how she uses the visual influences of photographs and drawings to determine a colour palette as the foundation from which her textile designs might develop. Inspired and motivated by the textile manufacturing industry in Yorkshire, she took many photographs of mill buildings and old machinery and this encouraged her to experiment with a variety of processes to create distressed looks in her fabric designs.

Alisha Simpson developing a colour palette directly from her own photos and drawings

Alisha Simpson's photo of wood panelling, revealing the history of the building

Colour forecasting

While colour theory and an intuitive use of colour can assist and often suffice for a successful colour outcome, design industries such as the fashion industry also use colour forecasting. This is a process in which specialist teams or individuals analyse and interpret data to anticipate colours, fabrics and styles at least two years in advance. Designers then use these predictions as a means of ensuring next season's colours in their designs. International colour and trend predictions can be found throughout the industry, from car interiors to interior design.

Conclusion

Someone once said to me that you can put any two colours together and they will look good, and, as this section suggests, many different combinations from the colour wheel appear to harmonise. Some combinations may look better than others, but generally speaking, as long as the colour scheme is limited to two or three, it works.

Myths about effective colour use surround the subject. For example, I have always found the saying, 'blue and green should never be seen' surprising; is it not the case that a large proportion of landscape scenes around the world comprise largely of these two colours? As a student starting out in the design field it is important not to be put off by such phrases and remember that any two of the 10 million discernible colours will work together to create different visual and psychological effects. Things become more complicated when more than two colours are added to the palette, but this only makes the possibilities more exciting.

Developing a greater understanding of the use of varying colour proportions is an important part of some designers' work in practice. For fabric designers, the control of colour is paramount to the success of the look of the cloth. For example, as a customer, if the garment wasn't attractive in some way the clothes simply wouldn't sell.

The sheer number of colours that we can see also means that many combination colour palettes will not yet have been discovered, and surely the very notion of exploring new colour ground is motivation in itself to start your own explorations! David Hockney uses colour in his own unique way, and when asked about the purples or blues that he paints to describe tree bark, for example, he talks about a 'way of seeing' or a 'way of looking'. It is as though you have to open your third eye, the one that looks for a more colourful world, to be able to see in this way. In the way that, as designers, we were once taught to draw, it is only a similar notion that we teach ourselves to really *see* colour.

Finally, like much of the creative process, rules are there to be broken and in doing so more exciting and innovative results may be achieved. Colours are there for the taking, to beacon, enliven, allude, symbolise and even to confound. But one thing is certain, people are drawn to colour, and used effectively it can make an otherwise adequate design sing. So what are you waiting for?

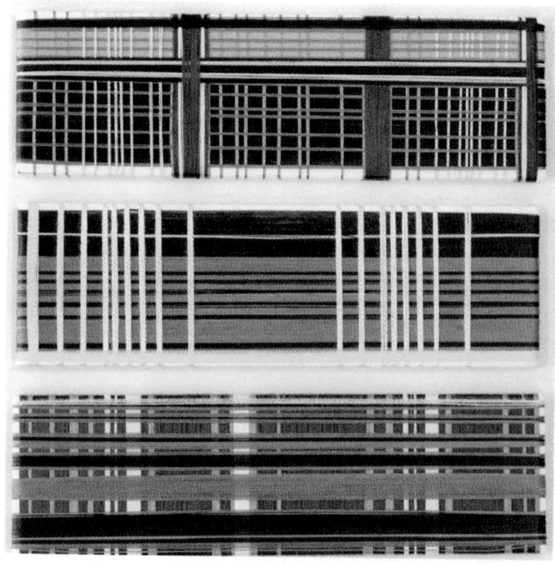

Colour proportion studies for woven textile designs by Anna Piper

Further resources

Books and magazines

Albers, J., *Interaction of Color – Revised and Expanded Edition*, Yale University Press (2006)

A real must for colour relativity and illusion, full of fascinating colour illustrations.

Chevreul, M.E., *The Principles of Harmony and Contrast of Colours*, Van Nostrand Reinhold Company (1967)

One of the principal colour theories expanding on Geothe and giving practical colour advice to painters and the textile industry.

Cole, D., *Textiles Now*, Laurence King (2008)

This book includes over 400 images from around 100 textile and fibre artists; an indispensable reference for anyone with a passion for textiles.

Cooper, E., *The Potter's Book of Glaze Recipes*, University of Pennsylvania Press (2004)

Including 400 recipes, this classic guide to making glazes covers all aspects of colouring, mixing, glaze materials and the application of glazes.

Diane, T. and Cassidy, T., *Colour Forecasting*, Oxford, Blackwell Publishing Ltd. (2005)

For fashion design students, this book discusses the driving forces of fashion, and the origin, development, terminology and processes of colour forecasting.

Gage, J., *Colour and Culture: Practice and Meaning from Antiquity to Abstraction*, Thames and Hudson (1995)

A thorough and diverse resource, taking the reader through the history of colour and colour theories from ancient Greece to the late twentieth century.

Hornung, D., *Colour: A Workshop for Artists and Designers*, Laurence King (2012)

This book introduces the application of colour to graphic design, illustration, painting, textile art and textile design. The author provides assignments that guide the student through a variety of colour experiences, moving logically from basic structural concepts to experiments with colour applications.

***Mix* magazine, Global Color Research, Mix Publications**

The leading quarterly publication for colour, design and trends, with accurate information about materials and products, helping you to apply colour trends to your business. www.globalcolor.co.uk/mix-magazine.php

Websites

www.aic-colour.org

The website for the International Colour Association: comprising of international research groups, the AIC hold regular annual conferences about the latest findings in colour research.

www.colour.org.uk

The website for The Colour Group of Great Britain: a group of artists and scientists who hold regular meetings and colour events throughout the year.

www.ncscolour.co.uk

The Natural Colour System website: NCS is a reference system for the accurate communication of the colours we see.

www.pantone.co.uk

The Pantone website includes products such as colour charts, the colour numbering systerm and Colour of the Year.

www.sdc.org.uk

The website for the Society of Dyers and Colourists: the world's leading independent, educational charity dedicated to advancing the science and technology of colour worldwide.

Career profile

Name:
Ptolemy Mann

Current job:
Woven Textile Designer and Colourist

What is interesting about colour from your point of view?

Colour has been the driving force in my work since the beginning; a complete devotion to the subject has meant that my name has become associated with a range of colour applications. There are three main areas of my practice, within all of which colour plays a major role – textile art, textile design and architectural colour consultancy. What came before any of these was an invaluable introduction to colour theory. At Central Saint Martin's College of Art in London, where I studied a BA in Textile Design in the early 1990s, everyone attended a one-day-a-week colour seminar by Garth Lewis, who introduced the theories and colour exercises of Josef Albers and Johannes Itten.

The other key factor was that the college only kept stocks of white yarn, so if any of us wanted to use colour in our weaving we had to dye the yarn ourselves. These two things inspired a lifelong exploration of the value and significance of colour. In my opinion, anyone interested in art and design should understand and learn at least basic colour theory; this will become the building block for everything you do thereafter, no matter what your subject or discipline... even if you then reject colour, you still need to understand what you are rejecting.

When making one-off artworks or designing a commercial fabric, how colours *interact* with each other becomes extremely important. Understanding complementary opposites, recognising the importance of not just the 'chroma' (shade) of a colour but its 'saturation' (depth and brightness) and 'value' (lightness or darkness), how they all come together and their balance, will determine a good design or artwork. My hand-dyed and woven artworks have become so dominated by their colour that the weave structure and fibre has become secondary; in fact, their plain woven surfaces are deliberately designed to show the colour in its most pure state.

The flat surface of the cloth can appear three-dimensional by interesting colour interaction choices. By applying simple colour theory based around complementary opposites you can make a completely flat surface appear three-dimensional and full of movement. Warm colours, such as yellow, orange and red, will appear nearer than cool colours, such as blue, green and purple, when placed next to each other. 'Violet Indigo Dynamic' illustrates this clearly, as the mustard yellow horizontal band appears in front of the purple; the complementary opposite of yellow is the indigo background. What adds to this sensation is the dynamic created by the *changing* coloured bands. By using a dip dyeing technique, inspired by the Ikat process, a wonderful sense of atmosphere and complexity can be achieved, even with simple fibres and techniques. Ikat describes a process where the thread is bound tight in certain places, dictated to by the pattern being created, to prevent certain sections of the yarn absorbing the dye when it is immersed in a dye bath. This is also referred to as resist dyeing. Once the threads have been dyed then they can be woven into cloth.

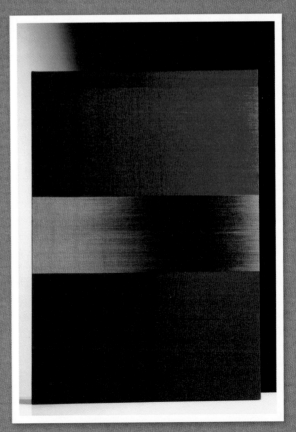

'Violet Indigo Dynamic' woven wall hangings by Ptolemy Mann

King's Mill Hospital. Architect: Swanke Hayden Connell International Limited; colour designed by Ptolemy Mann
Photograph: Timothy Soar

Tell us about an interesting project

In 2006 I was approached by an art consultant who wanted to appoint an artist to specify colour for the external façade of a hospital in Newark. My experience in working with architects meant I was well suited to the project. This then began a whole new way of working for me, using the analogy and methodology of weaving and colour in application on the façade of a building. It was also the first time I fully understood that my knowledge of colour was a skill in itself, and a valuable skill at that. Although the design of the hospital was complete, the architects had identified that there was an opportunity for colour; my brief was to reduce 'threshold anxiety' and make visitors (many of them children) feel more secure and even excited about entering the building, despite the circumstances. I also felt way finding was an extremely important issue, and that colour externally could help people navigate through the large building.

There is a tangerine orange colour that only appears at the entrance/exit and therefore is synonymous with this space, and as you move away from this hub the colour cools through the spectrum the further away you get. The north-east elevation is where the ambulances arrive (often referred to as the 'blue light' route) and where A&E is located (often symbolised by red); consequently, the external colours here are shades of blues, reds and magentas. The intention was that colour was placed across the façade with intelligent purpose rather than random selection.

King's Mill Hospital interior
Photograph: Timothy Soar

Several projects later and I have become a bona fide colour consultant who specifies internal and external colour on many kinds of buildings, although specialising in healthcare environments.

Do you have any advice for young designers?

Colour is a deeply emotive subject. For most of us it is also highly personal; we each have a unique response to colour that we develop internally through experience and association. How we feel about certain colours often has more to do with what hue our childhood bedroom walls were painted than anything else. Experiences, good and bad, associated with certain colours affect our response. For me, it is this more emotive understanding of colour, with a good basic knowledge of theory, that will help all of us to use colour in an intelligent and meaningful way across a range of disciplines. Colour, like any other subject, needs to be studied and understood and too often it is neglected and considered an afterthought. The significance of colour must be celebrated.

Project story

Designer:
Sara Moorhouse, Designer and Ceramicist

Project title:
PhD research project to determine how colour appears to change the space and shape of 3D form

'Oilseed Rape', 2008
Sara Moorhouse

Tell us about the project

Through my ceramic practice I examine the ways colour may appear to change the space and shape of three-dimensional forms. I initially recognised the relationship between colour and space while walking through the countryside in Nottinghamshire where I grew up. I was fascinated by the colour of arable, often linear, landscapes, as different fields appeared to change in size and volume depending on the colour of the crop and the time of year. Every year I watched fields transform in magnitude and grandeur; dark quiet fields clouded in mist throughout winter seemed close and oppressive, while the loud shining brilliance of yellow rape in spring dramatically magnified that same area of land. Such contrasts left me with an overwhelming vision of colour and its intrinsic relationship to space.

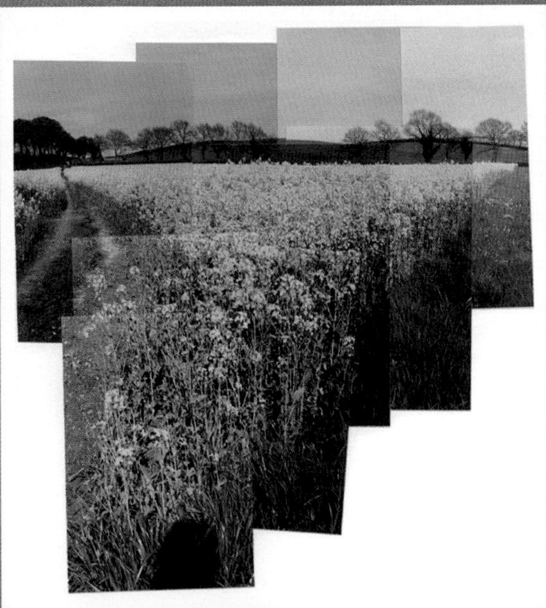

Oilseed rape field, Yorkshire, 2003
Sara Moorhouse

On a Master's course in 2003 I began to explore this relationship of colour and space through the medium of ceramics. I started by translating colours from landscape scenes into banding arrangements and applying these to the internal and external surfaces of ceramic bowl forms. After all, I thought, if farmers unwittingly, yet dramatically, altered the perception of landscape space on a massive scale, why couldn't coloured lines applied to a bowl alter the perception of form in the same way?

For my first series of forms, I referred to the close and oppressive appearance of a summer storm-drenched sky, specifically when a heavy and finite sheet of cloud was occasionally broken by bursts of yellow light, which splayed onto the ground beneath. My resulting ceramic artworks, dark blue/grey bowls pierced with narrow lines of vivid yellow, were entitled 'Storm'. These confirmed to me that colour imbues weight to an object: the dark blue hues lay heavy on the thinly-walled forms, pulling the rims downwards, as the 'shots' of yellow enlivened and lifted an otherwise densely weight-laden form.

'Storm', 2004
Sara Moorhouse

'Arable Landscape', 2004
Sara Moorhouse

The idea for my second series, 'Arable Landscape', came from a phenomenon that I had noticed within a wide and brightly coloured landscape, where individually coloured fields drew the eye from place to place in a very haphazard, almost zigzagging way. As the spectator I was no longer able to pass a gentle glance to the horizon and found myself gazing for prolonged periods across such arbitrary scenes. I transferred colours from a particular scene in a similarly haphazard way to the bowl form. I found that the colours dominated the space, encouraging the viewer to look firstly, and for longer, at the bright reds and oranges and then at the narrow blue details. I also noticed that paler hues made the internal rims appear wider and bases appear deeper, while darker hues appeared to pull the rims inwards and drew the bases closer. Not only, then, did colour articulate the space and direct the gaze, it also pushed and pulled it backwards and forwards in space, depending on the hue.

'Oscillation', 2004
Sara Moorhouse

'Beyond Form: Red', 2010
Sara Moorhoose

In other forms, such as 'Oscillation', I specifically used the complementary colours of green and red on the internal surface and blue and orange on the external surface. The juxtaposition of complementary colours creates vibration where the two colours meet. This, together with the arrangement of finer lines, generated such apparent movement that the entire form appeared to oscillate and hover.

Within the contexts of ceramics these illusory phenomena had not been explored or written about, and following my Master's a tutor suggested that I examine the idea in more detail through a PhD. I then won a bursary from the University of Wales Institute Cardiff and began a PhD in 2005 to further my enquiries into colour perception.

My most recent work from this project is 'Beyond Form: Red'. On this piece the inside red appears to raise up and over the actual rim so strongly that the entire form appears to reverse in shape. This has led me to believe that colour not only *affects* space, rather it can *control* space and, if used correctly, be a key protagonist to the apparent space, shape and form of an object.

What processes and skills were most relevant to this project?

An ability and confidence with the use of colour was necessary for me to develop my colour work on the bowl form. Although I had always worked intuitively with colour, my knowledge of colour theory helped me to develop my use of it in practice. Working with colour three-dimensionally, it is important to consider the form in relation to the colour arrangement and to make sure that the proportions of band-widths are sympathetic to the form. I have noticed, for example, that the rule of thirds, apparent in two-dimensional composition, also works very well on the bowl forms. This can be seen in the pieces 'Oilseed Rape' and 'Beyond Form: Red'.

What challenges did you face?

Working on the bowl form means that I have to be aware of and consider the whole form, i.e. the internal surface in relation to the external surface, as colour relationships from one to the other can greatly affect the spatial dynamics across the form. For example, a vivid red band on the external foot and one on the internal rim will create a relationship between these two points and draw the eye very quickly from the outside to the inside of the form, and vice versa. Unaware initially how to deal with both surfaces, I made a decision when I first started working with the bowl form not to be frightened of the spatial possibilities it offered, and instead to work with and utilise both surfaces to enliven and animate the space around the form.

'Arable Landscape: Spring', 2011
Sara Moorhoose

What were the highlights of the project?

One of the best things about working with colour and being inspired by landscape is seeing possibilities for new colour palettes. It can be a challenge in itself to know which combinations to work with next, as I often feel spoilt for choice! All too often a favourite sight is when I return home to Nottinghamshire and see the brilliance of the oilseed rape fields. Bursting with mouth-watering yellow, they seem to expand towards me time and time again. I made some work influenced by these landscapes in 2003 for my Master's and then in 2008 (see 'Oilseed Rape'), and have recently made a group of bowls called 'Arable Landscape: Spring', which are once again influenced by the yellow fields of Nottinghamshire. This time each bowl, dominant with yellow, shares the form with one other vivid hue found in the spring/summer landscape, and together as a group they reflect the colours across a vast panorama in the gently rolling arable landscape from my home county. Although I have also found inspiration from travelling to other countries, such as Switzerland and Australia, I doubt that I will ever be finished with this yellow from Nottinghamshire . . .

If you would like to see more examples of my work, then please visit my website:

www.saramoorhouse.com

What does it all mean?

Part three

13

Design history, culture and context

By Penny Sparke, Kingston University and
Juliette MacDonald, Edinburgh College of Art,
The University of Edinburgh

'Decode was a fantastic exhibition at the V&A. It helped me understand the potential of interactive technology – for me it was an important moment for inspiration in my ideas.'

Lior Smith, final-year design student

Creativity is spurred on by a thirst for developing innovative solutions for a particular customer and it is all about how you use and filter information from the things you see, experience and read and how you allow them to have an influence on your projects. Knowing what has been achieved by those who have gone before you and having a strong awareness of the world we live in today all contributes to your ability to produce great designs. Being culturally and contextually aware is going to give you an edge when you are discussing your work with future employers, so be comforted in the knowledge that this aspect of your studies forms an incredibly important part of your education as a designer.

This chapter provides an introduction to the history of design and takes you through all of the key art and design movements, with a focus on how this has impacted on the design profession. Then there is a detailed look at the type of work you might undertake as part of your course. This subject turns into a passion for some students and there are also career opportunities that have a basis in research and writing that are explored in the chapter.

The birth of design as a profession

By Penny Sparke, Kingston University

The desire to make useful things from the materials offered by the earth, and to decorate them so that they are not only functional but also beautiful, lies deep within the human psyche and determines both the nature and the appearance of the artefacts with which we surround ourselves. Design and designing have a long history. Indeed, they began when people began making and decorating ceramic pots on the banks of the rivers of ancient Mesopotamia 6000–2000 BC. The idea that we need to design and make things, to live, develop and improve ourselves, has always been there and has accompanied humankind's development over many centuries. Seen in this context, design and craft are interchangeable and are both highly skilled activities.

Mesopotamia bowl
Bridgeman Art Library Ltd

The Industrial Revolution and design

There is another story of design, however, that has a shorter timeline and that started far more recently. In this story, design owes its existence to the entrepreneurial energies of the men who transformed Britain's manufacturing industries at the end of the eighteenth and the beginning of the nineteenth centuries. In that context, design was a key by-product of the thinking that broke down the traditional craft process into its component parts and then reformulated them into a logical sequence of activities – in simple terms, designing, tooling, making individual components and assembling them. That led to the possibility of manufacturing large numbers of cheap, identical goods.

Within the new sequencing of tasks, the process of designing moved from being a spontaneous activity that happened along the way, as a craftsman worked intuitively with his material to create an object, to a highly orchestrated one that had to be completed before making even began. This meant that at the very outset of the manufacture of a product, the designer had already decided on its appearance and how it was to be made.

The advent of mass production

One of the eighteenth century's most successful entrepreneurs and mass-producers, **Josiah Wedgwood**, opened his first factory in Burslem in Staffordshire in 1759 in order to address the growing demand for domestic ceramics, brought about by the expanded interest in tea and coffee-drinking and the increasing popularity of hot, cooked meals. He worked on two levels, providing unique pieces for wealthy clients as well as quantity-produced functional wares. His mass-produced, utilitarian cream-coloured earthenware, for example, proved a huge success. New industrial

Wedgwood china
Frederica Cards Ltd

processes were introduced to make the pieces, and new classes of pottery workers (mould designers and carvers among them) emerged to transfer their designs into quantity production.

Wedgwood also manufactured one-off ornamental products for his wealthier clients in his Etruria factory in Stoke-on-Trent. He worked collaboratively with a number of successful artists of the day, among them the sculptor John Flaxman and the painters **George Stubbs** and **Joseph Wright** of Derby. Wedgwood would suggest to Flaxman the kind of work he wanted and the artist would produce drawings, wax models and occasionally plaster moulds as part of the development process.

A series of ceramic plaques depicting scenes from classical life, as well as the famous Portland vase, were among the many fruits of their close collaboration. Working with artists added value to his ornamental products and this also influenced developments within his utilitarian product range. This is very much what the Italian kitchen-utensil manufacturer, **Alessi**, does today with its commissioning of designs from high-profile architects such as **Zaha Hadid**. It enabled Wedgwood to manage the challenging tension between his desire to widen his market and still maintain the social cachet of his designs. To achieve this he also had to actively market his mass-produced wares and, in the words of the economic historian **Neil McKendrick**, to engage in what we now know as, 'inertia-selling campaigns, product differentiation, market segmentation, detailed market research and embryonic self-service schemes'.

So Wedgwood was a pioneer industrialist and a highly innovative marketeer, as he was one of the first to understand design and its relationship to the consumer in the modern sense.

It was at this time that the idea of a single, highly skilled designer-maker was generally replaced by a manufacturing team made up of less skilled workers, operating within a complex labour system with each person responsible for a small part of the process. In the nineteenth-century mass production of ceramic objects,

Wedgwood Portland vase
© Victoria and Albert Museum, London

Crevasse flower vase by Zaha Hadid for Alessi S.p.A
Alessi S.p.A

for example, no one single person took on the role of the 'designer'. Rather, a number of different people were involved in the processes of creating prototypes and moulds. As the design historian, **Adrian Forty**, has explained in his book *Objects of Desire: Design and Society 1750–1980*, the specialised workman who prepared ceramic prototypes for mass production, for example, was called a 'modeller'. In the British textile industry, mass production was undertaken by 'pattern-drawers' – fairly lowly factory employees. They were not paid much but were noted as highly important members of the team. As the design critic, **John Gloag**, has written,

> 'The designer was not regarded as a technician with authority. He was at best a pattern-maker, a malleable draftsman, the sort of man who could devise on his drawing board an infinity of variations upon a theme. Machinery could stamp out machinery by the mile. All that was needed to set the machine at work were drawings.'

While the financial advantages of mass production were clearly advantageous to the industrialist, it also meant that, unlike the people who commissioned customised craft-made artefacts and who knew exactly what they were getting, the consumers of the new factory-made goods did not know they wanted these new products until they saw them for sale. In order to stimulate demand, therefore, part of the designer's job was to make products visible, desirable and available. This fundamental transformation from craft to factory manufacturing, and the dramatic shift in the way goods were bought and sold, gave rise to what has become known as the modern design process, with which we are still familiar today.

Design reform

Wandle printed cotton fabric design by William Morris, 1884
Bridgeman Art Library Ltd

There was another aspect of nineteenth-century design, however, that has remained part of design culture since that time. Not all designers were happy to be passive members of the mass production team. Some sought to retain the design values that were linked with the original idea of craftsmanship and to create objects that would improve people's lives rather than merely feeding demand.

One such designer was the Englishman, **William Morris**. He was obsessed with the social and political implications of what he, like **Karl Marx** before him, believed to be the alienating effects of mass production. For Morris, unlike Marx, that alienation resulted not only in the despicable exploitation of the working class but also in what he believed to be the overly brightly coloured and 'unnatural' patterns on a carpet on a parlour floor. Indeed, for many of the design reformers, using nature as a source of pattern for textiles, such as carpets, and giving the impression that one was walking on real flowers, was the ultimate design crime. In 1856 the design reformer, Owen Jones, wrote in his influential book, *Grammar of Ornament*, that,

> 'flowers or natural objects should not be used as ornaments, but conventional representations founded upon them sufficiently suggestive to convey the intended image to the mind, without destroying the unity of the object they are employed to decorate.'

This must be one of the first statements about 'taste' in the household environment.

From the 1860s onwards, Morris' writings and work focused on the need to control ornament. Like those of many of his reforming colleagues, his early thoughts about design were stimulated by his exposure to the **1851 Great Exhibition of all Nations**, held in Hyde Park, London. In his view, the main problem was the process of industrial manufacture but he was not, as is often suggested, against the use of the machine *per se*: 'It is not this or that tangible steel or brass machine which we want to get rid of, but the great intangible machine of commercial tyranny which oppresses the lives of all of us.'

This could be one of the first statements against modern, capitalist consumerism.

Morris' ideas helped define modern design in another important way. He believed that decoration should not be applied as an afterthought or linked to social status, but rather as an intrinsic, defining property of objects – communicating their functions and their identities to their users – and he wanted people to understand this, to understand the merits of good, honest design.

Designing consumer machines

The very first consumer products that were mass manufactured in America looked like crude factory-made machines. The sewing machine manufacturer **Singer's** first sewing machine in 1851 was undecorated and highly utilitarian in appearance. Its own wooden packing case could even be used as a sewing table. It was not long, however, before the company realised that, while a housewife might be willing to buy a simple object as her first sewing machine, when she came to replace it she was more likely to want one that matched her other domestic possessions. She would also prefer to place it upon a polished wooden table that blended with her other furniture items.

With its broom handle, visible rivets and separate dust bag, the first suction sweeper produced by the **Hoover Company** in the early twentieth century was another domestic product that looked as though it had just come off an assembly line. In an attempt to redress that impression, a purple **Art Nouveau** pattern was applied to the surface of its metal body shell to ensure that it didn't look out of place in a domestic setting.

Early Singer sewing machine, c. 1850
NYPL/Science Source/Science Photo Library

Early Hoover vacuum cleaner, c. 1900
Popperfoto/Getty Images

1928 La Salle Phaeton
www.carphoto.co.uk

The early twentieth-century designers were now becoming interested in new machines for the consumer. A pivotal moment in the development of the design profession is the 1920s story of the competition between the two largest American car manufacturers, **Ford** and **General Motors** (GM). It served to shift design's centre of gravity away from the factory and towards the marketplace. As his famous statement that a car could be any colour, 'as long as it's black' clearly indicated, **Henry Ford** believed that all cars should look identical. Through General Motor's understanding of the importance of the needs of the consumer, and the employment of the automotive stylist **Harley Earl**, it demonstrated that, while designers were important to the production process, they were even more crucial in creating desirable goods that would sell. GM was one of the first companies to give the consumer things they didn't know they wanted.

The market-focused definition of design developed by General Motors was adopted by many companies from the 1920s onwards. The company's appointment of Harley Earl influenced the emergence of a trend for American industry to use design consultants, and the profession reached its peak of influence in the 1930s. Men such as **Norman Bel Geddes**, **Walter Dorwin Teague**, **Raymond Loewy** and **Henry Dreyfuss** designed a wide range of new consumer machines – automobiles, typewriters, refrigerators, etc. – in the highly commercial context of the USA. In the process they helped create a new, mid-twentieth-century modern landscape. This idea of consultant designers who worked on an independent basis for several companies continues to exist today.

By the early twentieth century, therefore, the modern concept of designing and the role of the designer, as we now understand the term, were both fully formed, although design must and does evolve to meet the changing needs of the market. Design was a clearly defined activity isolated from the rest of the manufacturing process and a cultural concept that carried the values of mass production into the consumer's daily environment. In all its various forms, design acted as an important agent of the modern world. For many people, designed images, goods and environments became the primary signifiers of the modern age in which they lived, as well as the means through which they continuously renegotiated and challenged their relationships with it.

Modernism and post-modernism

By the 1920s and 1930s a design movement known as modernism, which had its roots in a German design school called **The Bauhaus**, had become highly influential. Modern objects, the Bauhaus teachers proclaimed, needed to look modern, and therefore they looked to the world of the machine for their influence. Designers in a number of different countries, including Britain, took up the Bauhaus baton, creating objects that focused on function and materials and that displayed a minimal, geometric aesthetic.

1928 Cesca S32 chair by Marcel Breuer
Bridgeman Art Library Ltd

1981 'Casablanca' by Ettore Sottsass
Bridgeman Art Library Ltd/Studio Ettore Sottsass s.r.l

After the end of **World War II** in 1945, modernism was revived and reached a wider audience. It provided the basis for design education internationally and the majority of designers respected its rules of 'form follows function' and 'truth to materials', which had been formed by Morris a century earlier. As design influenced the lives of more and more people, designers began to free themselves from those rules and began to think that their role was to give people a new form of self-expression, rather than just useful objects.

The early twentieth-century, 'fully formed' profession of design began to question its assumptions. As this opening up of the concept and definition of design grew stronger, the term 'post-modernism' emerged and, by the 1980s, had become highly influential.

Designer culture

Burberry design
Mint Photography/Alamy

One of post-modernism's side-effects was a popular acceptance of using the name of a designer as a marketing tool and as a guarantor of 'added value'. It reflected the familiar practice of naming fine artists, which had been part and parcel of the way in which the market value of art works had been established, from the Renaissance onwards.

Within the market economy of the late twentieth century, individualism had become a sought-after commodity, and consuming a product with a designer's name attached to it represented a means for people to assert their own individuality (in spite of the fact that many people were, in fact, confirming their identities by purchasing the same things). By the 1980s and 1990s the label 'designer' had been attached to many objects, such as 'jeans'. These then ceased to be mere items of utilitarian dress and were now transformed into high-fashion goods, with the names of designers such as **Giorgio Armani**, **Gianni Versace** or **Donna Karan** attached to them.

The strength of the designer culture, of the popular idea that designers wield magic wands that can transform lives, reached a peak in the early 1990s, and is still with us today, as evidenced in this **Gucci** bag, pictured. In response to this phenomenon, however, following the lead of the nineteenth-century design reformers, many designers chose to step back. Several decades earlier the German

Gucci bag
Caroline Menne/Camera Press Ltd

'Quin Lamp' by Bathsheba Grossman, 3D digital printing
MGX Materialise

designer, **Dieter Rams**, had described himself as a 'humble servant' and now others followed his lead. The British designer, **Jasper Morrison**, for example, in sharp contrast to the extrovert Frenchman, **Phillipe Starck**, adopted a low-key approach.

In recent years there has been a return to the idea, promoted by the Italian 'anti-designers' of the 1960s (Archizoom and Superstudio among them), that designers should work anonymously in collectives. Today the Spanish duo, El **Ultimo Grito**, work successfully in this way, and there are many new collectives in the UK. It is recognised as a safer way to take control of your work – by teaming up with like-minded souls to share the experience of going it alone.

Now, in the early twenty-first century, in spite of the rise of post-modernism, design's relationship with manufacturing remains as strong as ever. However, the changing nature of the latter has inevitably meant that that relationship has been constantly evolving. Many models of mechanised manufacture co-exist in today's world – from high-volume mass manufacture, through low-volume batch production to 'rapid manufacturing', as in this example of the 'Quin Lamp' from **Bathsheba Grossman**. Manufacturing can be determined by designers, clients or even market forces. These are all accepted processes. Above all, though, the concept of design is still inextricably linked to industrial manufacture, not forgetting the pursuit of beauty, and it is likely this will be the case for many more years to come. So where is design going?

The future

It is possible to see where design is heading at this point in time. Recent new developments that have affected the way designers work include the increasing importance of multi-disciplinary teams and the application of what has come to be called 'design thinking' to challenges beyond the world of consumers and manufacturing. Designers are applying their skills to services and to a wide range of other areas of business and social enterprise. Recent **Design Council** initiatives to link the UK design industry with the National Health Service, for example, have looked at redesigning hospital equipment to make it easier to clean and a project to reassess the 'customer experience' of visiting an accident and emergency department.

In the post-modern era, design continues to be defined by characteristics formed back in the nineteenth century, i.e. its links with industry and the marketplace. However, designers are being pulled in new directions and longstanding disciplinary barriers have been eroded. Fashion designers **Ralph Lauren** and **Calvin Klein**, for example, have moved into homewares and interior design, while the gap between high and mass culture has also disappeared as fashion designers such as **John Galliano**, **Jean-Paul Gaultier** and **Vivienne Westwood** take their inspiration from street fashion and youth culture.

The long-protected divide between, on the one hand, design as a professional activity and, on the other, as an amateur undertaking, has been further challenged by the capacity of computer software programs that enable everyone to become his/her own designer. The computer has radically challenged the traditional world of the designer. The rapid growth of interaction design, for example, has taken design further into the world of the virtual and is eating away at design's boundary with psychology.

Of course, the biggest challenge we face is the growing vulnerability of the planet and the effects of over-consumption. Designers have become conscious that they can play a positive part in counteracting this by using sustainable materials, by ensuring that their products can be disassembled easily for recycling, by designing products that will last and by just being clever, as in this lamp made from plastic drinks bottles by **Sarah Turner**.

A new set of rules, or ethical guidelines, is being established by the ecological and financial challenges we all face in the early twenty-first century. In addition, on a public rather than a private level, we continue to need designers more than ever before to create our prosthetic limbs and aortic heart valves, to interpret for us the advances in complex technologies that are beyond our understanding, to help us to orientate ourselves as we move from place to place and to assist us in expressing ourselves and communicating with others. While on one level we can all be designers, on another we are as dependent as ever upon educated and experienced professionals with their highly developed creative and communicating skills. Designers are as important now to our lives as they always have been, and will undoubtedly continue to be so for the foreseeable future. Indeed, such are the complex demands on designers today that they need an ever broader knowledge and skills base.

Cola 10 recycled bottle lamp
Sarah Turner

We may no longer have a simple formula for 'good design' but the social, political, cultural, aesthetic and ethical ideas expressed by William Morris a century and a half ago are still ringing in our ears and will undoubtedly help us develop a definition of design relevant to the challenges of the twenty-first century. You are part of this too. You will be studying design culture and context in detail, and to start this journey the next section maps out the context and introduces you to the critical thinking that is necessary to be successful in the design world.

Design context and culture

Juliette MacDonald, Edinburgh College of Art, The University of Edinburgh

Appreciating the relevance of art and design history, as explained in the previous section, is an important part of your ability to be a successful designer. This section aims to explain the connections between being culturally and contextually aware and your own design practice.

Learning about the wider context and cultures associated with design will be a core part of your design education and it will continue throughout your degree course. Different institutions will have different names for this theoretical aspect of your course, such as 'Historical and Cultural Studies', 'Critical and Contextual Studies', 'Context', 'Design Studies' or 'Design History'. Whatever it is called, the aims of these strands will be broadly the same; that is, to introduce

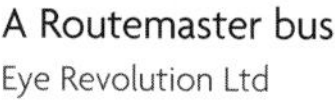

A Routemaster bus
Eye Revolution Ltd

A Dalek
Hugh Threllfall/Alamy

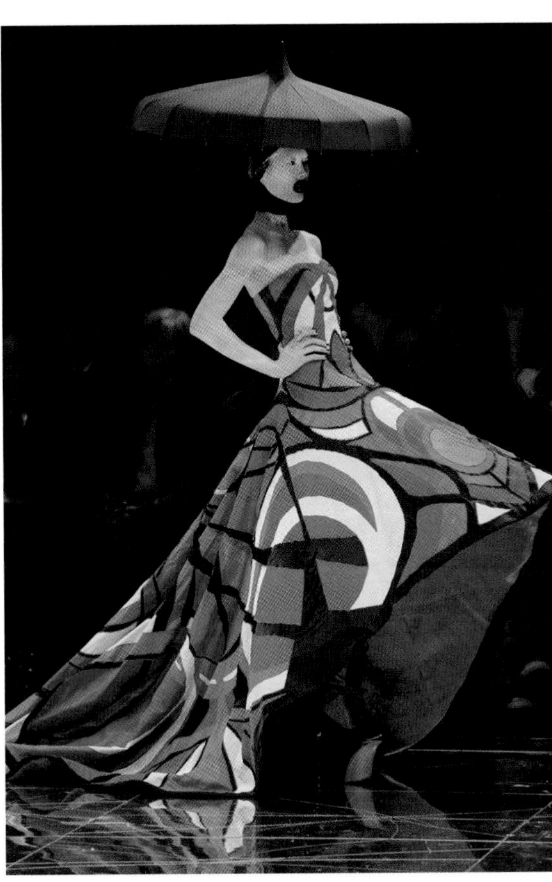

Alexander McQueen
Pascal Le Segretain/Getty Images

you to some of the key ideas and theories that have had an influence upon the discipline you are now studying.

This section will:

- relate the importance of context and culture in framing your work
- introduce you to some of the best critical thinking in the field of design
- empower you to question everything you come across during your education.

At times, especially to begin with, it may be difficult to see the need to spend time learning about the historical and theoretical context of design when you are busy with a project and have a presentation to prepare. However, examples of innovative thinking from **popular culture**, music or literature that you study as part of your contextual studies can in turn lead you to new and exciting insights within your own practice.

The use of materials and the way ideas manifest within an historical and cultural context are a rich source of inspiration for designers, and the study of these relationships is sure to be part of the introductory contextual themes you will encounter. The post-modern era we live in is characterised by a very different way of using history and culture, and the concept of **post-modernism** is not an easy one to grasp.

Popular culture: this reflects the trends, attitudes and tastes of a given society. So music, fashion and images from mass media, such as advertising, films and television, and the Internet, such as Facebook and YouTube, are all part of what is known as popular (or pop) culture.

Post-modernism: the post-modern condition is the term used to describe the economic and cultural state of Western society following on from **modernity**.

Modernity: is the term given to the historical period following the demise of feudal societies and is associated with a firm belief in progress and an attachment to the idea of the new.

Despite its slipperiness and contradictory nature, it is also an alluring topic that has challenged and informed designers and makers across many disciplines.

It is hoped that by thinking about post-modernism and some of the elements that are associated with it – irony, authenticity and retro – you will be able to see how useful and rewarding this knowledge can be and how it can, in turn, feed into your practical work.

Post-modernism

The architect **Charles Jencks** famously quipped that, 'modernism died at 3.32pm on 15 July 1972'. He was referring to the moment when the Pruitt Igoe housing scheme in Missouri, USA, was destroyed. For Jencks, the scheme's destruction was an exciting moment, marking the end of modernist architecture with its emphasis on rational forms and function and the beginning of a playful and eclectic post-modernism. In reality, such a change did not happen instantaneously. Post-modernism was the outcome of broader changes taking place in society as a result of, and a reaction to, political and economic events. Partly as a result of the Second World War (1939–45), there was a growing disillusionment with authority and alongside it a gradual breakdown of the distinction between 'high' and 'low' cultures.

At the same time, consumption, rather than production of goods, became the defining characteristic of Western society, with an associated increasing emphasis on the way things looked.

The term post-modernism gained currency in the early 1980s, and you can see from **Dick Hebdige's** quote below that, even while it was taking shape, it was not restricted to boundaries or certain disciplines.

> 'It becomes more and more difficult as the 1980s wear on to specify exactly what it is that "post-modernism" is supposed to refer to as the term gets stretched in all directions across different debates [. . .] When it becomes possible for people to describe as "post-modern" the décor of a room, the designs of a building, the narrative of a film, the construction of a record [. . .] or an arts documentary, or the intertextual relations between them, the layout of a page in a fashion magazine or critical journal [. . .] then it's clear we are in the presence of a buzzword.'
>
> Dick Hebdige (1988) *Hiding in the Light: On Images and Things*, London, Routledge, pp181–2

What you can notice in this quotation is the confusion created by the term, but in a way such confusion was why post-modernism was so appealing to many designers, because it allowed them to dismiss the formal qualities that had been at the heart of so many designs of the early to mid-twentieth century and to quite literally play with ideas, materials and processes in a resourceful and humorous way.

Post-modern culture embraces a mixture of cultural practices, influences and ideas. So a person or group can hold differing ideas or 'truths', as all are considered valid. This 'pluralism', along with the **decline of metanarratives**, resulted in designers abandoning long-held beliefs or rules for the way something should be made or should look.

Decline of metanarratives: the idea that one culture, religion or political system is more important than others is described by philosopher **Jean-Francois Lyotard** as 'metanarrative'. The blurring of boundaries between 'high' and 'low' cultures is a symptom of the decline of these metanarratives in the post-modern era.

In Italy in the 1980s a band of designers and architects formed a group known as **'Memphis'**.

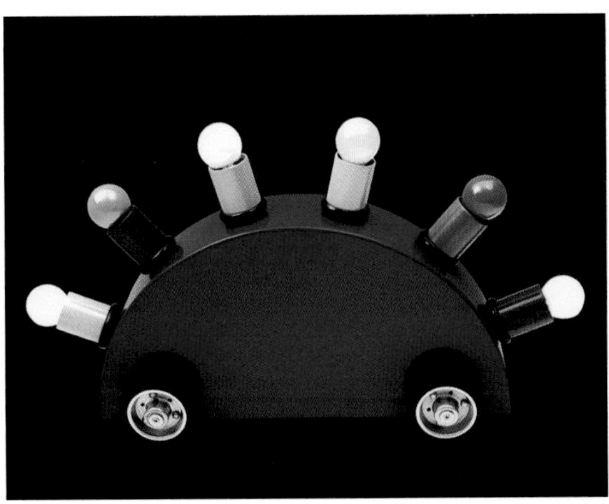

1981 Martine Bedin (for Memphis), 'Super Lamp' prototype
Memphis s.r.l.: Aldo Ballo, Guido Cegani, Peter Ogilvie/Martine Bedin

Martine Bedin, a member of the group, designed the 'Super Lamp' in 1981. Bedin focused on creating an object that was mobile, colourful and celebrates the bare bulbs in apparent rejection of the conventional idea of what a lamp should look like. This expensive, luxury item would perhaps look more at home in a fairground than a living room, but part of the intention of the designers in the Memphis group was to distance their designs from established rules and expectations.

The Memphis group of designers worked in a wide range of media: furniture, ceramics, glass, textiles and metal, but were united in their belief that design should imitate fashion in its rapid turnover of style, as fashion writer **Caroline Evan** comments.

> 'Since the decline in the 1960s of a seasonal "look" of which women could be sure, mainstream fashion has deliberately constituted itself as a variety of "looks". But in the 1980s the turnover of looks speeded up hysterically.'
>
> C. Evans and M. Thornton, *Women & Fashion: A New Look*, London, Quartet, 1989, p59

The 'Super Lamp' is a good example of the possibilities that present themselves when rules are challenged or deconstructed (taken apart to be understood in more detail). Post-modernism has been criticised and often dismissed as simply a style or fad. However, if we try to understand it as a description of the society we live in, it can help us comprehend the condition of pluralism that enables exciting new hybrid forms to take shape and flourish.

Finding different ways to think about the society in which we live and work is an important skill for designers to develop. It is this exploration that gives us starting points for our work – creativity. This is demonstrated here by the winner of the Stewart Parvin Award for Best Fashion Graduate 2011, **Felix Chabluk Smith**. His collection, 'Kin', for the Graduate Fashion Show in London, 2011, was the result of his exploration of family and ancestry. His inspiration was stimulated by his interest in a distant part of his family who lived in Luxembourg in the 1800s.

Chabluk Smith comments that:

> 'Rather than creating a tribute to my relatives, relying on hackneyed historicism and irrelevant costume, I wanted to convey a sense of the gulf that divides us all – temporally, socially, sartorially and geographically – from our ancestors, and the genetic links that bind us to the distant dead.'
>
> www.scottish.parliament.uk/visitandlearn/44333.aspx

'Kin' collection by Felix Chabluk Smith, 2011

He used fragments of a bespoke tailoring pattern from a gentleman tailor's pattern book, *Thornton's Sectional System*, published in 1895, to develop the garments in his collection. Again he notes that:

> '(b)y using these drafts for long-obsolete articles of male dress – clothes that would have filled the wardrobes of the Brandebourgs – and building up modern garments in a homeopathic process, this original genetic material has become watered-down, altered and in some cases almost entirely lost, but regardless of the final outcome, each garment will have a traceable link to its history, its family, its kin.'
>
> www.scottish.parliament.uk/visitandlearn/44333.aspx

Chabluk Smith clearly recognises that he wasn't designing 'authentic' garments. His very successful outfits suggest the past rather than recreate it. Through his research and his work he was able to make a connection historically and emotionally to his own family history, enriching his experience of the design process.

Practise analysing the work of others in a critical manner as this will help you develop your skills and build an understanding of how this relates to your own thinking processes.

Having a good understanding of what post-modernism meant to people in the past and what it might mean to you now could also get you thinking about what your work might mean to other people in the future.

Irony

The emphasis on irony, playfulness, humour and **kitsch** in design and other cultural practices, such as dance, music, literature and architecture, is another important characteristic of post-modernism. Modernist designers in the earlier twentieth century felt a need to create work where object, image, content and form fit together as an efficient whole.

The 'LC/4' chaise longue, below, created in 1928 by **Charlotte Perriand, Pierre Jeanneret** and **Le Corbusier**, is a good example of this interest in form and function, and Le Corbusier even referred to this chair as a 'relaxation machine' because of the way it mirrored the body's curves. For him and for other modernist designers, an object was considered to be worthy and beautiful when its form followed its function in a clear and rational way.

> 'A good modern machine is . . . an object of the highest aesthetic beauty.'
>
> Kurt Ewald, 'The Beauty of Machines' (1925–6), quoted by Paul Greenhalgh, in *Modernism in Design*, p11

The witty sarcasm or irony often found in designed objects from the later part of the twentieth century is far removed from these ideals. As noted earlier, post-modern designers were not interested in rationality and efficiency because of their experience of a society that is too complex and **multi-layered** to be addressed with a **single idea**.

As can be seen in Bedin's 'Super Lamp', designers not only dismissed the need for an object to be rational but, by being playful and ironic through the use of materials, in this case to create a lamp that looked more like a toy, they used their work to question modernist design and its conventions. This intelligent use of irony aims at raising a smile but also at encouraging the viewer or user to think about the work in more depth by making associations with unusual materials or subverting visual elements.

Memphis designers' ironic use of a combination of cheap and expensive materials and their references to popular culture worked well because they tended to use 'inappropriate' design language in a knowing way to create **contradictions**. In other words, they knew the rules of design and were then able to

Kitsch: the term comes from the German *verkitschen*, which means to make cheap. Kitsch gives value to things that might usually seem to be cheap or 'tacky'.

Multi-layered/single ideas: post-modernism moved away from the idea of one single explanation or story to explain how society works. Instead, ideas that had numerous layers, that could mean different things to different people or even different things to the same people, were embraced in order to demonstrate that society was complex, changing and could not be pinned down by just one person, one idea or one, straightforward connection.

Contradictions: using unusual materials and combining different visual styles, this bringing together of unusual objects to provoke a reaction is reminiscent of the **Surrealists'** aim of using chance encounters to create new objects.

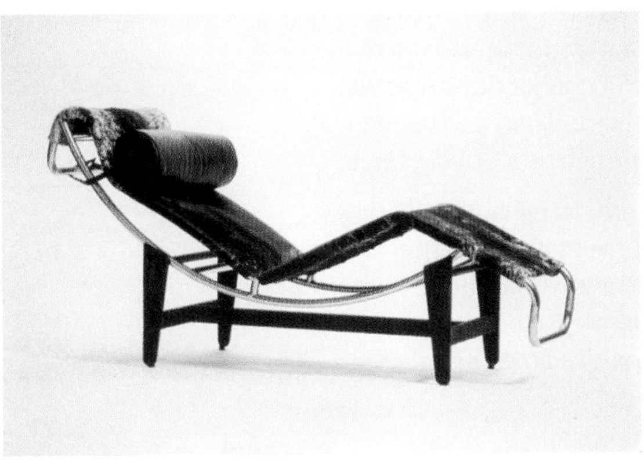

Charlotte Perriand, Pierre Jeanneret and Le Corbusier, 'LC/4' chaise longue, 1928

'Bottoms Up' doorbell by Peter van der Jagt for Droog, 1994

dismiss some of them, bend others and borrow new ideas or processes from other genres.

Through the 1990s irony was used by other designers, often in order to critique post-modernism and its self-referential focus. The Dutch design collective, **Droog** (probably the best-known group from this era), use ironic comment and a strong sense of playfulness and humour in their work. Some of Droog's simplest designs are the most eloquent. For example, the 'Bottoms Up' doorbell by **Peter van der Jagt** for Droog Design, 1994, is visually and aurally playful: the upturned glasses reference the phrase 'bottoms up' and when the button is pressed, the sound of the doorbell is of glasses clinking together as if someone has just raised a glass and clinked it against another in order to honour someone or wish them good health. As the design has been reduced to a minimum of components (two glasses, a clapper and some wiring), it can also be read as an example of the minimal design that was popular in the 1990s.

Irony should not be mistaken for playfulness or kitsch. Yes, it can have these elements too, but it is the knowing misappropriation and subversion of ideas that gives irony a complexity that playfulness and kitsch do not have. Alessi's kettles, by Michael Graves, fall into the playful category but are not necessarily ironic.

It is important to remember that, while irony often works best when used as a critical tool, not all ironic design has to be created as a reaction to a current trend or idea. What makes an object ironic can be the thoughtful connections that will take the viewer or user beyond the obvious pun or joke. The designer and lecturer, **Tim Parsons** makes a good point for the use of irony in design as a driving force.

> 'While we need worthy projects to take us forward, irony is a stabilising force. Every field needs internal mechanisms that facilitate reflection and evaluation – this is what satire offers to politics. Those who criticise ironic design for being "design for designers" would do well to consider that perhaps that is no bad thing and, if done well, provides an essential component in the debate about where we should be heading.'
>
> Tim Parsons, 'The Design Comedy: In Defence of Irony', *Core77*, 3 June 2010

Playfulness

A good example of a playful rather than an ironic object can be found in **Pablo Matteoda's** 'Sharky' tea infuser. In 2009 it won a prize in the 'Beyond Silver' competition by www.designboom.com, whose brief was for a functional and prestigious item of silverware with a new, innovative twist.

Matteoda's infuser meets the brief but is playing with us on a number of levels. First, the name of the object plays with the idea of it being a shark and a tea infuser (Matteoda is Argentinian, and in Spanish 'y' means 'and' – so the title is literally saying it is a shark and a tea infuser). Second, Matteoda makes reference to popular culture with the fin/sail, suggesting the iconic film *Jaws*. Finally, the trail of colour from the tea further reminds us of *Jaws*, giving the appearance of a severed fin, while also providing a playful element as the user waits for the tea to be fully infused.

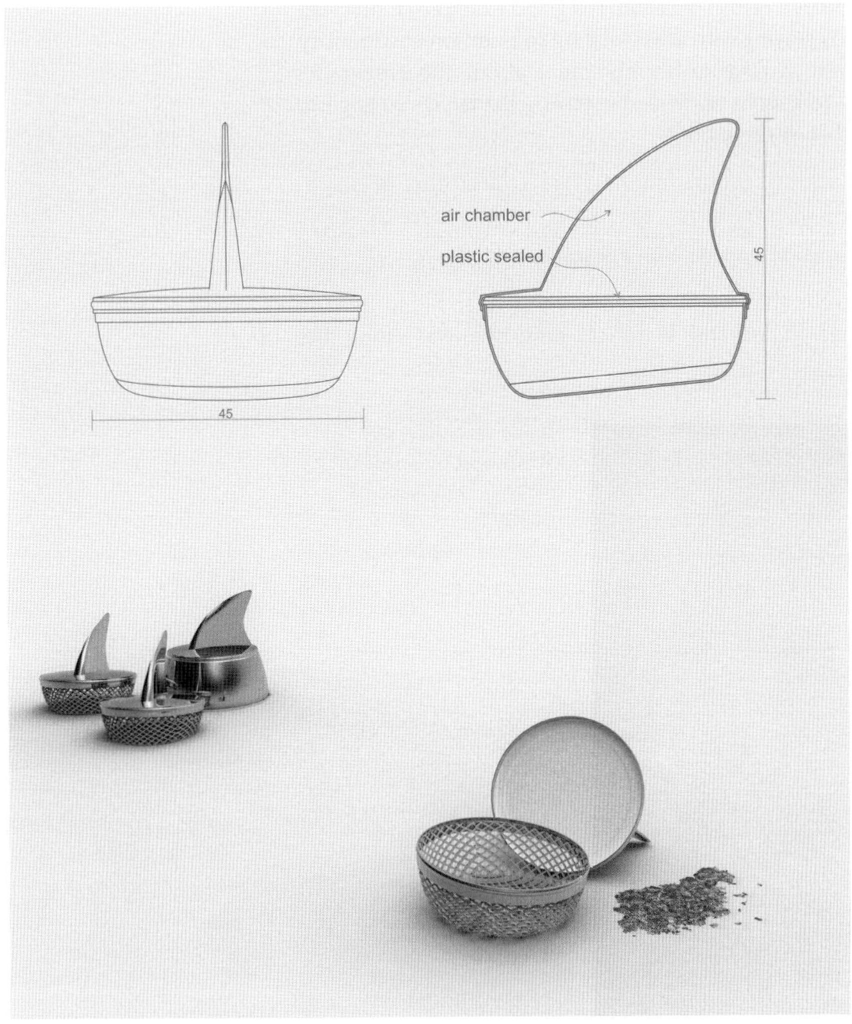

'Sharky' tea infuser by Pablo Matteoda, 2009

Authenticity

Before the post-modern era it would have been unthinkable to mix styles from different historical periods or cultures. Now that all times and cultures are considered equally important and easily accessible, it is perfectly acceptable that a rococo-style bodice could be matched with a 1950s polka-dot skirt because the wearer likes the idea of them together. The 'authenticity' or origin of an idea or style is no longer tied to its original context or meaning, and **pastiche** and **parody** have become important tools for designers.

Pastiche: described by Frederic Jameson as 'the random cannibalisation of all the styles of the past, the play of random stylistic allusion' in his book, *Postmodernism: Or, the Cultural Logic of Late Capitalism (Poetics of Social Forms).*

Parody: the imitation of style for humorous effect or ridicule.

A good example of a fashion designer's use of pastiche and parody can be found in the work of **Vivienne Westwood**, who uses historical costumes and aristocratic clothing in order to mock traditional notions of Britishness. Traditional Scottish materials and tartans are among Westwood's many sources of inspiration, and this portrait of Westwood as a British monarch is a good example of her poking fun at the establishment by playing with ideas relating to tradition and identity. Westwood's playful iconoclasm is clear in her comment about the crown: 'It's comic, but terribly chic. I like to keep it on when I'm having dinner. It's so English.'

Claire Wilcox writes that Westwood:

> 'brought in a gentle parody of establishment styles – the clothes of boarding schools, royalty and country wear – giving them a new lease of life.'
>
> Wilcox, C., *Vivienne Westwood*, London, V&A Publications, 2004, p21

Again, what can be seen here is that while Westwood was anarchic in her use and juxtaposition of clothes, styles and historical eras, she was using the

Dame Vivienne Isabel Westwood, by Angus McBean, 1988, National Gallery

language of design in an appropriate and knowing way to create her hybrid collections.

Jean Baudrillard, another post-modernist thinker, claimed that since post-modernism there were no longer originals, only copies or, to use his terminology 'simulacra' (a commonly used derogatory term from the nineteenth century for an image without the substance or qualities of the original). He claimed that the distinction between an original and a copy was diminishing, so that ideas of originality and authenticity were becoming meaningless. Our experience of the world no longer relies on a first-hand experience, indeed the fake or the simulacra could be more realistic than the real. For example, a Disneyland experience of a 'train journey on the Big Thunder Mountain Railroad' might be a better fit with our idea of a railway through the mountains to an old mining town in the American southwest, and therefore seem more 'realistic' than an actual journey.

Television, advertising and media also play an important role in our post-modern culture, to the extent that we no longer need a personal engagement with a place, object or product in order to believe in its worth or effectiveness. The result of this is that the phrase 'as seen on TV' can now be just as powerful an endorsement of a product as a first-hand experience of it. Evidence of this can be seen in sales figures that demonstrate sudden spikes in sales following mention of a certain product on a television programme by well-known people, from chat show hosts to TV chefs. For example, in 1998 **Delia Smith**, in one of her *How to Cook* programmes, described her omelette pan as a 'little gem'. This in turn prompted a huge leap in sales. The 'Delia Effect', as the phenomenon came to be known, is still effective as we continue to engage with ideas and objects mediated through advertising, film and television.

The change in attitude towards authenticity and our addiction to simulation has on-going implications for design. **Tibor Kalman** was a graphic designer who felt ill at ease with designers having to continually create **simulacra**. In an essay from 1990 called 'Good History/Bad History', he asked whether designers were just being lazy by reusing old ideas to sell to new clients. He was also unhappy with graphic designers' ability to make products appear to be something more than they were.

Simulacra: this might refer to an image as a representation of something that might be vague or superficial in its rendering.

One example that he used to demonstrate this was the packaging for a particular brand of spaghetti sauce, where he questioned if it was right to make it seem that the sauce had been made by someone's Italian Grandma somewhere in the Tuscan hills. The writer, **Michael Beirut**, explains that we are still willing to engage with communication design that creates a sense of the real or authentic, even though we know it is an imitation, and he quotes **Charles Eames** to support his claim:

> 'Innovate as a last resort. Simulation, evocation, contextualism: call it what you will, but this thing that we designers are so good at seems to serve a basic human need.'
>
> M. Bierut, 'Authenticity: A User's Guide', observatory.designobserver.com

At **Timorous Beasties**, a textile company formed in 1990 and based in Glasgow, the designer/directors **Alistair McAuley** and **Paul Simmons** focus on the exciting possibilities that emerge when historical knowledge and technical skill are combined. Together, they use their knowledge and skills to recreate handmade 'toiles' (eighteenth-century decorated cloths or other designs), but rather than draw their

'Glasgow Toile' by Timorous Beasties, 2004

inspiration from eighteenth-century visual language, they have looked to the everyday city scenes they see around them in order to update and 'urbanise' the images.

> 'By depicting uncompromisingly contemporary images on traditional textiles and wallpapers, Timorous Beasties has defined an iconoclastic style of design once described as "William Morris on acid". Typical [of our work] is the "Glasgow Toile". At first glance it looks like one of the magnificent vistas portrayed on early 1800s Toile de Jouy wallpaper, but closer inspection reveals a nightmarish vision of contemporary Glasgow, where crack addicts, prostitutes and the homeless are depicted against a forbidding backdrop of dilapidated tower blocks and scavenging seagulls.'
>
> www.timorousbeasties.com

Timorous Beasties, like Vivienne Westwood, use pastiche and parody in their designs in a knowing, yet anarchic, way. What the examples in this section demonstrate is that a dialogue with authenticity may be troublesome, but it remains an important element in post-modern design.

Retro

The word 'retro' became popular in the 1960s and is an abbreviation of 'retrograde', i.e. a movement towards the past. The term 'retro' refers to a style or fashion from the past, with its popularity beginning in the 1970s when it was used to describe the work of Parisian film-makers, writers and fashion designers. It was, however, often used as a derogatory label, as can be seen in the following quotation from 1979, when the fashion journalist **Bernadine Morris**, in an article called 'Will the "Retro" Look Make it?', reported that:

> 'After a number of years during which fashion was perceived as having made tremendous progress in the direction of freedom, comfort and style, the forward thrust of design was "aborted".'
>
> Bernadine Morris, quoted by Elizabeth E. Guffy in *Retro: The Culture of Revival*, Foci, 2006, p15

Her comment highlights her belief that the rising interest in retro was exploitative rather than innovative, that it would prevent fashion from continuing to develop and that it would not be so exciting or encourage the freedom many designers had been exploring.

Post-modern thinkers such as **Jean Baudrillard** and **Frederic Jameson** were also critical of retro style. Jameson argues that indiscriminate use of the past reduces history to a style, and that this results in a continuous reduction of our understanding of history itself. Jameson sees this as a serious problem, because rather than helping to keep alive a knowledge of the past, this use of history is contributing to a collective loss of memory.

> '. . . the disappearance of a sense of history, the way in which our entire contemporary social system has little by little begun to lose its capacity to retain its own past, has begun to live in a perpetual present and in a perpetual change that obliterates traditions of the kind which all earlier social formations have had in one way or another to preserve [. . .] The information function of the media would thus be to help us to forget, to serve as the very agents and mechanisms of our historical amnesia.'
>
> F. Jameson, 'Postmodernism, and Consumer Society' in *The Cultural Turn: Selected Writings on the Postmodern, 1983–1998*, p125

Jameson also saw such reuse of styles as an indication of a loss of creativity: 'In a world in which stylistic innovation is no longer possible, all that is left is to imitate dead styles, to speak through the masks and with the voices of the styles in the imaginary museum' (*The Cultural Turn*, p7).

However, retro should not necessarily be dismissed solely as a backwards-looking, nostalgic device that stifles all avenues of creativity. If used in a thoughtful and creative way it can promote a conversation between what might be thought of as old-fashioned and on-trend. Retro's disregard for 'high' and 'low' culture, its embrace of kitsch and luxury, irony and playfulness, its interpretation of the past and its questioning of authenticity and originality make it a powerful tool for post-modern designers. It continues to be popular, as evidenced in everything from typography, the popularity of certain colours, clothes, advertising images and household products, to cars and films, and designers with an informed knowledge of history and culture can ensure that retro is used innovatively and creatively, and not simply as an empty nostalgic device.

The 'Dharma' lounge chair that was shown at the Interior Design Show in 2008 is a good example of bringing together innovative ideas with retro styles.

'Dharma Chair' by Palette Industries, 2008 (www.paletteindustries.com)

The typeface, the choice of colour for the words and the structure of this chair by **Palette Industries** all draw upon past styles, which for some users will evoke particular memories. Even a glimmer of a memory raised by the style, colour or typeface, or even all three together, will encourage the user to perceive the design in a particular way and set up a stronger emotional attachment. Visual perception happens quickly, but it sets in motion a very complex set of interactions, such as the emotional connection between an object and a user.

These connections are important in the reception of a new design, and the development of such positive relationships can be crucial in the long-term success of an object. Your understanding of these aspects is also paramount to your long-term success as a designer. Your skills will develop in this area through lectures, seminars and coursework.

Course content

Undertaking research and writing during your course about design history, culture and context prepares you for your final year in which you will be expected to complete a larger, more complex piece of work, often referred to as a dissertation. This may consist of research projects, poster presentations, essays and visual essays.

Poster and visual essay

A poster is a one-sheet presentation of imagery and words; a visual essay is a combination of images and words, quite different to that of an ordinary essay.

Visual essays and poster presentations are helpful ways of thinking about the connections you see between your own work and that of other designers, and this visual format doesn't by any means have to be restricted to your own studio practice. It can help you to find a stronger sense of flow and a smoother communication of the relationship you might 'feel' you have to a person, idea or a period of time. Sometimes it can be hard to put into words what you can see quite clearly.

Often, the simple act of structuring a visual essay or a poster will help you to find ways to explain the process of your thinking without it seeming to be such a difficult obstacle to overcome. Even if you are asked to write a more formal-style essay or report,

> *'[The visual essay is] an interesting way to demonstrate a form of writing. Why as designers should we be limited to the traditional form of text, when we can show this in more playful ways? To write an essay can be formal and dull; when presenting an essay we can show all our enthusiasm.'*
>
> Anon

you might find that drawing your ideas in the first place or finding the images you think are important to your ideas are the best ways to start to construct your argument or opinion.

Dissertation

This is a final-year extended and research-based essay.

Students writing their dissertation are allocated a supervisor, or mentor, who meets with them regularly to discuss, debate and support the process of creating the work. These meetings might be on an individual or group basis.

Final-year student, **Rosalind Northern**'s dissertation submission was about military uniform and its influence on fashion and tailoring. Rosalind found it relatively easy to decide on her topic, as she explains:

'As part of your investigation into which dissertation topic you'd like to tackle, review previous well-graded dissertations in your subject matter to learn how high-achieving students structure their research and present their arguments. Ultimately, the aim is to gain insight into the methodologies of how good dissertations are constructed, and to use your new-found knowledge to guide the research and construction of your very own and "unique" dissertation.'

Lincoln Fan, graphic designer

> 'My interest in military clothing began while studying for A Levels, when I made a military-inspired jacket. My first-year design project was inspired by utility wear, and I then wrote an assignment on the evolution of the trench coat in my second year. So, for my dissertation I decided to delve deeper and really investigate military influences on fashion, past and present.
>
> I began by visiting the National Army Museum and was inspired by their collections and paintings of military uniform. I visited the study centre and received excellent support from the staff to access archives and their supporting data. This gave me the confidence to contact military tailors directly and focus my methodology on specific survey questions, which I elaborated on following personal visits to prestigious tailors, such as Gieves and Hawkes of Saville Row and Stones of Norwich.
>
> Deciding on the title for my dissertation and sticking to it was actually one of the most difficult tasks. I spent quite a while refining a title that was interesting, realistic and outlined what I truly wanted to discover. Once I had the title, the next challenge was picking out the most relevant historical data to match my dissertation theme from a mass of information on military uniform, and creating a logical order.
>
> My advice to students embarking on this task is to start everything early and just keep going. Set yourself a target to get something done each week. Record interviews and write up everything as you think of it – even scribbled notes on a train journey. Your first thoughts and instincts are often the best. Keep focused on your topic and title and really try not to digress too much. Don't be afraid to approach anybody (large companies, organisations, museums, etc.) to ask questions and arrange visits – I got some excellent direct feedback that I did not expect.
>
> The most rewarding aspect was seeing the final work printed and completed with plenty of graphic images to support the depth of text.'

Your ability to ask questions of what you do will help you to be a better designer. Making time for thinking about what you are creating is essential, and expressing yourself is an invaluable tool that will help with self-promotion for jobs and funding. In addition to helping you develop your ideas, being able to think and write about your own work and that of others also opens up other career opportunities.

Job opportunities

Your strengths in the area of design culture and context might lead to post-graduate study or working as a journalist, design commentator, critic or teacher (see Kirby Dowler's story on her journey to teaching design in Chapter 15, 'So, where is this going to take you?). Many designers opt to follow multiple paths through their careers and use their studio work to inform their critical writing, and vice versa. Through your studio work you have the opportunity to shape the world around you. By making the most of your exploration of design, culture and context you will be equipping yourself with essential tools for developing ideas that will continue to be of use to you long after you have graduated from your course.

Designers are often thought of as providers of style and trends, but they are much more than that, they are also cultural interpreters, broadcasters of ideas – even, as we have seen in this section, critics of society.

Conclusion

This section has introduced you to the history of our profession and examined post-modernism through irony, playfulness, authenticity and retro. It has provided you with a taste of the ideas, issues and themes that might be covered in your Critical and Contextual Studies lectures and seminars. Many of the ideas discussed will often be of ongoing interest in design studios as people continue to explore the possibilities created by post-modernism. Designers need to embrace the past and refer to this exciting and informative arena, where old ideas can be reused and playfully reinterpreted.

You will find that as you study contextual ideas and themes it can be an incredibly rewarding process. Readings and tasks will be set for you to help you understand and learn in more depth. You will find out again and again during your studies that the tasks you are set are there to help you develop your ideas and are often designed to enable you to see how the issues and themes might relate to your own practice.

The following project story demonstrates perfectly this joined-up thinking and shows **Katie Roberts**, who studied Fashion and Textiles at postgraduate level, using her dissertation research to inspire a project. It is followed by the career profile of **James Clegg**, who followed his undergraduate study of ceramics with a postgraduate course in Contemporary Art and Art Theory, and has made design culture and context his career choice.

Further resources

Books

Adamson, G. and Pavitt, J. (Eds), *Postmodernism: Style and Subversion, 1970–1990*, V&A Publishing (2012)

This catalogue from the exhibition of the same name, hosted by the Victoria and Albert Museum, was the last in a sequence of exhibitions that charted the development of design through the twentieth century. The essays and images focus on the radical ideas of post-modernism and the impact it had on all areas of art and design.

Breward, C. and Wood, G. (Eds), *British Design from 1948: Innovation in the Modern Age*, V&A Publishing (2012)

This catalogue from the exhibition of the same name, hosted by the Victoria and Albert Museum, offers you the chance to see the development of design from modernism to post-modernism. It focuses on design and designers, as well as providing essays by curators and commentators, and outlines the changes in approaches to design between the Olympic Games of 1948 and the Olympic Games of 2012.

Francis, P., *Inspiring Writing in Art and Design: Taking a Line for a Write*, Intellect (2013)

New guide aimed specifically at art and design students.

Harrington, C.L. and Bielby, D.D. (Eds), *Popular Culture: Production and Consumption*, Blackwell Publishers Ltd (2000)

This is a useful text because it covers a wide range of perspectives on how culture is produced and consumed. It uses examples from television, advertising and music.

Sparke, P., *A Century of Design: Design Pioneers of the 20th Century*, Mitchell Beazley (1998)

A definitive view of designers and their work over the last century.

Sparke, P., *The Genius of Design*, Quadrille Publishing Ltd (2010)

This book will put everything you see around you and everything you design in context. A 'must read' for any designer.

Storey, J., *Cultural Theory and Popular Culture: An Introduction*, Pearson (2012)

This book is wide-ranging so helpful for contextualising design in the twentieth and twenty-first centuries, and chapters include examples of popular culture as well as giving suggestions for further reading, which is helpful if you want to learn more about specific areas.

Ward, G., *Postmodernism*, Teach Yourself Books, Hodder Education (1997)

This is a really helpful and accessible book that will give you a good foundation in post-modernism, key theorists and ideas.

Places to visit

Museums and art galleries throughout the UK host exhibitions and events that are often related to design disciplines and are good to visit as they can help you to see how your own interest and practice has developed and expanded, and to see how these changes relate to broader social and cultural changes. The catalogues and talks often associated with the exhibitions will help you see how the work is contextualised, and demonstrate why it is important to be able to think and write about the ways your work might relate to the bigger social or cultural picture.

The following institutions are particularly helpful if you want to find out more, and you want to keep up to date with developments in your own field:

Bankfield Museum, Akroyd Park, Boothtown Road, Halifax HX3 6HG

If you are interested in costumes and textiles, this museum is a good place to visit as it regularly hosts exhibitions and events that use local textile manufacturing history as a starting place for the themes of its exhibitions.

Design Museum, 28 Shad Thames, London SE1 2YD

Designmuseum.org/exhibitions

The Design Museum hosts contemporary exhibitions on a wide range of design disciplines and issues. It also organises talks from designers, curators and critics, and if you can't get to London very easily it has a good website so that you can keep up to date with shows, talks and important issues relating to design.

Websites

www.craftscouncil.org.uk

The Crafts Council organises events, conferences, courses, talks, touring exhibitions and fairs throughout England, and its website gives you information on when and where the events will be taking place. For anyone interested in applied arts and crafts, the website is a good place to visit to keep up to date.

designobserver.com

This website provides a wide range of material, from essays by critics and designers on current design issues to jobs and events.

Project story

Designer:
Katie Roberts

Project title:
Dissertation on surrealist photographer Tony Ray-Jones/'Extraordinary' fashion and textiles project

Tell us about the project

'Extraordinary' was driven by my dissertation research into surrealist photographer, Tony Ray-Jones. Ray-Jones had the ability to seek out surrealist moments that were evident in everyday life, producing fascinating photographs that accentuated incongruous scenes full of strange juxtapositions. The unexpected combinations of imagery in surrealism form strange illusions and alternative realities, inviting the viewer to look further and to question and re-analyse what they see. I wanted to capture the idea of looking further into a design, to identify hidden features that we do not see on first glance.

What processes and skills were most relevant to this project?

Photography became an extremely important research tool for the project. It allowed me to gain a large amount of contrasting imagery and investigate colour and mood in greater detail. Initially, I used photography to generate research of the natural landscape and animals to generate my overall theme. I was able to capture the intense atmosphere of taxidermy to create a haunting range of imagery to further enhance the surrealist edge. I used more photo sessions to investigate creating a scene and setting a mood by photographing shadows, projecting patterns and working with models in styled shoots.

Dress with placement image

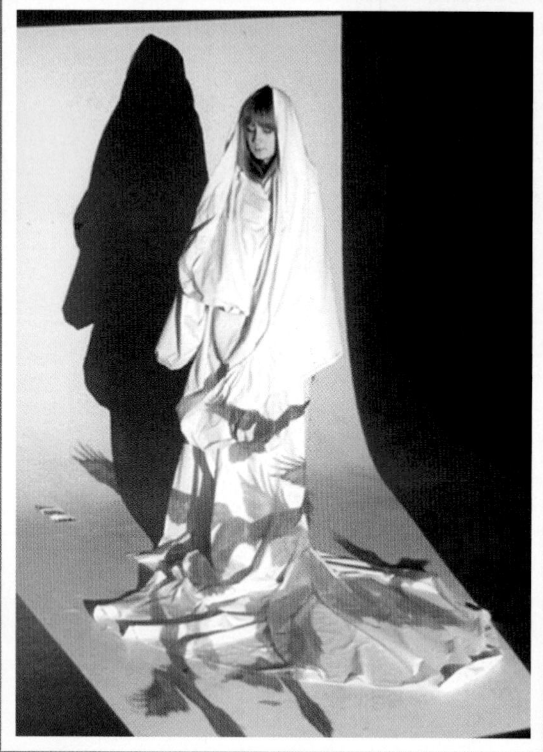

Design development – image manipulation through projection

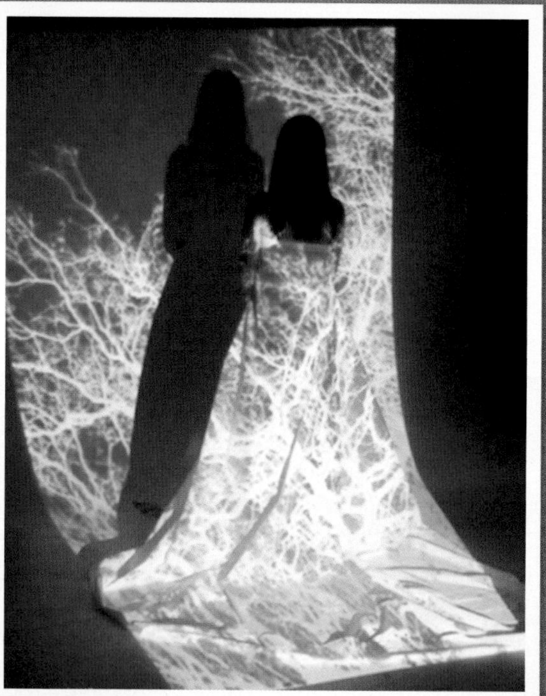

Projecting imagery to discuss scale

What were the highlights of the project?

Having the opportunity to project my imagery onto 3D form was extremely useful in helping me to debate which application of scale and placement of imagery worked the best. I found that the most successfully placed images were when the drawings were scaled-up large.

This distorted the imagery and offered another type of aesthetic, which was very intriguing.

The drawings of the faces of animals were particularly effective, as it appeared that the animals were emerging out of the fabric. The large scale of the prints also highlighted the detail of colour seen in the original drawings and this led me to develop large-scale digital prints.

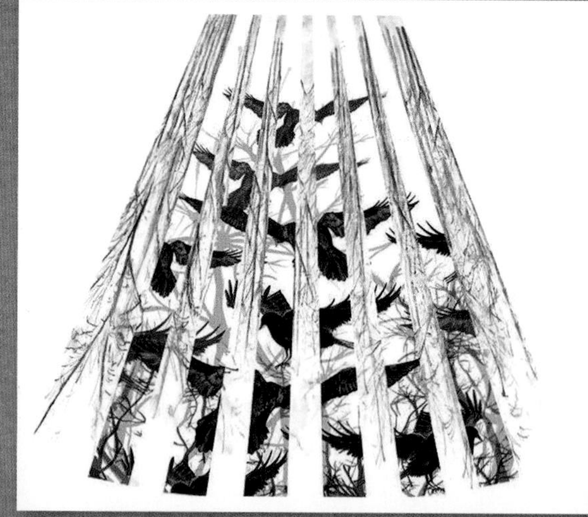

Placement digital print for fashion garment

Career profile

Name:
James Clegg

Current job:
Exhibitions Assistant at Talbot Rice Gallery, Art Critic for *Art Review* and *Art Monthly* and Lecturer.

Describe what you do

Working for an ambitious art gallery has given me a great appreciation for how much work goes in to a good exhibition. From the initial idea to the time you watch the artworks roll back out when it's all over, you are constantly negotiating with lots of different people and have to be good at solving problems. What I enjoy about this job is its variety, where in one day I can go from using my research and writing skills for creating press releases, to using my organisational skills to plan the transportation of valuable items. This kind of job provides some opportunities for travel because you work with people from all over the world, but in these times of recession most galleries try to avoid the need for constant jet setting.

Being an art critic draws heavily upon the contextual studies I undertook at art college, but it's not something I consider to be purely academic at all. In fact, for me, writing has always felt like a creative activity and I enjoy the challenge of trying to put into words the effects of an artist's work. Of course, you need to be able to provide a judgement of whether or not an exhibition is critically and/or artistically successful – and that takes confidence and experience. I review contemporary art exhibitions in Scotland, so I frequently travel from Edinburgh to Glasgow or Dundee.

I'm also a lecturer. Sometimes this feels like being a really, really dedicated student. You learn a lot doing it because you always have to be knowledgeable about your subject. And it's the intense learning you need to do when you're preparing lectures and writing courses, when you know you're going to be put under scrutiny by an audience, that can make it really tough; not to mention the marking and administration. One thing about this kind of job is that it is never part-time, even if that's what the contract says, because there's always so much to do. But then you're working with people wanting to learn, and if you put in the hard work it can feel really rewarding. I love the performance of lecturing too and because I get excited about art I've never found it hard to get excited when I talk about it.

Tell us about your career so far

I specialised in ceramics at art college. That's not a sentence I feel comfortable with. For me, 'ceramics' was just a name given to a course where you use a very versatile material to pursue your art, which I would have probably described as 'sculpture' or even 'conceptual art' at the time. The disadvantage of this open approach is that the good idea of 'specialising' loses its particular meaning. The advantage, however, is that you don't get stuck in a particular area. Throughout the course I enjoyed mixing art theory and practice, and my degree in Ceramics led neatly to a Master's Degree in Contemporary Art and Art Theory.

When I graduated from college I felt pretty lost, as I think many people do. Nothing clicked. Despite getting a first for my degree and winning awards – which seemed like a big deal at the time – nothing seemed to happen at all. I worked some pretty bleak, prosaic jobs, felt confused and cheated. At times I wondered why it had ever seemed like a good idea to do something called 'ceramics', but actually not really do ceramics but something much more mysterious relating to creativity. Was I really surprised I didn't have a job at the end of it? Then I had some kids, which added being knackered and screamed-at-in-the-face to the confusion.

Phew, this seems a bit melodramatic now, but it did work out OK in the end. I was lucky to fall into doing some lecturing, which gradually increased. Well, lucky and I had also impressed the right people as a student when I gave presentations. I started writing really short reviews for *The List* magazine, as well as doing lots of my own 'academic' and 'fictional' writing. Then I bullishly phoned an editor of *Art Monthly* out of the blue and said, 'I want to write for you'. Their Scottish critic was sick and so I was commissioned to write my first 'proper' review. I worked hard at honing my own style of critical art writing, took opportunities to get published – including entering competitions such as the Collective Gallery's New Writing Scotland Project – and suddenly I was writing reviews regularly. Can you make a living writing art reviews? No. I was being paid about £15 per review for writing for *The List* magazine, and now get about £150 per review for writing for magazines such as *Art Review*. The limiting factor is that you can only really get one review in a magazine each month.

As a lecturer I really started to find my feet and was suddenly being asked to generate new courses and lectures for both undergraduate and postgraduate programmes. This really helped my confidence. Can you make a living as a lecturer? Yes, you can. But it's hard getting a permanent contract unless you have a PhD, and even then it can still be hard. One job I applied for had over 50 applicants with PhDs.

I later got the job at Talbot Rice Gallery when lecturing became less certain, departments were being closed down and reshuffled and I was fed up with the ad hoc nature of lecturing (especially tough with a family). The job at Talbot Rice brought me into close contact with a much more tangible side of art. I'm now aiming to become a curator because I think that would bring together a good balance of the skills I've needed for these other positions: the voice of the critic, the knowledge of an academic and the practical understanding of exhibitions I've acquired over the last two years at Talbot Rice.

What advice do you have for students considering a similar career?

I've tried to be frank in my answers here because you pick up all kinds of expectations at art college that can really come back and bite you. It's probably good to know that other people go through frustration and have had to learn to be patient. Certainly I've never really felt there was a nice neat 'path' leading to what I am doing now. But what I want to say is that, even though I've found pursuing a creative career difficult, and even though I never expect to make much money, I've never lost my belief that being creative is pretty much unavoidable. I've never accepted easy answers about what art can be and I've worked really hard at everything I've done, even if I didn't know why at the time. Looking back I can see that was really important and that it's provided a foundation for me. I really appreciate what I'm currently doing and look forward to where I might be in a few years' time.

14

Future directions

By Casper Gray, Wax RDC Ltd,
Sally Halls, PearsonLloyd and
Amy Ricketts and Fan Sissoko, Innovation Unit

'Researching the concept of waste-free production really opened my eyes to the array of issues, debates, considerations and contradictions that surround eco-design and sustainability.'

Anna Piper, textile design student

Anna Piper's woven fabric, part of her project to minimise waste in the production of clothing
Photo: Anna Piper

Design is a constantly evolving process. As society, technology and our environment change, so does design's response to those factors. In fact, there are many directions design is moving in at this point in time. As a future designer you need to be aware of these and be prepared to take part in that change. These are exciting times.

This section covers the following topics:

- sustainable design
- inclusive design
- socially responsible design

After reading about these topics you will:

- realise the complexity and importance of sustainability to your work
- understand why your products must be accessible to anyone who wants them
- see that your design abilities can have an influence on society that can spread far wider than the disciplines outlined at the start of this book.

The directions we're introducing in this section inspire an increasing number of design students every year. What follows is certainly not a definitive list; there are many others that could be added, enough to fill a book, actually. Also, many experienced designers would argue that they've been practising these 'new' methods for years. This is true – they are often part of the sub-conscious skill-set of an experienced designer.

The following specific directions are quite evolved and mainstream already, and should be in your core body of design knowledge. The sections are written by people experienced in these fields. And remember, everything we design must be sustainable, inclusive and socially responsible. The design profession thinks these 'trends' are no longer 'trends'; they're here to stay.

Sustainable design:

Sustainability is a constant factor in all our lives now, and the design world is starting to settle down and take a stance on this aspect. The media might still squabble occasionally over the seriousness of climate change, but the world has now agreed that something must be done. Some people think design has solutions to sustainability problems. Some think that design is part of the problem, in that we promote new solutions, dismiss the old ones and encourage consumerism.

The author Casper Gray will explain this in more detail. His background is in design but he has moved on to advising companies about sustainable methods and practices.

Inclusive design:

Inclusive design is a legal requirement in certain parts of the world – we as designers must take account of the variation in the human population and design services, environments, information and products that are as accessible as possible to the widest section of society. This must not only take account of disability, it should also address aspects of ageing such as physical and psychological deterioration. Inclusive design has come out of the growing relationship between design and ergonomics, the science of how people interact with the world.

Sally Halls demonstrates in her writing the effectiveness of inclusive design. She has worked in this area for the Royal College of Art Helen Hamlyn Centre for Design, the Design Council and multinational companies, and is an expert in inclusive design methodologies.

Socially responsible design:

A socially responsible design outlook can be applied to many of the disciplines outlined at the start of this book; however, it may soon have to be categorised as a discipline of its own. There are principles of design process involved that have grown out of various other disciplines, and not just those in the design world. It tackles big, social issues and is practised by people from a wide range of backgrounds.

Amy Ricketts and Fan Sissoko from Innovation Unit demonstrate the scope of socially responsible design with a large project case study and a good description of the range of professionals who get involved in this expanding area.

These aspects of modern design add up to a complexity and seriousness that wasn't there 25 years ago. As the discipline of design matures, these new tools and methods give us the professional standing that the industry deserves.

Sustainable design

Casper Gray, Wax RDC Ltd

Hannah Lobley's 'Paperwork' bowls, made from old books and magazines
Hannah Lobley Paperwork

Design is amazing. As a designer you can be a part of creating the world that you want to live in. More and more people understand that our actions have consequences and that our traditional design techniques and business practices need to change.

Our awareness of the issues of sustainability (such as materials, processes and energy costs) and the importance we give those issues has changed over time and continues to change. We are heavily influenced by the media, and it is sometimes difficult to know whether an issue is happening more frequently or whether the media are just getting better at reporting it. The importance of these issues is also influenced by direct and visual effects, such as climate change, material scarcity and business performance in this changing world.

After reading this section, you should have an understanding of:

- the importance of sustainable design
- the relationship between sustainability and design of all disciplines
- the benefits and challenges of practising sustainable design
- common techniques for sustainable design

Sustainable design is simply good design, and good design takes some effort. It is sometimes easy to ignore sustainability for short-term reasons, such as cost. When you're designing you can easily be distracted by experimenting with materials and processes and not really think about the impact of your choices. Some designers (and/or their clients) think it will take up too much time and energy to undertake the relevant research, but others argue that the long-term benefits far outweigh the short-term costs. However, if you want to make an intelligent investment in your future career and life – then read on.

What is sustainable design?

There are many challenges and opportunities out there. When you consider how to protect our way of life for existing and future generations, there are a lot of issues that must be solved and you can play quite a part in this if you want to. Apart from the design and manufacture issues mentioned above (materia ls, processes, energy costs, life-cycle analysis), other, bigger and more important issues affecting all of us include:

- climate change
- social inequality and poverty
- extreme weather events
- education
- food shortages and loss of agricultural land
- access to healthcare

- pollution (of water, earth and air)
- ageing population
- material scarcity
- poverty
- homelessness and employment
- loss of biodiversity
- economics
- legislation

These issues are all interconnected and affect us all, whether it is directly through more extreme weather or indirectly through rising food or house prices. Businesses may be affected by issues such as material scarcity, increased environmental reporting, social reporting and increasing pressures from supply chains and customers, not to mention global economic factors. These are all issues that need to be managed if a business is to survive.

Regardless of the cause of these changes, they create new market, product and service opportunities for designers interested in providing sustainable solutions.

The aim of designing sustainably is to meet the needs of today without compromising the ability of future generations to meet their own needs. While the needs of sustainability change over time, there are a few basic targets that can always be agreed upon and the basic needs of food and water are likely to be most pressing, particularly for poorer markets. Our ability to meet these basic needs of food and water is increasingly put at risk by factors such as climate change, growing populations and the use of agricultural land for non-food crops such as bio-polymers and bio-fuels.

Why is sustainability important?

In 2006 UNESCO predicted that in the next 40 years 2–7 billion people will be without adequate access to water. The International Panel on Climate Change (IPCC) states that, in order to have a chance of avoiding the worst effects of climate change, we must keep the levels of carbon dioxide equivalents (CO_2-eq), otherwise known as 'greenhouse gases', below 350 parts per million (ppm). In 1988, CO_2 rose above 350ppm and in July 2012, it was already 392 ppm.

Richer markets may have more secure access to food and water and often take this access for granted. These markets have other needs and may be more concerned over issues such as availability of raw materials, which will in turn affect the availability of everything we buy, and even the Internet. To preserve access to materials requires us to create and sustain recycling societies. Richer nations are often considered to have had the most influence in accelerating climate change and so some believe they have the responsibility and the power to lead and develop solutions.

So, as you set off with your next project, what has sustainability got to do with your work? Approximately 80 per cent of the environmental impact of any product or service is fixed during the design phase. However, the designer is generally just one member of the team working to bring products and services to market – this means many different people influence the decisions that are fixed at the design phase. We need to be mindful of the effect we can have, which is not always positive.

> 'There are professions more harmful than industrial design, but only a very few of them.'
>
> Victor Papanek, *The Green Imperative*, 1995

Every design discipline, whether it be product, textile, graphic, craft, spatial, interactive or theatre, film and TV, has an impact on people, economies and the environment. It could be because of the location where products are manufactured or because of the materials they use, or for many other reasons. Designers of all disciplines have an opportunity to help make societies sustainable and thereby help to protect their own career and personal future. As demonstrated in this quote, there is a wide range of response to the sustainability problem.

> 'I'm interested in so much in the world, but specifically within design: what is worth bringing into existence amongst all the rubbish in the world? I'm trying to design non-material things to avoid part of that problem at the moment.'
>
> Lior Smith, final-year design student

There are many opportunities to practise sustainable design. The market for sustainable products and services is growing (albeit slowly) all over the world, however *all* products and services will need to consider how to protect material flows, human health and the surrounding environment. There have been waves of awareness about sustainability issues in the 1960s and the late 1980s and early 1990s, but perhaps now we are entering a period of more prolonged change, with businesses being more concerned about efficient use of resources.

Growth in this market has several drivers. Governments have binding commitments to reduce greenhouse gas emissions and increase recycling. Business is beginning to understand the need to establish reliable access to the materials they use in their products, the need to protect their brand reputation from negative reporting and to keep up with competitors. Digital media and social networking means that non-sustainable practices can easily and quickly be understood by the mass audience and this is also driving change towards more sustainable practices.

This all means that customers (whether the general public, business or government purchasers) are increasingly creating a demand for more sustainable products and services to protect their own beliefs, reputations and obligations. In turn, producers are required to deliver more sustainable products and services. This creates employment opportunities for designers with skills in sustainable design.

Knowledge of sustainable design techniques and commercial opportunities will give you an edge over your competitors when you complete your studies and need to search for a job; however, your general design skills must still be good. You need to be a good designer and in addition have a good knowledge of sustainable design.

Key sustainable design techniques

There are standard systems and methods used across industry and commerce that frame certain aspects of sustainable design. An awareness of these is important, although this is not an exhaustive list and obviously people are working constantly on new methods and tools.

Lifecycle assessment (LCA):

LCA is a strategy for assessing the environmental impacts of a product or service over its entire lifetime; this means design concepts can be compared according to their environmental impacts and the best option chosen. LCA is based on consideration of impacts during the product's lifecycle. The 'lifecycle' means the different stages that happen to create the product, e.g. material extraction, transport, storage, manufacture, retail, use and disposal at the end of its life. LCA is a useful tool for comparing design concepts because it identifies the impacts and this helps you to identify ways to reduce or remove those impacts. LCA can be costly and time-consuming, so it is often most effective to perform just a simplified assessment or to perform lifecycle thinking.

Lifecycle thinking (LCT):

LCT is a quicker, simpler and more common sense approach to LCA. Although LCT may be less scientifically precise than LCA, it can generally be done more quickly and cheaply – this makes it a very useful tool. For example, using LCT, we can conclude that generally speaking the environmental impacts of energy-using products will be in the production and use phase.

Financial return on investment:

Establishing the potential financial investment and returns of your concepts is key to giving comparisons that are meaningful and useful to decision-makers such as your boss, tutor or investors – you must prove your arguments and be sure of them in this area as much as, if not more than, you have to normally.

Service design:

Sometimes offering the service of a product can be more appropriate than offering the product itself. For example, Zipcar is an example of a service whereby individuals and businesses can have the service of a product (in this case a vehicle), but do not have to own one. In order to design a service like this, both product and business model must be designed together.

A couple of key benefits of this service are that the user does not have to invest in buying the product, thus freeing up cash-flow, and does not have to worry about any ongoing maintenance. This type of service means that fewer vehicles are likely to be required, therefore saving materials and energy.

In this type of system, a company may lease the product to the user, which means that the company keeps control of the product and is able to earn ongoing income from the product. Remanufacturing can often be a key element of this type of system.

Remanufacturing and reuse:

Intelligently designed remanufacturing and reuse systems can allow environmental impacts and business profits to be significantly increased. To achieve these benefits, design for remanufacture/reuse requires the product and the business model to be designed together.

Remanufacturing is the process of returning a used product to an 'as-new' condition with a warranty to match. This has traditionally only been commercially viable for products that can be reliably collected at the end of their life, those with high value and low turnover – cars for example.

However, as materials become more expensive and products such as small consumer electronics use more and more high-value materials, remanufacturing is becoming commercially interesting to a wider range of companies. It also creates a need for skilled, higher-wage employment. This is helping to drive industry interest in the use of smart materials and new techniques for automatic, cost-effective disassembly.

Use recycled content:

It is important not only to make products recyclable but to build the market for recycled content, so that there is a commercial reason to recycle materials. If you specify recycled content for your product this will create a demand, which means that the chances of that material being recycled are increased.

These bird boxes by Chandni Kumari are developed from waste sawdust and timber off-cuts from a furniture factory in East London.

The problem was how to produce a natural binding agent to bind the materials together. Chandni says:

'Waiting for everything to dry was torture! It would often take between four and seven days for everything to completely dry. I attempted to speed up the drying process once by placing my mixture in the oven on a fairly low heat, and when I came back to it 20 minutes later the whole thing had collapsed in the middle of the baking tray and the glue just oozed out onto the sides, so that was a disaster. From then on I just let everything dry naturally!'

Many people might have given up and used a basic wood glue at this point, but Chandni persevered and her wonderful research paid off.

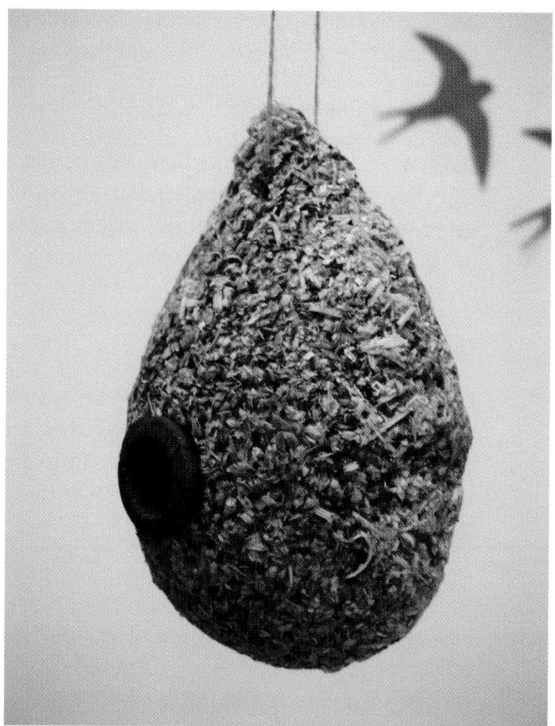

'Bird Box' by Chandni Kumari

Reduce energy consumption:

A key sustainable design technique when designing energy-using products is to reduce energy consumption during use, which is often the largest environmental impact for these type of products. This will often be the work of specialists, such as electronics engineers, but as a designer you can also influence the way people use the product to mean that energy is not wasted. An obvious, yet increasingly ignored technique is having a simple 'off' button on a product, to avoid the unnecessary energy and money wasted when a product is left on standby.

A sample of the test pieces for 'Bird Box'

Disassembly:

Most products are made of different components and different materials. In order to recycle these, they need to be disassembled. To maximise the profits that can be made from recycling products or reusing components, it is important to design products so they can be easily disassembled.

Active disassembly (AD) is an innovative method for automated non-destructive disassembly, thereby making it more cost-effective to reuse or repair certain components. An example of an AD technique is where a product casing's snap-fit is engineered to react in a certain way under controlled stimuli (e.g. a specific temperature). This would mean that the snap-fit ceases to hold when the stimuli is applied, thus releasing components for reuse or repair.

Understand users throughout the product's lifecycle:

When considering the lifecycle of a product, you will find there are many different organisations and people involved in making decisions that affect how the product is designed. It is important to extend your view of 'the user' to include all these organisations and individuals. For your plans to happen as you intend, you need to understand their needs and design your product and business model to meet those needs.

It is also very important to understand the barriers to practising sustainable design. While there are many opportunities for working in this area, there are many problems that are worth highlighting early on in a project. If you study or work in sustainable design, here are a few issues you are likely to face and some tips on how you might overcome them:

How do I deal with my own attitude towards sustainability and attitudes of others, such as fellow students, friends, family, tutors, employers or colleagues?:

Try to understand what you are doing, and why. Sometimes these reasons conflict with each other through personal, academic or professional causes. If this is the case, work through them, understand the priorities of sustainability (which mean the most to you, your life and career), analyse them and figure out which are most important. This will help you prioritise the issues so you can deal with the most pressing issues first. Sometimes this is 'quick and dirty' (there may not be enough time to get lots of clean data and your analysis is needed very quickly) and sometimes your analysis will have to be more structured (and hopefully your client will pay for the time this requires).

Make sure you can justify your actions, know your reasons and be able to communicate them. This is a key skill that will help you get support from others. If you are consistent in the ways you respond to designing sustainably this will help build credibility with others. Your confidence and knowledge might even influence the way others think!

There is so much to think about and do!:

Sustainability is a massive area and everything appears all-important. The more you understand, the more you will realise that there is a lot to learn and little time to make the changes that are required! Collaboration can help you learn quickly and achieve your aims. It is really important to not work in isolation. Build teams and networks around you so that you can complete more complex tasks. Social media and events are great tools for building your network of trusted collaborators.

There's a lack of information. Where do I find it?:

Access to quality data is often expensive. Universities often have subscriptions to commercial and scientific databases that companies may not have access to. Use professional social networks to gain from others' experience and improve your ideas. Collect information and datasets when you have the opportunity to access them. File them somewhere you'll be able to find them again in the future and use them to your advantage.

What are the employment opportunities and are there enough clients interested in sustainable design?:

In the past this has often been perceived as a big hurdle, but opportunities are on the increase. In order to make sure you can take advantage of opportunities in sustainable design, make sure you acquire sound design skills as this has to be the basis from which you tackle sustainability.

Who pays for research into sustainable solutions?:

Clients can be interested in the environmental and social impacts of their products but often don't have a budget to pay for any extra work to deal with the issues. When this is the case it can be a good idea to think of more innovative ways of working, e.g. using lower-cost LCA tools, such as Okala, when you are studying.

To summarise sustainable design

It is important to understand the reasons for wanting to study or work in this area because this will help you choose the most effective strategies. To help you in your academic and commercial career, it is particularly important to understand the differences and boundaries between 'ethics' and what is 'business'. Knowing the distinction between them will help you avoid some common pitfalls, such as forgetting business drivers and becoming fixated on options that are not realistic for you or your company, and this will help you know what design and communication strategies to use, with whom and when.

It is also important to keep up to date with relevant news and trends in business and design, and shifts in consumer behaviour. This means designers' skills in the analysis of both quantitative and qualitative data are essential.

If there is a sustainable focus to the brief, then designers and companies will have to interact with it. A brief may ask for consideration of particular sustainability issues, in which case this is the clear reason to apply sustainable design strategies; however, when it doesn't, it is likely that you will need to find a clear commercial driver, i.e. a business reason – the advantage to the project of investing this extra money in the design process – that the company accountants will understand.

Finally, to increase your chances of success, identify as many fellow students and business contacts who share this interest as possible, with whom you can discuss ideas, strategies and opportunities. These people will become useful allies in the future.

Inclusive design

Sally Halls, PearsonLloyd

> 'What you think a client would need, and what they actually want are often two completely conflicting opinions. It is important to take the time to talk to the users and find out what they hope to gain from your project.'
>
> Chloe Muir, Interior Architecture Design student

The way we use and interact with everything in our society, whether it's products, graphics, or even environments, is largely determined by designers and people like you. For example, have you ever picked up an object and found it difficult to understand what it is? Why was it so difficult to use and how did

Chloe Muir's final-year project – a bereavement centre for armed forces families

it make you feel? Did you get frustrated? Did you try and work out how it might be designed to be easier to use?

The way a handle is shaped, how a poster is illustrated, or even where a light switch is positioned, all affect how comfortable we feel in our surroundings and how happy we are to use them. The problem is that we don't all respond in the same way and therefore do not fit the normal stereotype of a 'user'. Most things are designed for the average user – typically someone that's able-bodied, right-handed and not old. So if, for one reason or another, you fall outside of this category, then things can start to get much more difficult. Inclusive design focuses on designing for *all* people in society, so that people with differing needs can use your designs.

By the end of this section you will understand that:

- the world and its population are changing at a rapid rate and designers need to be acutely aware of this
- it is a designer's responsibility to make sure that as many people as possible are able to experience what they design
- users can enhance the design process and make you a better designer.

You will also discover a range of methods and ways of working that can enhance someone's experience of your designs. Helping all people to feel happy using the things you design can have great benefits. Not only will you have a very happy customer, but making your design easier to use will provide you with a business advantage over your competitors, leading to more sales, a happier client and a better reputation for you as a designer.

What is inclusive design?

Inclusive design is the name given to an approach and methodology that helps ensure that your designs are easy to use for the greatest number of people. This means that people of all different physical abilities, sex, race, age, or even geographical location, can still use and be satisfied with your design. So your role as the designer is to understand what all these different types of people might need from your design and to respond to this challenge.

'Inclusive design' is the term used in the UK to describe this approach to design, but there are other similar movements around the world, which have different names. Originally it started in the USA and initially focused on the built environment to ensure that all people had the same unlimited access to buildings and spaces. It then became a broader term that enabled all people to use products, spaces and services on an equal level, making them 'barrier-free'. The term 'inclusive design' has now been taken up by Japan and the other Pacific Rim countries, although there are companies who use the term 'universal design'.

Across Europe and Scandinavia the term 'design for all' is used and has a similar meaning to those above. It similarly began from providing barrier-free accessibility for people with all disabilities, and the principles are now advocated by the European Commission to all designers.

Why is inclusive design important?

The world we live in is now changing at an incredible rate. We are staying healthier and living longer, with the result that, in the UK, life expectancy has reached its highest ever levels on record, with men living on average for 78.1 years and women for 82.1 years.

The increasing number of older people in our ageing population is the biggest social trend happening right now. The 'baby boom' generation, those people born in the years after 1945, have just started to retire, which means that the proportion of over 65s will only continue to grow over the coming years. In fact, the UN predicts that by 2045, there will be more people over the age of 60 in the world than under the age of 15.

An ageing population means that the way things are at the moment will have to change. Our society, goods and objects will have to adapt to accommodate the different needs of older users, especially as these new older users are expected to be far more demanding and expectant of good services than previous generations have been. And not only are they fitter, healthier and more active than previous generations, they are also much more affluent!

How do you become an inclusive designer?

Inclusive design means designing for users with different needs, but how do you find out what those needs are? There are a number of straightforward activities you can do to start understanding what they are. The great thing about inclusive design is that, actually, you just have to get out there and start talking to people . . . lots of them!

User research: involve people in your design process

As a designer you will often be given a brief about something outside your scope of knowledge. If that happens, the first thing you'd normally do is to start researching the topic and its background. Inclusive design is no different, it's just that the research is broader and more extensive. It involves engaging with people, learning from them and their experiences and gaining insights into the context and use of the product.

What is key to inclusive design, though, is the users you choose to talk to. Talking to an 'average' or 'normal' user is a good place to start, as it can increase your understanding of the way in which something is used on a day-to-day basis. This will give you an insight into the most common aspects of the project. This is useful if you aren't familiar with the situation, even though it's not particularly unusual. Say, if you are working on a bus ticketing system, but you never use the bus! Talking to 'normal' users, however, is unlikely to yield many new and surprising insights.

It is the users who push the existing design to the extremes that can provide you with deeper insights. They will be stretching the limits of use through things such as frequency of use, limited mobility, or other constraints. So with the bus ticketing system example, these might be people who use late-night buses in city centres, pensioners who use them often and for free, or people with limited mobility. These people can be called 'lead users', or 'extreme users', and they're the people that you need to find to give you the extraordinary insights you're looking for.

Product Design student Emily Boniface created models of her coffee set to test perceptions of it. The contrast between the dark and light areas of the mugs helps people with limited vision to see where the contents are by providing the right background colour.

Emily Boniface's 'Black Dot' tea and coffee set
Emily Boniface Design

Find ways to observe the environment you're working in

Observation is an invaluable tool that helps you understand how designs are used in reality. Often what people say they do is quite different to what they actually do. Usually this is because they don't realise there's a difference between what they say and do, or it may be that they just don't want to be 'different' to everyone else.

Whatever the reason, observing people in their natural environments can help you gain key insights and knowledge into ways that designs or environments might be improved. Often it's the way people work around a problem when something goes wrong that can point you in interesting directions. Look out for situations where someone's cobbled together a modification, or put up a hand-made sign – these are good indicators that things aren't functioning well and need improving through redesign.

Take a 360-degree view of the situation

You also need to 'zoom out' of the problem/product/system area and have a good look around to try and understand any issues surrounding it. This can mean various different things, depending on which design discipline you are working in. It can be:

- learning about the environment in which the product is used (3D design and craft)
- how a space is traversed by crowds of people (spatial design)
- what the big purpose of a piece of communication design is (graphic and interactive media design) or
- whether your design proposal is even communicating the right emotion (theatre, film and TV design, automotive design).

These things can all help you to understand the greater context of use, and therefore create a better design for it. For example, a recent project that PearsonLloyd was involved in looked at how design could help reduce violence and aggression in hospital accident and emergency departments. The reason we won the project in the first place was because of our preliminary research – we were trying to understand the big picture and learning about why so many people were becoming angry in these situations. We needed to make sure we were addressing the right problem, and not trying to 'solve' something that wasn't broken!

Organise a series of interviews over the course of the project

Interviews are a great way of gathering qualitative information. You can start with general interviews, letting the subjects take the conversations where they want to, then use a series of statements or questions to prompt them to talk. These might be something like:

- 'Tell me about your experiences of using this product/service/space . . .'
- 'Is this something you would recommend to a friend or family member (in a similar situation)?'
- 'If not, why not?'
- 'How do you think it could be improved?'

Be very wary of asking leading questions though. If a publisher asked a group of students, 'University text books are expensive, aren't they?', we can probably guess what the answer would be!

Run focus groups to encourage discussion between stakeholders

Sometimes talking to users on a one-to-one basis may not be practical or appropriate. In this case, you may wish to organise a focus group, to draw lots of users together in one place. This can have many benefits, as the users will each have their thoughts and opinions that they can bounce off each other and this will stimulate the conversation. In this case, your role becomes one of observer and facilitator.

Sometimes gathering a group of new people together can be an uncomfortable experience so it's up to you to help make everyone feel relaxed. You need to:

- keep it friendly, so that everyone feels comfortable
- make sure that everyone gets a chance to speak and nobody is too dominant
- keep the conversation relevant, directing it to areas of interest and away from controversial themes
- be flexible and ready to adapt – if you don't get a response, try rephrasing the question, or if an activity isn't working, move on to the next one
- note down or record all the key insights (you'll forget them otherwise!)

It may take you a while to get used to doing this, but keep practising as it will become much easier with experience. You may also find it easier to facilitate with a friend or colleague and divide up the tasks between you. You can then talk through your findings afterwards.

The main advantage of a focus group is that you have many more opinions to draw on, and areas of commonality (for example, areas/aspects that everyone likes or finds particularly easy to use) can help you build a strong argument to support your design. Of course, if everyone finds something particularly difficult, then you'll find out pretty quickly what you have to change!

TIP

Be aware of what you're trying to find out from the focus group, and make sure you've planned the session accordingly. Remember that you're trying to get everyone really engaged with the topic, so keep the activities fun and interesting and they should end up being productive!

When starting out, you may want to try and keep the numbers in the focus group quite small; three or four people makes it easier to engage everyone in the group, while still having enough people to stimulate the conversation. An hour is a good amount of time, but it means you need to be strict about sticking to your schedule!

Here's a quick example of how a focus group might run:

1. **Introduction** (five minutes) – who you are, why everyone has been gathered together, a quick summary of the project.
2. **Ice breaker** – try to make this fun, for example, ask people to tell a silly fact about themselves, or a secret; this is to help your participants feel comfortable talking in front of each other.
3. **Try to keep the sessions interactive** – when asking for responses to a question, write down all the answers on a flip chart on a whiteboard; everyone's opinion is equally important and should be acknowledged.
4. **Keep a stack of Post-it notes handy**; people can then write down and display their ideas/thoughts.
5. **You may want to break into smaller groups** to brainstorm particular questions/problems; each group can then present their thoughts back to the rest of the room.

Remember to take photos of everything you create – boards, notices, layouts – so that you can access them again at a later date.

Finally, if you want to take photos of your participants you must get their written permission, particularly if you want to use their image in a presentation or publication.

Monitor blogs/web forums, or even start one yourself

If your design is targeting a specific market, then there are likely to be many blogs and web forums dedicated to it. These will provide all the basic information you need to know, as well as having forums where interested people can discuss any questions or specific needs they might have. When you have an idea or a question to ask, these are a good place to get responses.

Design yourself a questionnaire or survey

Surveys are a useful tool to understand how a larger group of people think or respond to a topic. They generally provide a list of questions for a participant to fill out by themselves and this affects the quantity and the kind of questions that can be asked. For example, in order to encourage participants to complete the survey, it is best to keep it very short and easy to understand. It is a good idea if a number of the questions are multiple choice, as these require less time to complete and are easier to analyse. However, you need to consider the answer options carefully, as you need to prepare a tick box for each one. Getting responses back that have many 'don't knows' ticked, or 'other' boxes ticked, implies that your answer selection wasn't adequate.

Generally, the more people you ask, the more certain you can be of the accuracy of any trends that you find. Surveys can be conducted through emailing your personal network of family and friends, or there are now websites available, such as www.surveymonkey.com, that allow you to set up an online survey for free, or for very little cost.

Designing a questionnaire is a big job and requires a certain amount of research beforehand to ensure you're asking appropriately phrased questions and tackling the right issues. You must interview several people and discuss the area before you do a questionnaire. You might find that you have quite a bit of information after this and you might reconsider whether a questionnaire is worth doing. Don't assume they're the best solution. You also need to pilot the questionnaire with five or six people to make sure it works and there are no mistakes.

Research tools

You can also create your own set of research tools. These may be in the form of a 'probe pack'; this is a set of things that you give to a user to help understand how they live their daily life. For example, you may want to include a diary for them to plot out their activities or emotions, or you could create a visual map for them to fill out. Alternatively, you can create a 'design provocation', something that is designed deliberately to stimulate the conversation and provide responses to an idea in some way.

It's important to remember that the 'perfect' research tool doesn't exist. Each project is unique and will have its own needs and most appropriate research tools. Sometimes it is just a question of trial and error, developing your tool until you are able to generate useful results.

Organise regular user testing sessions throughout the project

Once you've developed your design to a first prototype, you will need to test it with your users to see if it actually works as you intended, and does what it's supposed to. Tests can be done in a lab, workshop or in the public domain. You need to mimic a real situation and be careful that your experiment set-up doesn't affect how people view your design and subsequently rate it.

Empower the ordinary person in the street with co-design

If you take inclusive design to its farthest extent, you then embark upon co-design. This is where you work collaboratively with a user, or community of users, to develop and design a solution together. It empowers the users, enabling them to get much more involved in solving their own problems. It can, however, be a very complex and time-consuming process to manage – finding subjects to use, organising facilities and resources such as rooms, paper, Post-it notes, etc. You also need to ensure that your design doesn't become customised to the very specific needs of your co-design partner (unless of course, that *is* the brief!). Its advantage is that it can often lead to new and unexpected insights, different to what any other methods might yield, and result in new and exciting designs.

The methods described above are suggestions to help you change the way you design and the way we experience our surroundings, whether they are objects, systems, resources or just fun stuff. These methods are applicable to all design processes, so have a go at building them into your own brief and see how useful they are.

To summarise inclusive design

Inclusive design research methods are certainly more difficult than sitting at a desk with a computer, a cup of coffee and a sketch pad just dreaming up ideas. However, the benefits are always worth it. As stated in the introduction to this section, these methods are not options any longer as consumers and clients are more discerning than ever. As designers you have a responsibility to embrace these and enlighten your tutors, employers and clients with your use of them.

Incorporating these methods into your designer's skill-set will help to differentiate you from other designers, and give your ideas a solid rationale. You may also find your designs heading into new and exciting territories too! And, of course, as you become more used to working with users, you'll discover that it becomes easier and easier!

So, read through the stories at the end of this section and then plan to integrate an aspect of this user research into your next project!

Socially responsible design

Amy Ricketts and Fan Sissoko, Innovation Unit

> 'With my major interior architecture project, I wanted to understand childrens' perception and interaction towards the built environment and aid their development. I wanted to come to an understanding as to how these experiences ultimately shape their future.'
>
> Anna Deery, final-year Interior Architecture student

Design can be said to be a desire to make things 'better'. Ask all the design students and professionals you know about their motivations to become a designer and you will probably hear an overwhelming diversity of stories. But dig a bit deeper into their responses and it is very likely that you will find a common desire to make things better. Designers tend to be motivated by the belief that, through the exercise of their creativity, they can generate more ingenious and more beautiful solutions to everyday problems – a more legible typeface, a more comfortable shoe, a more entertaining game, a smarter phone.

What 'better' means within the particular context of a professionally commissioned design project is defined by a whole range of factors, often driven by economic imperatives. This is due to the fact that design has emerged from the development of an industry for mass-produced goods: designers are often tasked with the mission of, not exactly making things 'better' under a general agenda of human development, but of improving existing designs so that they can be sold

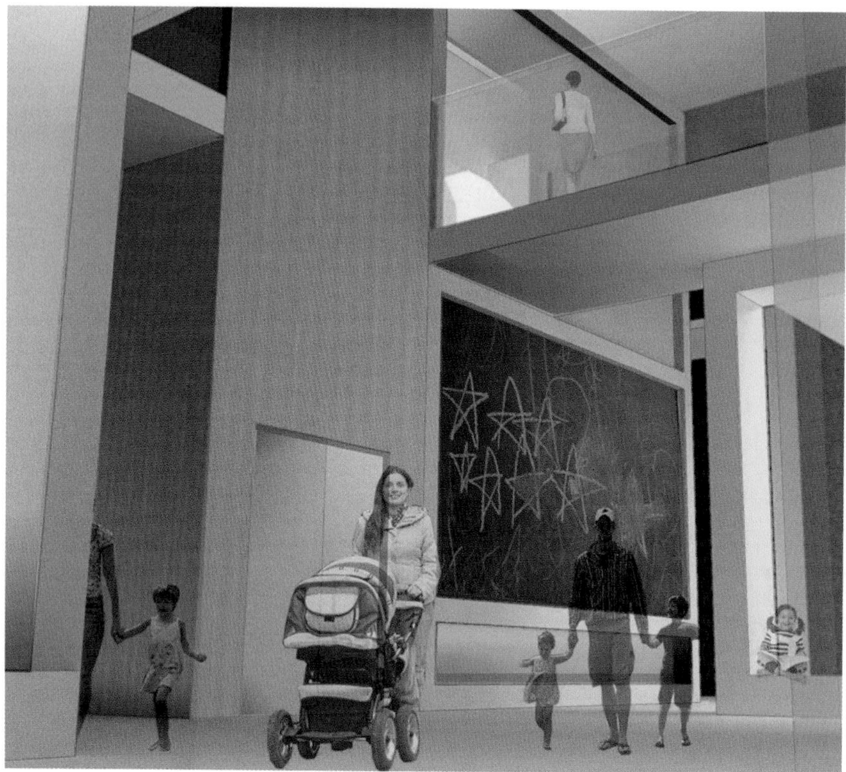

Child education centre – kidsPACE by Anna Deery (see Chapter 6, 'Spatial design', for her story)

to more people. They have been almost exclusively concerned with adding functional and aesthetic value to material goods, through manufacturing and advertising as opposed to responding to peoples' needs from a social perspective. In other words, their efforts to make things 'better' have been led by the values of markets and consumers.

However, this is starting to change. Designers are increasingly taking into account the wider social and environmental context in which they work, and demonstrating that the social implications of any product can be determined at the design stage. For example, some are directing their designs towards greener and less exploitative manufacturing processes, choosing longer-lasting and less wasteful materials, or ensuring that their design is inclusive and will not harm or cause prejudice to a particular group of people. This has been demonstrated in the previous sections on sustainable and inclusive design. This section hopes to demonstrate that, whatever their field of speciality, designers now have a range of options for putting social responsibility at the heart of their practice and having a truly positive impact on the world around them.

Final-year Interior Architecture student, Kristen Brice's major project in Kielder Water, Northumberland, exemplifies this ambition. The columbarium and promatorium aim to explore the stimulation and deterioration of memories in relation to death, remembrance and burial.

> 'The columbarium was a culmination of approximately seven months' work for my final degree project. My project developed into a concept about memory and how memory is affected by genetic spelling mistakes. I established the concept: the stimulation and deterioration of memory in the brain.
>
> The next step was exploring the concept in a spatial way. In crude terms, memory could be simply illustrated as two light sources touching one another: two electrical signals communicating with one another. However, throw in genetic spelling mistakes and the communication may start to

dwindle until finally it deteriorates and dies. I illustrated these theories through a number of light experiments, trying to achieve both stimulation and deterioration of communication. The concept quickly developed, but in some respects I was still designing blind.

At this point my tutor advised me to look at death. I wasn't certain. She maintained that death would lend itself very well to a concept about memory, but I didn't like the finality of it. It seemed too harsh. I researched crematoriums, churches, mausoleums, crypts and then finally came across a columbarium. I'd never heard of one before, but as soon as I saw photos of one I was inspired. It was like a beautiful library of ashes. I could suddenly see the project could be taken forward, focusing on the memory of the deceased.

I loved the idea of designing a columbarium, but it still needed, ultimately, to be just a library, therefore I set out to design both elements of the burial process. I would design a crematorium with a ceremony hall two, inspired by the deterioration of memory. The building would then flow into the columbarium, where memory would hopefully be stimulated.'

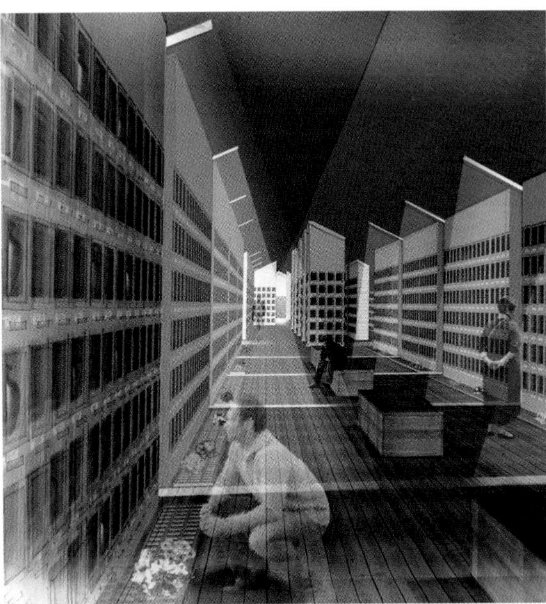

Computer-generated illustration of the lighting

The view over Kielder Water from the columbarium

One thing that is important to understand is that socially responsible design is not a traditional discipline in the same way that, for example, graphic or product design are. In other words, you can decide to be a graphic designer who only works for charities, or you can decide that you will always try to consider the environmental impact of the products you design, even though it may not be directly included in your client's brief.

After reading this section you will:

- understand how a design approach can be applied to solve complex social challenges
- realise that the influence you could have goes way beyond aesthetics
- hopefully be inspired and feel the need to respond to more meaningful problems.

What is socially responsible design?

Designers who identify themselves as socially responsible have started to demonstrate that the social implications of any product can be determined during the design process. Socially responsible design can be framed in very *proactive* terms. Designers are increasingly recognising that they can apply their creativity and ingenuity to problems that are not market-defined, but pressing social issues. For an example of this visit the Design Council website and see the results of a project to reduce violence and aggression in the accident and emergency departments of UK hospitals: www.designcouncil.org.uk/our-work/challenges/health/ae.

The challenges facing humanity

So, to begin with why don't you think about all the challenges humanity is currently facing, from environmental degradation to loneliness and isolation, from an ageing population to crime, from the impact of budget cuts on health services and the education system to social deprivation. If what motivates you is a desire to improve things, then it makes sense for you to take on a wider, social well-being agenda and look for opportunities to design solutions that are relevant to the needs of humans in all their complexity and diversity.

Recognising that you as a designer have a role to play in solving problems that are usually the responsibility of governments can seem quite daunting. However, let's consider the interventions that you might make that can have a significantly positive social impact, such as:

- **design a website** that will help a charity to raise funds to support homeless people into employment
- **design an accessible technological device** that helps isolated older people to connect with their neighbours
- **design a family game** that raises awareness of climate change
- **design an informative poster** to communicate a critical public health issue to people who lack literacy skills
- **design a special care environment** for children with specific needs.

As you can see, there is a variety of ways that you could use your skills and ingenuity to improve the society you live in.

Designing better solutions

Socially responsible design goes beyond designing environments, systems and objects, it looks for better solutions for end-users. Most social issues have systemic roots and cannot be solved by the design of a new App, game or poster. That's why designers are also starting to get involved in more ambitious programmes of social change and are now working in partnership with governments, local authorities and third-sector organisations. These solutions are often 'services', i.e. not a tangible object, space or piece of communication. This has led to the notion of a 'service designer': someone who designs experiences, interactions, information, systems, etc. Here are some examples that encompass this wider approach:

- partnering with a charity to design a service that helps homeless people to get into employment
- partnering with a local authority to design a strategy that will reduce social isolation for older people
- designing a programme of school-based activities that will support families to choose more environmentally-friendly lifestyles
- engaging a group of people who lack literacy skills in the design and delivery of a literacy programme.

In this context, you as a designer would not just contribute your technical, aesthetic and craft skills, you would also need to demonstrate that you can think laterally and tackle these systemic problems in a creative and innovative way. The story behind the project later in the section will give you a more detailed insight into some of these methods and processes, but for now let's focus on design as an approach and on why it is particularly relevant to social innovation projects.

Why is socially responsible design important?

From healthcare through to education, design is a way of getting to the heart of the issue. There are a number of general principles that are essential to creating social change – they all focus on enabling us, as designers, to understand more completely the problems we are trying to solve. This generally means spending more time carrying out research and thinking about people, rather than refining your technical skills. While you are studying, begin to think about how you can practise some of these skills in your own work.

Question everything:

The briefs you are working on will often require you to work in new environments, with new people and in new ways. This places you in an ideal position to question why things are done a certain way. You can approach projects and problems as someone who is very naïve, meaning that assumptions and ways of doing things are broken and seen from a fresh perspective.

- **Exercise part 1**: Think of a problem – it might be something big such as social isolation, or something more manageable such as local vandalism. Now write down everything you know about this problem; it is likely that there are huge gaps in your knowledge. These gaps can be filled by asking 'why' and 'how'. Make it your job to find out.

Empathy:

This involves being a good listener and observer and being able to communicate with different kinds of people. Empathy is about understanding the people you are designing for and 'stepping into their shoes'.

- **Exercise part 2**: You can now use your 'why' and 'how' questions. You have identified the gaps and possibly the people you need to speak with to fill them. Perhaps you know someone who experiences the problem; spend some time speaking with them about how they feel and exploring their ideas about how it could be solved.

Lateral thinking and problem solving:

Designers work with a range of people and perspectives, meaning that design outputs will be the result of a number of considerations. Thinking clearly about the problem and your research findings will help you to get to the solution in a clear and logical way.

- **Exercise part 3**: Now that you have some results from your questioning and empathic research, you can begin to draw some conclusions about what has been said. There will probably be some repetition and clues to bigger problems; group the findings and create a set of 5–10 insights (concluding headings). You now have some design principles to begin problem solving.

Design often deals with a great deal of ambiguity, where you may not know the answer until you have followed a lengthy and in-depth research process. It is important to refrain from jumping to conclusions and making quick decisions about what the most appropriate solution may be. Being comfortable with the fact that you might not know what the solution should be allows a project to grow and evolve as more information is uncovered.

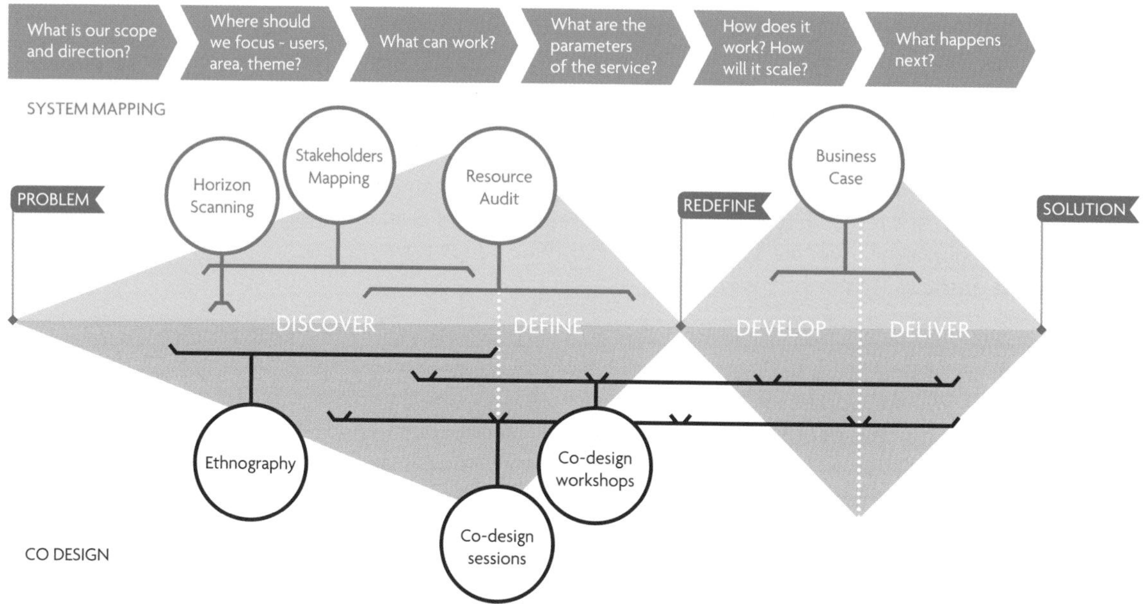

The 'double diamond' design process

The more traditional design skills, such as drawing and making, are great assets. There are, however, more key skills, such as visualising bodies of data, mapping people's attitudes or building system prototypes that are required as they add value by breaking down any barriers that might exist between people with different perspectives. Being able to explain visually, in a simple and clear manner, a range of often quite abstract or intangible information is a fantastic way of enabling everyone to imagine how a system, service or product could work. As a designer, all of your skills will enable you to tackle many of the really important problems facing society today.

Many excellent examples of socially responsible design have been promoted by the UK Design Council and their website, as stated earlier, has some very detailed reports on how those projects were planned, carried out and implemented. Through this pioneering work, the Design Council has also worked on promoting research and design methods and even specifying its own framework for these projects. They define this as a 'double diamond model' – Discover, Define, Develop and Deliver – to embrace many of the concepts, methods and tools that are central to this type of design process. The framework is useful for many types of projects – 3D, interior, graphics, etc. The principles are outlined here and then demonstrated in action in one of the project stories later in the chapter.

Discover

This is the first stage within most service design projects. This involves getting a deep understanding of what the problem is by gathering as much insight and intelligence as possible. Without this, your final design proposal could be inappropriate and not meet the needs of the users, or you might overlook a crucial element that makes it impractical.

Ethnography:

Ethnography is an incredibly powerful way of uncovering real insight into the lives of people for whom the service is aimed at. In a good design process ethnographic fieldwork involves spending prolonged periods of time with service users or staff, possibly in their homes, places of work or other environments in which you are both comfortable. This will enable a relaxed and informal conversation to take place. It is best to approach these meetings with an open mind – your preconceptions are often shattered as you find lovely people willing to open up their lives, revealing the most intimate human experiences that have led them to where they are today.

Ethnographic research
Innovation Unit

Data gathering:

This can be one of the most difficult but important tasks, where you try to get hold of the data you need to make your design proposition feasible and, crucially, to make sure it's an improvement on the current situation. In contrast to gathering qualitative, human data (as the previous methods described), here you will be looking to understand processes, practical limitations such as procedural restrictions and hard quantitative data.

Horizon scanning:

Horizon scanning is probably the closest method to traditional design research. As its name suggests, it's about having a good look around to get a good understanding of the problem and the kinds of things that may offer solutions. You should spend some time looking at what currently works well and find examples of the innovative ways that people are tackling similar challenges in other organisations and environments. This allows you to identify strong examples as inspiration and you might discover ways to improve the status quo.

Define

This is the stage where you will clearly define the problems and write them into the brief you have set yourself. Having gathered the insights and data, this stage can feel overwhelming, as it might be rich with options. This can be the most difficult but exciting phase in the project – knowing that the answer is somewhere in the room but still having to find it.

Co-design:

Co-design is a participative method, often run as a workshop, allowing lots of different people, typically with different perspectives, to come together and share ideas, work through a specific problem collaboratively and hopefully come to a consensus. This is a very democratic process, where each representative party is able to have a say in how they feel things could be improved. You will play a facilitative role here – having designed the workshop, you are simply on hand to steer groups towards the next phase in the process, encourage people to discuss things more deeply or to simply record what is happening through filming, photography or note-taking.

Prototyping:

Prototyping gives you the opportunity to mock up aspects of the service that you would like to test. This doesn't need to be expensive. There are a number of easy, cost-effective and quick ways of prototyping a new concept. Consider designing a *role play* session where different roles and scenarios are explored. This should give you a good insight into how each person might interact and respond to the concept. Another option would be to physically mock-up a part of your proposed design on paper – paper prototyping. This may be a phone application, website, poster or something else entirely. This allows you to see what might and might not work without spending lots of time or money. If you prefer to work in a hands-on visual way then this can also be done using Lego, plasticine or other props to help you to organise your thoughts and ideas.

What you are trying to do here is create ways in which you might be able to test and evaluate your ideas. All of these options should result in more robust solutions that are more realistic and achievable.

Develop

Here you are finalising the details of your project, identifying who will be involved and how it will work.

Service blueprint:

A blueprint details every aspect of the idea, from the roles of users and other stakeholders to the back-end processes where infrastructure and investment are required and data and money might be exchanged and processed. It gives an overview of how a 'journey' works – from when someone begins to use it, to where they might stop. Blueprints can be plotted

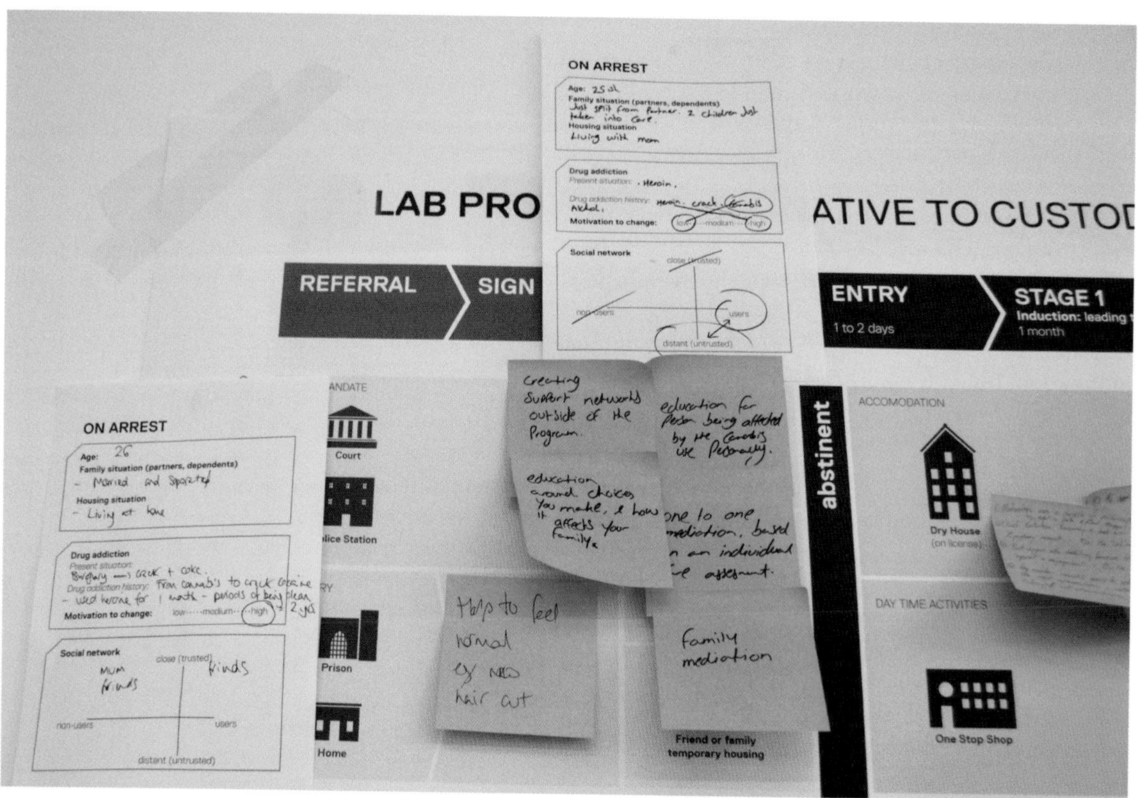

Analysing feedback from a co-design workshop
Innovation Unit

against time, split roughly into phases of encountering the system, using it and leaving it, or buying the product, using it and then disposing of it.

Business case:

Building a business case allows you to capture the viability of the proposition. It will include things such as the costs associated with delivery and the financial streams to show how it will be funded. If the project is part of the government's social service this might come from a government department. If it's a new craft workshop it might come from a European Union cultural budget. If it's a new product concept it might be from an entrepreneurial investor. (See Chapter 16 'Working for yourself', for more detail on this stage.)

Deliver

The final stage in the process does not involve a great deal of design, if any.

Piloting:

To pilot something is to try it and see what you can learn from the experience. You can then feed this knowledge back into the last stages of the design process to improve the final concept.

Piloting may take place as part of the delivery, or perhaps slightly earlier depending on the nature of the project. It follows on naturally from prototyping, as this should be a small-scale but 'real' test of how well it works in practice. In all likelihood problems will emerge, but so will other exciting and unexpected results!

Implementation:

Here the project will move forward to the point where the organisation is investing in the project or recruiting new employees to implement it. Designers do still have a role to play in ensuring that, as the project is implemented, the concept is protected from hard-nosed business decisions that risk overlooking the unique aspects of the project.

As well as championing the concept, it is also important for you to encourage ongoing development and experimentation. Especially in a service design situation, continuing to offer support and suggestions and adapting the service where necessary are important factors in arriving at a sustainable solution.

To summarise socially responsible design

Design can be used to help rethink how we might be able to improve *services*, making them not only cheaper to deliver but much better for users. These kinds of services touch the lives of almost everyone in the country at some point and are a vital aspect of our society, from something as important as helping us to feel better when we're unwell, down to collecting the rubbish each week. Currently, these services are facing a number of challenges, both financially and through an ever-increasing demand. This provides you, as a designer, with ample opportunity to make a real difference where it is most needed.

The pioneering examples of socially responsible design in this section and in the project stories by practitioners and students at the end of this section can inspire all of us, no matter what our discipline, to consider the wider social issues in all that we do.

Conclusion

This section on future directions has demonstrated various strands of design that give us hope and direction as we head into the twenty-first century. The new challenges we face are being tackled head-on by some inspiring people in our profession. We are also looking at the wider picture as designers and realising how we can get so much more involved in society in the larger sense. There is also a very good appreciation of these issues in the design students of today. Degree shows up and down the country and the work displayed every July at the graduate show New Designers in London exemplify this.

There are other challenges not covered here, challenges that we are grappling with now:

- **MIY – make it yourself**: designers and ordinary people are realising the value of their thinking and design skills and are sharing knowledge, ideas and skills via social networking. If someone creates something they can share the idea with others. This is not done for profit, it's done for true ethical reasons, for the wider good.
- **Emotional design**: designing things we can really connect to, work with, value and, yes, love. We start with a teddy bear when we are young and end up with a cappuccino machine, pair of shoes or a car to die for!
- **Critical design**: design for art's sake, as it is described by some. It puts design in a gallery, to provoke debate and to get us as designers to question what we do. Critical designers sometimes design things that don't work. They're not meant to. They are there to get us talking and arguing. Sometimes critical designers design things that do work – visit www.droog.com to be inspired by some of them.

We now have a wide range of stories to conclude this section that will give you a very good idea of the excellent work being done today in the name of sustainable, inclusive and socially responsible design. Given that quite a lot of it is student work, it also gives you an idea of how well the profession will meet these challenges in the future.

Further resources

Books

Berman, D.B., *Do Good Design*, Peachpit Press (2008)

A really easy-to-read account of why design matters and what designers can do to make the world a better place.

Black, S., *Eco-Chic: The Fashion Paradox*, Black Dog Publishing (2011)

Tackles the contemporary issues of sustainability in the field of fashion.

Braungart, M. and McDonough, W., *Cradle to Cradle: Remaking the Way We Make Things*, Vintage (2009)

An inspirational and authoritatively written introduction to sustainable design.

Clarkson, J., Coleman, R., Keates, S. and Lebbon, C. (Eds), *Inclusive Design: Design for the Whole Population*, Springer (2003)

A comprehensive introduction to the area of inclusive design.

Fletcher, K., *Sustainable Fashion and Textiles: Design Journeys*, Routledge (2008)

An encouraging and forward-looking approach to sustainability issues in relation to the manafacturing lifecycles in fashion and textiles.

IDEO, *Research Method Cards*, IDEO (2002)

A collection of 51 cards representing diverse ways that design teams can understand the people they are designing for.

Papanek, V., *Design for the Real World: Human Ecology and Social Change*, Thomes & Hudson (1985)

A pioneering book advocating socially and environmentally responsible design. First published in 1973, it is just as inspiring now as it was then.

Thackara, J., *In the Bubble: Designing in a Complex World*, MIT Press (2006)

A journey through the challenges faced by humanity today and how designers can address them.

Thorpe, A., *The Designer's Atlas of Sustainability*, Island Press (2007)

An informative book and guide to the need for sustainable design and design strategies.

Websites

www.350.org

350.org – an international campaign aiming to unite the world around solutions to climate crises.

www.designcouncil.info/inclusivedesignresource

www.designcouncil.org.uk/resources-and-events/Designers/Design-Glossary/Co-design/

The Design Council is a charitable centre that promotes the benefits of design to business, as well as helping designers to tackle new problem areas.

designmuseum.org/exhibitions

www.vam.ac.uk

A number of museums periodically have exhibitions on inclusive design or related fields, so it's worth checking their websites to see what's on.

www.ergonomics.org.uk

The Institute of Ergonomics and Human Factors often has links to interesting articles, as well as having up-to-date information on upcoming events and conferences.

www.hcdconnect.org

A website run by IDEO focusing on human-centred design.

www.hhc.rca.ac.uk

Helen Hamlyn Centre for Design – this centre pioneered inclusive design in the UK, and works with industry, academia and students to promote the uptake of inclusive design principles. It also hosts the Include conference every two years, which brings together the world's forward thinkers in inclusive design.

humancentereddesign.org

The Institute for Human Centered Design is an American non-governmental organisation (NGO) dedicated to advocating the benefits that design can bring in expanding opportunities for people of all ages and abilities.

www.ideo.com/work/human-centered-design-toolkit

Human-centered design toolkit – IDEO (2009)

An open-source toolkit to inspire new solutions in the developing world.

inclusivedesign.no

The Norwegian Design Council provides businesses with a comprehensive guide to inclusive design, so that they can understand its benefits and start to incorporate it into their companies.

www.inclusivedesigntoolkit.com

Developed by Cambridge University's Engineering Design Centre, this site gives information about inclusive design, as well as showing methods that can be used.

www.interfaceflor.eu/sustainability

Interface – a global leader in sustainable business practice and design.

www.nudges.org

An American blog on social well-being.

www.servicedesigntools.org

An open collection of tools to help you design.

www.storyofstuff.com

An inspirational, easy-to-watch piece of communication about the need to change the way we produce and use products.

www.sustainabledesignnet.org.uk

The Sustainable Design Network – an initiative bringing together like-minded people interested in sustainable design.

Places

The Centre for Sustainable Design

www.cfsd.org.uk

A centre that has helped business and organisations around the world understand the benefits of sustainable design and develop effective strategies.

The Design Museum

designmuseum.org/exhibitions

This museum has specific interest in show casing contemporary and historical design.

Design studios with a user-centred focus

Factory Design

www.factorydesign.co.uk

Harry Dobbs

www.harrydobbs.com

PDD
www.pdd.co.uk

Research Centered Design
www.researchcentereddesign.com

Rodd Design
www.rodd.uk.com

Smart Design
www.smartdesignworldwide.com

Sprout Design
www.sproutdesign.co.uk

Studiohead
www.studiohead.com

Vitamins
vitaminsdesign.com

We Are Human
www.wearehuman.cc

Career profile

Name:
Casper Gray

Current job:
Owner of Wax RDC Ltd

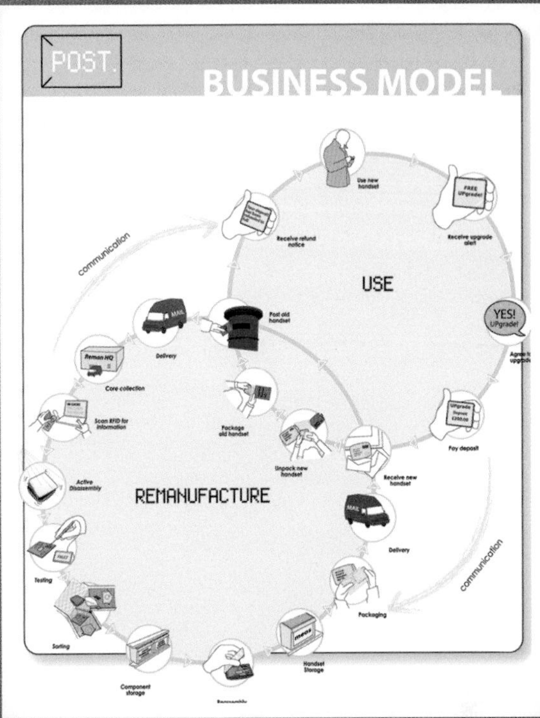

Business model diagram

Describe what you do

Since 2007 Wax RDC (Wax) has researched issues of practising design sustainably, in order to work out what designers can do, and how. We've designed products to help clients benefit from sustainable practices and worked in consultancies to share knowledge with others. This work has taken me to a wide variety of designers and manufacturers around the world.

Tell us about your career so far

At school and college I was interested in maths, art and languages. I wanted to go to university and follow these interests but didn't know what to study. During a 'year out' I travelled in Sweden, did administration work to save up some money and attended university open days. I visited different courses and eventually applied to a Furniture and Product Design course.

Going to university is a big investment and I was determined to make the most of it – I studied hard and had a lot of fun, I learned basic Italian and did a year work placement in Italy. Despite my interest, I was discouraged from specialising in 'sustainability' because of a lack of industry engagement at that time. I graduated with a good grade but, on applying for jobs, I realised that my tutor had been correct and sustainable design really was not on the design industry's radar. So I went back to university and studied for a Master's in Sustainable Product Design, where I met Damien Jones and we set up our design business, Wax.

Is there a particular project you would like to tell us about?

Exoteq, a small start-up firm, were interested in the potential financial and environmental benefits of remanufacturing their smart mobile handset, 'Meos'. Funded by The Centre for Remanufacturing and Reuse, Wax worked with Exoteq to design concepts for the product and how the business model would work if the product were remanufactured. We calculated the potential financial and environmental benefits and costs.

The concepts were evaluated through a simple **lifecycle assessment** (LCA). A selection of potential user scenarios was produced to cover a period of ten years.

> **Lifecycle assessment**: Okala is a simple, cost-effective LCA tool and guide that can be used for educational purposes.
>
> The Okala guide provides an introduction to ecological and sustainable design for practising and student designers.
>
> The guide is produced by and available from The Industrial Designers Society of America (IDSA).

This allowed multiple lifetimes of remanufacture and use to be simulated in an Excel spread sheet, and the environmental and financial benefits calculated and then visually communicated.

What advice do you have for students considering a similar career?

If you understand the issues of sustainability and can apply the techniques you will be an attractive potential employee. Your company will benefit from the insight you can bring to a project. Now, as a student in academia, you can often have more free access to information than when you are working in a company. During this time in your life, it can be a good investment to use LCA tools and remember the information so that you know it when you need it, because when you are working it can be hard to access this knowledge. So the start of your career is the perfect time to think a little about what you hope to learn and achieve from your studies.

Career profile

Name:
Sally Halls

Current job:
Designer, Design and Strategy, PearsonLloyd

Describe what you do

PearsonLloyd is a design consultancy based in London. I have a senior role in the company, working across projects to help shape the direction of the designs. I tend to work at the front end of projects, conducting research, analysing and dissecting results, exploring social trends and getting a good grasp of the context of the project. This helps the studio to create a coherent brief for the project, something that we believe is key to ensuring a good design. We're lucky enough to be able to work across a wide variety of interesting projects, from transport to healthcare to workplace design.

Tell us about your career so far

At school, I was always interested in design but combining it with double maths and physics for my A levels meant that I was recommended to study engineering at university. I spent my gap year working and travelling, and then went on to study Mechanical Engineering, with a year abroad in Japan. While studying, I heard about a course at the Royal College of Art (RCA), which focused on giving Engineering graduates the skills to become designers and fill that grey space between design and engineering. It sounded perfect! I pulled together a portfolio, applied and managed to get a place on a very intensive two-year course that gave me the basic skill-sets to be a designer.

It was while studying Industrial Design at the RCA that I became very interested in medical design and this in turn led me to engage in new design research methods. My graduation project looked at how to provide mums with better contact with their newborn babies while they are in incubators. This work brought me to the attention of the Helen Hamlyn Centre for Design, a design and research centre based at the RCA, working to promote inclusive design. After graduation I joined their research associates programme in their Design for Patient Safety team. The programme partners research associates with clients from industry and explores different problem areas through inclusive design methods.

Through my work here, I was able to work with the Department of Health, the National Patient Safety Agency and the Design Council and go on to win a number of design awards. It was a great experience, but after four years I felt ready to move on, and my experience in inclusive design enabled me to get a job at Panasonic, the Japanese consumer electronics manufacturer. Here I worked as a design manager, commissioning projects to design agencies across Europe, ensuring that the work and ideas were appropriate to the Panasonic brand. I eventually moved on to join PearsonLloyd in my current position.

Is there a project you'd like to tell us about?

I thought it would be interesting to tell you about some of Panasonic's work, so you can see how important inclusive design is in industry. I travelled quite extensively in this job, visiting design studios, conducting market research and doing lifestyle research in places such as Eastern Turkey, to understand better how people live, work and might use our products.

Panasonic is a multinational consumer electronics manufacturer, with its head offices based in Japan. It was rated the 89th largest company in the world by Forbes in 2009, with revenues of $104.88 billion in 2011. It makes a huge range of products, from audio-visual equipment to domestic appliances, to batteries and semiconductors.

The company has long been a proponent of what they term 'universal design principles', and all its products need to follow their six rules:

1. Easy-to-understand operation
2. Uncomplicated displays and indicators
3. Natural posture and ease of movement
4. Space to support easy access
5. Peace of mind and security
6. Consideration of how the product is used and maintained

When Panasonic decided to enter the European washing machine market for the first time, they knew that they'd have to have a strong product that stood out in the market.

It was through the consideration of inclusive design principles that the concept for the tilted-drum washing machine was born. Top-loading machines allow easy access into and good visibility of the drum, but they also make the laundry hard to reach at the bottom. Side-loading machines require the user to bend down to gain access to the laundry, as well as restricting sightlines into the drum. By taking the step of tilting the drum at an angle, the user becomes able to access the laundry easily, while still being able to see exactly what's inside and if any clothes are left.

The tilted drum also created further benefits to the user, as it was more water-efficient, especially for smaller loads. And it also meant that users were more able to add items, midway through the laundry cycle.

The designers further increased the usability of the product by considering how people load and unload their laundry, and how the design could better facilitate this. Firstly, the door height was raised from the position of a conventional side-loader, to ensure all users could easily access the door area. The size of the door opening was also increased, so that clothes could be loaded and unloaded through the opening far more easily.

The opening mechanism was changed to a push-button operation, so that you could open the door and load the machine with armfuls of laundry. Finally, the hinge was upgraded to a double hinge structure, so that the door was able to open fully through to an angle of almost 180 degrees. This enabled the washing machine to be approached from either side by wheelchair users, as well as helping prevent users from banging their elbows against the door – thereby making the drum easier to access.

The washing machine was a big success, with greater demand than anticipated across the European market. Universal design principles are employed with the same enthusiasm throughout the company's design facility.

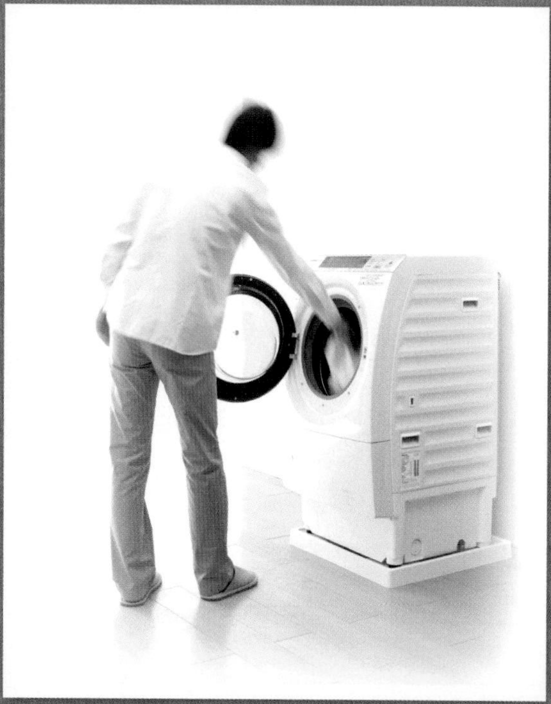

The benefit of the higher, tilted drum
Panasonic Corporation

What advice do you have for students considering a similar career?

Retraining to become a designer was the best decision I ever made! Every day is completely different and brings unique challenges that I wouldn't ever have anticipated! Of course it's hard work, and the hours are long, particularly when you're starting out. However, creating something from scratch, developing and growing it into something that can be held and used is an incredibly rewarding and satisfying experience.

To be a good designer, you need to have enthusiasm by the bucket load! As a designer, you're not only the creator, but also the marketer and the salesman for your design. So you need that enthusiasm to be able to sell your idea, and sell it well, ideally with a rationale behind why it's that shape/colour/design!

However you also need to have enthusiasm for learning, because there'll always be new software you suddenly need to learn, new project areas that will come in that you'll quickly need to become expert in and, most importantly, you'll be constantly learning how to be a better designer.

Career profile

Name:
Amy Ricketts

Current job:
Design Intern, Innovation Unit

Tell us about your career so far

I originally worked as a graphic designer upon graduation from a BA (Hons) course in Visual Communication. I worked with a fellow student on a variety of client-based and self-initiated projects, ranging from branding and website design, to album artwork and an interactive window display. I also assisted the teaching of a foundation course in Graphic Design at the University for the Creative Arts in Farnham. I then went on to study for a year on a Master's course in Service Design. This introduced me to the idea of design thinking – design as an approach to problem solving.

Since joining Innovation Unit, I have been involved with a range of projects in a number of ways, including conducting user research, producing learning materials and tools, writing case studies and supporting workshops.

Is there a particular project you'd like to tell us about?

During university, my final project looked to understand and communicate value within the National Health Service. I carried out research with patients and clinicians to capture the different kinds of 'value' through interviews and activities, asking people to draw, describe and rate aspects of the Health Service. This research phase gave me great opportunities to think in new ways about how to understand people and their behaviours.

What advice do you have for students planning a career in this area?

Be curious!

Be interested in things that are not design-related. Designers often seek inspiration by looking at the work of other designers and artists. This is great for your work to develop aesthetically, but if you want your work to also have a real social significance you will need to get curious about the social issues that affect people around the world.

Get passionate about those issues and try to understand as many perspectives as you can by keeping up to date with the political developments surrounding them.

Ask questions!

Suddenly having to work on really complex social issues (with people who have spent years trying to get their heads around them) can be a bit disempowering. But think of your naivety as an asset. One of your skills should be asking why things are the way they are and how would things be if they were different?

By asking questions you will push the experts you work with to express their views and the challenges they face in simple and refreshing terms. This is a very powerful intervention in itself.

Career profile

Name:
Fan Sissoko

Current job:
Service Designer, Innovation Unit

Tell us about your career so far

My background is in graphic design. Probably as a result of growing up in the multicultural suburbs of Paris, I have a strong interest in the blurry areas of cultural identity and social cohesion, and believe that communication design should move beyond public or corporate communication, to serve the wider purpose of enhancing communication between people and communities.

I feel lucky that my first job after graduating was with PCC, an Irish communication agency working exclusively with progressive, non-profit organisations. The clients for whom I designed logos, publications and websites ranged from small community groups to national charities and governmental organisations, such as the Department of Education.

This experience motivated me to explore how I could apply my creative skills to social and environmental challenges in more extensive ways. So I moved to London to complete a MA in Design for Development at Kingston University, which gave me some strong theoretical insight about social change and sustainability, as well as the confidence to define what I had to offer to organisations that want to change the world, beyond simply graphic design.

I now work for Innovation Unit, a social enterprise committed to using the power of innovation and creativity to solve social challenges. I am part of the service design team. Our aim is to design better public services, from education to healthcare. While my work mostly consists of visualising new services and making new ideas engaging through design, illustration, photography, film and animation, I also spend a fair bit of my time doing research with people who use and deliver services in order to understand the challenges we work on from their perspective.

Is there a particular project you'd like to tell us about?

My MA research focused on the power of food to bring people together, and to engage diverse communities on sustainability issues, so after I graduated I decided to set up a community kitchen using food surplus in South London, where I live.

What advice do you have for students planning a career in this area?

Every designer should be socially responsible. You might not decide to become a service designer, or you might not have the opportunity to work on social innovation projects, but that does not mean that social responsibility should be left out of your journey. Designers are in the privileged position to be able to transform the world around them, through objects, through spaces, through communication materials, through services. Every project has a potential social impact.

When you are about to begin a project, start by asking yourself if what you are designing could have harmful consequences to people or to the environment? Ask yourself if the consequences of your work are in tune with your own ethical values? Try to use your ingenuity to address ethical issues if they arise.

University is a great place because, to some extent, you are free to give your projects the direction you want them to take. Keep in mind that this will not always be possible in a professional context and make the best of this freedom. Use your course as an opportunity to shape your personal values around all that you do as a designer to explore social issues that you are passionate about.

Career profile

Name:
Hannah Lobley

Current job:
Environmental Paper Artist and Owner of 'Paperwork'

Adding personal value to old books

Describe what you do

After accidentally leaving a book out in the rain in 2002, I developed the internationally exhibited and award-winning 'Paperwork' – a unique recycling technique using the printed pages of unwanted books and paper. The pages are layered and transformed back into a solid, wood-like material. Traditional woodworking methods (lathing, sawing and drilling) are then used to create interior objects from that material. The surface patination of the paper when the objects are worked echoes wood grain. *Wood becomes paper becomes wood.*

'Paperwork' is committed to the sustainable agenda as it combines traditional wood working with recycling techniques; the material and objects produced are very tactile, open and adaptable.

As it is a recycled product, a customer can determine the book or papers from which the object can be made, so the product can retain the book's original sentimental value or context.

Tell us about your career so far

I studied Decorative Arts at Nottingham Trent University, graduating in 2001. This was a great course for me as it explored many different materials; I always worked three-dimensionally but hadn't found a material to specialise in, but eventually I found a passion for wood working. After this I went on to complete a Master's degree in 2004; during this period of study I developed

Paperwork bowls turned on a wood lathe

'Paperwork' and I established my business the following year, operating within the East Midlands.

'Paperwork' products have been recognised for their unique qualities, winning accolades including:

- 2010 Winner of Creative Nottingham Image Competition
- 2008 Best-in-Show Eco-Design Award, Liverpool
- 2007 Rowan Best Business Innovation Awards, third place
- 2004 Winner of Channel 4's Ideas Factory Creative Class

'Paperwork' has exhibited at some prestigious events, such as 100 per cent Design and Origin at Somerset House. A range of my products were showcased by the Crafts Council at the Victoria and Albert Museum shop and 'Paperwork' has also been on sale in Liberty in London. In 2008 I made my television debut after appearing on *BBC East Midlands Today*, where I was interviewed for a recycling feature.

I hadn't envisaged any of the projects or adventures I have encountered in my career but I wouldn't change them – they have been unique and interesting and my practice and clients continue to surprise me every year. Running a creative business you need to be adaptable and prepared for the unusual, it's great! You never know what or where the next project will be.

Is there a particular project you would like to tell us about?

As a result of the Asian interest in the products of 'Paperwork' and the concept behind them, I travelled to Tokyo again in September 2008 to complete a large-scale, site-specific commission. I created a huge suspended paper sculpture that is permanently featured in the largest shopping mall ever built in Asia. I received the commission because the concept of Paperwork is eco-friendly and sustainability was part of their philosophy also.

I have also recently completed an amazing project for an establishment within the UK Ministry of Justice in which I have been asked to develop a 'Paperwork' piece that will enhance the health centre, helping to make it more of a 'healing' environment. Considering the patients' well-being and responding to a socially responsible design brief has been an interesting challenge.

What advice do you have for students considering a similar career?

It's hard work but could be the best thing you ever do. It was for me. Be prepared that making your work is such a small part of running a creative business; you have to learn marketing, accounting, correspondence etiquette, sales, to name just a few. I would advise joining creative groups, going to all the networking events and meeting as many people as possible. The creative sector is a very generous sector and fellow business people are always willing to help and share knowledge.

A very important point for me is that I have two strands to my business – selling through retail and working to commission. The retail products I produce provide me with a regular income that sustains my business and help to promote my work nationally and internationally to new audiences, demonstrating the materials and processes I use. Working to commission is where I can be more creative and add innovative and exciting elements to my working practice. The sculptures can therefore become more daring.

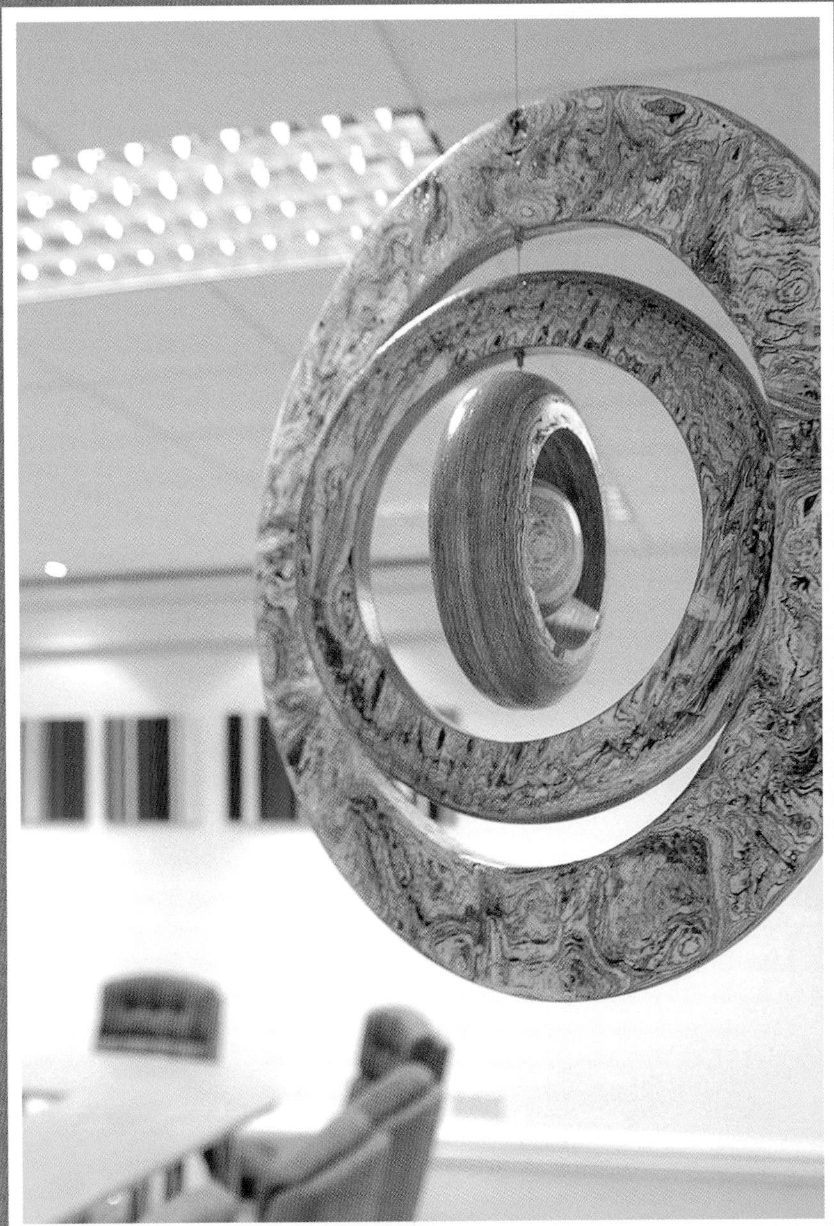

AEON Group commission – 'Paper Swaying'
Photograph by Dan Lane
www.dl-web.net

Project story

Designer:
Anna Piper, textile design student

Project title:
Visualising Music: Seeing Sound

Tell us about the project

For my final exhibition statement I really wanted to bring together all the areas that interest me and the skills that I had developed so far, to push them further, challenge myself and have fun. I wanted it to be a statement that represented me and the things that are important to me as a designer. A key part of this was to work towards a sustainable outcome. This shaped the project and sparked the idea – to produce engineered, waste-free, woven garments. Having made clothes in the past, it seemed feasible to combine dressmaking principles with the weave techniques I had explored in my second year to create a simple garment.

'Visualising Music: Seeing Sound' was inspired by the process of data visualisation and the concept that inspiration for design can come from and capture something that is non-visual. I had taken the decision to have one theme that tied together my two final-year projects, but to explore two very different aspects within each to give my portfolio both diversity and cohesion. Part one used colour, pattern and image to convey the structure and emotion of music. Part two focused on translating the music-inspired imagery into 3D form, exploring the relationship between form, surface pattern and colour to convey movement and rhythm. The challenge was to translate and recreate the shapes and structures generated through material manipulation into woven fabric.

The development process would allow me to experiment with material properties and weave structures to generate movement and form, and to translate my 3D visual inspiration into a 3D outcome. So the project had started to take shape, but first of all I had to establish what being sustainable really means!

What processes and skills were most relevant to this project?

Being technically minded I enjoy the process of research, experimentation and analysis, but in this particular case I had not anticipated the complexity of my undertaking. Not only did I have to test and control the shrinkage of yarns to pleat and gather the fabric, design garments and work out how to construct and fit them and develop pattern placement, colour proportion and yarn combinations, but I also needed to do this sustainably – working to eliminate waste.

For me, working sustainably is predominantly focused on finding ways of reducing the impact of my work on the environment, by choosing materials and processes responsibly. On the surface this appears to be relatively simple – achieved by selecting natural yarns, not being wasteful and sourcing responsibly. But what happens when natural materials don't meet your needs? What if you think your discipline is inherently wasteful or you are unable to establish the origin of your yarns? This can all be further complicated when you read and research more, as I found out!

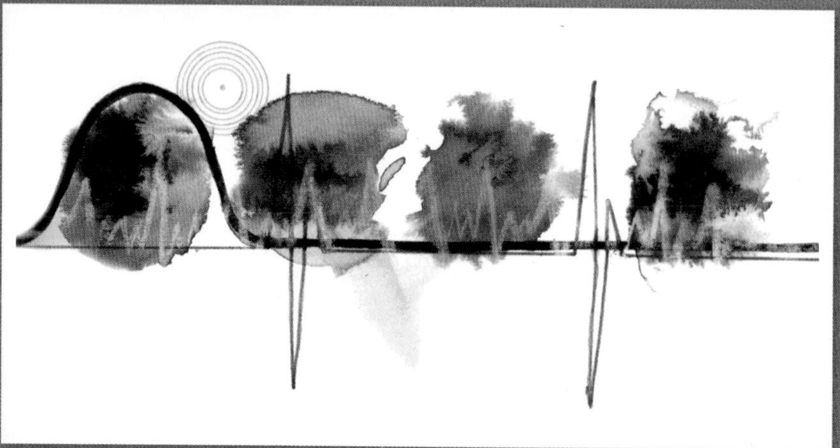

An early visualisation of sound waves

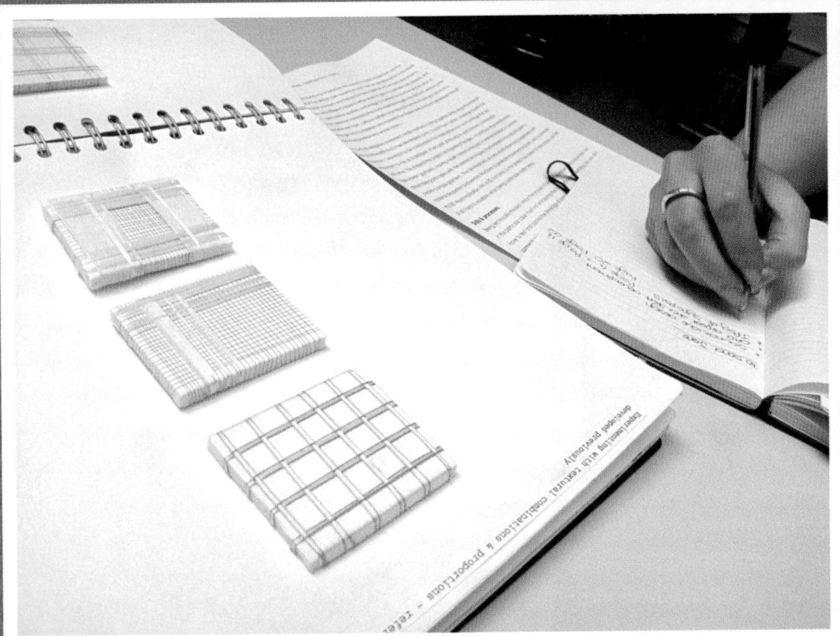

Translating patterns to woven designs
Photo: Anna Piper

Waste use & reduction – final warp & garments:

- The final warp was based on the last sampling warp (on which the prototype skirt was produced) & was used to produce all 6 garments.
- The block threading combining silk & monofilament provided versatility allowing 2 garments to be produced using the full warp (including monofilament). The monofilament was then removed for the final 4 garments. A much smaller monofilament warp was wound to accommodate this.
- The warp calculations were surprisingly accurate – the above photograph shows the amount of warp remaining on completing the final garment.
- Some of the remaining warp has been utilised in the finishing of the garments e.g. to hand stitch the seams.
- If the production & development process was to continue this

Conclusions & evaluation:

- Although the initial intention was to produce a waste-free garment, I soon realised that this was an unrealistic aim & that eliminating waste was a much more significant undertaking than anticipated.
- Working towards the elimination of waste as part of a waste-free process became the aim & I have developed a process that is moving in the right direction. A collection of garments & samples has been

A section from Anna's report into sustainable production

What challenges did you face?

With one research source championing the sustainability and biodegradability of natural fibres and the next highlighting the negative impacts and resource consumption of their processing, I was presented with questions, rather than the answers I was looking for:

- Does a waste-free production process really constitute a waste-free garment?
- What about the waste generated by the garment once it has been worn and discarded?
- How much waste is generated during yarn production?

. . . the list goes on.

Despite the barriers and confusion, decisions needed to be made to progress the project and I had to be realistic about what I could achieve while still adhering to my initial goal of a sustainable outcome. By setting parameters and identifying key principles to inform my decisions, I was able to balance aesthetics, technical requirements and sustainability.

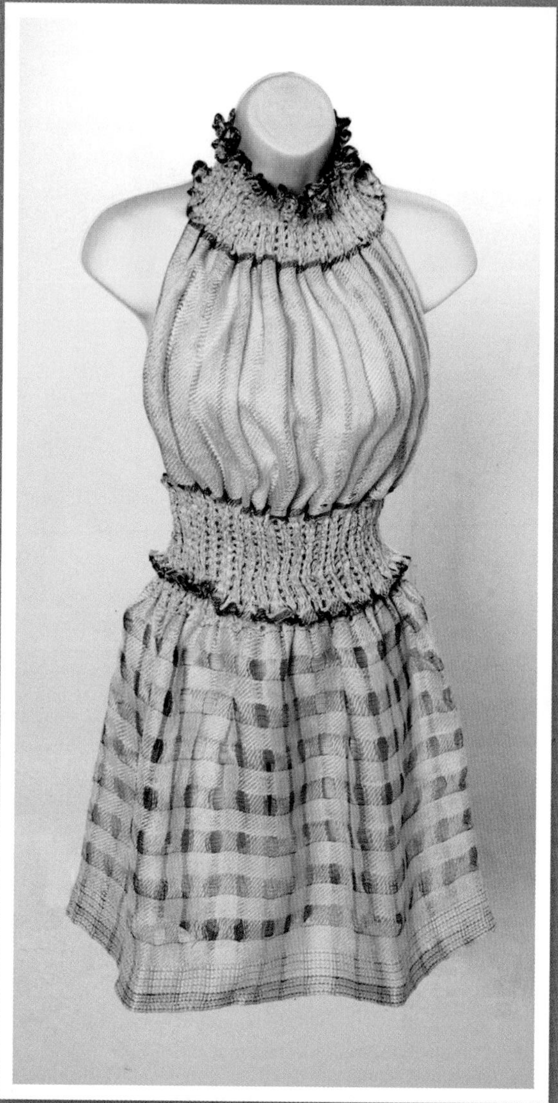

One of the final collection of garments

What were the highlights of the project?

I didn't achieve the waste-free garments I was aiming for, but have taken positive steps towards it, learning a lot along the way. I have developed a framework and processes that will guide my future work, they will continue to evolve into an increasingly sustainable practice and I am confident that my revised aim – to produce waste-free garments as part of a waste-free design and production process – will be achieved. In the meantime, I am satisfied that I made the best-informed and most responsible decisions possible for this project to ensure an effective balance was struck between good-quality design and sustainable practice. Most importantly, I produced an exhibition statement – a collection of six hand-woven engineered garments – that I am proud of.

Project story

Designer:
Sally Halls, National Patient Safety Agency/ Helen Hamlyn Centre for Design

Project title:
Inclusive design guidelines for the packaging of injectable medicines

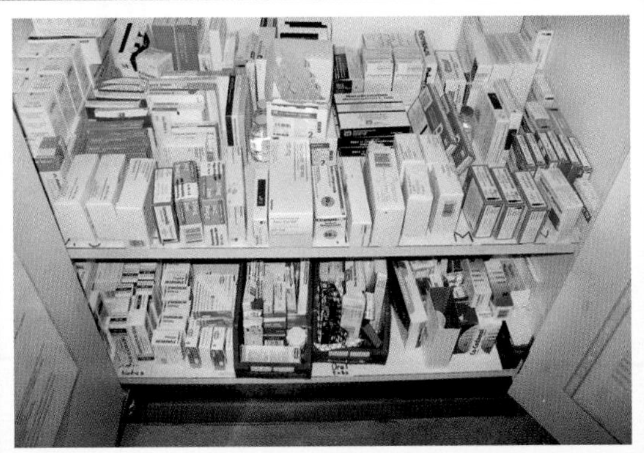

The problem: confusing but important information in a drugs cabinet

Tell us about the project

Every day thousands of patients will receive their medication, of which a small minority will receive them via the injectable format. But the National Patient Safety Agency (NPSA) found that a disproportionate number of accidents were happening with the injectable medicines, where, for example, the wrong drug was given, or it was given at the wrong dosage or concentration. The packaging design was thought to be a large contributing factor to this, so we were asked to create some design guidelines on how they should be designed.

What processes and skills were most relevant to this project?

The NPSA were able to put me in touch with hospital pharmacists, and the first weeks were spent visiting them. Medicines are normally administered to the patient by a nurse or doctor, so time was also spent on the wards observing them at work, understanding where the drugs were stored and talking to staff about the process they go through in order to administer medicines.

Many medicines are patented by the manufacturer to protect the idea, so no other company can make the same medicine until the patent expires. When it expires, any pharmaceutical company can make it, which means that the same medicine can be bought by a hospital from a variety of different manufacturers. The manufacturers are keen to differentiate themselves from each other, and they do this through developing distinctive branding on the packaging, just as we would if we were selling breakfast cereals.

Unfortunately, this often happens at the cost of highlighting the crucial medicinal information. So, for example, if the company's name is in a much larger font than the medicine's name or other information that a nurse or doctor might need in order to administer it, it will be more difficult to read and this is more likely to lead to mistakes.

Each issue was identified and illustrated in a guideline document for the NPSA, showing both the existing standard practice and also highlighting what the best practice should be. As there are a number of different packaging formats, a chapter was dedicated to each one. For example, the design guidelines for a 1ml ampoule, which is very, very small, will need to be different from that of the surrounding box.

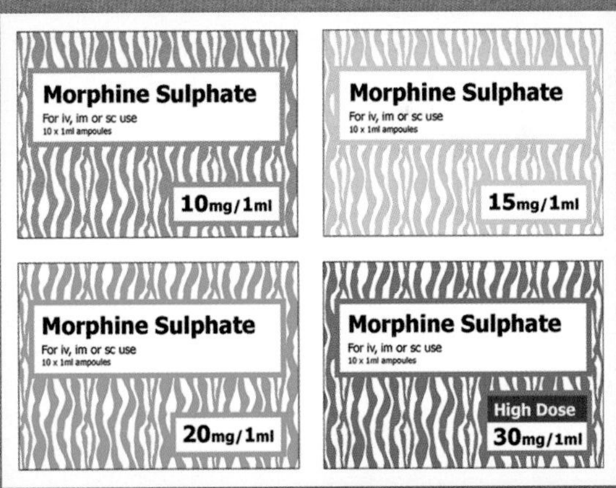

Clear, unambiguous information

What challenges did you face?

As the design guidelines would be issued by the NPSA and Department of Health to pharmaceutical companies operating in the UK, they had a very strong and vested interest in what the guidelines would be recommending. It was therefore important to liaise with them, and ensure that any recommendations I wanted to make were feasible and would actually be implemented by the pharmaceutical companies. This required a careful approach to the information hierarchy on the pack, with any branding information, such as manufacturer or branded medicinal name, balanced with the need for good, readable instructions and dosage information.

What were the highlights of the project?

The guidelines were issued as a publication and distributed to every single pharmaceutical company in the UK. They were also distributed to all the NHS purchasing teams, giving them the knowledge to be able to choose the medicine supplier with the safest packaging designs. By raising the expectations of the buyers it had a big impact on the pharmaceutical companies, and new packaging designs have improved considerably.

The guidelines can be downloaded from here: **http://www.nrls.npsa.nhs.uk/resources/?EntryId45=59831**

Recommendations

- Print the medicine name longitudinally, along the length of the ampoule.
- A good rule of thumb is: if the visible width of the label is less than the height of the label then the name should be printed longitudinally.
- The information that must be present on containers 10ml or smaller is:
  - Medicine name
  - Expression of strength (where relevant)
  - Route of administration
  - Posology (for self medication)
  - Warnings
  - Expiry date
  - Batch number
  - PL number
  - MA holder's name

Recommendation documentation

Project story

Designers:
Amy Rickets and Fan Sissoko, Innovation Unit service design team, UK Government/ Innovation Unit

Project title:
Service design for public services: designing an intensive, community-based service for re-offenders with a substance misuse problem, essentially an alternative 'path' that re-offenders might take, avoiding short-term prison sentences.

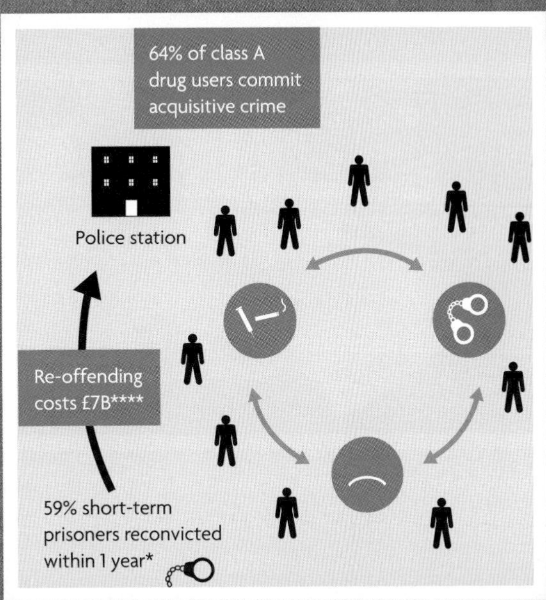

The problem we were trying to solve

Tell us about the project

Research suggests a high correlation between class A (heroin, ecstasy, cocaine, etc.) drug users and acquisitive crime (theft, burglary, vehicle crime, fraud, etc.), with 59 per cent of short-term prisoners being convicted again within one year, at an estimated cost of £7 billion to UK taxpayers.

As a result, the Ministry of Justice became interested in exploring how innovation might be used to reduce prison overcrowding, repeat offending and associated costs. Innovation Unit was commissioned to do this, working over a period of six months, in conjunction with West London Magistrates Court and the London Borough of Hammersmith and Fulham.

Our brief was to develop a new model of service that could be implemented and scaled right across the UK. As far as service design projects go, this is one of the most exciting and challenging projects we've been involved in. Exciting because what we've designed could have a massive impact on reducing the costs associated with prosecution and custody of re-offenders. But, more importantly, because of the positive impact a better model could have on the lives of offenders and their families in breaking the vicious cycle many of them are in.

What processes and skills were most relevant to this project?

Service design has its own methods that allow designers to work through often very complex logistical and human problems in order to find the solution. They involve focusing on both the technical aspects of how a service is delivered, as well as how people think, feel and behave.

This project followed a pattern of phases that are based on our own interpretation of the Design Council's 'double diamond model', described earlier in this section (Discover, Define, Develop and Deliver). Here we will explain our methods within the context of this particular project, in roughly the order that they were used:

Discover

For this project, we conducted a number of full-day ethnographies with users of class A drugs. As researchers, we created bespoke tools to build trust and uncover stories. These included mapping of personal journeys (what they do and where they go), income and spending and social network and neighbourhood mapping (who they are in contact with). This was carried out in hostels, drug treatment centres, cafés and the homes of their family and friends.

Define

Over a number of weeks we ran a series of intensive sessions to look at our findings from different directions, inviting different people to join us and using creative methods such as journey mapping, role playing, personas and brain storming to converge on a solution.

Co-design:

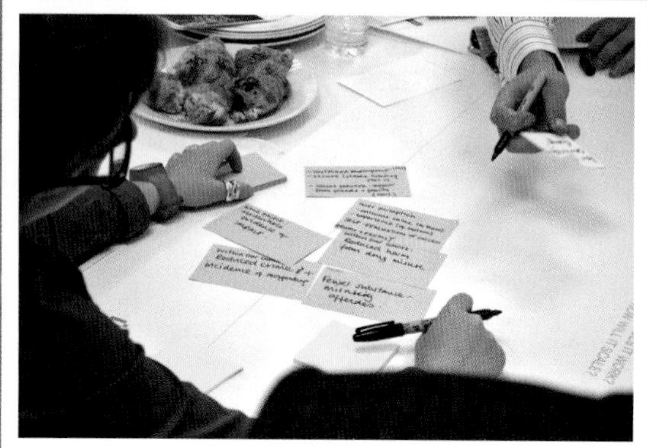

Brainstorming exercise during co-design workshop
Innovation Unit

One project involved a co-design workshop with a mixed group of service providers, including probation officers, members of the police, a district judge, drug workers, ex-users and carers. Other stakeholders, the leader of the host council and representatives from the Ministry of Justice and the government were invited to view the process so they could consider how the discussions might inform their wider policy decisions.

Prototyping:

Using Lego to prototype the funding process of the new service
Innovation Unit

Lego was used to visualise and analyse the relationships and issues within the funding of the service. It allows people to share understanding and get involved in modifying the prototype to see what the difference might be between different ways of implementing the concept.

Develop

Within this stage of the project, we organised a number of meetings with experts to begin to 'flesh out' some of the issues, including those around drug treatment, housing eligibility and the process of police referrals. This then allowed us to work towards creating a blueprint for a service that would provide holistic support to drug users coming out of prison and facilitate their journey into stable housing and employment.

Deliver

The final stage consisted of building the business case for the solution, and enabling partnerships between the different organisations that will deliver it.

We believe that the above story is a perfect example of the fascinating and incredibly challenging design process that can be followed, demonstrating an insight into how design really can play a wider, more important role within social change.

What challenges did you face?

We found particular difficulties in trying to uncover the kinds of data we were looking for, including which services were currently available, their cost to the taxpayer and impact on the users. Trying to get the information that allowed us to understand how each element linked together was difficult but nonetheless crucial to the success of the project.

The fact that we were trying to liaise with people who had extensive expertise in areas that we knew very little about provided a challenge for us, as we had to really prove to them the value of involving designers to improve and support the service.

What were the highlights of the project?

One of the most rewarding highlights of the project was to conduct and work through a rigorous research process that uncovered a fascinating number of insights, such as the importance of stable housing for people to break out of the offending cycle.

Another was the running of the co-design sessions and the outcomes. These involved a huge amount of preparation and build up, resulting in a fantastic buzz of energy during the workshops themselves.

Thirdly, as the project outcomes have been shown to people and organisations who are responsible for these services nationally, there is a real possibility of reproducing this work in other areas of the country. The potential for this to be rolled out to reach a wider audience is now a very real possibility and is proof of the project's success.

Project story

Designer:
Chloe Muir, interior architecture and design student

Project title:
Memory Lane

Tell us about the project

The direction and focus of the project is motivated by the facts and realities of war and the thousands of servicemen and women who put their life at risk every day in order to fight for their country. These individuals leave behind husbands, wives, partners and children and it is these people in particular that help form the basis of the scheme. Memory Lane aims to establish an intervention that helps to assist the ways in which families of the fallen are aided through the bereavement process. The scheme will provide an intimate welcoming environment, allowing those who wish to, to escape from the mundane activity faced within day-to-day life and enter into a new comforting family community.

Memory Lane will encourage social interaction, helping families to meet other individuals in similar situations, with the intention to assist one another through the various stages that losing a loved one can entail. A combination of spaces innovatively designed to cater for families at different stages of the bereavement process will assist in enabling families to deal with the stages of rehabilitation at their own individual pace and comfort.

The site, located in an industrial area of Birmingham, functioned as a gun proofing house, first established at the peak of the English gun trade in 1831. In contrast to its dominant façades and heritage, the new function allows the building to, in a sense, almost redeem itself by helping to re-piece the fragmented lives of the individuals within the scheme, who have indirectly been destroyed by war and ammunition.

When a loved one is lost, the individual's ability to deal with grief often becomes an insurmountable task, making many aspects of life difficult to handle. This is a process that is extremely personal, which can result in this grief being hidden away from the outside observer. These hidden elements can often cause the individual to feel as though they are existing between two worlds, neither belonging to the past nor the future. Over time these broken, missing pieces and faded memories within the individual find their way back together, initially helping to rebuild lives and the person's ability to cope with the bereavement process.

'Acceptance Chapel' illustration

The visual above portrays the Acceptance Chapel of the scheme. The building is merely a shell, stripped of all doors and windows, meaning the space is open to the elements and allows unrestricted views into the space. It has been designed this way in order conceptually to represent the idea of an individual opening up and allowing their emotions to be viewed by others – something that is often difficult during the early stages of grief. A new shell has been inserted into the existing façades, exposing half of the original brickwork on one side, representative of the past, and covering the remaining half, representative of the future. Large linear façades have been inserted drawing the light down into the lower space of the chapel, guiding the visitors up towards the acceptance bridge intervention over the canal.

What processes and skills were most relevant to this project?

It is extremely important to know your user and who you are designing for. Without knowing your target audience, you will end up designing a project that is completely unusable for the client. It is also important to know exactly what the user would need and want from the intervention you are designing. It is often easy to get caught up in what *you* think that the user would need and gain from your intervention. Without this first-hand, primary research, it is inevitable that you will design an intervention that will not be successful and that will appear to be very sterile and perhaps, however unintentionally, offensive.

What challenges did you face?

It was extremely important to design elements that didn't patronise the user or force them to feel a particular way at an inappropriate stage of the bereavement process. Instead, through conscientious design, the intervention was created to influence the user to open up to their emotions and feel that it is acceptable to feel the way in which they are feeling.

Bereavement affects individuals in different ways and there is no definitive route that people will follow. Therefore it was important to allow the user to develop their own journey and options within the intervention, enabling them to deal with the stages of bereavement at their own individual pace and comfort.

What were the highlights of the project?

There were many highlights of the project, the key one for me being the whole conceptual stage of the project. Without the conceptual stage, the project wouldn't have developed into the end result it did and would have merely ended up an obvious, uninventive, sterile project.

The conceptual stage of the design process enables the designer to think outside of the box and really be creative. A spark of imaginative thinking at the beginning of the project can really develop into a significant feature that can then underpin the whole project. Memory Lane incorporated the concept of broken, fragmented memories and how these memories often become faded over time.

So my key advice to anybody who is considering taking on a project like this is be conceptual. The small conceptual aspect that the designer may think is irrelevant could turn out to be a feature that can essentially underpin the whole project.

The presentation at the end of the project was extremely rewarding. To stand back and look at all you've achieved and see the elements that have evolved from small conceptual aspects is fundamentally rewarding.

What's next?

Part four

15

So, where is this going to take you?

By Jane Bartholomew, Nottingham Trent University

'After university, I decided to apply for jobs based on their job descriptions, rather than their titles (not letting these put me off) and I sent off endless application forms. It goes to show, though, that persistence does pay off – even when you get no reply from 80 per cent of the applications (which can be disheartening), getting that one reply saying you have an interview is amazing!'

Jenna Wright, assistant press officer, Next Plc

New Designers annual graduate exhibition at the Business Design Centre, London

If you are currently doing the first year of your degree, you may end up flicking through this section of the book thinking to yourself that graduation is a long way off and you don't need to concern yourself with this information right now. When the graduates from the course I teach on return to the university to give a talk to the current students about their careers so far, they often mention that they wish they had done more, during their studies, to find out about the wide variety of career possibilities that they now know exist in the industry. Even the graduates with a good number of relevant work experiences on their CV, and a good competition success behind them, found that they only had a fairly narrow view of the possible job opportunities open to them.

As first-year students, the majority of you will arrive with a clear goal to become a 'designer'. During the second year of your studies you will have many other potential career opportunities introduced to you as you become more aware of the types of roles within the industry. You will probably undertake live projects, attend trade fairs and exhibitions and learn from visiting lecturers. These experiences will provide you with a broader understanding of what else might be on offer. By the time you enter your final year you will have a fairly good idea of the specific skills and motivations that are really driving you and the way you work best. Spending time working out how these can be incorporated into your career decisions will result in you feeling comfortable with the choices you make.

> *'I don't think you can underestimate the importance of getting work experience. Employers look for relevant experience of working within design to show your passion and commitment alongside your studies. Without having various placements on your CV, it may prove difficult to find paid employment in this industry straight out of university.'*
>
> Adele Parsons, graduated from an MA in 2009

At the end of this section you will find career profiles and stories by recently graduated students and industry specialists who are all keen to give you an insight into what the industry is like from their perspectives. They are full of practical guidance that will capture your imagination and are sure to set you off investigating specific areas of the industry that you probably know little about at this time.

Design degrees can lead to a myriad of careers too numerous to mention, but what this section is attempting to do is offer you an insight into the types of roles and businesses within the creative industries and provide you with the tools to find out more about the areas that interest you.

Give yourself a head start by committing to finding out about your own skills, strengths and weaknesses, as this will help you understand what types of roles and ways of working you might be suited to in the future. Many degree courses are committed to supporting your understanding of what you might do when you graduate, and often have career development sessions built in to the curriculum or a career service keen to support you.

*'Starting my job was definitely one of the scariest days of my life – it's one thing selling yourself but an entirely different thing proving that you deserve the position. For me, I knew very little about **PR** and so felt like I was starting university all over again. But looking back to that first day, just three months on I'd learnt so much already and did realise that a lot of skills that I had learnt studying design were invaluable. For instance, I have many tasks to handle at once – including endless requests for product samples and images from journalists, along with having to be proactive in thinking up ideas for how to promote the ranges that I help look after. My role involves writing mini press releases, assisting with events and arranging photo shoots.'*

Jenna Wright, design graduate 2008

Allocate real time in your diary so that you can tackle the subject of 'your career', as it is not a quick task to work out who you are and what you might want to do in your future and who you want to become within the industry.

PR (Public Relations): this is a role that involves managing the communication of a company's or designer's image and ethos to the outside world.

Make the most of your tutors and visiting lecturers to find out as much as you can while studying for your degree. The majority are practitioners and specialists as well as teachers, and have more knowledge than you realise, so, when you can, pick their brains for advice and information.

If you are keen to continue studying and enjoy asking questions of your practice, materials and processes, then deepening and broadening your understanding by doing a postgraduate course might be something to find out more about by reading this section. Each of the following sub-sections will offer you many ways in which you could do further research into what you could do next:

- postgraduate study
- jobs market
- marketing yourself
- working as a designer in the industry

For those of you contemplating working for yourself in the future, then you will find some great advice and practical approaches in Chapter 16, 'Working for yourself', to help you understand what it takes to be an entrepreneur. It also includes a very useful guide on how to protect your design rights.

Postgraduate study

By Dr Lynn Jones and Dr Alison Shreeve, Buckinghamshire New University and Caroline Norman, Birmingham Institute of Art and Design, Birmingham City University

After graduating you may decide to get a job and gain industry experience, or you may prefer to continue your studies and apply for a postgraduate course. Some students find it very useful to have a couple of years in the industry before studying for an MA, as this provides an important context from which to undertake deeper levels of studies into a specific area of design practice.

For some students the experience of engaging in more in-depth questioning about materials, processes, aesthetics or social design issues may lead on to a desire to undertake doctoral study. These opportunities are often associated with larger research groups in educational institutions where particular themes are explored. For example, certain institutions within the UK have particular focuses of research interest, such as sustainability, graphics, textiles, furniture, interaction, transporting goods, and so on.

Sophie Alidina studied Smart Design at postgraduate level and worked on a collaborative project with industry. The project was called 'Living Lace' and her plan was to reinvent the lace curtain and explore new opportunities to reduce pollution in the home. The concept and rationale for the project was fuelled by the enhancement of people's well-being and involved detailed investigations into a range of technologies, processes and materials. Her first task was to identify and develop the concept.

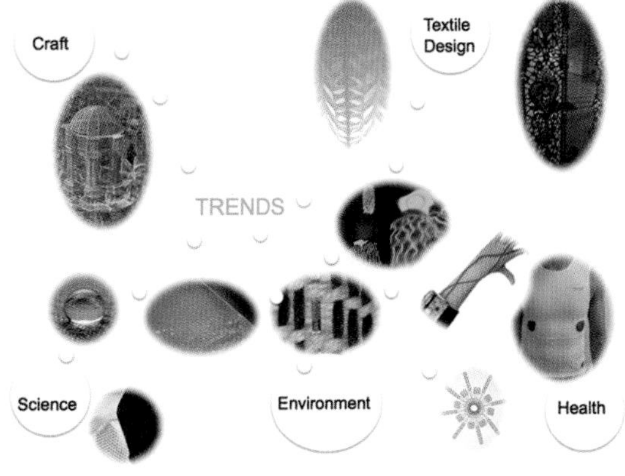

Visual representation of the trends underpinning the design project by Sophie Alidina

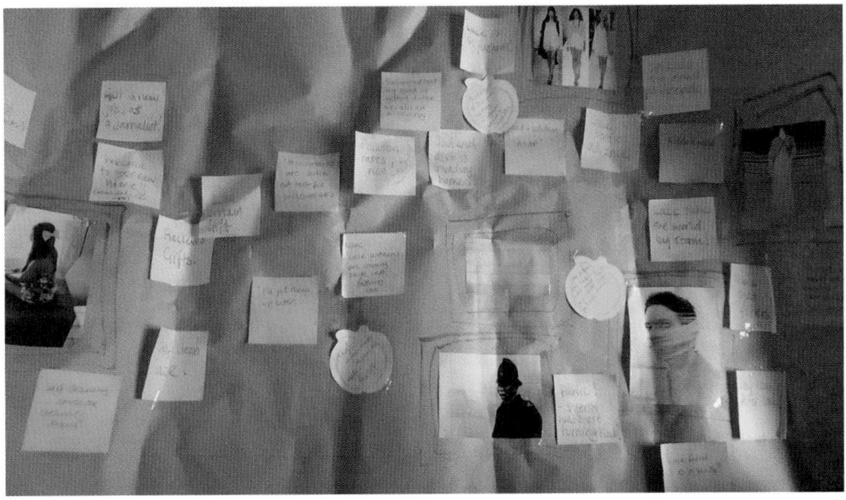

Concept and idea generation

Sophie notes:

> 'The project was based on a future-facing narrative that I wrote about lace curtains. In my narrative, lace had become a way to filter clean air into people's homes because our country has become over polluted. The story was a combination of utopian and dystopian ideas. The utopian aspect was lace benefiting people's well-being.
>
> I explored this narrative by creating laser-cut lace fabrics that represented those from my story. The fabrics were tarnished on the outside, having consumed pollution – however, they kept their filtering function. I used SMA (shape memory alloy) wire to explore the opening and closing action of the laser-cut designs. The wire was sewn into the fabrics. When the fabrics warmed up, the heat made the sections move slowly up and down.'

Postgraduate study is becoming more sought after as it encourages individuals to develop original ideas and undertake a deeper level of research into the subject area.

Wing patterns for flapping origami crane

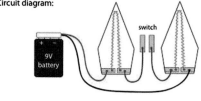

scotch tape
copper tape
muscle wire
crimp beads
thread

Circuit diagram:

9V battery
switch

SMA circuit

Applying for postgraduate study

In order to apply for postgraduate study, you will need to have an idea of what you are hoping to achieve by the end of your course. Some courses might ask you to submit a proposal of study that outlines the type of project that you want to do. Most courses also require you to have achieved a first-class or upper-second degree qualification, but this isn't always the case so check the entry requirements and application procedures carefully.

Postgraduate courses are available in both full-time and part-time modes at many institutions, and some may contain periods of time studying and practising abroad while others may have placement opportunities built in. Think carefully about the amount of time you can dedicate to studying, as you may have a part-time job or other commitments that will reduce the amount of time you spend studying.

Grey felt and gold polyester laser cut design by Sophie Alidina

Practice-based courses

Practical design courses are clearly focused around 'making', and your time in the workshop will be important to the development of your project.

James Poole studied MA Product Design and Innovation Management. His focus was researching perceptions of quality and longevity in his product and furniture designs. This examination led to the design of a chair that embodied these values. He realised the importance of the story behind an object – where it comes from, why and how it needs to be developed. He also embodied the importance of 'touch' within the chair – when the person sits and touches it, they can feel the care and attention that has gone into the fine detailing and level of finish in the frame.

If you are more drawn to the theoretical subjects at Master's level then you may find that a full-time route is manageable; there will be many more flexible hours outside of the taught time in which you can study. You need to look carefully for the kind of attendance that suits you and also the emphasis of the course. Many include other aspects of professional practice, such as management, technology or specific career opportunities, or may give you the chance to submit a project of your own choice that has an emphasis on researching a particular topic in depth.

Aesthetic qualities captured in the detailing of the chair by James Poole

Theresa McMorrow embarked on a creativity and innovation project designing furniture for children. After examining the research methodologies available, she put together a blueprint for her designs; this takes account of the essential differences between dealing with adults and children. Her personal strategy diagram demonstrates the control that the project needed to ensure a successful outcome.

When you apply for postgraduate study you need to make sure that you can finance your studies. In a few cases there will be bursaries available to help with your costs. These are well worth finding out about. Teaching approaches will be similar to your undergraduate study, but there will be more emphasis on you directing your own investigation and being prepared to undertake more rigorous research than previously. There will be introductions to different ways to research and greater guidance on how to analyse your progress. Read Adele Parsons' story about her postgraduate experience at the end of this section.

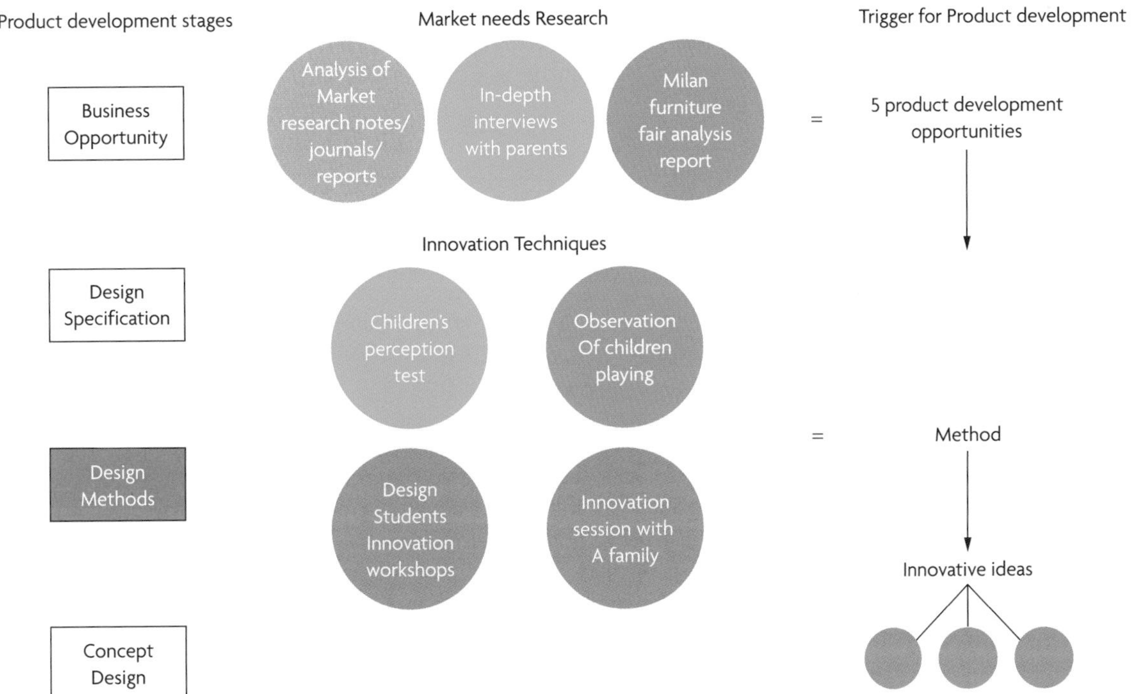

Innovation strategy by Theresa McMorrow, MA Product Design

Becoming a teacher

If you want to go into teaching as a career then you will need to undertake a postgraduate qualification referred to as a **PGCE**.

There are many different courses to choose from and they all specialise in teaching at different levels, from primary school to further and higher education. It is also now becoming commonplace to undertake a teaching qualification if you are considering lecturing as a career. The Design and Technology Association is the professional body for design and technology teachers in schools (www.data.org.uk). Read Kirby Dowler's story about choosing a career in teaching at the end of this chapter.

PGCE – Postgraduate Certificate in Education: if you are interested in becoming a teacher then undertaking a year's intensive training, with practical hands-on experience and the theory to underpin it all, is a requirement for industry.

Kirby Dowler, secondary school teacher
The Weston Road Academy 2013

Studying design management

You might well be wondering what design management is, as it's a relatively new discipline that is most often taught at Maste's level, however there are some undergraduate programmes that also cover the subject. Design management isn't simply about the management of design or designers, it is concerned with the management of the relationship between design and business and other organisations involved with design.

Master's programmes are usually aimed at both recent graduates and practitioners with several years' experience, who come from a wide range of disciplines.

Course content

Design management often involves working with students from more than one design discipline, so Design Management postgraduate courses might be described as multidisciplinary in their content. The types of content that you might experience on a Design Management course are as follows:

- the design industry and how it operates
- the role design plays within organisations and consultancies
- the nature of creativity and innovation
- the strategic role of design and how it contributes to organisations
- business and marketing strategy
- the practical aspects of design practice
- skills development in relation to interpersonal, team working and leadership

You will usually have the opportunity to focus your study in an area or discipline of your choice and, at postgraduate level, your work will involve a high degree of independent study that needs to have a clear set of goals determining what you want to achieve. Some courses will have a distinct academic and theoretical emphasis, whereas others will be more practice-focused.

Master's courses are usually studied on a full-time or part-time basis. Some courses are designed to be studied by distance or flexible learning, which means you can study whilst working.

The relationship between design and business

Shop display, Macy's, New York, 2012
Photograph by Linyang Zhang

The management of this relationship is important because design work involves people from a variety of disciplines who have different views and expectations to those of designers. Put simply, design can be seen from two different perspectives, that of the designer and that of the client (the person who is commissioning the work). Read about Linyan Zhang's career as a design manager for Calvin Klein at the end of this chapter.

The designer's perspective is concerned with the design itself; how successful is it in achieving its objectives and what does the designer need to do to support this? The client's perspective is more likely to start with the business or organisational objectives; how can the end result achieve these objectives and what are the costs associated with this?

The two different perspectives are underpinned by different motivators; the client is motivated by the value of the work to the company and the designer is motivated by solving problems to create the best possible solution to a given brief. In a design-led organisation these differences may not be so evident, however, many designers will find themselves working with organisations where design is not at the centre and they might experience that they have limited influence. These differences can be quite challenging for designers in their early careers, as they can undermine the process and ultimately the performance of the outcomes.

The role of the design manager

The role of the design manager is to understand and 'bridge' these differences and to bring them together, through insights into how to integrate the value of design with the management process. So, studying design management is about understanding the strategic value of design – how can you ensure there is a return on the investment in a design project and how can this most effectively be achieved? Studying design management will help you acquire these business and marketing insights and develop your ability to communicate effectively with clients.

Who works in design management?

Designers in the early part of their career often learn a lot about the management of design 'on the job', and there is no distinct career path into design management. However, practising designers with an interest in business may progress into management. Equally, individuals who don't have a design background but have a keen interest in design may gravitate towards design management. It is practised by a whole range of individuals working in consultancies, companies with in-house design departments and marketing departments, with job titles such as:

- designer
- design manager
- project manager
- account manager
- marketing manager

- buyer
- merchandiser
- freelance design consultant

What skills do you need to be successful in design management?

- To be a 'people's person' with good interpersonal skills.
- To be a good listener who can understand others' needs and nurture their skills.
- Have excellent presentation skills and be able to influence and persuade.
- Have an empathy with designers and their process.
- Be resourceful and have good organisation and project management skills.
- Have the ability and the motivation to champion the value of design.

Doctoral study

Many design students are not aware that it is possible to become a Doctor of Philosophy in design-based disciplines. This gives you the right to call yourself Doctor (Dr.) and to put the letters PhD after your name. Increasingly, this is becoming an important qualification if you wish to teach in higher education. Since the 1990s, in the UK, practice-based doctorates have been awarded that consist of your practical work and a written **thesis** that contextualises, justifies and validates it as a piece of original work.

There are a few universities that offer a practice-based doctorate without a written element, but the majority still expect a thesis to be undertaken.

Some institutions offer funding for doctoral study, particularly where they have a subject or project that needs researching. These opportunities are usually advertised in the relevant press (the 'Education' section in the *Guardian* newspaper being a good first point of reference). They can take between three and five years to complete.

Thesis: this summative piece of writing positions the practice-based work within existing debates and theories about art and design. It describes the way that the research has been undertaken (the methodology) and the analysis that has been done, and determines its level of success.

The jobs market

by Katie Dominy, Arts Thread and Jane Bartholomew, Nottingham Trent University

Planning your career – working backwards

Starting on your career path after completing your degree can be a daunting prospect. You may have another qualification and a new set of skills, but being able to tell people about yourself and find the right job is the next step. It is normal at this stage to not know exactly what you want to be doing, and hopefully this section will help you think through your options and give you some guidance on what to do next.

'I feel that I am gaining a lot of skills and work with a supportive team that I can, in turn, learn from. Meeting new industry contacts all the time and chatting to them makes me realise that my career path isn't set in stone and that one opportunity can lead to another.'

Jenna Wright, Assistant Press Officer, Next Plc

What steps do I need to take to achieve my career goal?

Have a go at developing a career plan to see how your career might develop in one, three, five and ten years, and consider where you will be and what you will be doing. Use a matrix like the one below and give yourself enough room to answer the questions. Then consider how you are going to work toward these goals by identifying the smaller steps that you need to take in order to reach your goals. The following table has been designed to help you plan ahead to achieve your goals.

Career Planning	1 year	3 years	5 years	10 years
Career: imagine what you want to be doing and how you will progress				
Destination: where in the world do you want to be based?				
Personal: lifestyle and family commitments need considering				
Interests: other aspects of your life may influence career decisions				
Goals: think about each period of time and write down things you need to do that are relevant to help you to achieve your vision	1. 2. 3.			

These goals may need to be broken down into even smaller goals to make them appear more achievable. You might even find it useful to create a further plan with your own deadlines, so that you meet your own targets.

It is quite likely that if you were to look back at all these plans after ten years, your career ideas would have changed considerably since your first thoughts. This is quite natural, but this exercise will certainly get you heading in the right direction at the time you do this planning. Obviously it and you can change, and it is a good idea to update it every couple of years.

Designers from many different specialisms speak about their 'chance meeting' with someone who gave them an opportunity and their 'lucky break'. This can certainly help, and the more networking you can do, the better the opportunity for conversations of this nature to arise. However, you can plan your way to success too!

'Even being asked to put my thoughts down on paper for you to read I jumped at the chance . . . I thought – great! Something else to put my name to. And that's the way you need to think about everything you do. So many doors can be opened through new opportunities; even after university it is about making the most of all meetings with people. I collect everyone's business card and always hand mine out. Always smile! That sounds a bit silly, but you have no idea who these people are or who they might represent.'

Martin Bonney, embroidery designer, graduated 2009

How do I achieve my goals?

If you have a goal and you need to work out how to get there, do some research:

1. **Find out more about the people you admire** and whose job or role you would like in time. Ask them how they got there? Don't be embarrassed to contact them and ask. Many people also have their curriculum vitae (CV) on public sites such as LinkedIn – so make the most of these, as the site includes a detailed breakdown of people's career paths.
2. **Speak to lecturers and university professionals** about their knowledge of the area in which you want to work. After speaking to different people, you should be able to build up a pattern as to how these various individuals arrived at their goals.
3. **Look at a range of university websites** as the pages explaining each course often include details of what their best graduates have gone on to do.
4. **Attend lectures or talks** given by people you admire – often there is time at the end to ask them questions about how they started out.

What are the different ways I could work within the industry?

Working for a company:

If you want to become an established employed professional then you need to get into the industry and move around until you find the role that best suits your skills, abilities and interests.

For some areas of design, it is natural for graduates to start their career by being employed by companies, with a view to gradually moving up within the organisation, or applying for a more senior position in another company. This applies largely to all areas of visual communication, fashion, industrial, spatial and product design. Once in your first job, do as much as possible to gain experience within the different areas of the design department and build an awareness of all of the different people you meet and what they do, as this will help you identify where you might be best placed.

Working for a small business:

Many of these people will have specialised in some way or another within a particular discipline, and you will need to develop a full awareness of how the business operates and get involved with many aspects of the day-to-day running of the business, as well as design. You will soon appreciate what it is like to run your own business and have a higher level of responsibility. Many businesses in the creative industries may have started out as 'craft' businesses, covering traditional making processes such as ceramics, glass, jewellery, furniture, textiles and the bespoke end of fashion design, but they could also be service-based, for example within the graphics and illustration fields.

Working as a freelancer with self-employed status:

As a freelance designer you will be working on projects set by others and getting paid for these designs or services. This may happen as a consequence of exhibiting your work at the end-of-year shows and simply saying 'yes' to providing something for a customer. You are then on your way to self-employment.

Starting your own business:

If you have become immersed in your practice and know that there is a market for your products or services, and have thought long and hard about the pros and cons of setting up on your own (refer to Chapter 16, 'Working for yourself', for plenty of information about this), then you will soon be in charge of running your own business. Most countries and regions view design as a very important part of their culture and offer some level of support to graduates from the creative disciplines.

Placement students Kelsey Pilgrim and Kirsty Armstrong working on business strategies for Hanna Francis Design

A first salary

As a graduate, you have to be prepared to be flexible to get where you want to go and to earn what you want to earn. Unless you take a job such as buying, marketing or sales, the wages for junior designers can start low and vary between £15,000 and £25,000 per annum, depending on where you are based and sometimes the size of the company. Smaller businesses in their first few years of operation may need an extra designer or someone to take on certain parts of the work. These businesses will probably not be making huge profit at this stage and not be in a position to pay much. The decision as to whether you work for low pay is up to you; however, the experience and insight will make it all worthwhile.

On placement doing real projects

Some graduates aim to get a placement when they finish so that they are at least doing something in the industry and gaining more contacts, even if the placement is unpaid. Or, you may find a part-time job with the design company of your dreams for three days a week, but if they cannot afford to pay a high enough salary, you may have to obtain other part-time work to become more financially stable.

International awareness

You need to consider where you would like to work! And, allied with this is the fact that certain areas of design are based in certain cities, countries or continents. It is important to find out from your own research, and by absorbing as much as you can during your studies, where the industry is placed within the world and identify who the main companies are that you might like to work for.

Work visa requirements

It is important to consider the work visa implications when choosing a different country to work in. You should check with your embassy and identify how long the process will take and whether there are any other issues that you might need to consider, such as having the right skill base. Large, multinational companies will be able to handle visa applications from a student from another country to spend time on a placement with them, but smaller companies may struggle with this as there is a fair bit of paperwork to do and they just might not have the resources. Some countries also offer overseas students the opportunity to stay in the country that they have chosen to study in and work for a short while – this is worth finding out about.

Marketing yourself

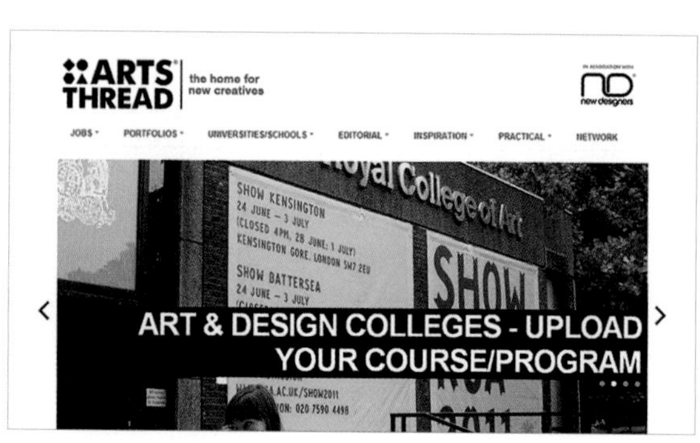

www.artsthread.com

Marketing yourself needs to start well before the end of your final year and in good time before the final degree show. Whether you want a job in the industry or to sell your own ideas, here is a list of what you will need to have ready by then:

- Portfolio (collection of work)
- Online portfolio
- Business cards
- Curriculum vitae (CV)
- Press release – about you and your major projects
- High-quality imagery of the work from your final year

There is plenty of advice about preparing for the industry in Chapter 9, 'How to succeed as a design student' and in Chapter 16, 'Working for yourself'.

Online portfolio – own website or a purposely-designed portal?

The key to online portfolios is not to have your work only in one place, so do have your own website if you wish, but there are also various different creative portals that are available to help get you promoted. Creative portals such as Arts Thread, Coroflot and Behance do all the work of marketing your work online alongside other work of a similar nature, so that people can search a variety of work in one go, make comparisons and find the work that they prefer (thus saving you time). Remember, you can be on as many sites as you want, the only thing you must make sure you do is to update them all regularly.

Image galleries, such as Flickr, also offer opportunities for you to showcase your work for free. Don't forget to use YouTube and Vimeo for your videos and slideshows of your work. And get involved in blogs too.

YouTube
Imogen Ransley-Buxton

To create your own website

Type your name into www.yourname.co.uk and you will find out if your name is available as a domain name. Buying domain names is relatively cheap. When you buy a domain name you can pay someone to design and run your website for you. If you do not want to choose a format from all of the templates available, you can design a website that suits your needs exactly. You can design your own or, if you don't know how to, enlist a friend to help you design and manage the uploading of the pages. You may also want them to continue to do this for you, unless you can learn how

Vimeo
NTU Industries

to do it yourself at some point. Be mindful that if your friend moves away or changes jobs then it might be costly to enlist a new person to undertake the management of a website. So, the advice might be to work with an established company who have preferential rates for new businesses.

TIP

It is important to think through in advance how exactly you would want someone to be able to navigate the pages of your website. Think about being a client yourself and try and imagine what it would be like to view the information. Time spent doing this in advance is better than having to find time to do lots of reworking.

Image requirements

A creative portal or website has certain requirements that you must meet to make sure your work is presentable and loads quickly.

Think hard about your choice of images and spend time cropping them and presenting the content in the best way possible to suit the templates you are given. You should resize your images to 72 dpi for the web and make sure they are saved as RGB colour (not CMYK) and as a jpg (the easiest file format for all sites). Always present your best work on the main homepage. Text should be kept to short sentences and use bullet points for clarity.

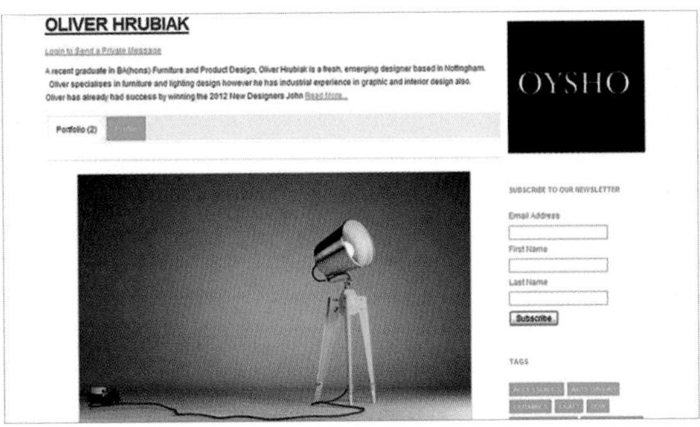

www.artsthread.com portfolio with quality images

Business cards

Most courses advise that you and your fellow graduates order your business cards as one big order from a printing company, thus saving you all money. Add an image that acts as a visual *aide-mémoire* of a piece of work that you want to be remembered for. This will help people recognise your work. Remember that people often like to make their own notes on a business card – so leave some blank space (i.e. not black on both sides!) and avoid laminating them. Add your name, email address, mobile number and postal address. Include a key phrase that sums up the project, or you.

Example of an Arts Thread graduate portfolio

Press release – for your final year's work

A press release is for you to hand out to journalists and possible employers or potential customers, etc. at the degree shows. It is written in a descriptive manner and briefly outlines all of the key features of the project and identifies the type of customer the work is aimed at. Include what makes the project 'newsworthy' by highlighting the unusual and new aspects of the design – the materials, the technique, the inspiration.

It should also contain brief answers to the following questions:

- what is the new product/exhibition/event?
- who is involved in the design?
- where and when is the event that the press release is promoting?

It can also include a personal statement similar to something you might write on your CV. Focus on what makes you unique and different.

Keep it brief and design it to look good on one side of A4 – half a page about your work and a half-page about you. Choose a typeface that is easy to read and set your text with plenty of room for the journalists to write notes alongside the text, because if you get as far as handing one of these over then you are sure to be having a conversation and they will probably want more information. Keep any explanations short and concise, use straightforward language and put the most important information near the top of the page.

Put your name, email address and all phone numbers on it, together with the date and a title of the press release. You may wish to create a separate email address just for press – something like press@mycompany.com – as a journalist may contact you at short notice in order to meet a deadline, perhaps for a photo shoot or article.

Pay careful attention to spelling and grammar. Print it out and read it through. It is a good idea to ask a friend to read through it as well, as a fresh pair of eyes can often see an obvious mistake. Mistakes are rarely forgiven at this level, so check and check again!

High-quality imagery of your final year's design work

This is critically important, especially if you are considering starting your own business. There are some useful rules:

- If your institution does not provide a photographer, then it is worth paying to have your work professionally photographed – this is vital for any smaller pieces where light, surface and close-up details are important, such as pieces of jewellery and ceramics.
- Have your images saved as both high-resolution (tiff files) and low-resolution (jpg files) images. Tiff files are used for printing posters, flyers and for magazines; jpgs are used for printing and for the web.
- Only put low-res images on a CD to give to the press (not high-res ones). You ideally want them to make contact with you for the high-res versions, which will then enable you to keep a record of where your work is being published.

Showing your work at an exhibition or trade fair

Before you part with money, always visit the exhibition or trade fair in advance to see if it is the right venue and customer base for your products, and whether it is well attended. Ask other exhibitors whether they have got a lot out of the show and whether it has been well organised. See how well the event is being covered in the press. Have material for the press ready as soon as possible, as often journalists need images and press releases about four months before the event.

You should be sent an exhibitors pack of information and, once committed to exhibiting at a show, develop a list of deadlines to include your own manufacturing deadlines and those for press.

Purchase the right insurance, whether it be public, product or professional insurance. Make sure you leave yourself enough time to design your stand well and ensure you have asked people to help you look after the stand so that you can take breaks or spend a bit more time with an important potential customer.

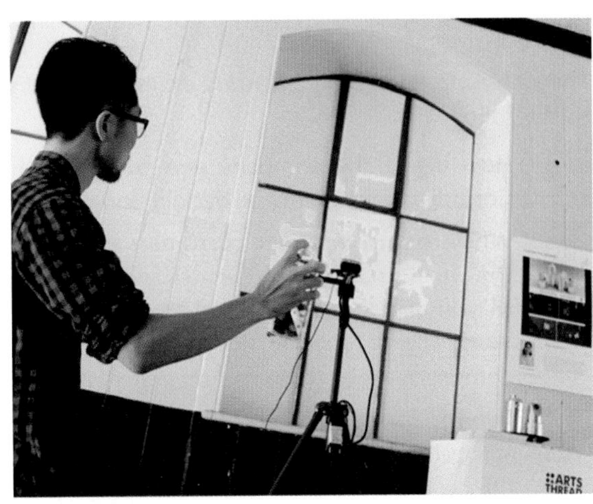

Satoru Kusakabe at Designersblock
Photo © artsthread.com

How do I find a job?

Once you have a fairly good idea of the type of jobs you are looking for, then you can begin the search. The following routes are all possible ways to do this:

- Arts and creative media job-related websites.
- Specialist and generic recruitment agencies – online and office-based. Some will actually want to view your portfolio so that they get to meet you and can make better matches for companies when they send in their vacancies.
- Newspapers – particularly the *Guardian* on Monday (arts) and Tuesday (creative and education – you might want a job as a researcher or to apply for a postgraduate course at a university). If you have decided where you are going to be living, then look up the local paper and their website so that you can be searching them regularly for possible jobs before you move.
- Networking – ask around and make sure you are letting university lecturers and other contacts know what you are looking for.
- Join Facebook groups and LinkedIn so you are able to find information quickly about possible jobs, and so that people can find you too!
- Actively introduce yourself to the companies that you would like to work for by offering to do some work experience for them for a couple of weeks, or by asking them for 20 minutes of their time to review your portfolio.
- Attend trade fairs and events where you have opportunities to meet people in the industry. If you are still at university then contact your careers office and check you know the dates for all of the career and recruitment fairs.

TIP

Have everything ready to go – an up-to-date CV and key statements about skills and abilities.

When the perfect job comes up you can apply immediately by referring to the content of those to complete an excellent application form.

Preparing for an interview

Think about preparing for an interview in advance by writing down your answers to the following:

1. **How big is the company and what are their products and services?** Consider doing plenty of background research on the company.
2. **Who are their intended customers and market competitors?** Understand the way the company operates and who their customer base is. Find out all you can about them and try and gather this intelligence by tracking down how they market themselves and any publicity they have had, as this demonstrates to them at interview that you are interested in them.
3. **Can you foresee any new potential markets for them?** They might find this an interesting topic to discuss at interview but take care how you approach this, as they might not want to employ someone who is too opinionated.
4. **What strengths do you have in relation to the job description?** Try and find ways to lead them to your strengths through your answers to all of their questions.
5. **Why do you want to work for them?** This might appear an obvious question but it is not easy to answer – think of their needs and yours when you respond to this.
6. **What can you bring to the company that is different to the other applicants?** This is your opportunity to shine and point out your unique qualities and skills that are matching up well with what they are looking for.

An interview is a very good way for a company to assess your abilities to cope under pressure and to determine whether you can verbally and visually present your thoughts in a logical and succinct manner, so, think before you speak.

After being successful in obtaining a job, Elouise Holland, a design graduate, wrote these prompts for interview technique:

- try to come across as someone everyone would love to work with
- ask yourself the question – what have the other applicants got that I haven't?
- remember small companies offer big challenges
- don't be demoralised by knock backs, someone will give you a break eventually
- always address your letter and CV to the right person
- always call to see if they have received your information
- if you are not successful, always call after the interview to thank them for the opportunity and get an extra brownie point! (everyone knows everyone else in the industry!); remember to be brave and ask for feedback
- take any job within the industry that you want to work, and work your way up.

Working as a designer in the industry

By Jane Bartholomew, Nottingham Trent University

This section explains the various ways you might become a designer in the industry and looks in depth at the skills and abilities required to succeed through an analysis of some job descriptions.

The size of a company and where they are positioned within the supply chain are two factors that can influence how you might operate as a designer. Something else to be aware of is the market level and subsequent price points that will influence the types of processes and materials that you will be working with. Much of the design sector is governed by events in the calendar that heavily influence the pace of the specific industries and determine where the busiest points of the year fall.

'It's amazing how it's worked out really. If it wasn't for the live project in my final year, I wouldn't be where I am now, because it was with those designs that I won the Bradford Textile Design competition and consequently won a one-week work experience with Johnstons of Elgin. Later that summer the design director offered me a one-year placement, which I gladly accepted. They've now offered me a permanent contract! So, I now have my dream job in my favourite part of the world!'

Jen Smith, designer, Johnstons of Elgin, Scotland

Whatever the brief and whoever the customer is, attention to detail, working to deadlines and communicating your ideas are all fundamental skills required to be a successful designer.

Accepting a junior design role and getting into the industry is a fantastic step following graduation. It is also important to look ahead and see yourself in ten years' time and imagine the type of lifestyle and job you would like to have. You can take steps toward it by making sure you stay up to date with your own training needs and by keeping abreast of the changes occurring in the wider industry. Another factor will be the size of the team involved in the design process and the way the tasks are divided among the members of the team, based on their skills and abilities.

By taking each of these factors in turn now, and exploring them in more depth, you will gain a greater understanding of the skills required and the variety of ways you might work as a designer in the industry.

The role of the designer working in companies of different sizes

If your design role is well defined and you are working as part of a design team in a large organisation, with departments for each range of products, then it is likely that you will be focusing on the design process for your specific product or service. You will need to know your customer and market very well and be able to do the necessary research into trends, which will probably include travelling to relevant trade events across the world. It is also probable that you will source raw materials for the products, liaise with the industry and contribute to design development meetings within the company.

It is unlikely that you will get involved with other design projects happening within other departments, but if the company is committed to the development of their staff and they can see you have emerging interests in other areas then they are usually keen to talk this through to see if there is a way they can support these interests, as they would much rather have a happy team with the right people in the right jobs.

In a smaller company with perhaps just six employees, for example, your design role will certainly include all of the above, but you could easily be expected to interface with other aspects of running a business.

'Working in a small office of ten people does mean that I have a lot more responsibility than you would normally find in equivalent graduate roles.'

Jenna Wright, Assistant Press Officer, Next Plc

The most likely areas that you will be expected to get involved with might be promotion and marketing, and even supporting the overall management of the company to include answering the phone and responding to customer enquiries. This type of role would require you to be very flexible. You would also need to be an excellent team player and be willing to learn in order to get the most out of a position of this nature.

The 'supply chain'

The customer for the products or services that you are designing might be the next company in a line of manufacturing processes, or might be an individual making a direct purchase of one of the products in your range sold via a website. To understand these routes more clearly, let's look at a supply chain for a product made by a large company and think about the comparative journey within a small business.

Whether you are designing props for a theatre production or packaging for an international brand, ideas for the product must start somewhere and involve a variety of people and a range of processes along the way. The scale of the process – for example, the manufacturing of a car in contrast to creating an illustration for a book cover – will determine how long the process will take, from initial idea to the customer receiving the finished product or design. This is referred to as the 'critical path' in some industries, or the 'workflow' in others.

Understanding that there might be any number of different companies (or departments) in the supply chain, and the different roles that they play, will help you become more aware of which roles you might like to explore further. Whether you decide to go into business as a designer-maker or to develop designs and have them put into production, it is likely that you will have to deal with all aspects of product development, from initial research to identifying how you are going to get your product to the intended customer, often referred to as 'route to market'.

Different roles within the industry

The role of the designer varies from company to company, but the liaison with suppliers and being able to keep abreast of all market and customer needs is something that all designers will have in common, no matter how big or small the business. The impact you might have as a designer for a particular company on some sections of the supply chain may be zero, but this is simply because you are employed by a company that only manufactures one part of the product and therefore cannot influence what happens in other parts of the production process.

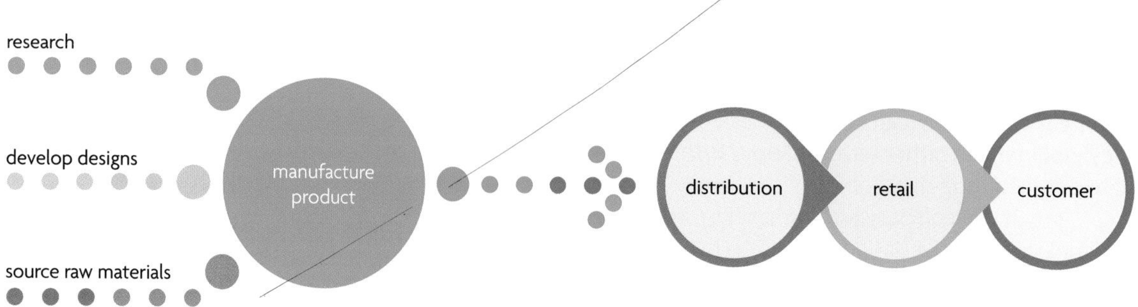

The manufacturing supply chain

RESEARCH concept, market, customer, trends

DESIGN prototype using CAD, design boards & present ideas

REFINE design ideas

SOURCE raw materials, components & manufacturing processes

COMMISSION & monitor manufacturing of product

MARKETING sales, promotion

DISTRIBUTION & packaging

CUSTOMER retail, gallery, website

Understanding the design roles within the manufacturing process

An important part of the job is to maintain an up-to-date knowledge of new materials, processes, approaches and trends so that you are designing from an informed knowledge base. Taking time to visit trade fairs and exhibitions and keeping a close eye on your competitors are fundamental to ensuring you understand the potential market for your product and can fulfil the customer's needs.

As a designer you may have plenty of time to develop prototypes in order to refine the design, or you may find that you have a very tight deadline to work to and all that you can do in that time is draw on your knowledge and experience and develop sketches or virtual ideas using CAD software to simulate potential ideas. As your career as a designer gathers pace you may be required to work more and more like this. Recent graduates are often quite shocked at the speed in which they need to develop designs in industry, as there is generally a more leisurely pace to complete projects during their degree. However, design itself is not your only option, as you will have realised having read about the various design disciplines.

So, are you great at negotiating, good with numbers and keen to travel? If yes, then you might consider becoming a buyer. A buyer would source products and raw materials and keep an eye out for new developments and trends within the industry and liaise very closely with the designer (See Katie Roberts' career profile at the end of this chapter about what it takes to become a successful buyer.)

Or, are you interested in the ever-changing customer, able to think creatively and wanting to drive the success of the business? Then perhaps you would enjoy being part of the sales and marketing team. Or, do you instead like the idea of supporting the future vision of a company and are able to see projects in their entirety? Then maybe you might consider training as a manager.

Understanding different market levels

Let's look at understanding more about designing for different levels of the market. While studying design you might have become particularly motivated by new materials, processes and approaches. Perhaps you have really embraced 'risk-taking' – playing and pushing your creative abilities to the limit. From this start point, you then need to determine what sort of a designer you want to become and what level of the market you might want to be positioned in.

As a student you may be of the opinion that to be a successful designer-maker, working with new or expensive processes and materials and launching your own products or services, is the only way to be sure that you have truly succeeded as a designer.

Self-employment does not suit everyone, so it is important at this point to attempt to ask yourself serious questions. You need to consider what you have learnt about your strengths, weaknesses, abilities and skills and have analysed what type of activities you might prefer to spend your working week doing.

An example of this might be: as a fashion designer you have been working with a lot of very expensive fabrics as part of your final collection, which were perhaps given to you to experiment with by a company and you enjoyed developing ranges for *haute couture*, high-end fashion. You might then perceive that you will struggle to feel comfortable working with cheaper fabrics if you accept a design job working for a fashion company positioned at the lower end of the high street. Clearly you will have different questions to ask of your materials at a lower level of the market place where goods are cheaper and the addition of interesting trims and finishes would have to be within the budget determined by the overall recommended retail price (RRP).

This, for most designers, is where it gets interesting! Being faced with a budget and an urge to create innovative ideas gives the designer a problem to solve. This is something that you will now recognise as one of your best skills, as you have been trained during your studies to problem solve and you're really good at it!

On the other hand, if you are completely inspired by specific processes and working with certain types of material for a particular level of the market, then you need to be aware that you are reducing the number of opportunities that will come your way. If this does describe you, then you will need to remain focused and driven and will perhaps have to wait longer for the perfect design job to come up. It is noted that students have succeeded using both of these visionary approaches, but it is also worth noting that others have become confused along the way and have had to think again about their goals.

The designer's lifestyle

Tight deadlines, fluctuating work patterns, opportunities to travel and liaising with many different people are all part of the lifestyle of being a designer.

The fashion industry, for example, is governed by the calendar of the London, Milan, Paris, New York and Tokyo fashion weeks, where the top fashion designers launch their new ranges. This, together with new seasonal trends, sends the design teams within the high-street stores into a whirlwind of activity, racing against the clock to bring the most up-to-date looks into their stores. This will have an effect on you. As a fashion designer you will be expected to contribute, and that may result in working long hours at certain points in the year.

'I design sets for TV. I am often expected to work late around a deadline, which is totally normal for all of us in the team. I really love my job so it would be weird if I didn't want to stay on when I needed to finish a project.'

Liam Sands, one year into his first job, having studied Theatre Design

All design disciplines have this 'lifestyle' aspect of tight deadlines, some more than others. One common theme is the accompanying feelings of excitement and elation when the work is finished. It doesn't have to be something that you worry about, as luckily most designers really enjoy their jobs!

Skills and abilities

Designers need a range of specialist and generic skills to be proficient and work well as part of a team. To create a perfect design team, each member needs to feel at ease with all of the other members so that honest, full discussion and the giving and receiving of critical appraisal can drive forward the outcomes of a project. Your tutorials and seminars at university go some way to preparing for this. A professional attitude, having a flexible approach and being a good listener are all key attributes that potential employers are looking for.

Being interviewed for a design job is not only about finding out whether they want to employ you from a creative perspective, but whether you and your personality will complement the existing team or not. It is not unheard of for the designer with the best portfolio at interview not to be offered the job, the reason often being that their personality and ways of working were too similar to others already employed in the team! This can be very frustrating if you really felt the job was right for you.

We all bring different strengths to a team, and it is gaining an understanding of the strengths that the individuals can bring that often results in the best design outcomes. Some people have natural abilities to initiate ideas and conceptualise, while others are able to map out a project over time and ensure deadlines are met. Abilities to problem solve, apply research, remain objective, delegate, stay committed, check that work is completed to the highest standard and bring relevant specialist knowledge are all sought for in the make-up of a design team.

You will also need to have vision and excellent visual and verbal presentation skills. (Refer back to Chapter 9, 'How to succeed as a design student', for improving your verbal presentation skills and entering competitions. And chapter 10, 'Being creative and innovative', looks at practical ways in which you might focus on improving your design process skills.).

This doesn't mean that a design team needs to consist of a certain number of members, as individuals can obviously have more than one of these skills. For those considering self-employment, it does, however, raise an interesting question about the range of skills required of an individual person. To find out more about this and to see if this is something you might consider in the future, refer to the information about self-employment later on in this section. The theory behind team-working forms the basis of the work by Dr Meredith Belbin, and her website gives a clear overview. If you want to find out more, refer to the further resources information at the end of this chapter.

As you prepare for your first job, it is now that you probably wish you had spent more time thinking about and improving these skills while you were studying. A practical suggestion would be to start looking at and analysing job descriptions during the second year of your degree onwards, and building a list of the type of skills and abilities you might need to succeed as a designer in the industry.

Job descriptions – looking at the details

If you are in the middle of your course, then looking at some job descriptions now might trigger some new motivations to ensure that you make the most of your final year and guide you to make conscious decisions about the skills and knowledge you need to improve your chances of getting the job you desire. If you are in your final year, or have just finished your course, then use this section to prepare for your next phase – job hunting!

Let's analyse different descriptions that appear within the advert for three creative design roles as a way of understanding the types of skills and knowledge that are required to undertake certain roles. This is a practical activity that you could do for every job you are aspiring to, or going to apply for.

To understand what a company is looking for in a successful candidate, various statements have been unpacked to tease out important information and placed under the headings a) **skills**, b) **knowledge** and c) **experience**. An explanation then follows, with some useful advice and suggestions that will help you follow up if you want to evaluate yourself against the skills, knowledge and experiences mentioned.

Job description 1: Graphic Designer

'We have an opportunity for a designer to work within a fast paced in-house marketing team . . . fully proficient in Quark, Photoshop & Illustrator and a knowledge of Dreamweaver and Flash would be advantage'
a) **Skills:** confidence, market awareness, communication and organisation.
b) **Knowledge:** ability to competently use the majority of software.
c) **Experience:** examples of projects completed under pressure.

a) **Skills**: the only way to develop true confidence is to believe in yourself. Prepare well for presentations so that they become second nature to you and you enjoy delivering them. Then see how appreciative the audience is when they are delivered well.

b) **Knowledge**: it is unlikely that all of the software listed will be formally taught during your studies, so here are some practical suggestions for you:

   - find out if they are available and set yourself time to understand them, or force yourself to use them during your next project as it will be quicker to pick up if there is a purpose to your quest
   - when looking for a placement, ask which software they use as this might be time well spent and they won't mind if you are still learning
   - find a local college that offers a short course in the software

c) **Experience**: within every project, practise working under pressure in a focused, relaxed manner so that your creativity is not stifled, but understand the deadline requirements and meet them! Remember, you will need to rely on a reference from a lecturer to get your first job, so you won't want to give them the wrong impression and not demonstrate what you are really capable of.

Job description 2: Product Designer

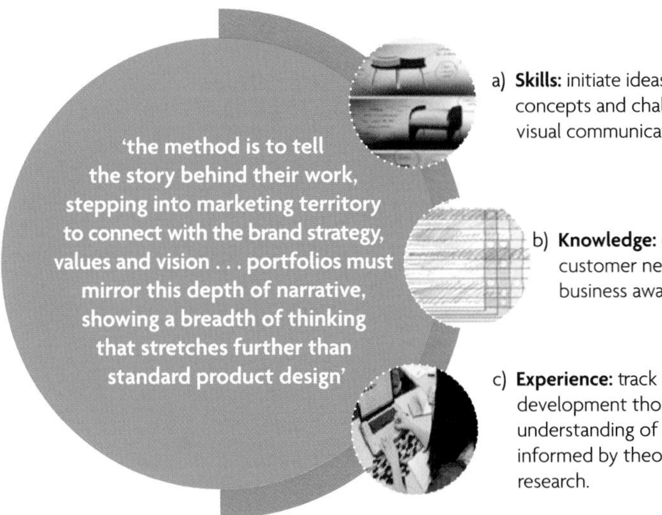

'the method is to tell the story behind their work, stepping into marketing territory to connect with the brand strategy, values and vision . . . portfolios must mirror this depth of narrative, showing a breadth of thinking that stretches further than standard product design'
a) **Skills:** initiate ideas, develop innovative concepts and challenge perceptions, visual communication.
b) **Knowledge:** deep awareness of customer needs and market level, business awareness.
c) **Experience:** track own design development thoughts and gain understanding of how designs are informed by theoretical and market research.

a) **Skills**: push yourself to take the lead on group projects or to engage in discussions with others about different ways to start a project, and find a challenging angle to come into it from. For example, take a new material or concept and research the possibilities of applying it to a new product area, then undergo the relevant market and customer research to see if you can provide an innovative addition to the market place.

b) **Knowledge**: train yourself to think like the customer! See the product from their perspective and ask yourself the following questions:

- What do I want from a product like this?
- Where else could I shop for a similar product?
- What is it made of and how long will it last?
- How much is it, and is it a fair price for what it is?
- If I can afford to spend more, where else might I shop for one?

c) **Experience**: you are sure to uncover key reasons to support your designs if you keep focused all the way through the project on the person for whom the product is intended. This will also help you avoid designing for your own taste instead of concentrating on fulfilling the customer's expectations.

Understanding the connections you are making between your own ideas and the all-important culture and context in which you are being asked to design, is something that students in the first half of their degree find a challenge.

Slow down your thought process and spend time thinking through the research and the design process you are undertaking. Capture the ideas on paper! This should help you reflect on the reasons for the way you made your decisions and this should uncover concepts that inspire your designs. Then you can build the research around them and this will make the purpose of your project more meaningful.

('Analyse' and 'reflect' are the words you might see regularly in module information, and this is what it is referring to!)

Job description 3: Women's Footwear Designer

'you will be an ambassador for the brand, you will articulate the company and brand vision clearly and will have the ability to predict emerging trends and best-sellers by analysing the markets. You will be able to translate print and colour trends into the brand handwriting'
a) **Skills:** verbal presentation, numeracy and able to work with data, highly creative.
b) **Knowledge:** footwear market, trend forcasting.
c) **Experience:** placement experience to understand the inner workings of a company's ethos.

a) **Skills**: practise beforehand any opportunities you have to present your ideas verbally. Even prepare ahead for a tutorial where you are bound to be asked to talk about your work and think ahead about the type of terminology you might need to put your thoughts and questions across succinctly. Set yourself some investigation time to understand basic analysis of data so that you can apply the information and talk knowledgeably about it. Even a short course tackling basic spread sheet operations can help. In an interview situation this demonstrates that you are willing and able to learn about business and financial matters.

b) **Knowledge**: the majority of fashion-related courses take their students to trade fairs, and it is here that you will learn directly about the importance of trend prediction. There are hundreds of companies all specialising in particular aspects of the market, from interior colour and pattern trends to yarn and texture trends for knitwear. A designer's role, depending on how their company operates, will either be using specific trend information that the company has decided to buy in, or being given the task, perhaps as part of a team, to identify trends as part of the job.

c) **Experience**: placements are one of the best ways to become immersed in a company's vision and to understand the importance of the decisions that are made in relation to designing a range for a specific market. Every company will have a different set of goals, and the more you can do to utilise a work experience or placement to understand company ethos, then the more you have to talk about at interview.

Job specifications

In addition to this, the job specification will contain even more information and statements about the ideal candidate and these might include the following:

- to be able to work in a fast-paced multicultural environment
- to produce accurate technical information
- to liaise with factories in the development of prototypes
- to have had experience working with overseas providers

Let's think about these in more detail. The design industry spans the entire globe. Working successfully within a multicultural environment means you need to develop an understanding of people from different cultures by recognising their differences and accepting their preferences. Respect for others' ways of working and appreciating that there is always more than one way of doing something is crucial to be able to design effectively in teams made up of people with different backgrounds. Empathy is an important quality for a designer: to understand the motivations and needs of your colleagues and customers.

The final point to remember is why lecturers often request that you keep an accurate record of all of the processes you use while you undertake practical work; companies need to be sure that the successful candidate for a job has a keen eye for detail and can maintain accurate technical records – this is to ensure that a product or process can be recreated exactly if another batch needs to be made, or replica versions constructed.

Conclusion

This chapter has offered you an opportunity to interface with what might happen next in your career. Thinking ahead to the sort of career you might want can have a direct impact on the sorts of projects that you might set yourself in the final year of your course. The portfolio that summarises your practical, theoretical and contextual understanding of your subject will therefore be more appropriate for your initial career aspirations.

If you have worked through this chapter at a 'reflective' pace, then this will have enabled you to do some thinking and planning about your future. It is likely that it will have prepared you to take steps in the right direction to becoming a highly employable design graduate. If you have scanned through it quickly for now then, perhaps unknowingly, you will have taken a number of points on board and maybe this will influence some of the decisions that you make, as you journey through your course.

The following career profiles and stories illustrate some of the career opportunities open to you. Their personal thoughts will help you appreciate the sort of preparation that you need to think about.

Further resources

Books

Belbin, R.M., *Team Roles at Work*, Butterworth-Heinemann Ltd (2010)

If you are interested in the theory behind successful team-working then this is one of the core texts on this subject.

Heller, S. and Fernandes, T., *Becoming a Graphic Designer: A Guide to Careers in Design*, John Wiley & Sons (2010)

This provides a comprehensive survey of the graphic design market, including complete coverage of print and electronic media and the evolving digital design disciplines.

Mainstone, J. and Reynolds, K., *The Careers Directory 2012/13: The One-Stop Guide to Professional Careers*, COA (2012)

This is a very useful guide to the variety of careers available, with practical tips and advice as to how to progress with them.

Websites

www.belbin.com

Team-working information website by Dr. Meredith Belbin.

www.csd.org.uk

The Chartered Society of Designers is a professional organisation that promotes the design profession and provides accreditation for higher education courses, and is a useful website for competition information, etc.

www.dba.org.uk

The Design Business Association promotes professional excellence through productive partnerships between commerce and the design industry to support all aspects of effective design.

www.direct.gov.uk/en/YoungPeople/index.htm

The original connections website is now part of this overarching government site to help young people prepare for their careers.

www.prospects.ac.uk

Careers website with plenty of general advice.

www.thedesigntrust.co.uk

Help for designers and crafts people to develop and run their own businesses.

www.tda.gov.uk

Government website about becoming a teacher.

www.totaljobs.com

Alphabetical listing of all job roles, making it a useful site to find out about the variety of jobs available as well as search for possible jobs to apply for.

www.web.data.org.uk

Great for guidance on getting into teaching.

www.yourcreativefuture.org

Information about events, books and general advice for all the design disciplines.

Project story

Designer:

Adele Parsons, Print Designer for Wacoal Eveden, graduated from MA in 2009

MA project title:

'Does the Person Make the Journey or Does the Journey Make the Person?'

What were your reasons for studying at postgraduate level?

After spending a few years in industry working as a buyer, I decided to return to education to study for a Master's in Textile Design. I needed to retrain in an area of textile design that suited my skills and get back in touch with print design. Early on in the course, I was fortunate enough to win the prestigious Paul Smith Scholarship to study in Tokyo at Bunka Women's University. I spent this time absorbing the Japanese culture and being inspired by the fashions and textiles, and gathering research to support the formation of the concept base for my Master's degree.

My final collection consisted of three printed textile design ranges for womenswear, aimed at the exclusive end of the high street. My work encapsulated the essence and processes of Japanese Shibori textiles and also captured influences from my recent 'round the world' travel experiences. This idea of making the actual journey inspired me so much that it led me to undertake deeper research that helped formulate the question that framed the brief: 'Does the person make the journey, or does the journey make the person?'. Setting the scene for the brief contained the following statement, to place the project in context:

> 'Exploring the notion of the migration of swallows and their bird's eye view of the world, my practice considers the idea of, "does a person make a journey or does the journey make a person?". Using a combination of traditional patterns and Google maps of the 21 countries I have visited, my printed textiles encompass cultural and geographical impression.'

Adele Parsons' MA show, 2009

What processes and skills were most relevant to this project?

To achieve this I experimented with a range of textile processes in the print studio, including Japanese Shibori tie-dying and heat transfer processes with inks. This was so I could develop the initial mark-making process before I manipulated the imagery in a more controlled format. I then scanned the imagery into Photoshop and worked on developing the marks and visual effects created by these processes to create a series of both abstract and geometric designs.

What challenges did you face?

Learning how to work and understand the Photoshop programme was a huge challenge, as this was new to me. However, despite this I managed to master the fundamental skills to create a comprehensive body of work that resulted in an innovative and contemporary collection.

What were the highlights of studying at postgraduate level?

My Master's rewarded me with a great sense of achievement as I had gained a number of new skills and trained myself to become much more proficient in Photoshop. I had a wonderful research term in Tokyo and the entire set of experiences gave me a great foundation to launch my career as a printed textile designer. The highlight for me was becoming a Master of Textile Design!

Career profile

Name:
Katie Roberts

Current job:
Assistant Fashion Buyer

What do you do?

I began my first job as a trainee buyer after graduating with a first-class honours degree, and after 18 months was promoted to assistant buyer, responsible for buying 7–13-year-old's girlswear fashion. I now manage the buying for blouses, skirts, shorts and dresses and am responsible for a total 'buy' of over 8 million pounds-worth for a spring/summer season, for example.

Working as an assistant buyer in a high-street, 'fast-fashion' brand means that I need to work at a very fast pace and therefore need excellent time-management skills. A buyer needs to be able to work under high levels of pressure and always deliver results – you can't miss a deadline. No working day is the same and I enjoy the variety that this brings. I can be working on 15 to 20 projects simultaneously across a single day, so being able to multi-task is essential. I need to maintain a strong commercial focus in order to maximise sales and be able to react to changes on the high street dictated by our customers and our competitor stores. This job is very much about being a team player, you need to be clear and concise in all communications in order to work effectively with colleagues and suppliers and you must be able to take on other people's points of view/ideas and constructive criticism.

My work involves travelling to Europe, the USA and the Far East. The focus for these trips might be for inspiration, or to gather intelligence about emerging fashion styles and to acquire information about the latest trends and key shapes for the forthcoming seasons. Travelling to India and the Far East is a crucial part of our business' development as we buy samples and work directly with our suppliers, who are predominantly based out there. We visit the factories and business offices to ensure we have a clear understanding of how things are manufactured. Being able to travel and see the world is a fantastic part of the job, but it is also tiring and the working hours can be very long. It is not unusual to work over weekends on trips and you are often surviving on very little sleep. This can be tricky when you still need to remain focused and professional in all that you do.

What has your career consisted of to date?

In the final year of my Textile Design degree, I took the initiative to gain some first-hand experiences and organised five work placements within the fields of interiors and fashion. Although I learnt a huge amount while on these placements and gained a great deal of confidence in my design skills, I concluded that a career as a designer was not for me. I knew that I did not have the desire to become self-employed and knew that I worked best in a team, as I had always enjoyed working collaboratively and sharing ideas with others.

I wanted to use the skills I had learned while studying design and apply them to a more corporate role within the commercial fashion industry. I had no previous buying experience, so the role of 'trainee buyer' suited me perfectly as I could be trained on the job – this has proved to be the best way to learn quickly. The knowledge acquired through studying textile design, together with my work experience in fashion and retail, gave me a great basis from which to become a buyer. At interview, I was able to show my prospective employers that I was passionate about the industry by drawing on the experiences and varied challenges that I had already undertaken while studying for my degree.

Is there a particular project you would like to tell us about?

In the fashion and textile industry, the year's structure is built around the buying seasons, and different times of the year have different focuses within them. In preparation for the forthcoming season, we work with a trend prediction company that helps us identify the key colours, textures, silhouettes and themes for the upcoming season. We then undertake a lot of broad research and create mood boards. These are visual displays that capture the thoughts and vision for the season and include fabric and yarn samples. We then develop these themes by making decisions on the focus for the season and develop a strategy plan for each product area so that product development can get underway.

During this development stage, we agree on the cost prices with the suppliers, deciding on a pricing structure and work out which stores will have which products. We also work on the fitting of the garments and approve the 'lab dips' (dyeing process to determine colour matching) and select prints and graphics. Towards the end of this period, we would present to the CEO so that the designs and production process is signed off. Later on in the season the buyers are also involved in the presentation of the new range to the store managers. This means you do need to be confident in speaking to large groups of people and be able to show your passion and enthusiasm for the product.

The range is also promoted via press shoots – I have been lucky enough to be chosen to be the press coordinator for my department. In this role I am expected to help coordinate the selection and send the promotions company the samples, and I also have the extra opportunity of attending the press shoots, which is another aspect of the job that I really enjoy.

What advice do you have for students considering a similar career?

Buying is a very rewarding job – there is nothing better than seeing your garments in store and worn by customers in the street. It is also fantastic and very rewarding achieving fantastic sales on a product that you have worked very hard on!

Career profile

Name:
Kirby Dowler

Current job:
Secondary School Teacher – Subject Leader for Design and Technology

Describe what you do

I lead a small team of staff to deliver creative and innovative design and technology education. I work hard to motivate and engage 11–16-year-old pupils in a classroom environment, delivering projects that are relevant, up to date and challenging to all. Since taking on this teaching role, the GCSE pass rate at A*–C has risen by 30 per cent in three years. I am committed to ensuring that all pupils succeed and come out of my class with a GCSE that will give them life skills that are relevant for the rest of their lives!

I also deliver projects to primary school pupils to aid in the transition up to high school. As a member of the Design & Technology Association's James Dyson Foundation Innovation Group, I work on curriculum tasks to share with other teachers across the UK.

Tell us about your career so far

I have a BTEC National Diploma in Art and Design. I did this as I couldn't decide whether I wanted to be a designer or a teacher, so I had an extra year building up my skills while I made a decision. I then decided

Life skills as well as technical skills are an important part of design and technology education
The Weston Road Academy 2013

to keep my options open and study for a design degree as I felt I would gain more skills that would be relevant to either career.

I chose to study Furniture and Product Design and opted for a four-year sandwich course. I knew having time in the industry would give me invaluable skills that would ensure I became a more rounded designer. I also knew that all of these experiences would translate well to the classroom.

Following my degree, I did a PGCE at Loughborough. I worked in two different schools, at opposite ends of the spectrum, and this was important to me as it meant I could work with children from all walks of life. I learnt a lot about how their personal family situations affected how they wanted to learn.

Is there a particular project you would like to tell us about?

The 'Interactive Buddy Project' is a PowerPoint file (with hyperlinks) that is used in a classroom to assist in the teaching of design and technology subjects. It has been designed to help not only the higher-ability pupils progress through their project at a faster pace, without much physical teacher support, but also the lower-ability pupils who perhaps go at a slightly slower pace but still want to complete the project on their own. The Interactive Buddy can be used to check the processes and techniques by watching video-clips of specific processes.

It allows me more time to be a better facilitator in the classroom and encourages pupils to learn more independently. The Interactive Buddy helps them through a project but doesn't tell them how to solve problems and avoid mistakes. Pupils do make mistakes, but it is important for me as a teacher to work with them so that they understand what has happened through reflecting on how they did things. My favourite quote is taken from Edwin H. Land: 'The essential part of creativity is not being afraid to fail', and that is the environment we are trying to build in the design and technology department at my school. We are now taking this Interactive Buddy into many design and technology subjects (food, product design and electronics), as all teachers have now seen the value of this resource.

What advice do you have for students considering a similar career?

You need to be enthusiastic about your subject and about working with pupils of different ages. There is a lot of different work to do in a school. Teaching, writing reports, marking work and doing assessments can take up a lot of your time. You need to be passionate about giving something back to the community and young people, and enjoy watching them grow and develop into talented young designers/ adults. It is this that makes it all worthwhile.

I was swaying between whether to become a designer or a teacher but, as a teacher, I now get to support the development of many design projects everyday. My creative mind never stops! Enthusiasm for your subject gets the children engaged in your subject, they will work for you as they can see how passionate you are. They think at times I am a little bit crazy as I get so passionate, but they tell me they love my lessons.

I feel it is the best career ever and I can't imagine doing anything else. I kept my options open for as long as possible and I am grateful for all the experiences I picked up on the way. I am now able to share these with the children, so they can relate what they are doing in my classroom to what happens in industry.

Watching pupils' growing awareness of design is very rewarding
The Weston Road Academy 2013

Career profile

Name:
Linyan Zhang

Current job:
Design Manager for Calvin Klein, China District (Mainland China)

Describe what you do

I recently joined China Ting Group Holdings as a design manager. China Ting Group Holdings is involved in the manufacture, retail and export of garments and branded fashion apparel and owns many international brands. I am responsible for the retail, merchandising and promotional planning of the Calvin Klein Performance brand in China. Both this job and my previous job have involved a lot of travelling to research the latest fashion trends. I travel to Hong Kong, Korea and Tokyo twice a year, which is totally different to travelling as a tourist. You work very long hours and need a lot of energy!

Tell us about your career so far

I graduated in Fashion Design and first worked in both fashion and fashion retail design. After almost three years, I felt I had learned a lot and was doing well in my career but I wanted to get more involved in fashion management.

I travelled to the UK to study the Master's in Design Management at Birmingham City University. I studied full-time there for a year so that I could get the management knowledge and skills I needed to progress my career.

Following my postgraduate studies I worked for Rmeo and RAPA (local fashion brands) as assistant designer, working with the design director. I learned a lot there about the design process. As a designer your job isn't just about designing, you need to be able to cope with all the different types of fabrics, trims, accessories and manufacturing considerations. In this job I was also responsible for the design of retail displays, which are important to make sure the customer gets to see the new product ranges.

I was then fortunate enough to be invited by a friend to join China Ting Holdings Group as a design manager. I think both my previous experience and my postgraduate study helped equip me for this new job. When I went to New York to buy 2012 autumn/winter products for the Chinese market, I knew how to conduct the research necessary to adjust for the difference between the US and Chinese markets. I have also learned about managing people. I have built my department with young fashion designers. This works well day-to-day, but I feel I still need to learn more; there is a lot of pressure to grow, which is good for me I suppose!

Company showroom in LinPing Head Office
Photograph by Linyang Zhang

Is there a particular project you would like to tell us about?

I was once asked to arrange the display for a new showroom in one day. This was a real challenge, as both the samples and the showroom were not finished! I had to get people to help me with the project and arrange all the samples from another department to be sent as soon as possible. We had to work all through the night! I had to be very organised; whenever I had a free moment I was planning in my mind how to arrange the samples. I felt very satisfied when the show room was completed.

The Master's in Design Management taught me well how to balance time, what needs to be achieved and available resources, all of which are essential to successful completion of a project.

The images demonstrate differences in display design for the US and Chinese markets. In this case we adapted the US display design to offer more styling guidance, showing the Chinese customer how to match separate styles together, which proved very effective for this market.

What advice do you have for students considering a similar career?

Your personal character is the crucial element for your future career. No matter what background you have, when you are involved in a fast-moving environment you need to be adaptable. It is important to build a professional reputation for yourself, the people you work with and the company. As a manager you need to be patient, kind to your team members and always full of energy. Lastly, learn how to be a highly efficient project manager; the ability to keep projects on time is very useful for both your working and personal life!

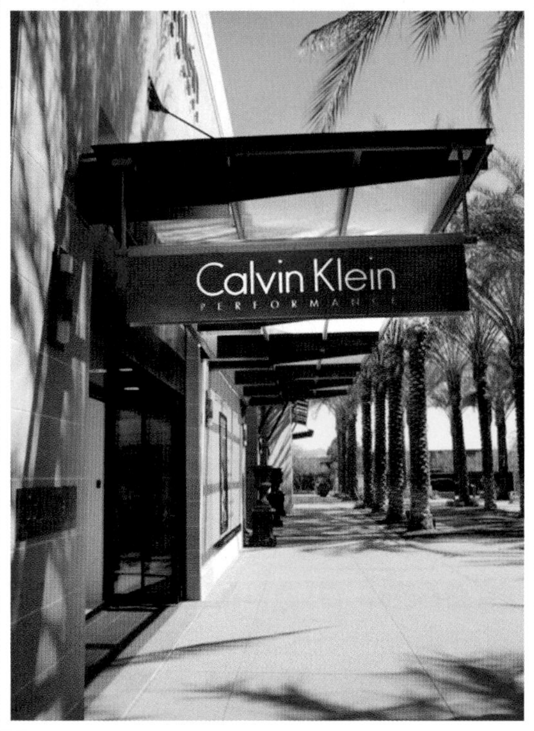

Scottsdale USA, flagship shop, 2012
Photograph by Linyan Zhang

Interior layout of Scottsdale flagship shop, 2012
Photograph by Linyan Zhang

Career profile

Name:
Edward Hollis

Current job:
Head of Interior Design, Edinburgh College of Art

Describe what you do

My job falls roughly into three categories:

1. I run a suite of degree programmes and this part of my job requires skill in organising people, timetables and resources, like any management job. It also requires me to be attentive and responsive to the needs of a diverse group of students on the one hand and the requirements of a university on the other.
2. The skills I need as a lecturer and studio tutor are rather different. It is not enough, as a teacher, to know a lot about one's subject. It is also very important to know how to communicate it to others and how to enthuse them about it. This is a real skill, and one I really enjoy practising and improving all the time.
3. All academics are required to undertake research as part of their job and I have chosen to pursue this through writing about buildings and interiors. Although this part of my job involves working alone, I love reading and writing and I find it feeds positively into my teaching.

Tell us about your career so far

I studied Architecture and worked as an architect for around six years before entering academia, as a year tutor on an Interior Architecture programme. After five years I was made head of that programme, and then the Interior Design course I now run.

Is there a particular project you would like to tell us about?

My first book, *The Secret Lives of Buildings*, was published in 2009 and I am now working on a sequel, *The Memory Palace: A Book of Lost Interiors*, which explores the ways in which we arrange the rooms we live in to remind us of what we own, how to behave and even who we are. Like *The Secret Lives of Buildings*, it's written as a series of short stories that take the reader from the cave where Romulus suckled the wolf to the last boudoir of Marie Antoinette.

What advice do you have for students considering a similar career?

If you want to be an academic in the design world it is very important, in my opinion, to have some sort of experience in practice under your belt before you start teaching. Designing is very, very difficult and it is as well to remember that from your own experience when you are teaching someone else how to do it.

TIP

If you want to be a writer . . . write! Seriously. Don't worry if you don't have anything to say. Just try to knock out a page a day, or even 100 words. Don't try to spend more than 45 minutes doing it, or you'll put yourself off. Writing is like baking or dancing or playing the recorder: a little practice once a day makes perfect.

www.edwardhollis.com

16

Working for yourself

By Jane Bartholomew, Nottingham Trent University

> *'I hadn't envisaged any of the projects or adventures I have encountered in my career but I wouldn't change them, they have been unique, interesting and my practice and clients continue to surprise me every year. Running a creative business you need to be adaptable and prepared for the unusual, it's great, you never know what or where the next project will be coming from.'*
>
> Hannah Lobley, owner of the company Paperwork

Ceramic tiles by Suet Yi Yip
Suet Yi ceramics

There are lots of success stories out there, but don't underestimate how much effort is involved in setting up your own business. You will never have worked as hard before in your life, but it'll be for you and no one else!

Many students have the ambition to start their own studio. This is understandable. It implies you want to be in control of what you do and have the recognition for your work that not all designers who work for other companies can have. However, although it is very easy to set yourself up with a studio, it is much more difficult to make it a successful venture. There's a lot to think about.

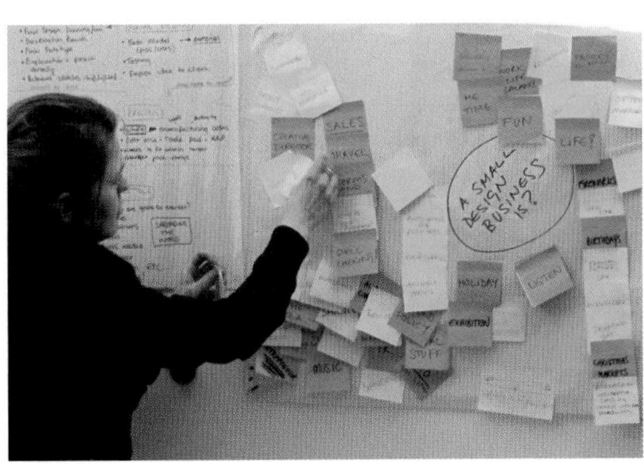

Scoping a new business, Kelsey Pilgrim on placement at Hanna Francis Design

There is no typical process to follow if you are to start up your own studio. Your networking skills should enable you to discover people who have travelled this road before you; you can learn from their experiences. You might even work for, or with them at some point. Sometimes design companies are created when one or more designer leaves a company to set up their own business.

Starting up on your own will rely on many of the personal skills you learn during your studies. This chapter will help you understand what it means to work for yourself. It will introduce you to the various business and financial considerations and offer you prompts to think things through in more detail before you commit. You can find out about intellectual property rights (IPR) and read stories and career profiles of those who have gone before you. You will certainly have a response to the notion of self-employment, either way, once you've read this section. And good luck if you go for it!

If you are dipping into the text at this point and considering self-employment then it will be worth your while referring back to the following sections for further advice and information about being a designer, not to mention reading to the end of this section to find out more about intellectual property rights:

- Chapter 10: Being creative and innovative
- Chapter 14: Future directions
- Chapter 15: So, where is this going to take you

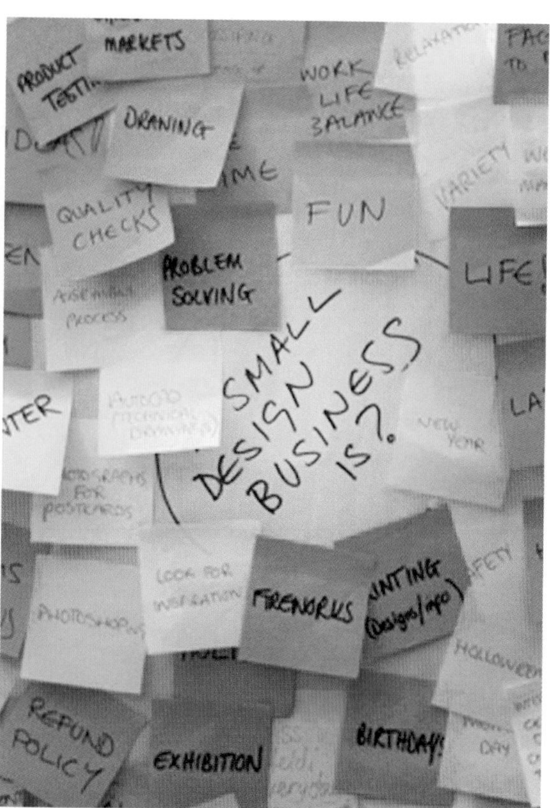

Hanna Francis' design business ideas board

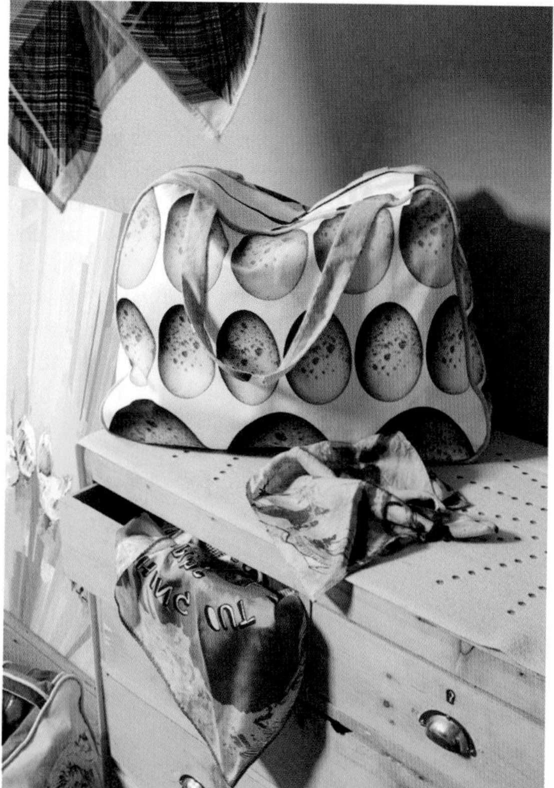
'Bad Egg' weekend bag by textile designer Craig Fellows

So, what does it take to be a successful self-employed person?

Take the test! Be honest with yourself and respond to the questions below.

You may have to think about what fellow students and tutors have said about you in the past to ensure you remain as impartial as possible when responding to this list of questions. Better still, ask a good friend to have a look at your answers – they may know you better than you do.

Self-employment – is it for you?

Statements	Tick	Actions
I am confident		
I can motivate myself		
I enjoy my own company		
I enjoy problem solving		
I work well to deadlines		
I like to meet people		
I feel comfortable presenting my ideas		
I can take advice		
I can handle stressful situations		
I am not impulsive		
I can make careful, appropriate decisions		
I do not give up when things aren't going well		
I am patient		
I can handle numbers		
I am adaptable and flexible		
I can plan ahead		

If the majority of the responses to these questions are positive then you are well on the way to identifying yourself as the right type of person to become self-employed. If there are quite a few un-ticked boxes then you need to take stock of the situation and carefully consider committing to self-employment at this time, and instead focus on spending some time gaining the skills you are missing – put some suggestions as to how you might do this in the 'actions' column. If you can work out a plan as to how to improve the situation and find ways to gain the skills or attributes that are missing at the moment, then this is still something you could consider for your future. You might find you would be better entering into the industry and gaining more experiences first before going it alone.

Expectations

Many students believe that self-employment will be great as they will be able to carry on being creative and develop whatever they want. In some ways this is true, but you also need to be able to promote, sell, market, research, negotiate, work to deadlines, keep financial records and basically be an excellent administrator!

It is useful first to do some detailed research about who your customer is and whether the products and services you want to sell are those actually sought after by potential clients, or whether they are items that you just want to keep on designing because you enjoy it!

Find out about the short courses or business services in your area so that you can get some advice about preparing to set up in business – they're invaluable.

'Running your own business can be liberating, but working on your own can also be isolating and you need to break out occasionally and explore your work in different contexts.'

Craig Fellows, self-employed textile designer, graduated in 2007

'I am still learning and developing my skills in my own business. For example, handling tax, VAT, manufacturing and working wholesale, etc. are all things that I've had to learn after graduation. By joining the organisation Design Factory, I now have the tools support and opportunities to help me further develop my business.'

Craig Fellows

Setting up in business isn't something that can be done quickly, and it is wise to uncover the true facts about the product, customer, market and financial implications to really understand whether it is going to make any money, as there is little point in going into business if it doesn't.

You might want to try it anyway, but you still need to understand that you need to generate enough sales to cover your overheads (rent, heating, etc.), raw materials (the things you need to buy in to make your products) and to pay yourself a wage (because you need to eat!). You also need to keep some money in the business account in order to move business forward so that you can keep developing new ideas.

Craig Fellows, 'Beetle Bum TWIRL' silk scarf

Business awareness and financial issues

'My advice would be to make a well-informed and intelligent appraisal of the proposition and realise there is no career, only vocation. There can be no career where there is no industry. The only way to make a living in this area is to work hard on the basics, define a product, set and find a market, wherever that might be. Understand the numbers involved – know your production costs and costs to get your product to market. I have always been a great believer in studying business practice to support my creative endeavours!'

Gill Wilson, paper-maker and gallery director

Throughout the next few pages, the same example will be used to help you understand and illustrate the basic financial issues that are common to all businesses. This is based on a small design business during the first few years of operation.

Overheads

Overheads – what are they?

Costs	Annual cost (£)	Assumptions (useful to keep notes here to explain cost in more detail, especially if you are wanting to borrow money or apply for a grant)
Rent and business rates	2,500	
Electricity and heating	1,600	
Telephone	600	
Legal and accountancy fees	1,000	
Interest on loans	330	
Advertising and marketing	700	(magazine and online promotion)
Selling	500	(travelling to clients)
Exhibiting	4,000	
Stationery	450	
Raw materials	5,000	(anything you buy in to make from)
Your ideal salary*	15,000	
Total	**31,680**	

* It is a good idea to do a separate list of all the personal costs you might incur over one year (living costs, running a car, food, entertainment, Christmas presents, etc.) so you can appreciate what you need to earn as a basic salary. In the first few years it is unlikely you will earn £15,000, but if you don't include a proper salary then you are being unrealistic from the start.

When you estimate what your overheads for running your studio might be, there are some fundamental questions that you need to ask yourself, such as where will the workshop be based? A word of caution here: if you start out by setting up your studio at home you might calculate the final cost of your products with the minimal amount of overheads that this will bring. If, however, within a couple of years you decide that you need to expand the business and move to a larger studio, you may find that if you try to increase the costs of your products, this will probably not be received well by your customers. Good advice would be to not undersell yourself in the initial stages – imagine your business expansion possibilities within the first five years and develop prices that cover all eventualities. This will be more sustainable in the long run.

How do you cost and price a product or service?

Keep a record of the time you spend developing your prototypes, even in your final year, including the cost of all of the materials, any tools that are used up during the production process (e.g. sewing machine needles, sandpaper, etc.), packaging and distribution costs. You can use this to work out costs for a batch of products, or for a single item. If you are selling a service instead of a product, then adjust the items list accordingly to include travel costs, etc.

Costing a product or service

Items to include	Cost in £	Assumptions
Raw materials		
Expendable tools		
Designer's labour costs		number of hours × your hourly rate (see later in text for details)
Employee's labour costs		if applicable
Industry's labour costs		if applicable
Courier costs		
Packaging		
Profit (usually 100%)		
Total (wholesale price*)		

* The wholesale price is the price you sell your goods on to a retailer for. Therefore, the final price to the customer will be at least double this, plus VAT (value added tax).

Remember to complete the costing table as if you were proficient at producing many products already. Allocate your time by working out how long it will take you to make the next one (or batch), as opposed to the first prototype.

What will the customer pay?

After you have analysed the product costs, if it is not worth the amount you need to charge then you need to consider whether you can do any of the following:

- find cheaper suppliers of the raw materials
- increase the perceived value by making it look more expensive so that it sells within the given market
- lower the 'time' costs and analyse whether it would be more cost-effective to employ someone on a lower wage to undertake certain parts of the making
- make more than one at a time
- look at ways to speed up the process
- buy in ready-made parts to incorporate into the product

How many hours a week will I be able to 'make' for?

You have to appreciate that much of your time is spent doing all of the other tasks that makes running a business viable, this means that only half the hours per week are given to the wonderful tasks of designing products and services. Look at the information in the table to gain a better understanding of the reasons behind this statement.

How many 'making' hours are there per year?

How many weeks a year do you work?	How many hours per week do you work?	How many of those hours do you spend designing and making?
(allow for bank holidays, annual holidays, sickness)	(be realistic – you will always work longer hours when there is a deadline)	(this allows for 16 hours for research, development time, sales, marketing, business development, administration, making cups of tea, etc.)
48	**40**	**24**
So, to work out the total number of making hours per year, the sum is **48 × 24, which equals 1,152 (total number of hours for 'making')**		

You may not want to agree that, on average, 16 hours a week is spent on other business-related activities, but this is part of general business advice and based on the experiences of other self-employed designer-makers. Generating more sales by improving promotion, marketing and selling strategies can soon take over. It is creating a balance between the two that will ensure a successful, profitable business.

> *'It's hard work but could be the best thing you ever do – it was for me. Be prepared that making your work is such a small part of running a creative business; you have to learn marketing, accounting, correspondence etiquette, sales, to name a few. I would advise joining creative groups, going to all the networking events, meeting people. The creative sector is a very generous sector and colleagues are always willing to help and share knowledge.'*
>
> Hannah Lobley, owner of the company Paperwork

Your hourly rate

Now you know what your annual overheads and your annual total designing and making hours amount to, you can work out what your hourly rate is. Taking the existing example, let's complete the sums.

Calculating your hourly rate

Overheads		Total designing and making hours		Hourly rate
£31,680	÷	1,152	=	£27.50

The hourly rate may be only an estimate, but can be very useful as you can begin to appreciate what your time is now worth and cost products and services correctly. By understanding money and costing issues it is likely that you will be in business longer than someone with limited awareness of these issues.

How do you start your own business?

Katie Dominy, Arts Thread and
Jane Bartholomew, Nottingham Trent University

Starting your own business in the UK is similar to starting to practice as a freelance designer. You inform the tax authority, start work, keep your receipts and get an accountant. If you do this from home then you're freelance. If you do it from a rented office space then you have started your own studio.

The simplest form of this is to operate as a sole trader in your own name. This doesn't require company registration, premises, etc., but you will have to fill in a tax return every year and submit accounts that detail your income and expenditure. You can claim certain expenditure against tax – so, for instance, you won't pay income tax on money you spend on train fares visiting a client. You can also claim certain other costs against tax, such as the costs of running a vehicle and a certain portion of your home, including heating and lighting, used primarily for your business. You should engage an accountant to manage this aspect of your business, no matter how small a business you are. It's the most 'value for money' cost you'll ever incur!

If you are considering renting premises from which to work, employing others or are involved in big sub-contracting projects, then there are other ways to set up in business. This includes limited company status, which means that, as a director of the company, you are not personally responsible for the liabilities of the company but rather the company remains responsible.

The UK Government offers excellent advice on the types of businesses you can operate and how to start operating legally . . . (www.hmrc.gov.uk/startingup).

The first practical thing you need to do is to create a company name and register it. Check all of the names already taken and any other companies with similar names, both with Companies House and online, so that you avoid any overlap or marketing problems. Companies House is the government organisation that regulates the registration of companies in the UK . . . (www.companieshouse.gov.uk). For some reason, most types of fruit have already been used as names for design consultancies. Don't ask me why. No one knows!

What to do next?

First, seek advice from:

- regional business links
- university lecturers and careers departments
- specialist organisations, such as the UK's Crafts Council, that offer advice or can be a platform for services available to graduates
- local specialist studios or workshops, set up to encourage creative businesses, who often offer start-up packages for graduates, including lower-cost rental.

One example is the Hothouse scheme from the UK's Crafts Council. It runs right across the UK, offering three-month business support and mentoring schemes.

> 'Hothouse provides up to 40 emerging makers (those defined as being within two years of setting up a practice) with a programme of focused, intensive business skills and creative development, complemented with one-to-one support over a six-month period.'
>
> www.craftscouncil.org.uk

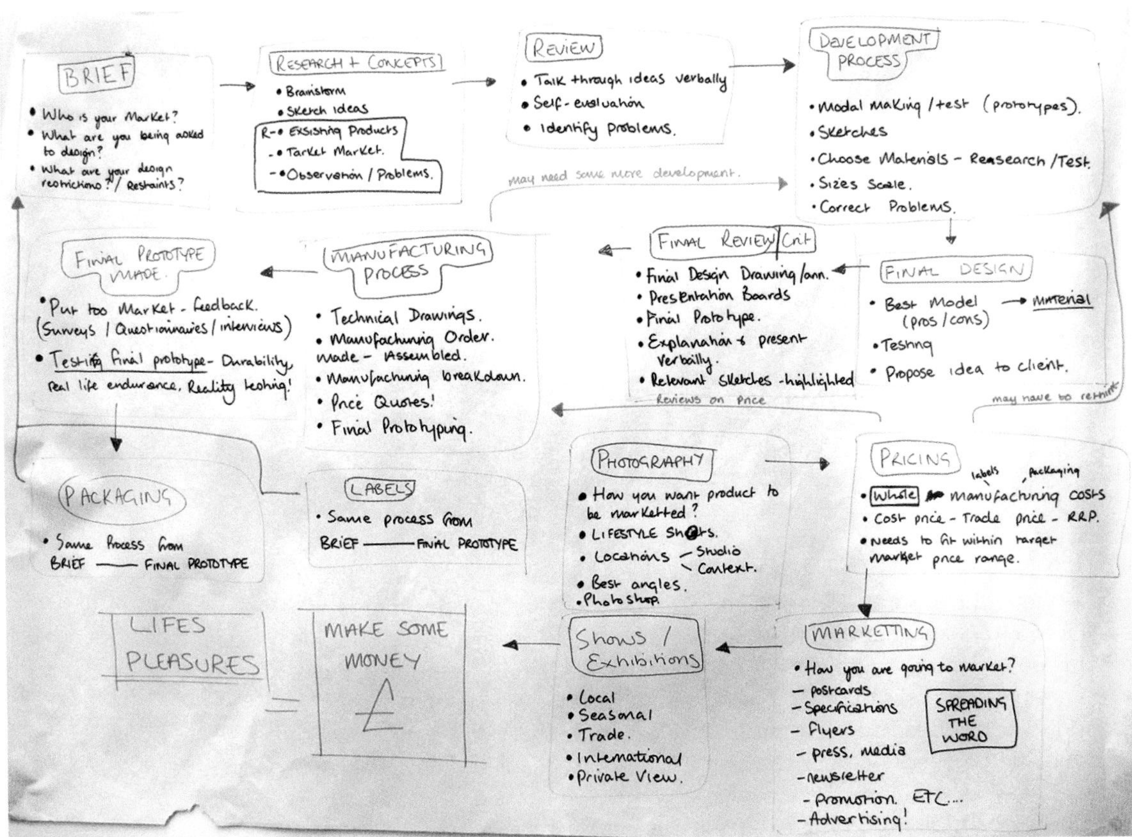

Business planning at Hanna Francis Design

A word of caution, though. James Coleman, Managing Director at Supercool, a graphic design and branding company, is now experiencing great success with his business, but it wasn't like that at the start! Looking back, he tells it as it was, when he and some friends on his course were considering going into business.

They soon realised that the success would be in the detailed understanding of what they were trying to achieve, and in the planning of it all.

> *'Many of our tutors were quite pessimistic about our chances of surviving in the big wide world by ourselves. One tried to persuade me that I'd be better getting a job first and then look at setting up by myself in a few years' time. Another said that it all depended on the profile and the network of people we could build up while still at university, and he was right. A year later we graduated and it became obvious that we'd failed miserably at building up any kind of network, let alone potential clients. One of us dropped out and we were down to two. We had virtually no work for the first six months, and the only thing to show for it was a fine from HMRC for failing to submit a tax return.'*
>
> James Coleman, Managing Director at Supercool

Intellectual property rights

By Steve Rutherford, Nottingham Trent University

As a designer, some of your work is protected automatically through intellectual property rights (IPR) legislation. However, with some of your work you may need to protect it yourself. This is dependent on what kind of work it is:

- **Patents** cover functional inventions. This definition is very wide and includes certain animals bred for research, drugs, aspects of computer interaction, as well as products and processes. Although it is cheap to apply for this, in the end it can become expensive.
- **Design Right** is free, automatic protection for 3D forms.
- **Registered Design** is a higher level of protection for 3D and 2D designs, for which you must apply and pay.
- **Trademark** protection allows you to protect your trademarks, logos and icons. You must apply and pay for this.
- **Copyright** is free, automatic protection for your art, drawings, photographs, patterns, writing, music and even software codes.

So, what can you do? Be aware of the legislation, as many designers aren't. Let people you meet in business know that you know about it and possibly have applied for protection for one of your ideas. This will gain you respect and might put people off stealing your ideas. Search the online databases of intellectual property and see what people in your industry protect (this is explained later).

If you are a student, your institution will have a policy on student IPR and this will state the exact position you are in. Some educational institutions retain the IPR and will want your help to exploit any good ideas you may have, and will reward you for doing so. Some state that you retain the IPR for your ideas; however, most of them have business development units that will provide invaluable support to help you exploit those ideas. Others will state that there is joint ownership and will be specific about the split in any proceeds if the IPR generates income. These, too, will have the expertise to help you take the ideas further. Educational institutions' positions on this subject might vary, and also might change with time. Always be aware of your own situation.

What types of IPR protection are there?

(The details in this section are subject to change and were correct at the time of writing.)

Patents – the protection of an invention or inventive step:

This can be expensive, and gets more expensive as you renew it every year on the basis that you will by then have started to make money from your idea. However, to apply for a patent is £30 and the initial stages are not prohibitively expensive, should you prepare the patent application yourself. If your idea is simple then this is achievable. If it is not, then you will need the services of a patent lawyer or consultant and this will be more expensive.

Costs: patent costs are complex. Application is cheap; however, to get the application finally granted involves other processes such as searches for similar patents. The Intellectual Property Office states that it normally charges £230–280 to process an application, but if your patent application is complex and requires the services of patent experts these costs will increase hugely. Renewal fees are on a sliding scale every year, going up from £70 in the fifth year of the patent to £600 in the twentieth year.

Designs – the protection of the visual design of something:

There are two stages to this protection. The first is Design Right. This is free and automatic – you do not need to apply for it. It protects the internal and external configuration of something, i.e. a three-dimensional object. Protection lasts 15 years from when you designed it, or 10 years from when it was first marketed, whichever is earlier. It will protect you from someone directly copying your work in every detail. Unfortunately it is very simple for a manufacturer to change one or two details of your design to get round this and manufacture close copies. Therefore there is also Registered Design protection. This protects not only the three-dimensional shape and form of an object, but also two-dimensional designs and surface patterns. It protects the general look of the design, not its function, and will help prevent the production of near copies. This is easy to apply for – you can produce the images and write the description yourself. This is a very affordable type of protection (see Paul Dack's project story at the end of this chapter for an example of how a second-year student did this).

Costs: £60 for the first year, rising every year to £450 for the fifth year, with 25 years' maximum protection.

Trademarks – the protection of logos, icons, names, etc.:

This protection is not automatic and costs; however, it is easy to apply for. This will offer you protection for the graphical and textual elements of your business: a company name in a certain typeface; any 2D graphics involved in the branding of the product and company; product names; etc.

Costs: £200 to apply, £200 to renew it every year.

Copyright – art, drawings, photographs, patterns, writing, music, software:

This is automatic and free, which sounds ideal! However, if anyone does steal your ideas you will have to go to court and prove that you had the idea first. This could be complex, expensive and fruitless. If you go up against the might of a multinational company with a huge copyright law department you should not be surprised if you lose. The element of evidence is what will win your case. Check the reference to ACID (Anti-Copying In Design) later in this chapter for advice on this.

It is not possible to include all of the important detail here regarding applications, international aspects, etc. All of the advice, detail, costs and forms you require to use IPR legislation are provided on the UK Government's excellent Intellectual Property Office website: www.ipo.gov.uk. All of the protection cited above will not make you immune to fraud. Some companies will take the chance that you will not find out, or will try to change your design just enough to get round the legislation.

Understanding IPR in more detail

Whether something you design is yours to protect in the first place has also to be examined. For instance, if you are employed as a salaried member of staff in a company, then your contract will probably state that everything you produce is the intellectual property of your employer. In this case you do not need to worry about IPR, it is the company's responsibility.

If you are an independent designer then it is something you need to be very aware of. If you act as a consultant to a client they will have a view on whether they want to own the IPR or whether they are happy for you to. As this is often linked to how you are paid for the design, then this should be clarified before entering into a contract. For instance, if a company is going to make something you have designed and **you retain the IPR** in the design, you might be paid only on a royalty basis. If they pay you when you hand over the designs they might be **selling the IPR** too. Clarify this before you start working!

There are no hard and fast rules on IPR and there will always be exceptions; however, these are two very common patterns of working in the design industry.

You retain the IPR: you do the design, they make it and they pay you a percentage of what they get for it. The money will start to trickle in. If it's a hit you might be very lucky and make it rich. If the design has a long life the money will continue to roll in for years. The down side is that you aren't paid at all during the design phase and you have to invest your own time and money on realising the design. Of course, it's a gamble for you and them as to whether sales take off or not.

Selling the IPR: you do the design and the company pay you for it (probably at the end of the design stage or, if it's a big job, in stages throughout the process) and take over the IPR. You get the money much earlier but, as you've sold the IPR, there's no more money coming in, no matter how successful it is in the future. This is less of a gamble than retaining IPR and receiving royalties.

To bother with IPR or not?

Not all designers or manufacturers put their faith in the IPR legislation. Many just 'go for it' and get their designs into production before anyone else and base their reputation on this innovative approach. They know people will follow and imitate, but by that time the true innovators will have moved on to the next big thing. Funnily enough, one of the disadvantages of protecting your work is that, in order to stop others copying you, your ideas must be published.

Anyone can go to the IPO website and search patents, designs and trademarks. Indeed, some companies do exactly that, looking for ideas to exploit – sometimes contacting the owners of the IPR and dealing with them up front, and sometimes waiting for the patent or registered design protection to expire before using the idea (not stealing it, because it is now in the public domain!).

Searching the databases is something you should do too if you have designed or invented something specific. Make sure you find out early in the project whether anyone has already done it and protected it, otherwise you may be wasting your time working on it and they might come after you!

There is a that sending copies of your work to yourself and not opening the envelope will somehow prove that you produced the idea before someone else. This may be because the UK Government's Intellectual Property Office (IPO) states that, in order to support your automatic design right protection, you may:

> 'take certain steps to provide evidence that you are the first owner of the Design Right. You could, for example, deposit a sample or a copy of your design drawings with a bank or solicitor. Alternatively, a designer could send himself or herself a copy by special delivery post (which gives a clear date stamp on the envelope), leaving the envelope unopened on its return. However, there is no guarantee that this will prove establishment of Design Right before the courts. A number of private companies operate unofficial registers, but it would be sensible to check carefully what you will be paying for before choosing this route.'

Therefore, do not assume that what is in the envelope will protect you if someone takes you to court for copying their work.

One of the 'unofficial registers' that the IPO refer to in the above quote is ACID (Anti-Copying In Design www.acid.eu.com). This is an organisation that will help you compile evidence. They work extensively with the design and craft industries, and their reputation in the industry is wide-spread. They also have links to a couple of law firms who deal with IPR on their website.

To sum up intellectual property rights

Why do it?

- It makes you look professional
- It can be easy and cheap
- You can often do it yourself
- It puts people off stealing your ideas

It also gives you something tangible to sell – the right to the IPR completely, or the right to manufacture via licensing or royalties.

Why shouldn't I do it?

- It puts your ideas in the public domain
- Lots of companies search through IPR databases to copy ideas
- Some manufacturers don't like dealing with it
- If your work is already in the public domain (degree show) it's too late!

And you might still get your ideas copied . . .

Remember that . . .

- Depending on what kind of work you are doing, you may have automatic rights over your designs (design right or copyright), that you don't have to apply or pay for.
- If, as a student, you produce a wonderful idea you will find that your institution is very willing to help you, but remember they'll want something out of it, obviously. Trust them.
- If you're going on *Dragon's Den* (©BBC), owning the IPR is a really good idea! Some of those promising ideas and deals fall apart when the gory detail of the IPR (or lack of it) comes to light.

Finally, you cannot protect an invention or a design if it has already been seen in the public domain. For the purpose of IPR, your degree show or any other event or competition external to your institution is considered the public domain. This also includes having your designs available via an online portfolio.

Once an invention or design is in the public domain, no one can protect the idea, not even you.

Conclusion

Working for yourself is a challenging and rewarding career choice but often financially difficult at the start. If you intend to start a business then you might need to find a part-time job to support yourself, so try and choose something that complements wherever possible – for example, working in a related retail environment such as a jewellery boutique three days a week and working on your own fledgling jewellery business the other two days (plus weekends!). Self-employed designers and makers find that they often need to work long hours.

The undergraduate dissertation 'Is Becoming a Designer-Maker a Financially Sustainable Career Choice?', by Sophie Minal, summarises the pros and cons of setting up a creative business. In this dissertation she undertook broad research, including a survey of a number of existing designer-makers, to arrive at this conclusion:

> 'The responses indicate that determination, time and effort, flexibility and being commercially realistic are necessities to survive. Factors such as where and how you sell your work, developing a strong brand identity and considering other sources of income are crucial.'

Her variety of experiences during her studies, including placements and high levels of experimentation with materials and processes, has led her to this concluding statement:

> 'These experiences made me realise two things: I want to create unique and innovative designs and I want to have full control over my designs and keep them as my own under my own name. Knowing this, and the fact that I have a true passion for being practical and making things, I decided to seriously investigate the possibilities of becoming a designer-maker.'

Good luck! The advice is to keep networking, as it is the support from others doing the same that will keep you focused and determined.

Read the following set of career profiles and project stories about self-employment to find out how it really feels.

Further resources

Books

Airey, D., *Work for Money, Design for Love: Answers to the Most Frequently Asked Questions About Starting and Running a Successful Design Business (Voices That Matter)*, New Riders (2012)

This book is inspired by the author's communications with up-and-coming designers on his blogs (www.davidairey.com, www.logodesignlove.com, and www.identitydesigned.com).

Benun, I., *The Designer's Guide to Marketing and Pricing: How to Win Clients and What to Charge Them*, How Design Books (2008)

A great book from the USA, it talks you through every step of marketing and financing. This is from the USA so there will be certain legal aspects that don't apply in the UK.

Branagan, A., *The Essential Guide to Business for Artists and Designers*, A & C Black Publishers Ltd (2009)

This has been designed for artists and designers who want to succeed at making money from their work. It covers business, promotion and legal money matters. This is a useful resource, with invaluable information for all creative practitioners.

Gauntlett, D., *Making is Connecting: The Social Meaning of Creativity, from DIY and Knitting to YouTube and Web 2.0*, Polity Press, USA (2011)

This book brings together the subjects of craft and creativity and places them in the context of the twenty-first century in relation to technologies and the community.

Perkins, S., *Talent is Not Enough: Business Secrets for Designers*, New Riders (2010)

A great book – however, the fact that it's from the USA means there are certain legal aspects that don't apply in the UK.

Williams, S., *The Financial Times Guide to Business Start Up 2013: The Most Comprehensive Annually Updated Guide for Entrepreneurs (The FT Guides)*, FT Publishing International (2012)

Understanding the business decisions you make in terms of financial implications is crucial to your business succeeding. This contains all of the important broader contextual information you need to understand to be successful in business.

Websites

www.acid.eu.com

Anti-Copying in Design is a membership organisation committed to raising awareness of intellectual property rights.

www.artsthread.co.uk

Arts Thread media is a social enterprise designed to connect designers to the industry. It includes new portfolios, courses, student and graduate shows, design exhibitions and competitions.

www.creativebarcode.com

Helps protect you as a designer and deals with matters relating to intellectual property rights.

www.creative-choices.co.uk

Part of the Government's creative and cultural skills intiative, this website is packed with interesting information, tips and profiles of practising designers, artists and other creatives.

www.designcouncil.org.uk

Looks at design in the context of all of the industries and reports on news matters, events, debates, etc. in the wider design community.

www.designersmakers.com

A not-for-profit agency for UK design and craft. Within the shop you can find work from over 60 members. Also useful information about past and future craft/design events throughout the UK.

www.ideastap.com

A creative network for young creatives, primarily in the areas of drama, film, photography and art.

www.mycake.org

An online toolkit to help you manage the financial side of your business simply and easily. So if you see commerce as a valid output for your creative endeavours, then MyCake is here to help you make the most of it.

Organisations supporting business start-up

This is not an exhaustive list and there are many more start-up support organisations across the country, so do plenty of research into what is available in your area.

Business Link

A government resource providing attentive information and grants, as well as business planning advice.

Capital Enterprise

This organisation runs various enterprise programmes for fashion, designers and freelancers.

Cockpit Arts

This organisation supports craftspeople who have chosen to follow a career in making and designing. They have incubator spaces for over 150 designer-makers and they offer help and support to accelerate businesses.

Cultural Enterprise Office

Scottish organisation that provides specialist business support.

Design Factory

Design Factory works to raise the standard of craft and design and to commercially support and develop the very best designers and makers who have developed a career as a practitioner.

The Princes Trust

This has its own enterprise programme aimed particularly at people aged 18 to 30. The trust offers legal advice and continued mentoring support, as well as start-up loans.

Shell liveWire

This is for young entrepreneurs wanting to start their own business. They can pitch their business idea to win cash prizes each month (www.shell-livewire.org).

Yorkshire Art Space

They run a studio start-up programme in Leeds for silversmiths, jewellery-makers and ceramicists.

Project story

Designer:
Paul T. Dack, furniture and product design student

Project title:
Registering my design

Front perspective – view from one side
Paul T. Dack

Tell us about your project

In my second year of university I had to design a bookcase for my fellow classmate, Lucy. The project encouraged me to use qualitative research methods that helped me build a picture of Lucy's practical and emotional needs. The research gathered information about Lucy's personal interests, hobbies, lifestyle and daily routine. This research included informal interviews with Lucy and observations and analysis of her living environment.

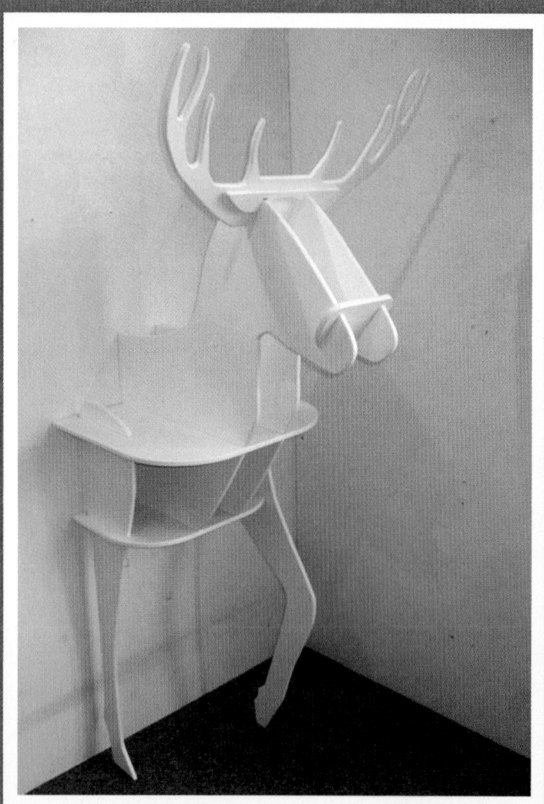

Bookcase
Paul T. Dack

The research highlighted Lucy's passion for animals. It was also clear to me that she was a friendly, fun-loving girl who liked quirky, witty design. As a student she lived most of her life from her bedroom, so a bookcase was more than just a place to store books, it was a place to display her most beautiful and treasured belongings. Lucy lived a very mobile lifestyle, having to move between rented student accommodation and her parents' house several times a year.

The final design came in the form of a floor-standing, wall-mounted, flat-pack deer. The design met with all of Lucy's aspirations, tapping into her love of animals and reflecting her fun personality. Being able to erect and dismantle the bookcase and move it easily made it great for moving house often.

Why did you think you needed to protect it?

The design is very quick, cheap and easy to manufacture. I began to get very excited about the commercial opportunities for my design. I needed to show my design to retailers so that I could get an idea of their interest and whether any changes needed to be made to the design to make it more appealing to them. I also needed to show it to manufacturers so that I could get an idea of the costs involved.

Before I could do this, I needed to get some design protection so that my design would not be copied or stolen. Sometimes as a student I have found it difficult to see the commercial value of my design work. However, the reality is that it is a very valuable service that should be rewarded fairly. I made a decision that I was not studying design to work for free!

Was it easy?

After a lecture on intellectual property rights I turned to the Intellectual Property Office to see what protection I could get. The government website **www.ipo.gov.uk** was very helpful and informative. As it was the look and feel of the design I wanted to protect, I applied to register my design. The application process was surprisingly easy and straightforward. I downloaded and printed the application form from the IPO website and filled in a few details. It was as simple as filling out my name, address and the word 'bookcase'. I also had to include illustrations of my design. It is important to clearly illustrate the look of the product, so I included perspective line drawings showing the bookcase from all sides. I also included plan views of the product from all angles. I included a cheque for £60 and posted it. Within six weeks I had received my 'Certificate of Registration of Design' – it was that easy!

Was it worth it?

Now I am free and safe to show people my design without worrying about having it stolen. I am considering developing the idea into a range of animals of different sizes, particularly for use as fun children's bedroom furniture. I am also looking into developing the design into a Christmas cardboard pop-up display with a red nose. I have since won a design competition with this idea. Once I had registered my design, I was confident that I could display my work publicly and online without having it stolen, thanks to my registered design status.

Career profile

Name:
Anna Glasbrook

Current job:
Self-employed designer of 'art for architecture'

Describe what you do

I'm a self-employed designer, making art for architectural settings. I create stitched architectural textile installations suitable for wall hangings, screens, panels and space dividers for interior and exterior spaces.

Being self-employed is both exciting and scary. It means that so far there is just me, working in my studio on my own and carrying out all the tasks. Many of these are unrelated to the creative part of my practice but are totally necessary in running any small business. Keeping the books, writing press releases, promoting myself (so much harder when it's just you), answering endless emails and fixing the telephone and other technology that goes down are just some of these tasks.

However, there are plenty of upsides to being self-employed, one of which is that I can decide when I want to work – often I'm at my most creative and have my best ideas pottering in the studio when it's dark, late in the evening. Or, I can decide to go to London for the day to get some new inspiration, although I still have to put in long hours in order to build my business. Being self-employed also means I am in control of the jobs I take on. So, one day I might be making a small domestic commission, the next designing screens for a show garden at Chelsea Flower Show, or creating a huge purple corset for Sir Peter Blake to 'unzip' for the grand reopening of the Holburne Museum in Bath.

I like to work on projects in collaboration with architects, engineers and interior designers, ideally creating large-scale, site-specific work that responds to the space and the architecture of the building and its surroundings. Of all the skills you need to work in this way, by far the most important is resilience. Both resilience and a determination to succeed, even when you are faced with rejections, need to be present within you.

'Purple Corset' for the reopening of the Holburne Museum in Bath, England

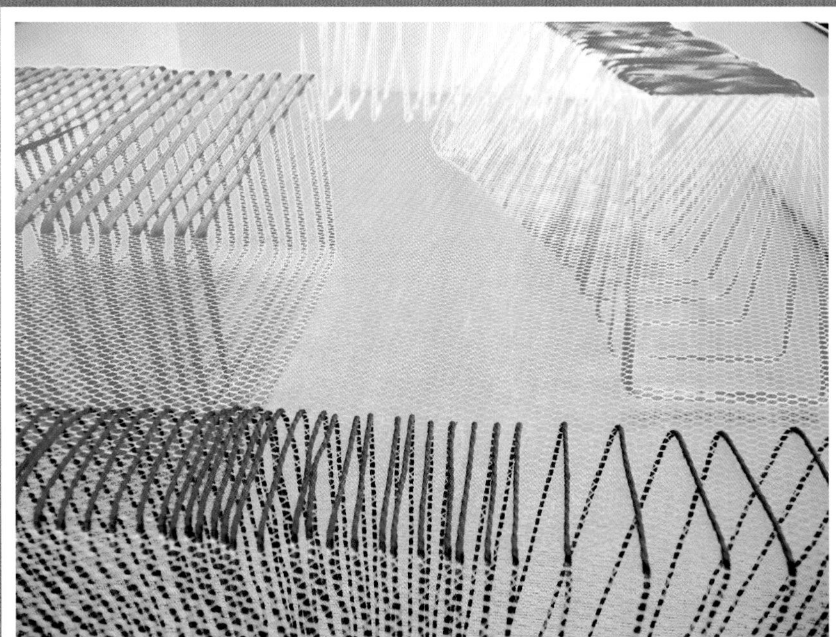

'A light exists in Spring'

What inspires and motivates you?

I always enjoy pushing the boundaries of textiles. I use ribbon and mesh and often work between two or more transparent layers to create an infinite weaving in space, designed to entice the viewer to investigate further and experience the colour, light and form of the work. My technique allows me to explore the space between the lines, rather than the line itself.

My sculptural installations capture colour and movement and play with perspective and the perception of space. Working in three-dimensions is exciting as it allows me to express a combination of delicacy and strength in my pieces.

I like to work big and I like to work small. Often an idea begins by stitching through clear plastic packaging – ready-made poppadom packets are great, but I end up having to eat a lot of them. Obsessed with transparency, I collect plastic packaging destined for the recycling bin; the rigid plastic that covered a tube of glue, little scraps of colour still clinging to the plastic where the cardboard ripped off. I stitch into these while at the breakfast table or watching a film, using that part of the subconscious that comes into play when you are not fully focused on a task – not trying too hard, but reacting to the object intuitively.

These ideas act as my sketchbook, encouraging free experimentation and the development of fresh new ideas. I now have a large collection waiting to be scaled up and developed into large pieces of work.

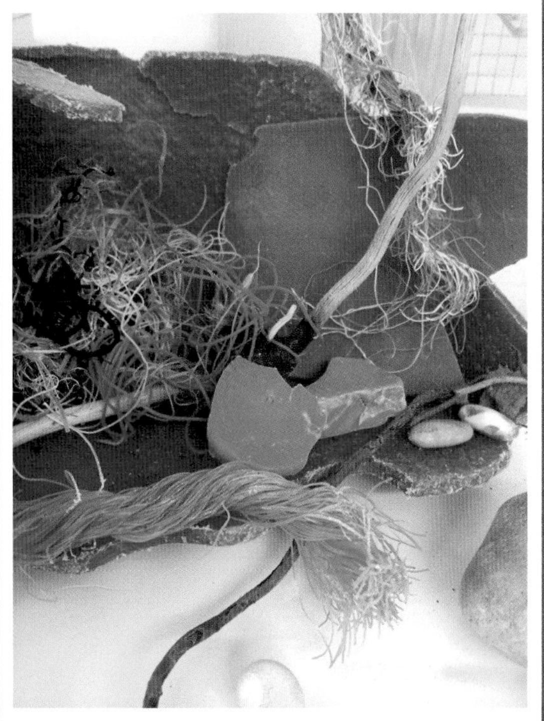

Inspirational materials

Tell us about your career so far

Although I originally trained as a speech and language therapist, I have always taken endless photographs of things around me and collected stuff – bits of coloured plastic I find on the beach, scraps of fabrics, interesting yarns and ropes. Eventually I realised that I needed to do something with this so I decided to start at the beginning again and do an Art and Design foundation course at Trowbridge College.

This was the best decision I ever made! It was experimental and a lot of fun and we got to try out and play with lots of techniques and materials. It is run diagnostically, and geared towards finding out which area you are most suited to, from print, textiles, 3D and illustration. This was when I really discovered a passion for textiles and got the confidence to apply for a degree course.

A year of a foundation course allowed me to get a strong portfolio together and I was accepted onto Bath Spa University's degree in Fashion and Interiors. As on foundation, the first year was about exploring lots of specialist areas – print, knit, weave and embroidery – before narrowing down to specialising in one area. The course allowed me to develop the way I wanted to. I was interested in weaving but wanted to do it my way (involving as little maths as possible). The technique I developed came about during the final year of my degree and grew out of my interest in weave. I managed to get into an exhibition of well-known textile artists and challenged myself to make a large-scale work for it.

The work was received very well and led to commissions that, in turn, gave me the confidence I needed, on leaving university, to book a stand at Tent London – an exhibition for new creative business ideas. This in turn led to more great opportunities and the all-important press, resulting in my business growing from strength to strength.

Is there a particular project you would like to tell us about?

One client I met at Tent went on to commission me to make a series of three large screens for his major show garden at Chelsea Flower Show. The screens were to be part of a striking circular pavilion in the garden and this meant working not only with the

Screens for the Homebase Garden at the Chelsea Flower Show

garden designer but also with the architect, joiners, fabricators and landscape contractors. Working with other designers involved being flexible and adaptable in my approach. Self-employment can mean that you are in your own company a lot, so it's great to get out of the studio and work with others whenever possible. I designed the screens in response to the theme of the garden, but also with reference to the architecture of the pavilion and the planting in the garden.

It was exciting to be part of such a high-profile project. I even got to help with some of the planting!

What advice do you have for students considering a similar career?

During your studies, talk to everyone and anyone and enjoy listening to people talking about themselves. Find out as much as you can about the areas you're interested in, and then do as much relevant work experience as possible before you graduate. Nothing compares to actually experiencing what goes on with real clients in a real studio. Then, probably most important of all, ask yourself what you are passionate about, what excites you and what's in your heart; follow that and you can't go wrong.

Career profile

Name:
Alexander Taylor

Current job:
Furniture and Lighting Designer

'Fold' lamp
Photographer: Peter Guenzel; manufactured by Established and Sons

Describe what you do

I have been running the studio since 2003 and am pleased to say that business is going well. I'm still heavily involved in production and create prototypes and models, although now I license my design to manufacturers, which allows me to be involved in many different projects. However, keeping a studio means I need more work as there are constant costs associated with running a business space, even before you consider employing staff!

I have worked with a leading design gallery, David Gill, creating limited-edition work for my collectors and I also consult for the sports brand Adidas. I have also been invited to lecture and run workshops in some of the world's best design schools.

Clients include some of the world's most respected furniture brands, including Established & Sons, who produce my 'Fold' light – now part of the permanent collection in the Museum of Modern Art in New York. (See the project story in Chapter 8, 'Three-dimensional design'.)

'Antlers' coat hanger
Image Courtesy of Thorsten van Elten; manufactured by Thorsen van Elten

Tell us about your career so far

I graduated in 1999 with a Furniture and Product Design degree from Nottingham Trent University. This career profile will give a brief indication of how determination and hard work can lead to a career doing something hugely satisfying and enjoyable.

My placement year, year three of my four-year course, was in the kitchen industry, primarily as an opportunity to live in London and work on the King's Road. It wasn't my ideal job but it was a start, and later gave me a way, after college, to earn enough to live in London and seek an unpaid placement with a design studio.

When I graduated I presented my work with the rest of the college at the New Designers Exhibition, the national showcase for graduate talent that occurs every July in London.

I had many encouraging conversations and my work was selected for publication in one of the best design journals. However, I must be absolutely honest that, at this point in time, I had no idea where I was heading.

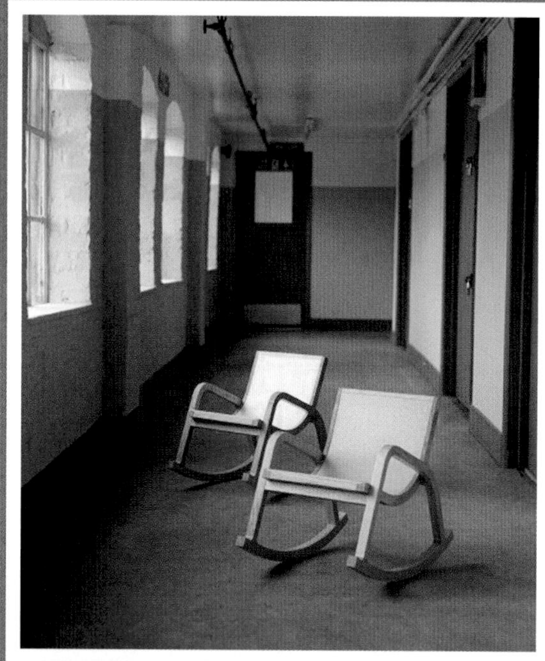

'Kids Rock' chair
Image Courtesy of Thorsten van Elten; manufactured by Thorsten van Elten

Once the New Designers show was finished and the summer was over, I had to do some work!

I went back to my placement company and worked part-time for a high-end kitchen showroom dealing with large kitchen contracts (and being shouted at by clients!). I happened to meet another architecture firm while working on one of their kitchens, which got me thinking. After some cold calling to practices whose work I had been interested in, I was offered an internship with a small, energetic and highly creative architecture practice. I was set to work creating independent pieces of furniture, later to be exhibited in Milan during the International Furniture Fair (annually in April). This proved invaluable experience for my future. I was then offered a full-time position designing furniture – and a salary! This allowed me to gain more experience and contacts in the industry until 2002, when I decided I wanted to realise my own designs and work from my own studio.

I set myself a five-year plan to be in a situation whereby I would be making enough money to survive and run a studio. I worked on many different jobs, not just designing my own work but creating furniture to sell from my studio in the Oxo Tower in London. This I sold privately or supplied to retailers at wholesale price. Sometimes I had to let pieces be sold to retail at a loss just to get my work in the shops, so that I could get noticed.

I spent only one or two days a week actually doing my design work, but I was able to keep my studio running. I had a couple of products put into production by London-based producer Thorsten van Elten, most significantly the 'Antlers' coat hanger. This really gave me the start that I had been looking for.

What advice do you have for students?

It's this variety of work and the daily challenges that make working for yourself so enjoyable. With a clear vision and plenty of confidence in your own ability to be original and work hard, anything is possible!

www.alexandertaylor.com

Career profile

Name:
Deryn Relph

Current job:
Self-employed textile designer and maker

'Retro Rainbow' collection 2011
© Alick Cotterill

Describe what you do

Although I specialised in knitted textiles for interior use during my BA (Hons) Textile Design at Winchester School of Art, I merged this with previous upholstery and furnishing skills to create unique and contemporary solutions.

Working from a studio on the south coast in Hampshire, I often reinvent discarded furniture and lightshades, making them desirable for a contemporary home setting. Inspiration for my work is often taken from aspects of nature, particularly structures or scientific imagery.

Knitted light shades
© Alick Cotterill

My work exploits the unique properties of knitted textiles to create an innovative range of interior textile products in rainbow-colour combinations. With a retro, yet contemporary feel, the pieces evoke a nostalgic happiness, hopefully encouraging people to keep things they love for longer. Embracing the power of colour as a sustainable design solution, the products are made using quality factory surplus or UK-sourced yarns and some carefully selected recycled objects. Inspired by natural structures, microscopic imagery and childhood memories, products combine machine knit, hand knit and crochet, with a quirky play on scale and an element of fun.

The 'Retro Rainbow' collection was inspired by the 'future potential' of seeds, the work of Rob Kesseler in association with the Millennium Seed Bank Project and my own emotional response to seeds, fruits and flowers.

The 'Uplifted' lampshades explore the emotive power of colour and the positive effect it can have on emotions.

Nostalgia and evoking an emotional response to engage the user with an object play a part; if something makes us happy, or we love it for the memories it brings, I believe we will treasure it for longer rather than send it to landfill.

Retro influences are often evident in my designs, inspired by my own childhood memories and happy times.

Is there a particular project you'd like to tell us about?

My most recent collection, 'Buttonbox', took its initial inspiration from my Nanna's button box. Favourite buttons, clothes I remember her wearing and that thriftiness within her that bothered to keep the buttons, zips and buckles. A box is like a time capsule of treasures, and this collection aims to capture some of that nostalgic spirit that is once more so relevant in current times.

'Buttonbox' cushions
© Alick Cotterill

What advice do you have for students considering a similar career?

Versatility is something I believe to be very important and I am keen to think 'outside the box' and take on any challenges for which my background might hold the key!

Issues affecting the environment and sustainable design underpin my design ethos, alongside the consideration to 'Reduce, Reuse, Recycle'. It is wise for you to consider your position on these important issues as a designer.

Career profile

Name:
Sarah Turner

Current job:
Self-employed Designer-Maker

Describe what you do

I design and make lighting from recycled plastic drinks bottles. I'm also involved in education in various different capacities.

Recycled plastic bottles turned into desirable lighting
Sarah Turner

Tell us about your career so far

It all started when my dissertation tutors seemed convinced I could take my project further after graduation. I was studying Product Design and I had written my dissertation about recycling and decided I wanted to make some products to go with the project made from everyday waste materials. So I set about collecting my and my housemates' rubbish and found we discarded a lot of plastic bottles.

After a little research I found that only a small percentage was recycled in the UK, so I decided that I would save a few of these bottles from the landfill sites and make something useful out of them. I experimented in the workshop with a variety of bottles of different shapes and colours and used as many different methods to manipulate them as possible: I melted them with paint strippers, sandblasted them to make them opaque and hacked them up with saws, to name a few. I managed to transform the bottles into beautiful forms that were totally unrecognisable from their original state and looked great as decorative lighting.

So much seemed to happen after I graduated from my course. The day after graduation we had our exhibition at Freerange in London, which was a great success. An interior design company attending the exhibition saw my recycled drinks bottle lights. They ended up commissioning me to make a large light-fitting for their showroom in the same style. This seemed to act as an early confidence boost; it showed me that people were actually interested in my work.

From this first exhibition I was selected for several smaller graduate shows in London, which helped me to promote both my work and myself as a designer. I then struck lucky when a talent scout from the Ideal Home Show saw my recycled lamps at one of these shows and selected me for the Innovation Nation Competition. By then I had designed a small range of lights made from plastic bottles. I and 25 other design graduates exhibited at the show in Earls Court and the public voted for their favourite. I ended up coming second in the competition, which was such an honour, and I also ended up making my first sale!

This early interest in my work made me realise that starting up my own business designing and making eco products was a real possibility. I loved the idea of being my own boss and I thought making and selling my own work sounded so satisfying. I thought the world of nine to five could wait while I had a stab at going it on my own!

I had no idea how to actually start up a business. Questions were occurring such as: where do you register? How do you pay tax? How do you calculate a wholesale and retail price? I just didn't know where to start, and it all seemed a little daunting.

I was fortunate, however, that my university had a business incubation centre that ran a free course to help start up a new business. I had to pitch my idea to the business team and try my best to show them that I had a good business idea. It was a little nerve-racking but all those presentations I had to do as part of my degree course were great practice. I was lucky enough to be accepted and, 12 weeks of business lessons later, I was equipped with some essential knowledge and a business plan.

It was pretty scary starting out, and to be honest it still is scary at times. I do have days where I think

Light design from recycled bottles
Sarah Turner

that this is ace and I'm doing well, but then there are days when I think, 'You are trying to make a living out of cutting up plastic bottles . . . are you crazy girl?!'.

Now I have designed three ranges of lighting, all handmade from waste drinks bottles. Local cafés and friends and family members collect bottles for me to use. I taught myself how to make a website and now my products can be bought online. I stock my work in shops and galleries too.

As my work involves recycling I seem to get a lot of freebies. For example, organisations want to exhibit my work for me and magazines keep asking to publish articles, giving me free publicity as if they also want to support the cause. The best 'freebie' to date was being offered the chance to exhibit as part of the Milan Furniture Fair, also referred to as the Milan Design Week. My work was to be in one of the many satellite shows that take place around the city.

The final exhibition looked great and I was showing my work among well-known, established designers. As a result of the show, many design magazines saw my products and several articles were published. Also, a few online retailers and international retailers saw my work and my products are now being sold with two retailers in Italy, and I am in talks with others.

Looking back to when I first visited Milan for the Furniture Fair with my fellow students during my studies in second year I never thought, in my wildest dreams, that I would be back a few years later exhibiting my own work!

Is there a particular project you would like to tell us about?

Another part of my business that has just developed in the past couple of months is teaching school children about creative recycling. A couple of schools found my website through Google searches and invited me to talk to their students and run a workshop for the day. The first was a class of five-year-olds down in London. I had some previous teaching experience in schools but not with an age group this young, so I didn't know what to expect really. The kids were great though, and were so keen to find out about what I did and how I made my lamps. They ended up making a simple lamp between the class, using coloured plastic bottles, and it looked great.

The second time involved a year 7 class in a school in Yorkshire. They were having a recycling week and the theme was fashion, so they each made a fashion accessory. They really got into it and came up with some fantastic ideas. We held a fashion show at the end of the day for the whole year group, and I had to introduce myself and talk about what I did in front of the 300+ staff and students . . . pretty nerve-racking! The show was great fun and the students were obviously proud of what they had achieved in one day. Teaching is definitely something I love to do, and I am now invited back to where I studied to give lectures about my work and to work with the students on their projects.

Glossary of terms

Ableton Live: a music sequencer, loop-based software tool for composing and arranging as well as enabling live performances; it also allows for the mixing of tracks.

Aesthetics: the study of form, proportion and beauty (see the book *Beautiful Thing* by Robert Clay).

Arduino board: an open-source hardware and software microcontroller board that makes the use of electronics easier in art and design interactive installation-type projects.

Augmented reality (AR): a live, direct or indirect view of a physical, real-world environment whose elements are augmented by computer-generated sensory input, such as sound, video, graphics or GPS data. It is related to a more general concept called 'mediated reality', in which a view of reality is modified (possibly even diminished rather than augmented) by a computer.

Boning: a technique used in 'contour' fashion to stiffen the seams of a bodice or corset. In Victorian times boning was made from whalebone, but modern boning is made from lightweight, flexible vinyl.

Branding: the study of a company's or product's identity: how this can be designed in relation to the market and the consumer (see the book *Wally Olins on B®and*).

Bread and butter: refers to work that forms the staple of the designer-maker's income; this work is often produced in relatively high numbers and sold for a reasonable price and could include perfume bottles, wine glasses, small ornaments, stained glass for the home, functional ceramics and souvenirs.

Built environment: refers to spaces and buildings, rooms and gardens, shops and hotels; in general, all architectural forms and their surroundings.

Choreutics: thought processes, or 'paths', are tracked spatially and are represented as a series of locations, creating flat-sided shapes that track the movement.

Commission: when a customer likes the work of an individual designer-maker and desires something unique so asks the designer to make a specific piece for them.

Contradictions: using unusual materials and combining different visual styles, this bringing together of unusual objects to provoke a reaction is reminiscent of the Surrealists' aim of using chance encounters to create new objects.

Creator's block: a term used to describe the moment in a project when your brain grinds to a halt because it has overheated! In this situation you will need a proper break – a long walk or some form of exercise; this will reset your brain. Leave the work alone for at least a few hours, or days if you have to, to unblock your creative flow.

Cross-disciplinary: a designer moving into another discipline; for example, a graphic designer might become interested in designing products.

Designer-maker: a self-employed designer who sells their work through specialist trade fairs, independent retailers and galleries, usually producing handmade products of high quality (sometimes the products are unique 'one-offs' but more usually they are limited edition pieces from small production runs).

Dissertation: this piece of work is completed in the final year of a degree course and is often illustrated and about 6–10,000 words in length. A specific interest in a concept or subject is at the heart of the piece. This is a critical and theoretical piece of writing, offering personal insights into your chosen theme.

Environmental Impact Assessment: a systematic approach to predicting the potential repercussions of the changes proposed within a design.

Epic theatre: a political form of theatre where everything presented has an objective basis and is devoid of illusion, made popular by Bertolt Brecht.

Ergonomics: the study of people's physical and psychological interaction with objects and systems – how we see, understand and use things (see the book *BodySpace* by Stephen Pheasant and Christine M. Haslegrave).

Ethical fashion: these brands form a small but influential sector of the industry. In recent years many small brands have emerged with ethical credentials that appeal to customers who are concerned about the impact on the environment of mass-production processes and the constant fuelling of our consumerist society.

Ethnographic research: the scientific study of human culture, involving contact with and in-depth study of people, society, working relationships, etc.

Fashion forecasting: a trend forecasting process that involves gathering research and information from a wide range of diverse aspects of modern life and world influences, including the global economic climate, modern technologies and the natural world, combined with textures, colour, art, design, literature, music and film references. This information is used to depict a set of conceptual themes for the fashion and interior industries.

Fast fashion: the term used to describe the speed at which new fashions can be conceived/sketched out and turned into collections and be ready for sale in the shops or on a website. Many successful retailers have refined and revised their systems, by using new technologies and advanced computer programs, but also by building flexibility into their production processes. This enables them to switch manufacturers if they have to, depending on what is being requested, and, in some cases, can reduce lead times to as little as three weeks.

Freelance designer: a designer who works either from their own studio or an agent's studio creating designs for many different clients. The key to successful freelancing is a full book of contacts – something you are unlikely to have when you have just graduated. However, some graduates do find regular work from specific clients through their graduate show and are able to build up their contacts over time. The designs are sold to the client, who can use them in any way they wish.

Grids: a system for arranging blocks of type and imagery to create an underlying consistency in page layout, pioneered by Josef Muller Brockmann.

Information design: the skill of preparing and presenting information for efficiency and effective understanding of the user; this may involve using visual representations of data, such as graphs, pie-charts, photographs and illustrations.

Intellectual Property Rights: retaining the IPR: you do the design, a manufacturer makes it and they pay you a percentage of what they get for it. The money will start to trickle in. If it's a hit you might be very lucky and make it rich. If the design has a long life the money will continue to roll in for years. The down side is that you won't be paid at all during the design phase and you have to invest your own time and money on realising the design. Of course, it's a gamble for you and them as to whether sales take off or not.

Intellectual Property Rights: selling the IPR: you do the design and a manufacturer pays you for it, probably at the end of the design stage or, if it's a big job, in stages throughout the process. You get the money much earlier but, as you've sold the IPR, there's no more money coming in, no matter how successful it is in the future. This is less of a gamble than retaining IPR and receiving royalties.

Investors: people who specialise in taking risks on new products, providing money for production, to build stock for sale or for marketing costs.

Kitsch: the term, from the German *verkitschen*, that means 'to make cheap'. Kitsch gives value to things that might usually seem cheap or 'tacky'.

Licence: this denotes ownership of an idea and allows that person to decide who will produce it.

Lifecycle assessment: Okala is a simple, cost-effective LCA tool and guide that can be used for educational purposes. The Okala guide provides an introduction to ecological and sustainable design for practising and student designers, and is produced by and available from The Industrial Designers Society of America (IDSA).

Master-class: when a 'master' of a particular material or technique passes on some of their specialist knowledge and experience – often to other professional artists, makers or designers.

(Decline of) metanarratives: a 'metanarrative' is the idea that one culture, religion or political system is more important than others, as described by philosopher Jean-Francois Lyotard. The blurring of boundaries between 'high' and 'low' cultures is a symptom of the decline of these metanarratives in the post-modern era.

MIDI (Musical Instrument Digital Interface): an electronic industry specification that allows digital instruments, computers and other devices to connect with each other.

Modernity: the term given to the historical period following the demise of feudal societies; it is associated with a firm belief in progress and an attachment to the idea of the new.

Multi-disciplinary: the opportunity for a designer to work across two different disciplines; for example, a fashion designer designs the clothes and then also designs the print for the fabric (therefore working across fashion and textiles).

Objective viewpoint: a statement, more of fact than opinion, that is difficult to argue against.

Parody: the imitation of style for humorous effect or ridicule.

Pastiche: described by Frederic Jameson as 'the random cannibalisation of all the styles of the past, the play of random stylistic allusion'.

Peer review: this is when fellow students give you their opinion about your work in a formal situation, probably as part of a taught session. It might be carried out as a group or individual exercise, and could form part of an assessment process. It is a very useful way of finding out what others think about your work and can help develop your projects enormously, but you do need to be receptive.

PGCE (Postgraduate Certificate in Education): if you are a designer interested in becoming a teacher then undertaking this one-year course of intensive training, with practical hands-on experience and the theory to underpin it all, is necessary.

Popular culture: this reflects the trends, attitudes and tastes of a given society – music, fashion and images from mass media, such as advertising, films and television and the Internet (Facebook and YouTube), are all part of what is known as popular (or pop) culture.

Post-modernism: used to describe the economic and cultural state of Western society following on from modernity.

PR (public relations): a role that involves managing the communication of a company's image and ethos to the outside world.

Première Vision: an annual Paris trade fair; one of the smaller exhibitions within, entitled Indigo, comprises many smaller companies, often referred to as 'swatch studios', that specialise in developing textile designs, which are then sold on to textile manufacturers to put into production or to fashion houses to be used as inspiration for forthcoming collections.

Price point: the price that consumers will pay in sufficient numbers to justify the production run and generate a profit.

Processing: an open-source computer programming language that makes it easier to create interactions, images and animations. It was designed to teach computer programming in a visual context.

(Low-volume batch) production: a batch of products is produced at the same time for distribution to a range of different clients or shops; there are economies of scale in manufacturing a certain number at the same time.

(One-off) production: a designer might produce a 'one-off' piece for a particular client, to that client's own brief. The client will pay more for the 'exclusive' right to the design and appreciate the fact that no one else owns the some piece.

Pro Tools: a widely used professional tool for recording and editing in film, television and music production.

Royalty: a payment from the manufacturer to the designer for every item sold in the shops, normally expressed as a small percentage of the retail price.

Scenography: an artistic perspective concerning the visual, experiential and spatial composition of a performance. It is an holistic approach to theatre performance that embraces all key artistic disciplines associated with theatre production.

Semantics: the study of the meaning of words, it deals with the language used to achieve a desired effect on an audience (as in advertising or political propaganda), especially through the use of words with novel or dual meanings.

Semiotics: the theory of visual signs and symbols and how they are used as representations with a number of alternative meanings and constructions.

Simulacra: an image as a representation of something that might be vague or superficial in its rendering.

Special effects: live action footage is combined with computer-generated imagery (CGI), which is added or manipulated to create environments and effects that would be too expensive, impossible to shoot or dangerous to capture on film.

Style guide: a graphically oriented publication that shows the development of a logo, for example, and how that logo should be used across divergent media; or a series of rules of use. The publication is created for the client to use when establishing or updating a brand.

Subjective viewpoint: a personal opinion that could be argued with.

Thesis: this summative piece of writing positions the practice-based work within existing debates and theories about art and design. It describes the way that the research has been undertaken (the methodology) and the analysis that has been done, and determines its level of success.

TouchDesigner: a software tool by Derivative for creating visual real-time projections, live music visuals, interactive systems or rapid prototyping – any type of rich user experience.

Trade fairs: large-scale, organised exhibitions where companies from a specific industry sector come together to launch their new products and services. This makes it easier for trade customers to compare the products available from different suppliers. The design sector has many of these events across the world and throughout the year.

Typography: the craft of arranging type using typefaces, point sizes, line lengths and spacing to communicate language.

User: anyone who is using a digital service, product, computer or any other digital device.

User experience interactions: looking at all aspects of the experience, including the aesthetics, interface and physical interaction.

User survey: anything from an interview-based discussion with a selection of potential or actual users of an environment to a full investigation of the purpose and requirements of a new multimillion-pound investment, such as a hospital.

USP (unique selling point): the one thing that will mark your idea out from the rest of the market.

Virtual reality: an environment created by software where the user is expected to suspend their disbelief and accept it as a real experience that involves responding to sight and sound. The environment can be a simulation of a real environment for education and training purposes or an imagined environment for an interactive story or game. The user can interact with the environment.

Way marking: used to help people navigate their way around and learn how to use the spaces. Main paths will be wider or colour-coded or perhaps brightly lit, while secondary paths may be narrower with more subtle lighting. Signs and symbols may be used, as well as key design elements such as gateways, entrances, exits and landmarks.

Web information architecture: the design of the organisation and labelling of websites and online communities using primarily 'wireframe' diagrams.

Zones of Visual Influence: an appraisal of how people will view the interior and exterior of a project.

Zoning: the way that designers plan the spatial arrangement of the proposed project into discrete areas of similar use, scale and appearance. For example, an active space (playground) may be planned separately from a quiet space (a library room). The two areas may have a close relationship (i.e. be linked by a door or corridor) but may be quite different in terms of space materials (colour, scale, etc.).

Index

D

G

H

I

J

K